HUMAN DEVELOPMENT 82/83

Hiram E. Fitzgerald, *Editor*
Michigan State University

Thomas H. Carr, *Editor*
Michigan State University

Cover Credit: NATIONAL MUSEUM OF AMERICAN ART, Smithsonian Institution; Roszak, Theodore, "Construction in White", 1937.

82-3232

ANNUAL EDITIONS

The Dushkin Publishing Group, Inc. Sluice Dock, Guilford, Ct. 06437

Volumes in the
Annual Editions Series

Abnormal Psychology
- Aging
- American Government
- American History, Pre-Civil War
- American History, Post-Civil War
- Anthropology
Astronomy
- Biology
- Business
Comparative Government
- Criminal Justice
Death and Dying
- Deviance
- Early Childhood Education
Earth Science
- Economics
- Educating Exceptional Children
- Education
- Educational Psychology
Energy
- Environment
Ethnic Studies
Foreign Policy

Geography
Geology
- Health
- Human Development
- Human Sexuality
- Management
- Marketing
- Marriage and Family
- Personal Growth and Behavior
Philosophy
Political Science
- Psychology
Religion
- Social Problems
- Social Psychology
- Sociology
- Urban Society
- Western Civilization,
    Pre-history – Reformation
- Western Civilization,
    Early Modern – 20th-Century
Women's Studies
World History
- World Politics

- Indicates currently available

©1982 by the Dushkin Publishing Group, Inc. Annual Editions is a Trade Mark of the Dushkin Publishing Group, Inc.

Copyright ©1982 by the Dushkin Publishing Group, Inc., Guilford, Connecticut 06437

Tenth Edition

Manufactured by George Banta Company, Menasha, Wisconsin, 54952

Library of Congress Cataloging in Publication Data
Main entry under title:
Annual editions: Human Development.
    1. Child study—Addresses, essays, lectures.
2. Socialization—Addresses, essays, lectures. 3. Old age—Addresses, essays, lectures. I. Title: Human Development
HQ768.A55   155   72-91973
ISBN 0-87967-401-6

ADVISORY BOARD

AND STAFF

CONTENTS

1

Perspectives

2

Development During the Prenatal Period

3

Development During Infancy

4
Development During Childhood

5
Education and Child Development

6

Child Rearing and Child Development

7

Development During Adolescence and Young Adulthood

8
Development During Maturity and Old Age

TOPIC GUIDE

This guide can be used to correlate each of the readings in *Human Development 82/83* with one or more of the topics usually covered in human development books. Each article corresponds to a given topic area according to whether it deals with the topic in a primary or secondary fashion. These correlations are intended for use as a general guide and do not necessarily define total coverage of any given article.

TOPIC AREA	TREATED AS A PRIMARY ISSUE IN:	TREATED AS A SECONDARY ISSUE IN:
Aggression	41. Go Get Some Milk and Cookies	7. Just How the Sexes Differ 38. Suffer the Children 42. How I Stopped Nagging
Anthropological/ Cross-Cultural Perspectives	4. Ethnic Differences in Babies 5. How Children Influence Children 6. The Japanese Brain 29. They Learn the Same Way 38. Suffer the Children	
Attachment	16. Biology Is One Key to the Bonding 19. Attachment and the Roots of Competence 37. When Mommy Goes to Work . . .	4. Ethnic Differences in Babies 10. Pregnancy 15. Premature Birth 18. What a Baby Knows 31. Infant Day Care 36. A New Look at Life with Father
Behavior Disorders	21. If Your Child Doesn't Get Along with Other Kids 39. The Curse of Hyperactivity 40. The Children of Divorce 52. Stress	9. The More Sorrowful Sex 22. Myth of the Vulnerable Child 32. Diet and School Children 42. How I Stopped Nagging 46. Why Johnny Can't Disobey 48. Single Parent Fathers
Behavior Modification	42. How I Stopped Nagging	
Biological/Genetic Factors	1. The Unique You 2. The Instinct to Learn 3. Twins 8. Why Women Live Longer 10. Pregnancy 39. The Curse of Hyperactivity 55. Living Longer	4. Ethnic Differences in Babies 7. Just How the Sexes Differ 11. Embryo Technology 12. Test Tube Babies 16. Biology Is One Key to the Bonding 17. Importance of Mother's Milk 50. Late Motherhood
Brain Development	6. The Japanese Brain	
Child Abuse	38. Suffer the Children	
Child Rearing	4. Ethnic Differences in Babies 21. If Your Child Doesn't Get Along with Other Kids 22. The Myth of the Vulnerable Child 36. A New Look at Life with Father 42. How I Stopped Nagging 48. Single Parent Fathers	14. Biology Is One Key to the Bonding 15. Premature Birth 17. Importance of Mother's Milk 19. Attachment and the Roots of Competence 31. Infant Day Care 37. When Mommy Goes to Work . . . 41. Go Get Some Milk and Cookies 44. Pregnant Children 46. Why Johnny Can't Disobey

TOPIC AREA	TREATED AS A PRIMARY ISSUE IN:	TREATED AS A SECONDARY ISSUE IN:
Cognitive Development: Infancy/Early Childhood	20. Are Young Children Really Egocentric? 26. Piaget 27. Learning About Learning 31. Infant Day Care	15. Premature Birth 18. What a Baby Knows 19. Attachment and the Roots of Competence 24. Mood and Memory 29. They Learn the Same Way 30. Navigating the Slippery Streams of Speech
Cognitive Development: Childhood/Adolescence	24. Mood and Memory 26. Piaget 27. Learning About Learning	7. Just How the Sexes Differ 9. The More Sorrowful Sex 22. The Myth of the Vulnerable Child 28. 1,528 Little Geniuses 29. They Learn the Same Way 30. Navigating the Slippery Stream of Speech 34. Learning Right from Wrong 35. Toward a Nonelitist Conception of Giftedness 41. Go Get Some Milk and Cookies 44. Pregnant Children 46. Why Johnny Can't Disobey
Cognitive Development: Adulthood/Aging	26. Piaget 28. 1,528 Little Geniuses	51. The Dynamics of Personal Growth 52. Stress 54. Must Everything Be a Midlife Crisis? 56. Coping with Death
Conception	1. The Unique You 10. Pregnancy 11. Embryo Technology 12. Test Tube Babies 14. A Perfect Baby	44. Pregnant Children
Divorce	40. The Children of Divorce 48. Single Parent Fathers	44. Pregnant Children
Emotional Development	9. The More Sorrowful Sex 13. Prenatal Psychology 16. Biology Is One Key to the Bonding 19. Attachment and the Roots of Competence 24. Mood and Memory 25. Erik Erikson's Eight Ages of Man 37. When Mommy Goes to Work 40. The Children of Divorce 47. The Sibling Bond 51. The Dynamics of Personal Growth 52. Stress 53. Coping with the Seasons of Life	31. Infant Day Care 38. Suffer the Children 41. Go Get Some Milk and Cookies 43. Adolescents and Sex 46. Why Johnny Can't Disobey 48. Single Parent Fathers

Left section

TOPIC AREA	TREATED AS A PRIMARY ISSUE IN:	TREATED AS A SECONDARY ISSUE IN:
Emotional Development (cont.)	54. Must Everything Be a Midlife Crisis? 56. Coping with Death in the Family	
Family Relations	23. American Research on the Family 36. A New Look at Life with Father 37. When Mommy Goes to Work... 40. The Children of Divorce 48. Single Parent Fathers	38. Suffer the Children 41. Go Get Some Milk and Cookies 42. How I Stopped Nagging 44. Pregnant Children 50. Late Motherhood 51. The Dynamics of Personal Growth
Fathers	36. A New Look at Life with Father 42. How I Stopped Nagging 48. Single Parent Fathers	22. The Myth of the Vulnerable Child
Intelligence	26. Piaget 28. 1,528 Little Geniuses 33. The Mismatch Between School and Children's Minds 35. Toward a Nonelitist Conception of Giftedness	7. Just How the Sexes Differ 32. Diet and School Children
Language	29. They Learn the Same Way 30. Navigating the Slippery Stream of Speech	6. The Japanese Brain 31. Infant Day Care 33. The Mismatch Between School and Children's Minds
Learning	2. The Instinct to Learn 24. Mood and Memory 27. Learning About Learning	
Learning Disability	32. Diet and School Children 39. The Curse of Hyperactivity	
Mothers	16. Biology Is One Key to the Bonding 17. The Importance of Mother's Milk 18. What a Baby Knows 37. When Mommy Goes to Work... 50. Late Motherhood	13. Prenatal Psychology 15. Premature Birth 22. The Myth of the Vulnerable Child
Nutrition	10. Pregnancy 17. The Importance of Mother's Milk 32. Diet and School Children	39. The Curse of Hyperactivity
Peers	5. How Children Influence Children 21. If Your Child Doesn't Get Along with Other Kids	24. Mood and Memory
Perceptual Development	18. What a Baby Knows 31. Infant Day Care	
Personality Development	22. The Myth of the Vulnerable Child 25. Erik Erikson's Eight Ages of Man 45. The Many Me's of the Self-Monitor 46. Why Johnny Can't Disobey 49. Does Personality Really Change After 20? 51. The Dynamics of Personal Growth	15. Premature Birth 21. If Your Child Doesn't Get Along with Other Kids 31. Infant Day Care 38. Suffer the Children 39. The Curse of Hyperactivity 40. The Children of Divorce 47. The Sibling Bond

Right section

TOPIC AREA	TREATED AS A PRIMARY ISSUE IN:	TREATED AS A SECONDARY ISSUE IN:
Personality Development (cont.)	53. Coping with the Seasons of Life 54. Must Everything Be a Midlife Crisis? 56. Coping with Death in the Family	
Prematurity	15. Premature Birth	14. A Perfect Baby
Prenatal Influences	10. Pregnancy 11. Embryo Technology 12. Test Tube Babies 13. Prenatal Psychology 14. A Perfect Baby 15. Premature Birth	1. The Unique You
School Influences	31. Infant Day Care 32. Diet and School Children 33. The Mismatch Between School and Children's Minds 34. Learning Right from Wrong 35. Toward a Nonelitist Conception of Giftedness	39. The Curse of Hyperactivity
Self-Concept	45. The Many Me's of the Self-Monitor 53. Coping with the Seasons of Life	19. Attachment and the Roots of Competence 40. The Children of Divorce 55. Living Longer
Sex Differences	7. Just How the Sexes Differ 8. Why Women Live Longer 9. The More Sorrowful Sex	
Sexuality/Sexual Identity	7. Just Why the Sexes Differ 8. Why Women Live Longer 9. The More Sorrowful Sex 43. Adolescents and Sex 44. Pregnant Children 48. Single Parent Fathers	11. Embryo Technology 12. Test Tube Babies 13. Prenatal Psychology 41. Go Get Some Milk and Cookies
Sibling Relations	47. The Sibling Bond	
Social Behavior/Development	21. If Your Child Doesn't Get Along with Other Kids 23. American Research on the Family 25. Erik Erikson's Eight Ages of Man 44. Pregnant Children 46. Why Johnny Can't Disobey 50. Late Motherhood 52. Stress 56. Coping with Death	5. How Children Influence Children 15. Premature Birth 19. Attachment and the Roots of Competence 26. Piaget 31. Infant Day Care 34. Learning Right from Wrong 41. Go Get Some Milk and Cookies 43. Adolescents and Sex 44. Pregnant Children
Sociobiology/Psychobiology	2. The Instinct to Learn 4. Ethnic Differences in Babies	3. Twins 16. Biology Is One Key to the Bonding
Twins	3. Twins	

PREFACE

During the past two decades many changes have occurred in the field of human development. At the risk of over-simplifying matters, these changes can be stated briefly as follows:

● Theory and research in developmental psychology have expanded to include the full span of life, from conception to death. In the past, developmentalists neglected the periods of adulthood and aging.

● Human development now is firmly an interdisciplinary field. This is a simple acknowledgement that no single discipline can comprehend the complexity of the living organism. Thus, the study of human development literally demands an interdisciplinary, team approach.

● Developmentalists have rejected the static "machine" view that emphasizes the dynamic and active nature of the living organism and the environments to which the organism adapts.

● Developmentalists have rejected the view that all behavior is continuous through the life span. Both continuities and discontinuities exist. Thus, the developmentalist must try to understand which characteristics of the individual change and which remain stable through the life span.

● Developmentalists are interactionists. Heredity and environment influence one another; nature *and* nurture contribute to the organization of behavior. The developmentalist tries to understand how factors blend together to affect the mutual interaction of organism and environment.

● Developmentalists tend to view the organization of behavior as occurring in various stages, such as those proposed for cognitive, social, moral, emotional, psychosocial, and psychosexual development. Implicit in the stage approach is the idea that each successive stage brings about qualitative changes in behavior that are not fully predictable from preceding stages.

● Developmentalists have emerged from the laboratory. Extensive laboratory research under highly controlled conditions has given rich information about the organization of behavior and it will continue to do so. However, knowledge obtained in the laboratory often does not generalize well to other settings. Today, developmentalists are conducting research in hospitals, mental health clinics, schools, and homes.

● Developmentalists have also emerged from their respective cultures. Information about human development in different cultures helps the developmentalist to understand which behavior patterns are culture-specific and which are species-characteristic. Moreover, cross-cultural research provides important information about individual differences.

The study of human development is exciting. New findings, new theories, new debates, and new problems come forth each day. It is our hope that much of this excitement will come through to you as you sample the theory and research in human development represented in this volume. The articles in this tenth edition of *Annual Editions: Human Development 82/83* were selected, in part, with the advice of the readers and with that of the members of the advisory board. The articles, written by professional sciences writers and researchers, originally appeared in popular magazines or newspapers. Collectively, they provide a sample of the issues and topics of historical and contemporary interest in the field of human development.

We thank the authors and publishers who have made this reader possible by permitting the reprinting of their works. Which articles did you find to be most informative and/or interesting? Which would you exclude if you were designing a revision? What topics would you have included that have not been included? Please return the article rating form located on the last page of this book and participate in the design of the next Annual Editions. Your opinions will be considered carefully.

Hiram E. Fitzgerald
Thomas H. Carr
Editors

Perspectives

1

The multidisciplinary field of human development is concerned with stability and change throughout the life span. At present, no single theory gives comprehensive direction to the study of human development. Instead, developmentalists typically are guided by one of four theoretical perspectives: biological-evolutionary, environmental-behaviorist, psychoanalytic, or cognitive-developmental. The biological-evolutionary approach emphasizes genetic aspects of psychological and behavioral characteristics. The environmental-behaviorist position emphasizes learned aspects of behavior, giving little attention to genetic or mentalistic (cognitive) explanations. Psychoanalytic theorists emphasize the importance of the early years for the formation of the ego, and conceptualize personality as being organized in a series of stages. Cognitive developmentalists propose a series of sequential stages that describe the various levels we achieve in our understanding of the events in the external, physical world.

Each of these traditions has its strengths, but the inherent limitations of each prevents any from achieving the status of a general theory of human behavior and development. As a result, many developmentalists have turned to general systems theory in the hope that it can provide unity for the psychological, biological, social, and physical sciences. The systems view of organization is an outgrowth of a general paradigmatic change taking place in all sciences. The word "paradigm" refers to the basic assumptions upon which theories are based. Such assumptions dictate how a theory will be applied to the study of natural phenomena and how they will be interpreted. Prior to Copernicus, vitalism was the paradigm of science. Vitalism's subjective dualism gave way to a mechanistic dualism which stressed an objective, empirical, reductionist science. Relativity theory and discoveries in quantum mechanics prompted the downfall of the mechanistic paradigm, while simultaneously giving birth to general systems theory.

The systems approach emphasizes the dynamic, active, transformational, interdependent, probabilistic, and informational aspects of natural phenomena. Thus, the family is a system composed of a number of individuals, but it is not the simple additive sum of these individuals. A family generates characteristics of its own which are different from the characteristics of its individual members. The interrelationships of family members change during development and in so doing transform the structure or organization of the family itself. Thus, whereas the form of the family may remain stable over time, transformations brought about by information exchange among family members will produce qualitative changes in its systemic character. The matter is even more complex when one begins to take into account social and cultural influences on the family system.

Articles in this unit of *Annual Editions* reflect many of the changes in human development referred to in the Preface. As you read each article try to determine whether it reflects a systemic approach to human development, or whether it reflects one of the four major theoretical traditions. How might the article differ if a systems approach was used?

Looking Ahead: Challenge Questions

Why is a comprehensive theory of human development necessary and how does one determine whether a theory adequately accounts for the phenomena of development?

How might cultural definitions of "childhood" affect the way in which child development experts interpret the results of their studies?

How can studies be designed to tease out the relative contributions of genetics and environment for sex differences in behavior? Or, can these two influences be separated?

How can the study of development in non-Western cultures contribute to our understanding of child rearing and development in our culture?

THE UNIQUE YOU

RICHARD HUTTON AND ZSOLT HARSANYI

We are entering the Age of Anonymity. As we struggle to remain individuals, our complex society increasingly identifies us by faceless numbers. We may soon live as units within an immense equation, only a digit, decimal, or binary zero removed from the next guy.

This slow slide toward sameness has compelled many people to search for new ways to reaffirm their individuality. Some go off to seek their roots—the number of genealogical organizations in the United States has grown from 200 in 1966 to more than 800 at present, and mail-order business in family coats of arms is booming. Other people, according to Queens College sociologist Paul Blumberg, try to manufacture individuality by wearing rhinestone tuxedos and running shoes or by rummaging around for distinctive status symbols and esoteric hobbies. For them, possession offers a claim to uniqueness: Their identities derive from their ability to own something handmade, homemade, self-made—something, in short, that no one else has.

The situation may seem pretty grim, but our salvation lies close at hand. It has, in fact, been here all along. Nature has wisely endowed each of us with unique characteristics that not even the postindustrial age can steal. And science is getting ever better at measuring and highlighting them.

The late Theodosius Dobzhansky, of Rockefeller University, noted the staggering number of possibilities that exist in the human gene pool. If all the imaginable characteristics are added up, he said, "the number of potentially possible gene constellations turns out to be greater than the number of subatomic particles estimated by physicists to exist in the universe."

So while society pushes us toward conformity, nature keeps pulling for variation and change. And while manmade institutions persist in trying to transform us into social clones, our internal blueprints ensure that we remain unique and uncopiable, as original, exceptional, and singular as flakes of snow.

Dozens of characteristics contribute to the uniqueness that distinguishes us from one another. Possible identifiers range from facial profiles and hand shapes to electrocardiographs, even the patterns of blood vessels in our eyes. Other traits may not become significant for decades, but these are in use today or are right on the horizon. They can identify one John as Travolta and another as Olivia Newton as easily as we can tell grease from water.

Our bodies are composed of many materials, and primary among them are proteins, the structural and functional building blocks of cells. Proteins come in all shapes and forms, and most proteins display a range of clearly discernible variations.

Because many tribes, races, and religious communities around the world tend to isolate their gene pools by inbreeding, protein variations can now be used to identify members of these groups. For example, red blood cells contain a protein, acid phosphatase, that is a blend of three structural variants known as p^a, p^b, and p^c. Native Greeks, as a group, have a distribution of 22 percent p^a, 71 percent p^b, and 7 percent p^c. The Yanomana Indians of Venezuela have only p^b, and Alaskan Eskimos have 56 percent p^a and 44 percent p^b.

This might seem important only if a destitute, mute, Asiatic-looking man were found wandering through the streets of Juneau, Alaska, but anthropologists have used these and other biological markers to trace migration, breeding, and evolutionary patterns in human populations. In addition, protein typing has been used to determine parentage in many paternity suits.

Because every person is a unique island floating in a sea of microorganisms, germ types can also be used to tell us apart. From birth we are surrounded by bacteria, fungi, and viruses of all kinds. A few are parasitic or exploitative and cause disease, but most consider us their homes and breed happily and harmlessly in colonies throughout our bodies.

Different populations of microorganisms get established in various parts of a human body. Because our body surfaces and orifices vary in their acidity, texture, and secretions, our glandular "signatures" actually determine which microorganisms will take up residence where. The remarkable constancy in the way certain microorganisms are distributed on the body of each individual has led Dr. Melvin Gershman, of the University of Maine, to suggest the development of a science of forensic microbiology, based upon germ types.

Since microorganisms are living creatures, each species has individual variants within it. Though most people might think that a Corynebacterium xerosis by any other name is still a xerosis, each germ has its own recognizable characteristics. So we can be identified not only by the distribution of our germ populations but also by the individual microorganisms themselves.

YOUR ODOROUS IDENTITY

One common microorganism activity offers another way to identify us—smell prints. An individual not only carries an identifiable odor but leaves a trail of it behind him—a fact that every dog's nose knows. We differ from one another across an entire spectrum of odors that emanate from our bodies; more than 100 constituents make up our unique olfactory signature. Engineers have reasoned for years that if a dog can differentiate individual smell patterns, a machine could do the same, without having to be fed.

People emit both general and specific odors. Women, for example, emit volatile substances that vary with their monthly menstrual cycles. One investigator discovered that if he gave German shepherds the general scent of progesterone (a female hormone), the shepherds could identify rods that had been held by pregnant women or by women in the latter stages of their menstrual cycle. Other specific odors that make us attractive or repulsive to mosquitoes are now being identified so that effective repellents can be manufactured.

An individual's smell can be distinguished from the smell of his clothes, residence, or diet, though all these factors can superimpose their own characteristics, at least in part. The olfactory sense is one of the least developed among mammals, however. Most of us have noses that can tell us little more than whether another individual smells fresh or foul. Still, individuals with highly refined senses of smell do exist. Freud reported the case of the "rat man," who was able to identify all the people he knew from their smell. One young boy was not only able to detect individual body odors, but his olfactory sense was so keen that he could even tell which part of the body a particular odor came from.

Such observations have led scientists to theorize that humans might even communicate subconsciously by means of subtle volatile substances. After all, most animals use chemical messengers, which are known as pheromones. These compounds tell an animal when to fight, move to a new territory, become friendly, or copulate. If we can respond to the chemical emissions of others, we might actually be doing ourselves and our ability to communicate a great disservice by bathing frequently and by coating ourselves with perfumes, lotions, and deodorants.

Machines under development today use gas chromatography to analyze and record the patterns of airborne organic vapors from humans. Analysis can now identify chemicals in concentrations below one part per billion — far beyond our own ability to detect odors. Though the technique is still in the experimental stage, it seems likely that we will soon be able to identify individuals by their olfactory signatures.

BLOOD BONDS

An old saying holds that a person's character is written in his blood. Scientists are finding that, in a sense, this is true. Blood varies from person to person.

When most people refer to "blood type," they are thinking of the most common designations: Rh positive or negative, or A, B, AB, and O. Hematologists, however, type people's blood into much finer categories. The A-B-O and Rh systems, for example, are but two of more than two dozen blood groups that differ among us.

Knowledge of blood types is important for safe blood transfusions, for preventing hemolytic disease in the newborn, and for transplantation surgery. It has also been used to clear up conflicting claims in cases of doubtful paternity, kidnapping, inheritance, and the inadvertent switching of infants in hospitals.

One of the most useful applications of blood typing is analyzing reproduction and childbirth. Couples with incompatible blood types, studies indicate, have a higher incidence of childlessness, since women of one blood type can produce antibodies against incompatible sperm cells. Prostitutes, who encounter dozens of different blood types in their careers, have been found to secrete similar sperm-inhibiting antibodies into their cervical fluid. This biological contraception has led some investigators to think antibody-antigen reactions may be the ideal method of birth control. In their thinking, the sperm become an invading organism, the pregnancy a preventable disease. If the sperm are considered a "germ," vaccination is the obvious "cure" for fertilization.

Virtually every human cell carries markers of biological individuality on its surface.

These markers, known as transplantation antigens for their role in sabotaging surgical transplants, are divided into four major groups, with as many as 22 different molecules possible in each group. Since each individual inherits eight of these antigens (four from each parent, one from each of the four groups), the total number of combinations is approximately 20 million.

Transplantation antigens function as passwords. This is why surgical transplants are so often unsuccessful; unless the transplantation antigens of the donor and of the recipient are identical, the body will recognize the transplanted tissue as foreign and will reject it. Some tissues, such as spleen cells, elicit extremely powerful rejection reactions from the body. It is ironic that the tissues that trigger the weakest reactions and that therefore would be the easiest to accept as transplants are the very ones that would be least desirable and least possible to transplant: the fat cells and brain cells, respectively.

Recently transplantation antigens have been put to another use in society, not in the biology lab, but in the courtroom. Because two random individuals are very unlikely to possess identical antigens, transplantation types are ideal for testing claims of paternity. All an attorney has to do is prove that a child does not have a combination of four antigens that his purported father would have supplied.

In a recent case, tried in Austria, the use of these antigens went a step further. A woman claimed that her child was an heir to the estate of a dead man. But the estate's lawyers rounded up the dead man's children and checked their antigens. From them, they were able to deduce his transplantation type and prove that the plaintiff's child could not have been sired by him.

No one knows why we are born with transplantation antigens in the first place. Surely nature could not have anticipated the increasing sophistication of the surgeon's scalpel. One theory comes from Nobel laureate Sir Macfarlane Burnet, who suggests that the antigen code was developed to enable the body to recognize and reject any cell that had mutated and become cancerous. According to his theory, cancer arises when the body's surveillance mechanism is unable to detect and kill an abnormal cell. Burnet's theory is one avenue scientists are exploring as they try to unravel the foundations of cancer.

GLEANED FROM YOUR GENES

The ultimate type of identification may be the karyotype. This isn't a biological group, but a picture of the chromosomes, the tiny, rod-shaped cellular structures that carry our genes. Ordinarily a karyotype is displayed as a photograph in which the 46 human chromosomes, stained for microscopic examination, are lined up in 23 pairs, according to size and shape. The

longest chromosome is designated number 1, and the shortest is number 22. In females, all the chromosomes normally form matching pairs, but in males, the twenty-third pair, which contains the X and Y sex chromosomes, is noticeably mismatched. Karyotypic analysis therefore readily distinguishes between the two most important human subgroups, males and females. It is a fact of which the International Olympic Committee is fully aware.

In normal individuals, there is little variation in size between chromosomes in a pair, except in the case of the Y chromosome. The length of this chromosome varies significantly among different groups of males; the Japanese have the longest and Caucasians the shortest. Whether the length of this male-determining sex chromosome is significant remains unknown, but it does not seem to correlate with variation in related anatomical features.

Thus far karyotype studies can detect only gross characteristics and abnormalities, but they are still accurate enough to reveal chromosomal alterations in about 2 percent of the population. New staining techniques and improved methods to identify the staining patterns, called chromosomal bands, will undoubtedly improve our ability to detect minor differences in gene structure. Someday we may be able to analyze chromosomes one gene at a time, the ultimate in genetic identification.

While full utilization of these techniques still lies a short distance down the road, scientists are already using identification methods that go far beyond fingerprints.

On September 24, 1974, for instance, a Torrance, California, man was convicted of manslaughter after it was shown that the pattern of his teeth fitted the bite marks on a slain woman's nose. The prosecution estimated that the five most distinctive marks in the nose could have displayed 59,049 possible combinations of alignment and tooth rotation; when 710 randomly selected models of adult dentition were compared to the damning marks, only the suspect's fitted every characteristic.

The dental pattern compares favorably with fingerprints in its uniqueness. Because each human tooth has five distinct surfaces, 160 different dental areas can be classified by hereditary formation, developmental alteration, natural or traumatic changes, and the extraordinary combinations of procedures used to repair or replace teeth. According to computer models, more than 2.5 billion possible permutations of these characteristics can be found in a dental print. X-ray analysis of the jaw can add yet another dimension to the process of identification.

Analyzing dental patterns is by no means a new science: Paul Revere, known more for his ability to ride a horse, relied on dental patterns to identify Joseph Warren, an American killed at Bunker Hill. The iden-

1. PERSPECTIVES

tity of John Wilkes Booth, Abraham Lincoln's assassin, was confirmed by dentition when his body was exhumed.

Today's more sophisticated analysis has given rise to an entirely new specialty, forensic dentistry, which works admirably when trauma to the rest of the body makes other kinds of identification impossible. Because tooth enamel is one of the hardest minerals known, the teeth are the body's most durable parts. They alone can survive fiery disasters where other identifiable remains are destroyed.

Forensic dentistry has been used rather recently to identify the remains of passengers in the crash of a DC-10 at Chicago's O'Hare Airport, in the investigation of the murders of young men linked to John Gacy near Chicago, and in the aftermath of the Jonestown community massacre in Guyana. In the San Diego plane crash 21 months ago, in which 144 people died, 140 of the victims were positively identified, 50 of them by their dental prints alone.

Working on the premise that no two mouths are alike, at least one nation, Norway, now requires that identification teams in disasters include law-enforcement officials, physicians, and dentists as well.

Thirty to 40 different methods are used to chart the mouth, but all have the same basic features. Each record notes missing or extra teeth, prosthetic devices, and the type, size, and location of fillings. Other measurable characteristics include the shape of the jaw and the position, shape, and relationship of the teeth.

Even this method of analysis has its limitations, though. Teeth change over the years, and dental records are often years out of date. Furthermore, investigators of a recent air crash in Europe found it impossible to identify five of the victims, because they had all been wearing complete sets of dentures.

SIGN IN, PLEASE

If no two mouths are the same, neither are any two signatures. One newly designed system has made the signature a more accurate means of identification than a page of handwriting, and more foolproof even than a fingerprint.

Handwriting analysis has always been a painstaking and questionably accurate process. Chance factors like mood changes or different writing instruments could disguise almost any hand, and forgers could copy any style so closely that finding the genuine document among ten purported wills by Howard Hughes could become an exercise in creative logic.

Now, however, Jacob Sternberg, a former college professor and founder of Conversational Systems Corporation, has developed a signature-analysis method that can avoid the pitfalls of chance and outwit the most skillful forgers. His electronic system measures not the appearance of the signature but its "pressure pattern"—the variations in pressure on the writing surface that occur as the signature is being written.

Sternberg found out that, although handwriting and signature are both complex patterns of learned behavior, the signature is repeated so often that it becomes almost a pure reflex action; its patterns are ingrained in the "memory" of the muscles. As a result, even when we sign something in the dark or in a rush at the bank, the pressure pattern remains distinctive.

When the U.S. Air Force tested Sternberg's system as a way to control access to highly restricted areas, its low error ratio pinpointed the Sternberg system's incredible potential for military and commercial use. The reason for this success lies in what the system identifies: It judges neither a reproducible characteristic (a fingerprint can be copied onto a rubber glove) nor some static photograph or reproduction, but the subconscious actions of an individual performing a complex task. In this way it identifies a combination of our unmistakable characteristics, personality, physical nature, and pattern of action.

Ever since humans first began to gather in groups, they have recognized the problem of identity. Early on, they learned to endow objects and individuals with names. For a long time these unique verbal signatures were enough to give each person a mark of his own within the community.

In these depersonalizing days, however, names have come a cropper. Anyone unfortunate enough to be dubbed Smith, Jones, or Johnson is lost amid a faceless army of identically named people. Names have lost their meanings. Our government, like most computers, much prefers to label us by number. This system works, but there is something depressing and a little scary about basing our uniqueness on arbitrary strings of digits.

Biological tools are on their way to changing all this. Nature's IDs can be as specific as numbers and more personal than names. Already they are employed in catastrophic conditions when nothing can visibly distinguish one body from another. Already they are establishing individual identities at some of our most sophisticated centers of research and learning.

Instead of forcing us toward anonymity, biological identifiers effectively celebrate our absolute specialness. It won't be long before checking a chromosome or tissue type will be as common as checking a signature is today. The results will be both more certain and more satisfying.

THE INSTINCT TO
LEARN

Birds do it, bees do it, perhaps even humans are programmed to acquire critical information at specific times.

James L. and Carol Grant Gould

James L. Gould, professor of biology at Princeton University, studies the navigation and communication of the honey bee. Carol Grant Gould is a writer and research associate in Princeton's biology department.

When a month-old human infant begins to smile, its world lights up. People reward these particular facial muscle movements with the things a baby prizes—kisses, hugs, eye contact, and more smiles. That first smile appears to be a powerful ingredient in the emotional glue that bonds parent to child. But scientists wonder whether that smile is merely a chance occurrence, which subsequently gets reinforced by tangible rewards, or an inexorable and predetermined process by which infants ingratiate themselves with their parents.

If this sounds like another chapter in the old nature/nurture controversy, it is—but a chapter with a difference. Ethologists, specialists in the mechanisms behind animal behavior, are taking a new look at old—and some new—evidence and are finding that even while skirmishing on a bloody battleground, the two camps of instinctive and learned behavior seem to be heading with stunning rapidity and inevitability toward an honorable truce.

Fortunately for the discord that keeps disciplines alive and fit, animal behavior may be approached from two vantage points. One of these sees instinct as the moving force behind behavior: Animals resemble automatons preordained by their genetic makeup to behave in prescribed ways. The other views animals as basically naive, passive creatures whose behavior is shaped, through the agency of punishment and reinforcement, by chance, experience, and environmental forces.

In the last few years, however, these two views have edged towards reconciliation and, perhaps, eventual union. Case after case has come to light of environmentally influenced learning which is nonetheless rigidly controlled by genetic programming. Many animals, ethologists are convinced, survive through learning—but learning that is an integral part of their programming, learning as immutable and as stereotyped as the most instinctive of behavioral responses. Furthermore, neurobiologists are beginning to discover the nerve circuits responsible for the effects.

Plenty of scientists are still opposed to this new synthesis. The most vociferous are those who view the idea of programmed learning as a threat to humanity's treasured ideas of free will. However, it now appears that much of what we learn is forced upon us by innate drives and that even much of our "culture" is deeply rooted in biology.

As though this were not enough of a shock to our ingrained ideas of man's place in the universe, it looks as though the reverse is true, too: Man is not the sole, lofty proprietor of culture; "lower" animals—notably monkeys and birds—also have evolved various complicated ways of transferring environmentally learned information to others of their own kind.

The honey bee provides entrancing insights into the lengths to which nature goes in its effort to program learning. These little animals must learn a great many things about their world: what flowers yield nectar at what specific times of day, what their home hives look like under the changes of season and circumstance, where water is to be found.

But new work reveals that all this learning, though marvelous in its variety and complexity, is at the same time curiously constrained and machinelike. Certain things that bees learn well and easily, they can learn only at certain specific "critical periods." For example, they must relearn the appearance and location of their hives on their first flight out every morning; at no other time will this information register in the bee's brain. Bee-keepers have known for centuries that if they move a hive at night the bees come and go effortlessly the next day. But if they move the hive even a few meters at any time after the foraging bees' first flight of the day, the animals are disoriented and confused. Only at this one time is the home-learning program turned on: Evidently this is nature's way of compensating for changing seasons and circumstances in an animal whose vision is so poor that its only means of locating the hive is by identifying the landmarks around it.

Since bees generally harvest nectar from one species of flower at a time, it seems clear that they must learn to recognize flower species individually. Karl von Frisch, the noted Austrian zoologist, found that bees can distinguish any color, from yellow to green and blue and into the ultraviolet. However, they learn the color of a flower only in the two seconds before they land on it. Von Frisch also discovered that bees can discriminate a single odor out of several hundred. Experimentation reveals that this remarkable ability is similarly constrained: Bees can learn odor only while they are actually standing on the flower. And finally, only as they are flying away can they memorize any notable landmarks there might be around the flower.

Learning then, at least for bees, has thus become specialized to the extent that specific cues can be learned only at specific times, and then only in specific contexts.

The bees' learning programs turn out to be restricted even further. Once the bits of knowledge that make up a behavior have been acquired, such as the location, color, odor, shape, and surrounding landmarks of a food source, together with the time it is likely to yield the most nectar, they form a coherent, holistic set. If a single component of the set is changed, the bee must learn the whole set over again.

In a very real sense, then, honey bees are carefully tuned learning machines. They learn just what they are programmed to learn, exactly when and under exactly the circumstances they are programmed to learn it. Though this seems fundamentally different from the sort of learning we are used to seeing in higher animals such as birds and mammals—and, of course, ourselves—careful research is uncovering more and more humbling similarities. Programmed memorization in vertebrates, though deceptively subtle, is widespread. The process by which many species of birds learn their often complex and highly species-specific songs is a compelling case in point.

Long before the birds begin to vocalize, their species' song is being learned, meticulously "taped" and stored somewhere in their memory banks. As the bird grows, the lengthening days of spring trigger the release of specific hormones in the males which in turn spur them to reproduce first the individual elements of syllables and later the sequence of the stored song. By a trial and error process the birds slowly learn to manipulate their vocal musculature to produce a match between their output and the recording in their brains. Once learned, the sequence becomes a hardwired motor program, so fixed and independent of feedback that if the bird is deafened his song production remains unaffected.

This prodigious feat of learning, even down to the regional dialects which some species have developed, can be looked at as the gradual unfolding of automatic processes. Peter Marler of the Rockefeller University and his students, for instance, have determined that there are rigorous time constraints on the song learning. They have discovered that in the white-crowned sparrow the "taping" of the parental song can be done only between the chicks' 10th and 50th days. No amount of coaching either before or after this critical period will affect the young birds. If they hear the correct song during this time, they will be able to produce it themselves later (or, if females, to respond to it); if not, they will produce only crude, vaguely patterned vocalizations.

In addition, the white-crowned sparrow, though reared in nature in an auditory environment filled with the songs of other sparrows and songbirds with rich vocal repertoires, learns *only* the white-crowned sparrow song. Marler has recently been able to confirm that the parental song in another species—the swamp sparrow—contains key sounds that serve as auditory releasers, the cues that order the chicks' internal tape recorders to switch on. Ethologists refer to any simple signal from the outside world that triggers a complex series of actions in an animal as a releaser.

Here again, amazing feats of learning, particularly the sorts of learning that are crucial to the perpetuation of an animal's genes, are rigidly controlled by biology.

The kind of programmed learning that ethologists have studied most is imprinting, which calls to mind a picture of Konrad Lorenz leading a line of adoring goslings across a Bavarian meadow. Newborn animals that must be able to keep up with ever-moving parents —antelope and sheep, for example, as well as chicks and geese— must rapidly learn to recognize those parents if they are to survive. To achieve this noble aim evolution has built into these creatures an elegant learning routine. Young birds are driven to follow the parent out of the nest by an exodus call. Though the key element in the call varies from species to species—a particular repetition rate for one, a specific downward frequency sweep for another—it is always strikingly simple, and it invariably triggers the chicks' characteristic following response.

As the chicks follow the sound they begin memorizing the distinguishing characteristics of the parent, with two curious but powerful constraints. First, the physical act of following is essential: Chicks passively transported behind a calling model do not learn; in fact, barriers in a chick's path that force it to work harder speed and strengthen the imprinting. Second, the cues that the chick memorizes are also species-specific: One species will concentrate on the inflections and tone of the parent's voice but fail to recall physical appearance, while a closely related species memorizes minute details of physical appearance to the exclusion of sounds. In some species of mammals, the learning focuses almost entirely on individual odor. In each case, the critical period for imprinting lasts only a day or two. In this short but crucial period an ineradicable picture of the only individual who will feed and protect them is inscribed in the young animals' memories.

By contrast, when there is no advantage to the animal in learning specific details, the genes don't waste their efforts in programming them in. In that case, blindness to detail is equally curious and constrained. For instance, species of gulls that nest cheek by jowl are programmed to memorize the most minute details of their eggs' size and speckling and to spot at a glance any eggs which a careless neighbor might have added to their nest—eggs which to a human observer look identical in every respect. Herring gulls, on the other hand, nest far enough apart that they are unlikely ever to get their eggs confused with those of other pairs. As a result, they are unconscious of the appearance of their eggs. The parents will complacently continue to incubate even large black eggs that an experimenter substitutes for their small speckled ones. The herring gulls' insouciance, however, ends there: They recognize their chicks as individuals soon after hatching. By that time, their neighbors' youngsters are capable of wandering in. Rather than feed the genes of their neighbors, the parents recognize foreign chicks and often eat *them*.

The kittiwake gull, on the other hand, nests in narrow pockets on cliff faces, and so the possibility that a neighbor's chick will wander down the cliff into its nest is remote. As a result kittiwakes are not programmed to learn the appearance of either eggs or young, and even large black cormorant chicks may be substituted for the small, white, infant kittiwakes.

Simply from observing animals in action, ethologists have learned a great deal about the innate bases of behavior. Now, however, neurobiologists are even tracing the circuitry of many of the mechanisms that control some of these elements. The circuits responsible for simple motor programs, for example, have been located and mapped out on a cell-by-cell basis in some cases and isolated to a single ganglion in others.

A recent and crucial discovery is that the releasers imagined by ethologists are actually the so-called feature detectors that neurobiologists have been turning up in the auditory and visual systems. In recent years, neurobiologists have discovered that there are certain combinations of nerve cells, built into the eyes and brains of all creatures, that respond only to highly specific features: spots of a certain size, horizontal or vertical lines, and movement, for example. In case after case, the basic stimulus required to elicit an innate response in animals corresponds to one or a very simple combination of discrete features systematically sought out by these specialized cells in the visual system.

The parent herring gull, for instance, wears a single red spot near the tip of its lower bill, which it waves back and forth in front of its chicks when it has food for them. The baby gulls for their part peck at the waving spot which, in turn, causes the parent to release the food. First, Niko Tinbergen, the Dutch Nobel Prize winner and co-founder of the science of ethology with Lorenz and von Frisch, and later the American ethologist Jack Hailman have been able to show that the chicks are driven to peck not by the sight of their parent but at that swinging vertical bill with its red spot. The moving vertical line and the spot are the essential features that guide the chicks, which actually prefer a schematic, disembodied stimulus—a knitting needle with a spot, for example.

Though the use of two releasers to direct their pecking must greatly sharpen the specificity of the baby gulls' behavior, chicks do quickly learn to recognize their parents, and the mental pictures thus formed soon replace the crude releasers. Genes apparently build in releasers not only to trigger innate behavior but, more important, to direct the attention of animals to things they must learn flawlessly and immediately to survive.

Even some of what we know as culture has been shown to be partially rooted in programmed learning, or instinct. Many birds, for instance, mob or attack potential nest predators in force, and they do this generation after generation. But how could these birds innately know their enemies? In 1978 the German ethologist Eberhard Curio placed two cages of blackbirds on opposite sides of a hallway, so that they could see and hear each other. Between the two cages he installed a compartmented box, which allowed the occupants of one cage to see an object on their side but not the object on the other. Curio presented a stuffed owl, a familiar predator, on one side, and an innocuous foreign bird, the Australian honey guide, on the other. The birds that saw the owl went berserk with rage and tried to mob it through the bars of the cage. The birds on the other side, seeing only an unfamiliar animal and the enraged birds, began to mob the stuffed honey guide. Astonishingly, these birds then passed on this prejudice against honey guides through a chain of six blackbirds, all of which mobbed honey guides whenever they encountered one. Using the same technique, Curio has raised generations of birds whose great-great-grandparents were tricked into mobbing milk bottles and who consequently teach their young to do the same.

What instigates the birds—even birds raised in total isolation—to pay so much attention to one instance of mobbing that they pass the information on to their offspring as a sort of taboo, something so crucial to their survival that they never question if or why these predators must be attacked? The mobbing call, it turns out, serves as yet another releaser that switches on a learning routine.

Certain sounds in the mobbing calls are so similar among different species that they all profit from each other's experience. This is why we often see crows or other large birds being mobbed by many species of small birds at once. So deeply ingrained in the birds is this call that birds raised alone in the laboratory are able to recognize it, and the calls of one species serve to direct and release enemy-learning in others. Something as critical to an animal's survival as the recognition of enemies, then, even though its finer points must be learned and transmitted culturally, rests on a fail-safe basis of innately guided, programmed learning.

The striking food-avoidance phenomenon is also a good place to look for the kind of innately directed learning that is critical to survival. Many animals, including humans, will refuse to eat a novel substance which has previously made them ill. Once a blue jay has tasted one monarch butterfly, which as a caterpillar fills itself with milkweed's poisonous glycosides, it will sedulously avoid not only monarchs but also viceroys—monarch look-alikes that flaunt the monarchs' colors to cash in on their protective toxicity. This programmed avoidance is based on the sickness which must appear within a species-specific interval after an animal eats, and the subsequent food avoidance is equally strong even if the subject knows from experience that the effect has been artificially induced.

But what is the innate mechanism when one blue tit discovers how to pierce the foil caps of milk bottles left on doorsteps to reach the cream, and shortly afterwards blue tits all over England are doing the same thing? How are theories of genetic programming to be invoked when one young Japanese macaque monkey discovers that sweet potatoes and handfuls of grain gleaned from a sandy shore are tastier when washed off in the ocean, and the whole troop (except for an entrenched party of old dominant males) slowly follows suit? Surely these are examples pure and simple of the cultural transmission of knowledge that has been environmentally gained.

Perhaps not. What the blue tits and the monkeys pass on to their colleagues may have an innate basis as well. The reason for this precocious behavior—and we say this guardedly—may be in a strong in-

The cells that bring you the world

There was a time when the visual system was thought of as little more than a pair of cameras (the eyes), cables (the optic nerves), and television screens (the visual cortex of the brain). Nothing could be farther from the truth. We now know that the visual system is no mere passive network of wires but an elaborately organized and highly refined processing system that actively analyzes what we see, systematically exaggerating one aspect of the visual world, ignoring or discarding another.

The processing begins right in the retina. There the information from 130 million rods and cones is sifted, distorted, and combined to fit into the four or so million fibers that go to the brain. The retinas of higher vertebrates employ one layer of cells to sum up the outputs of the rod-and-cone receptors. The next layer of retinal cells compares the outputs of adjacent cells in the preceding tier. The result is what is known as a spot detector: One type of cell in the second layer signals the brain when its compare/contrast strategy discovers a bright field surrounded by darkness (corresponding to a bright spot in the world). Another class of cell in the same layer has the opposite preference and fires off when it encounters dark spots.

The next processing step takes this spot information and, operating on precisely the same comparison strategy, wires cells that are sensitive only to spots moving in particular directions at specific speeds. The output of these spot detector cells also provides the raw material from which an array of more sophisticated feature detectors sort for lines of each particular orientation. These feature detectors derive their name from their ability to register the presence or absence of one particular sort of stimulus in the environment. Building on these cells, the next layer of processing sorts for the speed and direction of moving lines, each cell with its own special preference. Other layers judge distance by comparing what the two eyes see.

The specific information that cells sort for in other retinal layers and visual areas of the brain is not yet understood. Research will probably reveal that these extremely complex feature detectors provide us with what we know as conscious visual experience. Our awareness of all this subconscious processing, along with the willful distortions and tricks it plays on us, comes from the phenomenon of optical illusions. When we experience an optical illusion, it is the result of a particular (and, in the world to which we evolved, useful) quirk in the visual mechanism.

Feature detectors are by no means restricted to the visual system. In birds and bats, for instance, specialized cells have been found that recognize many nuances in sound—locations, repetition rates, time intervals, and precise changes in pitch— that allow the creatures to form an auditory picture of the world.

There is every reason to suppose that our experience of the world is based on the results of this massive editing. Since neural circuits differ dramatically from species to species according to the needs of each, the world must look and sound different to bees, birds, cats, and people.

—*J.L.G. and C.G.G.*

stinctive drive on the part of all animals to copy mindlessly certain special aspects of what they see going on around them. Chicks, for instance, peck at seeds their mother has been trained to select, appar-

ently by watching her choices and copying them. In the case of many mammals, this drive is probably combined with an innate urge to experiment. The proclivity of young animals, particularly human children, to play with food, along with their distressing eagerness to put virtually anything into their mouths, lends support to the experimentation theory. Perhaps it is the young, too naive to know any better, who are destined by nature to be the primary source of cultural innovation. The more mature become the equally indispensable defenders of the faith, the vehicles of cultural transmission.

Patterns, then, however subtle, are beginning to emerge that unify the previously unrelated studies of instinct and learning. Virtually every case of learning in animals that has been analyzed so far depends in at least some rudimentary way on releasers that turn on the learning routine. And that routine is generally crucial to the perpetuation of the animal's genes.

Even the malleable learning we as humans pride ourselves on, then, may have ineradicable roots in genetic programming, although we may have difficulty identifying the programs, blind as we are to our own blindness. For example, you cannot keep a normal, healthy child from learning to talk. Even a child born deaf goes through the same babbling practice phase the hearing child does. Chimpanzees, by contrast, can be inveigled into mastering some sort of linguistic communications skills, but they really could not care less about language: The drive just is not there.

This view of human insight and creativity may be unromantic, minimizing as it does the revered role of self-awareness in our everyday lives, but the pursuit of this line of thinking could yield rich rewards, providing us with invaluable insights into our own intellectual development. The times we are most susceptible to particular sorts of input, for instance, may be more constrained than we like to think. The discovery of the sorts of cues or releasers that might turn on a drive to learn specific things could open up new ways of teaching and better methods for helping those who are culturally deprived. Best of all, analyzing and understanding those cues could greatly enrich our understanding of ourselves and of our place in the natural order.

TWINS
R·E·U·N·I·T·E·D

Constance Holden

Anyone who knows identical twins has undoubtedly experienced profound paradoxical reactions: astonishment at their similarities and amazement at their differences. Twins have always been regarded as special; in some aboriginal societies they are venerated, in some they are slain. Identical twins also have figured as protagonists in one of the oldest and bitterest disputes among scientists—the nature/nurture controversy.

It has been almost a century since Sir Francis Galton, the founder of the science of eugenics, first proposed studying identical twins to determine the relative influences of heredity and environment on human development. Now a group of researchers at the University of Minnesota are involved in the most comprehensive investigation ever undertaken of twins raised apart. Initiated by psychologist Thomas Bouchard, the study consists of exhaustive physical, psychological, and biographical inventories of every twin. So far, 15 pairs, each of whom spent a week at the university, have been tested. Bouchard has located 18 additional pairs with the help of the publicity the project is getting.

It will take at least five or six years to analyze definitively the masses of data accumulated so far. But after a year and a half of examining twin pairs, the 17 members of the research team, which includes six psychologists, two psychiatrists, and nine other medical experts, have been overwhelmed with the similarities of the participating pairs. Some similarities, of course, are clearly coincidental, such as the twins who are both named Jim. Others may only seem to be: Twin sisters who had never met before each wore seven rings on their fingers. They may not have inherited their fondness for rings; perhaps only the same pretty hands prompted the fondness. And still other similarities, like phobias, which were long thought to be learned, may turn out after all to be hereditary.

"I frankly expected far more differences than we have found so far," says Bouchard. "I'm a psychologist, not a geneticist. I want to find out how the environment works to shape psychological traits." Bouchard has encountered some hostility on the Minneapolis campus because he remains aloof from the ideological fashion, which holds that behavior is largely shaped by environmental influences. Student hostility is ironic in view of Bouchard's political activism at the University of California in the radical 1960s.

Actually Bouchard's quest is considerably more sophisticated than critics perceive. To discover how the environment shapes behavior, one first must have an idea of how innate tendencies work to select a particular environment. The same surroundings can be interpreted very differently by two individuals. One person may find a library a good place to read, for example, while another may regard it as an excellent hunting ground for members of the opposite sex.

In the hope of making some of these distinctions, Bouchard has drawn upon his extensive experience in personality testing to develop the battery of tests for each twin pair. In addition, the scientists take detailed medical histories that include diet, smoking and exercise habits, electrocardiograms, chest X rays, heart stress tests, and pulmonary exams. They inject the twins with a variety of substances to determine allergies. They wire them to EEG machines to measure their brain wave responses to various stimuli.

During the six days they devote to each twin pair, the team intersperses the physiological probes with several dozen written tests, which ask some 15,000 questions. These cover family and childhood environment, fears and phobias, personal interests, vocational aptitudes, values, reading and television viewing habits, musical tastes, and aesthetic judgments. Each pair of twins undergoes three comprehensive psychological inventories. In addition each takes ability tests: the Wechsler Adult Intelligence Scale, the main adult IQ test, and numerous others that reveal skills in information processing, vocabulary, spatial abilities, numerical processing, mechanical ability, and memory. Throughout the week there is a good deal of overlap in an attempt to "measure the same underlying factor at different times," says Bouchard.

No scientific conclusions can yet be drawn from the masses of data collected, but the team has made a number of provocative observations.

The "Jim twins," as they have come to be known, have histories that are riddled with bizarre coincidences. Jim Springer and Jim Lewis were adopted as four-week-old infants into working-class Ohio families. They never met each other until they were 39 years old. Both had law enforcement training and worked part time as deputy sheriffs. Both vacationed in Florida; both drove Chevrolets. Much has been made of the fact that their lives are marked by a trail of similar names. Both had dogs named Toy. They married and divorced women named Linda and remarried women named Betty. They named their sons James Allan and James Alan. While the laws of chance dictate against such an unlikely string of coincidences, Bouchard has noted that twins seem to be highly subject to such strange similarities.

Other similarities, however, are probably more than coincidental. In school both twins liked math but not spelling. They currently enjoy mechanical drawing and carpentry. They have almost identical drinking and smoking patterns, and they chew their fingernails down to the nubs. Investigators thought their similar medical histories were astounding. In addition to having hemorrhoids and identical pulse, blood pressure, and sleep patterns, both had inexplicably put on ten pounds at the same time in life. Each suffers from "mixed headache syndrome," a combination tension and migraine headache. Both first suffered headaches at the age of 18. They have these late-afternoon headaches with the same frequency and same degree of disability, and the two used the same terms to describe the pain they experienced.

The twins also have their differences. One wears his hair over his forehead; the other has it slicked back with sideburns. One expresses himself better orally, the others in writing. Even though the emotional environments in which they were brought up were different, still the profiles on their psychological inventories were much alike.

Another much publicized pair is 47-year-old Oskar Stöhr and Jack Yufe. These two have the most dramatically different backgrounds of all the twins studied. Born in Trinidad of a Jewish father and a German mother, they were separated shortly after birth. The mother took Oskar back to Germany where he was raised as a Catholic and a Nazi by his grandmother. Jack was raised in the Caribbean as a Jew by his father and spent part of his youth on an Israeli kibbutz. As might be expected, the men now lead markedly different lives: Oskar, an industrial supervisor in Germany, is married, a devoted union man, and a skier. Jack runs a retail clothing store in San Diego, is separated from his wife, describes himself as a workaholic, and enjoys sailing as a hobby.

Their families had never corresponded, yet similarities were evident when they first met at the airport. Both sported mustaches and two-pocket shirts with epaulets. Each had his wire-rimmed glasses with him. They share abundant idiosyncrasies: The twins like spicy foods and sweet liqueurs, are absentminded, fall asleep in front of the television, think it is funny to sneeze in a crowd of strangers, flush the toilet before using it, store rubber bands on their wrists, read magazines from back to front, dip buttered toast in their coffee. Oskar did not take all the tests because he speaks only German, but the two had similar profiles on the Minnesota Multiphasic Personality Inventory. Oskar yells at his wife, which Jack did before he was separated. Although the two were raised in different cultures, investigator Bouchard professed himself impressed by the similarities in their mannerisms, the questions they asked, their "temperament, tempo, the way they do things." He also thinks the pair supply "devastating" evidence against the feminist contention that children's personalities are shaped differently according to the sex of those who rear them, since Oskar was raised by women and Jack by men.

The Bouchard team has enjoyed a run on female British twins in their late 30s, all separated during World War II, and all raised by people who did not know each other. Although Bridget and Dorothy, the housewives sporting the seven rings each, were reared in different socioeconomic settings, the class difference was evident only in that the twin raised in modest circumstances had bad teeth. Otherwise, the investiga-

Though Barbara and Daphne were raised separately, their handwriting is remarkably similar.

The Coffman Gallery area of the Coffman Union Program Council is responsible for the visual arts programs and exhibitions which are presented in its three formal gallery spaces and other facilities

The Coffman Gallery area of the Coffman Union Program Council is responsible for the visual arts programs and exhibitions which are presented in its three formal gallery spaces and other facilities.

tors conclude, the twins share "striking similarities in all areas."

Another pair, Daphne and Barbara, are fondly remembered as the "giggle sisters," because they were always setting each other off. There are evidently no gigglers in their adoptive families. The sisters both handle stress by ignoring it. Both avoid conflict and controversy; neither has any interest in politics. This similarity is particularly provocative since avoidance of conflict is "classically regarded as learned behavior," says Bouchard.

Irene and Jeanette, 35, who were brought up respectively in England and Scotland, turn out to have the same phobias. Both are claustrophobic, and balked when invited to go into a cubicle for their electroencephalograms. They independently agreed to enter the cubicle if the door were left open. Both are timid about ocean bathing; they resolve the problem by backing in slowly. Neither likes escalators. Both are compulsive counters, of everything they see, such as the wheels of trucks. Both count themselves to sleep.

Other similarities turned up in a number of twin pairs. Tests of vision, for example, showed that even in cases in which one twin wears glasses and the other does not, both require the same type of correction.

Bouchard says it is commonplace for identical twins to engage in coincidental behavior: Both will buy the same gift for their mother, or even select the same birthday card. But he is finding that such coincidences often crop up with twins raised apart. A favorite episode involves two middle-aged women in his study. Once, as children, they were brought together briefly to meet each other. Both turned up wearing the same dress.

As for IQ, a hotly controversial area of psychological testing, the Minnesota study confirms what other researchers repeatedly have shown: that of all psychological traits measured in identical twins, this one shows the highest degree of similarity. Bouchard, mindful of charges of investigator bias that are often leveled at IQ testers, arranges for outside contractors to come in to administer and score the Wechsler intelligence test. He wants to avoid imputations of fraud such as those leveled recently at the eminent British educational psychologist, the late Sir Cyril Burt, who some accuse of "cooking" his IQ data in order to buttress his belief in the heritability of intelligence. In the case of the Minnesota study, most of the scores do not differ any more than those of two tests taken by the same person at different times. In the few that vary considerably, the variance appears to reflect large differences in education.

Psychological histories also correspond well to studies of twins reared together. Psychiatrist Leonard Heston, the father of identical twin girls, is particularly well suited to studying these histories. Scientists have known for some time that there is a genetic component, often involving chemical imbalance, in many mental illnesses. For example, if one twin suffers from depression or schizophrenia the other stands a 45 percent chance of succumbing as well. Heston was surprised by the extent to which twins raised separately tend to share emotional problems—"things such as mild depressions and phobias that I would never have thought of as being particularly genetically mediated," Heston remarks. "Now at least, there are grounds for a very live hypothesis."

Psychologist David Lykken finds that the brain waves of the twin pairs in the Minnesota study resemble each other in the same way as those of twins who have been raised together. Moreover, Lykken finds that tracings from each twin of a pair differ no more than tracings from the same person taken at different times.

To be sure, identical twins differ in myriad ways. One of the most common is in tendencies toward introversion and extroversion. Another common difference, according to the English researcher James Shields, is dominance and submissiveness. Bouchard and his colleagues note that one twin is likely to be more aggressive, outgoing, and confident.

But they find twins reared together usually differ the most in that respect. In other words, dominance and submission seem to be traits that twins use to assert their individuality. Because twins brought up together often feel compelled to exaggerate their differences, David Lykken thinks it is possible that twins reared separately may actually have more in common with each other than those raised together.

Because of the comparatively small numbers of twins they have studied, the investigators face difficulties in proving that their results are more than a random collection of case histories. What they would like to do, according to Tellegen, is "invent methods for analyzing traits in an objective manner, so we can get statistically cogent conclusions from a single case." This will require first establishing what is to be expected on the basis of chance alone. For example, how likely are two randomly selected IQ scores to be as similar as those of two identical twins? This method of analysis will be crucial in weeding out similarities between twins that may be no more common than coincidences that occur between randomly paired people.

Of all the members of the Bouchard team, Lykken is the most willing to entertain ideas that so far are only supported by subjective impressions. He says, "Looking at these 15 pairs of identical twins, I have an enhanced sense of the importance of the genes in determining all aspects of behavior." But he acknowledges the importance of the environment as well. "What is emerging in my mind," Lykken concludes, "is that the most important thing to come out of this study is a strong sense that vastly more of human behavior is genetically determined or influenced than we ever supposed."

ETHNIC DIFFERENCES IN BABIES

Striking differences in temperament and behavior
among ethnic groups show up in babies only a few days old.

DANIEL G. FREEDMAN

Daniel G. Freedman, *professor of the behavioral sciences at The University of Chicago, spent last fall in Australia as a visiting fellow in the department of anthropology at the Australian National University in Canberra. There he extended his research into the newborn capacities of Australian aborigines. His doctorate in psychology from Brandeis University was followed by a postdoctoral fellowship at Mt. Zion Psychiatric Clinic and the Langley Porter Neuropsychiatric Institute in San Francisco. Much of the information in this article appears in an expanded form in Freedman's book,* Human Sociobiology, *published by the Free Press. With Fred Strayer and Donald Omark, Freedman edited* Human Status Hierarchies.

The human species comes in an admirable variety of shapes and colors, as a walk through any cosmopolitan city amply demonstrates. Although the speculation has become politically and socially unpopular, it is difficult not to wonder whether the major differences in physical appearances are accompanied by standard differences in temperament or behavior. Recent studies by myself and others of babies only a few hours, days, or weeks old indicate that they are, and that such differences among human beings are biological as well as cultural.

These studies of newborns from different ethnic backgrounds actually had their inception with work on puppies, when I attempted to raise dogs in either an indulged or disciplined fashion in order to test the effects of such rearing on their later behavior.

I spent all my days and evenings with these puppies, and it soon became apparent that the breed of dog would become an important factor in my results. Even as the ears and eyes opened, the breeds differed in behavior. Little beagles were irrepressibly friendly from the moment they could detect me; Shetland sheepdogs were very, very sensitive to a loud voice or the slightest punishment; wire-haired terriers were so tough and aggressive, even as clumsy three-week-olds, that I had to wear gloves while playing with them; and finally, Basenjis, barkless dogs originating in Central Africa, were aloof and independent. To judge by where they spent their time, sniffing and investigating, I was no more important to them than if I were a rubber balloon.

When I later tested the dogs, the breed indeed made a difference in their behavior. I took them, when hungry, into a room with a bowl of meat. For three minutes I kept them from approaching the meat, then left each dog alone with the food. Indulged terriers and beagles waited longer before eating the meat than did disciplined dogs of the same breeds. None of the Shetlands ever ate any of the food, and all of the Basenjis ate as soon as I left.

I later studied 20 sets of identical and fraternal human twins, following them from infancy until they were 10 years old, and I became convinced that both puppies and human babies begin life along developmental pathways established by their genetic inheritance. But I still did not know whether infants of relatively inbred human groups showed differences comparable to the breed differences among puppies that had so impressed me. Clearly, the most direct way to find out was to examine very young infants, preferably newborns, of ethnic groups with widely divergent histories.

Since it was important to avoid projecting my own assumptions onto the babies' behavior, the first step was to develop some sort of objective test of newborn behavior. With T. Berry Brazelton, the Harvard pediatrician, I developed what I called the Cambridge Behavioral and Neurological Assessment Scales, a group of simple tests of basic human reactions that could be administered to any normal newborn in a hospital nursery.

In the first study, Nina Freedman and I compared Chinese and Caucasian babies. It was no accident that we chose those two groups, since my wife is Chinese, and in the course of learning about each other and our families, we came to believe that some character differences might well be related to differences in our respective gene pools and not just to individual differences.

Armed with our new baby test, Nina and I returned to San Francisco, and to the hospital where she had borne our first child. We examined, alternately, 24 Chinese and 24 Caucasian newborns. To keep things neat, we made sure that all the Chinese were of Cantonese (South Chinese) background, the Caucasians of Northern European origin, that the sexes in both groups were the same, that the mothers were the same age, that they had about the same number of previous children, and that both groups were administered the same drugs in the same amounts. Additionally, all of the families were members of the same health plan, all of the mothers had had approximately the same number of prenatal visits to a doctor, and all were in the same middle-income bracket.

It was almost immediately clear that we had struck pay dirt; Chinese and Caucasian babies indeed behaved like two different breeds. Caucasian babies cried more easily, and once started, they were harder to console. Chinese babies adapted to almost any position in which they were placed; for example, when placed face down in their cribs, they tended to keep their faces buried in the sheets rather than immediately turning to one side, as did the Caucasians. In a similar maneuver (called the "defense reaction" by neurologists), we briefly pressed the baby's nose with a cloth. Most Caucasian and black babies fight this maneuver by immediately turning away or swiping at the cloth with their hands, and this is reported in most Western pediatric textbooks as the normal, expected response. The average Chinese baby in our study, however, simply lay on his back and breathed through his mouth, "accept-

ing" the cloth without a fight. This finding is most impressive on film.

Other subtle differences were equally important, but less dramatic. For example, both Chinese and Caucasian babies started to cry at about the same points in the examination, especially when they were undressed, but the Chinese stopped sooner. When picked up and cuddled, Chinese babies stopped crying immediately, as if a light switch had been flipped, whereas the crying of Caucasian babies only gradually subsided.

In another part of the test, we repeatedly shone a light in the baby's eyes and counted the number of blinks until the baby "adapted" and no longer blinked. It should be no surprise that the Caucasian babies continued to blink long after the Chinese babies had adapted and stopped.

It began to look as if Chinese babies were simply more amenable and adaptable to the machinations of the examiners, and that the Caucasian babies were registering annoyance and complaint. It was as if the old stereotypes of the calm, inscrutable Chinese and the excitable, emotionally changeable Caucasian were appearing spontaneously in the first 48 hours of life. In other words, our hypothesis about human and puppy parallels seemed to be correct.

The results of our Chinese-Caucasian study have been confirmed by a student of ethologist Nick Blurton-Jones who worked in a Chinese community in Malaysia. At the time, however, our single study was hardly enough evidence for so general a conclusion, and we set out to look at other newborns in other places. Norbett Mintz, who was working among the Navaho in Tuba City, Arizona, arranged for us to come to the reservation in the spring of 1969. After two months we had tested 36 Navaho newborns, and the results paralleled the stereotype of the stoical, impassive American Indian. These babies outdid the Chinese, showing even more calmness and adaptability than we found among Oriental babies.

We filmed the babies as they were tested and found reactions in the film

we had not noticed. For example, the Moro response was clearly different among Navaho and Caucasians. This reaction occurs in newborns when support for the head and neck suddenly disappears. Tests for the Moro response usually consist of raising and then suddenly dropping the head portion of the bassinet. In most Caucasian newborns, after a four-inch drop the baby reflexively extends both arms and legs, cries, and moves in an agitated manner before he calms down. Among Navajo babies, crying was rare, the limb movements were reduced, and calming was almost immediate.

I have since spent considerable time among the Navaho, and it is clear that the traditional practice of tying the wrapped infant onto a cradle board (now practiced sporadically on the reservation) has in no way induced stoicism in the Navaho. In the halcyon days of anthropological environmentalism, this was a popular conjecture, but the other way around is more likely. Not all Navaho babies take to the cradle board, and those who complain about it are simply taken off. But most Navaho infants calmly accept the board; in fact, many begin to demand it by showing signs of unrest when off. When they are about six months old, however, Navaho babies do start complaining at being tied, and "weaning" from the board begins, with the baby taking the lead. The Navaho are the most "in touch" group of mothers we have yet seen, and the term mother-infant *unit* aptly describes what we saw among them.

James Chisholm of Rutgers University, who has studied infancy among the Navaho over the past several years, reports that his observations are much like my own. In addition, he followed a group of young Caucasian mothers in Flagstaff (some 80 miles south of the reservation) who had decided to use the cradle board. Their babies complained so persistently that they were off the board in a matter of weeks, a result that should not surprise us, given the differences observed at birth.

Assuming, then, that other investigators continue to confirm our findings, to what do we attribute the differences on the one hand, and the similarities on the other? When we first presented the findings on Chinese and Caucasians, attempts were made to explain away the genetic implications by posing differences in prenatal diets as an obvious cause. But once we had completed the Navaho study, that explanation had to be dropped, because the Navaho diet is quite different from the diet of the Chinese, yet newborn behavior was strikingly similar in the two groups.

The point is often still made that the babies had nine months of experience within the uterus before we saw them, so that cultural differences in maternal attitudes and behavior might have been transferred to the unborn offspring via some, as yet unknown, mechanism. Chisholm, for example, thinks differences in maternal blood pressure may be responsible for some of the differences between Navahos and Caucasians, but the evidence is as yet sparse. Certainly Cantonese-American and Navaho cultures are substantially different and yet the infants are so much alike that such speculation might be dismissed on that score alone. But there is another, hidden issue here, and that involves our own cultural tendency to split apart inherited and acquired characteristics. Americans tend to eschew the inherited and promote the acquired, in a sort of "we are exactly what we make of ourselves" optimism.

My position on this issue is simple: We are totally biological, totally environmental; the two are as inseparable as is an object and its shadow. Or as psychologist Donald O. Hebb has expressed it, we are 100 percent innate, 100 percent acquired. One might add to Hebb's formulation, 100 percent biological, 100 percent cultural. As D. T. Suzuki, the Zen scholar, once told an audience of neuropsychiatrists, "You took heredity and environment apart and now you are stuck with the problem of putting them together again."

Navaho and Chinese newborns may be so much alike because the Navaho were part of a relatively recent emigration from Asia. Their language group is called Athabaskan, after a lake in Can-

ada. Although most of the Athabaskan immigrants from Asia settled along the Pacific coast of Canada, the Navaho and Apache contingents went on to their present location in about 1200 A.D. Even today, a significant number of words in Athabaskan and Chinese appear to have the same meaning, and if one looks back several thousand years into the written records of Sino-Tibetan, the number of similar words makes clear the common origin of these widely separated peoples.

When we say that some differences in human behavior may have a genetic basis, what do we mean? First of all, we are *not* talking about a gene for stoicism or a gene for irritability. If a behavioral trait is at all interesting, for example, smiling, anger, ease of sexual arousal, or altruism, it is most probably polygenic—that is, many genes contribute to its development. Furthermore, there is no way to count the exact number of genes involved in such a polygenic system because, as geneticist James Crow has summarized the situation, biological traits are controlled by one, two, or *many* genes.

Standing height, a polygenic human trait, can be easily measured and is also notoriously open to the influence of the environment. For this reason height can serve as a model for behavioral traits, which are genetically influenced but are even more prone to change with changing environment.

There are, however, limits to the way that a given trait responds to the environment, and this range of constraint imposed by the genes is called a *reaction range*. Behavioral geneticist Irving Gottesman has drawn up a series of semihypothetical graphs illustrating how this works with regard to human height; each genotype (the combination of genes that determine a particular trait) represents a relatively inbred human group. Even the most favorable environment produces little change in height for genotype A, whereas for genotype D a vast difference is seen as nutrition improves.

When I speak of potential genetic differences in human behavior, I do so with these notions in mind: There is overlap between most populations and

the overlap can become rather complete under changing conditions, as in genotypes D and C. Some genotypes, however, show no overlap and remain remote from the others over the entire reaction range, as in genotype A (actually a group of achondroplastic dwarfs; it is likely that some pygmy groups would exhibit a similarly isolated reaction range with regard to height).

At present we lack the data to construct such reaction-range curves for newborn behavior, but hypothetically there is nothing to prevent us from one day doing so.

The question naturally arises whether the group differences we have found are expressions of richer and poorer environments, rather than of genetically distinguishable groups. The similar performance yet substantial difference in socioeconomic status between Navaho and San Francisco Chinese on the one hand, and the dissimilar performance yet similar socioeconomic status of San Francisco Chinese and Caucasians on the other favors the genetic explanation. Try as one might, it is very difficult, conceptually and actually, to get rid of our biological constraints.

Research among newborns in other cultures shows how environment—in this case, cultural learning—affects reaction range. In Hawaii we met a Honolulu pediatrician who volunteered that he had found striking and consistent differences between Japanese and Polynesian babies in his practice. The Japanese babies consistently reacted more violently to their three-month immunizations than did the Polynesians. On subsequent visits, the Japanese gave every indication of remembering the last visit by crying violently; one mother said that her baby cried each time she drove by the clinic.

We then tested a series of Japanese newborns, and found that they were indeed more sensitive and irritable than either the Chinese or Navaho babies. In other respects, though, they were much like them, showing a similar response to consolation, and accommodating easily to a light on the eyes or a cloth over the nose. Prior to our work, social anthropologist William Caudill had made an extensive and thorough

1. PERSPECTIVES

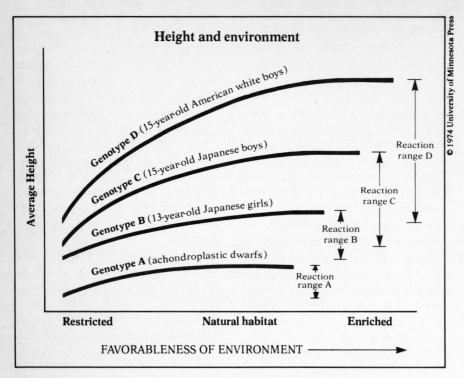

Height and environment

Average Height

Genotype D (15-year-old American white boys)

Genotype C (15-year-old Japanese boys)

Genotype B (13-year-old Japanese girls)

Genotype A (achondroplastic dwarfs)

Reaction range D

Reaction range C

Reaction range B

Reaction range A

Restricted   Natural habitat   Enriched

FAVORABLENESS OF ENVIRONMENT ⟶

© 1974 University of Minnesota Press

The concept of reaction range shows clearly in this comparison of adolescent groups: the better the environment, the taller the person. Although some groups show considerable overlap in height, no matter how favorable the environment, height cannot exceed the possible reaction range.

study of Japanese infants. He made careful observations of Japanese mother-infant pairs in Baltimore, from the third to the twelfth month of life. Having noted that both the Japanese infants and their mothers vocalized much less to one another than did Caucasian pairs, he assumed that the Japanese mothers were conditioning their babies toward quietude from a universal baseline at which all babies start. Caudill, of course, was in the American environmentalist tradition and, until our publication appeared, did not consider the biological alternative. We believe that the mothers and babies he studied were, in all probability, conditioning each other, that the naturally quiet Japanese babies affected their mothers' behavior as much as the mothers affected their babies'.

With this new interactive hypothesis in mind, one of my students, Joan Kuchner, studied mother-infant interactions among 10 Chinese and 10 Caucasian mother-infant pairs over the

first three months of life. The study was done in Chicago, and this time the Chinese were of North Chinese rather than South Chinese (Cantonese) ancestry. Kuchner started her study with the birth of the babies and found that the two groups were different from the start, much as in our study of newborns. Further, it soon became apparent that Chinese mothers were less intent on eliciting responses from their infants. By the third month, Chinese infants and mothers rarely engaged in bouts of mutual vocalizing as did the Caucasian pairs. This was exactly what the Caudill studies of Japanese and Caucasians had shown, but we now know that it was based on a developing coalition between mothers and babies and that it was not just a one-way street in which a mother "shapes" her infant's behavior.

Following our work, Caudill and Lois Frost repeated Caudill's original work, but this time they used third-generation Japanese-American moth-

ers and their fourth-generation infants. The mothers had become "super" American and were vocalizing to their infants at almost twice the Caucasian rate of activity, and the infants were responding at an even greater rate of happy vocalization. Assuming that these are sound and repeatable results, my tendency is to reconcile these and our results in terms of the reaction-range concept. If Japanese height can change as dramatically as it has with emigration to the United States (and with post-World War II diets), it seems plausible that mother-infant behavior can do the same. On a variety of other measures, Caudill and Frost were able to discern continuing similarities to infant and mother pairs in the old country. Fourth-generation Japanese babies, like babies in Japan, sucked their fingers less and were less playful than Caucasian babies were, and the third-generation mothers lulled their babies and held them more than Caucasian American mothers did.

A student and colleague, John Callaghan, has recently completed a study comparing 15 Navaho and 19 Anglo mothers and their young infants (all under six months). Each mother was asked to "get the attention of the baby." When video tapes of the subsequent scene were analyzed, the differences in both babies and mothers were striking. The Navaho babies showed greater passivity than the Caucasian babies. Caucasian mothers "spoke" to their babies continually, using linguistic forms appropriate for someone who understands language; their babies responded by moving their arms and legs. The Navaho mothers were strikingly silent, using their eyes to attract their babies' gaze, and the relatively immobile infants responded by merely gazing back.

Despite their disparate methods, both groups were equally successful in getting their babies' attention. Besides keeping up a stream of chatter, Caucasian mothers tended to shift the baby's position radically, sometimes holding him or her close, sometimes at arm's length, as if experimenting to find the best focal distance for the baby. Most of the silent Navaho mothers used only

18

subtle shifts on the lap, holding the baby at about the same distance throughout. As a result of the intense stimulation by the Caucasian mothers, the babies frequently turned their heads away, as if to moderate the intensity of the encounter. Consequently, eye contact among Caucasian pairs was of shorter duration (half that of the Navaho), but more frequent.

It was clear that the Caucasian mothers sought their babies' attention with verve and excitement, even as their babies tended to react to the stimulation with what can be described as ambivalence: The Caucasian infants turned both toward and away from the mother with far greater frequency than did the Navaho infants. The Navaho mothers and their infants engaged in relatively stoical, quiet, and steady encounters. On viewing the films of these sequences, we had the feeling that we were watching biocultural differences in the making.

Studies of older children bear out the theme of relative unexcitability in Chinese as compared to Anglos. In an independent research project at the University of Chicago, Nova Green studied a number of nursery schools. When she reached one in Chicago's Chinatown, she reported: "Although the majority of the Chinese-American children were in the 'high arousal age,' between three and five, they showed little intense emotional behavior. They ran and hopped, laughed and called to one another, rode bikes and roller-skated just as the children did in the other nursery schools, but the noise level stayed remarkably low, and the emotional atmosphere projected serenity instead of bedlam. The impassive facial expression certainly gave the children an air of dignity and self-possession, but this was only one element effecting the total impression. Physical movements seemed more coordinated, no tripping, falling, bumping, or bruising was observed, nor screams, crashes or wailing was heard, not even that common sound in other nurseries, voices raised in highly indignant moralistic dispute! No property disputes were observed, and only the mildest version of 'fighting behavior,' some good-natured wrestling among the older boys. The adults evidently had different expectations about hostile or impulsive behavior; this was the only nursery school where it was observed that children were trusted to duel with sticks. Personal distance spacing seemed to be situational rather than compulsive or patterned, and the children appeared to make no effort to avoid physical contact."

It is ironic that many recent visitors to nursery schools in Red China have returned with ecstatic descriptions of the children, implying that the New Order knows something about child rearing that the West does not. When the *New Yorker* reported a visit to China by a group of developmental psychologists including William Kessen, Urie Bronfenbrenner, Jerome Kagan, and Eleanor Maccoby, they were described as baffled by the behavior of Chinese children: "They were won over by the Chinese children. They speak of an 'attractive mixture of affective spontaneity and an accommodating posture by the children: of the 'remarkable control of young Chinese children'— alert, animated, vigorous, responsive to the words of their elders, yet also unnervingly calm, even during happenings (games, classroom events, neighborhood play) that could create agitation and confusion. The children 'were far less restless, less intense in their motor actions, and displayed less crying and whining than American children in similar situations. We were constantly struck by [their] quiet, gentle, and controlled manner . . . and as constantly frustrated in our desire to understand its origins.' "

The report is strikingly similar to Nova Green's description of the nursery school in Chicago's Chinatown. When making these comparisons with "American" nursery schools, the psychologists obviously had in mind classrooms filled with Caucasian or Afro-American children.

As they get older, Chinese and Caucasian children continue to differ in roughly the same behavior that characterizes them in nursery school. Not surprisingly, San Francisco school-teachers consider assignments in Chinatown as plums—the children are dutiful and studious, and the classrooms are quiet.

A reader might accept these data and observations and yet still have trouble imagining how such differences might have initially come about. The easiest explanation involves a historical accident based on different, small founding populations and at least partial geographic isolation. Peking man, some 500,000 years ago, already had shovel-shaped incisors, as only Orientals and American Indians have today. Modern-looking skulls of about the same age, found in England, lack this grooving on the inside of their upper incisors. Given such evidence, we can surmise that there has been substantial and long-standing isolation of East and West. Further, it is likely that, in addition to just plain "genetic drift," environmental demands and biocultural adaptations differed, yielding present-day differences.

Orientals and Euro-Americans are not the only newborn groups we have examined. We have recorded newborn behavior in Nigeria, Kenya, Sweden, Italy, Bali, India, and Australia, and in each place, it is fair to say, we observed some kind of uniqueness. The Australian aborigines, for example, struggled mightily against the cloth over the nose, resembling the most objecting Caucasian babies; their necks were exceptionally strong, and some could lift their heads up and look around, much like some of the African babies we saw. (Caucasian infants cannot do this until they are about one month old.) Further, aborigine infants were easy to calm, resembling in that respect our easy-going Chinese babies. They thus comprised a unique pattern of traits.

Given these data, I think it is a reasonable conclusion that we should drop two long-cherished myths: (1) No matter what our ethnic background, we are all born alike; (2) culture and biology are separate entities. Clearly, we are biosocial creatures in everything we do and say, and it is time that anthropologists, psychologists, and population geneticists start speaking the same language. In light of what we

know, only a truly holistic, multidisciplinary approach makes sense.

For further information:

Caudill, W., and N. Frost. "A Comparison of Maternal Care and Infant Behavior in Japanese-American, American, and Japanese Families." *Influences on Human Development*, edited by Urie Bronfenbrenner and M. A. Mahoney. Dryden Press, 1972.

Chisholm, J. S., and Martin Richards. "Swaddling, Cradleboards and the Development of Children." *Early Human Development*, in press.

Freedman, D. G. "Constitutional and Environmental Interaction in Rearing of Four Breeds of Dogs." *Science*, Vol. 127, 1958, pp. 585-586.

Freedman, D. G. *Human Infancy: An Evolutionary Perspective*. Lawrence Erlbaum Associates, 1974.

Freedman, D. G., and B. Keller. "Inheritance of Behavior in Infants." *Science*, Vol. 140, 1963, pp. 196-198.

Gottesman, I. I. "Developmental Genetics and Ontogenetic Psychology." *Minnesota Symposia on Child Psychology*, Vol. 8, edited by A. D. Pick. University of Minnesota Press, 1974.

How Children Influence Children:
The Role of Peers in the Socialization Process

Emmy Elisabeth Werner

Emmy E. Werner, Ph.D., is a Professor of Human Development and Research and Child Psychologist at the University of California, Davis.

I n his review on family structure, socialization and personality, John A. Clausen points out that an older sibling may be caretaker, teacher, pacesetter or confidant for a younger one. Yet such a reference work as the *Handbook of Socialization Theory and Research* by D. B. Goslin includes few references about caretaking of children by agemates. What cross-cultural evidence we can find, however, indicates that caretaking of children by siblings, cousins or other peers is a significant phenomenon in most societies of the developing world.

Sibling and Child Caretaking

Child caretaking is widespread cross-culturally, but relevant material about this topic is scattered throughout many ethnographic studies, which makes comparative analysis difficult. In one cross-cultural survey on the age of assignment of roles and responsibilities to children, based on ethnographies of 50 cultures, pancultural trends in the age of assignment of child care roles were observed.[1] These centered on the 5- to 7-year-old period, when care of siblings, peer play and the understanding of game rules are most frequently initiated. This is the same age when Western societies introduce formal schooling.

The most common worldwide pattern is informal child and sibling care that is part of the daily routine of children within the family and that is carried out without formalized organizational rules. Under these circumstances, child caretakers frequently operate under two simultaneous sets of pressures: one from their small charges, the other from their parents. In all non-Western societies in-

vestigated by the Whitings in their Six-Cultures Study, children were expected to do some child tending.[2] However, there were striking differences in the types of caretaking mothers were willing to delegate, the age at which they considered a child competent, and the amount of supervision considered necessary.

One question on the mothers interview indicates the value placed on the help given by the child nurse. When the mothers were asked who had helped them care for the sample child when he or she was an infant, 69 percent of the mothers from the East African community, 41 percent of the mothers from the Mexican barrio, 25 percent of the Filipino, 21 percent of the North Indian, but only 12 percent of the New England and Okinawan mothers reported having been helped with an infant by a child. The three highest-ranking societies, the East African, the Mexican and the Filipino communities, were also the societies that ranked highest in the nurturant behavior of their children, as observed independently in naturalistic settings.

There is usually a strong contrast between infancy and young childhood in terms of child and adult caretaking practices. The care of toddlers requires different skills and behaviors on the part of child caretakers than the care of an infant. Observations of the interaction of children with 2- to 4-year-old siblings in the Six-Cultures Study indicate that caretakers of toddlers were comparatively more apt to reprimand, criticize and punish. This is in some contrast to the predominant nurturant and responsive attitude shown toward the infants in these societies. Thus, the role of the child caretaker is a function of at least three factors: the physical maturation of the child; the availability of different caretakers; and the differing cultural conceptions of maturity of a child, which, in turn, leads to different patterns of caretaking by children.

Antecedents of Child Caretaking

The residence and size of a family, as well as the daily routines, subsistence

economy and maternal workload, are related to the frequency of child caretaking in the developing world. Sibling caretaking is more common in societies where women have more work to do, where the work takes the mother from home or is difficult to interrupt, and where circumstances of residence, birth order and family size make alternative caretakers available. A domestic group with a large number of kin and cousins present, a mother with many offspring and a daily routine that keeps brothers, sisters and other adults available for caretaking would be the optimal situation for the development of nonparental and sibling caretaking.[3] . . .

Consequences of Child Caretaking

There is a great need for additional data that document the possible effect on a child of either providing or receiving child caretaking. Most of the data available deal with attachment behavior and differences in affiliation versus achievement motivation.

Sibling caretaking seems to be of special importance in cultures that are polygynous. Africa leads the world in polygynous societies. A study of Kikuyu children illustrates the importance of the sibling group in socialization.[4] After the child is raised by the mother for the first one or two years and is given a great deal of maternal care that fosters strong attachment, he or she will move in with the siblings when the mother is pregnant again. The sibling group is mostly responsible for the socialization of the young child and becomes the main source of the child's emotional involvement.

In a short-term longitudinal study of 3-year-old American preschoolers, one researcher found that secure attachment may play a dual role in children's relationships with other children.[5] It may directly promote peer competence by encouraging a positive orientation toward other children, and, insofar as mothers who foster secure attachment also encourage expanded interactions, it may indirectly promote social competence by giving children the opportunity to learn from peers.

1. PERSPECTIVES

Studying urban Japanese families, Caudill and Plath were impressed with the role of siblings in the instruction and care of the younger babies, and by how this responsibility for parenting appeared to diminish any sibling rivalry and to create close bonding between brothers and sisters.[6] They ascribed the strong affectionate bond in interdependence between different members of the family to the sleeping arrangements. When the baby is new, he or she sleeps with the mother; when another baby comes, the child sleeps with an older brother or sister. Sleeping with another member of the family apparently strengthens family bonds and expresses a strong nurturant family life, at the same time lessening the sexual aspects of sleeping together.

In their Six-Cultures Study, the Whitings found that children who interact with infants were more nurturant and less egotistic than children who did not care for infants. These authors suggest that caretaking of infants appears to affect overall interaction with peers.[7] This becomes quite apparent when we take a look at the consequences of sex differences in child caretaking. B. Whiting and C. P. Edwards compared boys and girls in seven societies, the six cultures mentioned earlier plus the Kikuyu of Kenya, and observed incidences of nurturant and responsible behavior.[8] Older girls, aged 7 to 11, offered help and support to others more often than did boys. There were no such sex differences for children aged 3 to 6. The authors interpret the increased nurturance of older girls as due to the assignment to girls of increased childrearing duties, particularly infant caretaking. Another researcher observed Luo boys in Kenya who were expected to perform child-caretaking chores usually assigned to girls.[9] Such boys displayed more feminine social behaviors than boys not needed for such tasks.

Thus, it appears that sex differences in nurturance and responsible behavior may only occur at particular ages and are not uniform across all cultures. The critical factor for the development of nurturant behavior seems to be the demand for child care tasks within the home. It would be interesting to see whether similar findings could be replicated in our own culture, as sex-role expectations change in the wake of a more egalitarian type of childrearing.

Several authors, including R. Levy and J. E. Ritchie, have attempted to generalize about the effects of child caretaking on the development of individual differences in children.[10] These authors have dealt with ethnographic accounts of child caretaking in Polynesian societies and have argued that sibling caretaking restricts the development of individual differences in both children and adults. The possible effects of child caretaking were presented by Levy as the development of an easygoing or apathetic "you can't fight City Hall" orientation to life. Weisner and Gallimore suggest that these consequences need to be interpreted in terms of the social context in which the child will live as an adult.[11] The socialization goal of the societies in which these observations were made is the integration of the child into the social context, rather than fostering individual achievement and independent skills. Thus, it may be that children raised in a sibling-caretaking system develop psychological and behavioral characteristics that are adaptive in some settings and not in others. Systematic differences can be expected in the learning experiences of young children when taught by siblings rather than parents.[12]

From a brief overview of the rather scarce data available on sibling caretaking in the Six-Cultures Study and in Polynesian groups in Hawaii, New Zealand and Tahiti, it appears that sibling caretaking in extended families anywhere in the world may be a functional adaptation of low-income groups that allows economically marginal families flexibility in coping with crises and increases the number of potential resource contributors. A case in point is the American Indian family in the American metropolis society, which differs significantly from the nuclear-family, conjugal pair and single-person types that predominate in White America.[13] The less stable and the lower the amount of income among American Indian families, the larger the household. Brothers or sisters of the husband and wife, and nieces and nephews, all join together to pool their meager resources. This grouping together also characterizes other poverty-stricken households among other ethnic and racial minorities within the industrialized urban West.

It remains to be seen what positive roles siblings can play in helping the younger child adapt to changes brought about by modernization and industrialization, since older children appear more open and exposed to modern influences than younger ones.

Affiliation Versus Achievement Motivation

Evidence of the effects on motive development of sibling caretaking is either severely limited or indirect . . . In the Polynesian, African, Asian and Latin American societies where child caretaking has been studied, it appears that early parental demand for non-dependence serves, in part, to shift independence training to older siblings. Thus, refusal of help by parents redirects the child's overtures to siblings, who provide nurturance and training and, in turn, pressure for independence.

Of critical importance is the fact that this shift from adult to sibling caretaking can occur without the toddler learning self-care skills, which may impose a rather strong burden on the young caretaker that may lessen the child's achievement motivation at the crucial age when it tends to crystallize. Given the mother's behavior, the child has no alternative but to turn to siblings; thus, achievement motivation may be sacrificed for the sake of affiliation.

Reliance on sibling caretakers as a factor in the development of affiliation motivation has been suggested by studies of Hawaiian-Americans.[14] The pattern of being interdependent and affiliating with others is a significant feature of Hawaiian life and may cause problems in the classroom. Accustomed to sibling care, Hawaiian children are inclined to attend to peers rather than teachers and individual work, behavior that is often interpreted by teachers in terms of motivational and attentional deficits.

On the positive side, S. MacDonald and R. Gallimore found in a number of classroom studies that Hawaiian-American students perform at high levels if allowed to interact or affiliate with peers in team work or in the sharing of earned privileges.[15] Whether peer interaction is more motivating for those from families in other cultures where there is a great deal of sibling care of children is a hypothesis that has not been directly tested.

To sum up, child caretaking appears to be an important antecedent to nurturant, responsible behavior and to behavior that leads to affiliation rather than achievement motivation. Though it is presently preponderant in the non-Western world, child caretaking may in the future play an important role as an alternative to maternal caretaking in the West. P.M. Greenfield suggests that day care centers should involve older children and siblings in child care and that schooling or tutoring of primary school children should involve children as well as adults.[16] The Whitings argue that whether a child is told to take care of younger siblings or whether he or she is sent to school instead may have a more profound effect upon the profile of the child's social behavior than the manipulation of reinforcement schedules by the parents. Thus, major attention needs to be paid in future cross-cultural studies to the role of child caretakers as transmitters of new social values and as links between the family and the rapidly changing outside world.

Children's Play Groups and Games

Children's play groups are not necessarily dependent on caretaking patterns, but the two variables are frequently closely related. Child caretaking affects the sex composition of play groups and their physical and social mobility. Where caretaking is not limited to one's own siblings, it shapes contacts with children not in one's immediate family.

In a review article on exploration and play, Weisler and McCall trace the developmental sequence in the nature of children's social play.[17] At first the child plays in isolation, without reference to what other children are doing. The first indication of a social element is the occurrence of parallel playing, in which the nature of the child's behavior is influenced by and may be similar to that of nearby children, but there is no direct social interaction. Subsequently, there may be short interactions between children consisting of socially instigated but not truly interactive play, as when the behavior of another child is imitated. Later, full-scale group play can be observed, in which one child interacts verbally and physically in a prolonged sequence with other children.

Several theoretical orientations have emphasized the role of play as a means of reducing tension and anxieties. Cross-cultural comparisons might reveal, in addition, different social and cultural values that are infused in the play of young children. . .

Children's games around the world appear to play an important part in resolving conflicts over socialization pressures and teach social values and social skills essential for successful adaptation in a given society, whether through

physical skills, taking chances or making rational choices in a deliberate strategy.

A number of ecological and child-training correlates appear associated with assertive, competitive and rivalrous behavior in children. Increasing modernization, as measured by exposure to school and city life, and the opportunity for social mobility and the removal of the inhibiting effects of traditional stress on obedience and control of aggression seem to contribute to an increase in these behaviors, more so among boys than girls and among older than younger children. Extrafamilial socializers can counteract this trend in societies where cooperation is stressed as part of a deliberate philosophy.

Summary

In sum, the influence of peers as mediators of social change cannot be underestimated. The results of studies of child caretaking, of games and of competitive versus cooperative behavior in the classroom seem to indicate that peers, with and without the direct support of teachers and the sociopolitical system of a given society, transmit social values that are important in the process of modernization. The young, as role models for still younger children, become important pacesetters in the developing world and in human cultural evolution.

[1]B. Rogoff, M.J. Sellers, S. Piorrata, N. Fox and S. White, ''Age of Assignment of Roles and Responsibilities to Children: A Cross-Cultural Survey,'' *Human Development*, 1975, 18.

[2]B. Whiting and J.W. Whiting, *Children of Six Cultures*, Cambridge, Mass., Harvard University Press, 1975.

[3]L. A. Minturn, ''A Survey of Cultural Differences in Sex-Role Training and Identification,'' in N. Kretschmer and D. Walcher (eds.), *Environmental Influences on Genetic Expression*, Washington, D.C., U.S. Government Printing Office, 1969.

[4]J. I. Carlebach, ''Family Relationships of Deprived and Non-Deprived Kikuyu Children from Polygamous Marriages,'' *Journal of Tropical Pediatrics*, 1967, 13.

[5]A.F. Lieberman, ''Preschoolers Competence with a Peer: Relations with Attachment and Peer Experience, *Child Development*, 1977, 48.

[6]W. Caudill and D.W. Plath, ''Who Sleeps By Whom?: Parent-Child Involvement in Urban Japanese Families,'' *Psychiatry*, 1966, 29.

[7]Whiting and Whiting, 1975, op. cit.

[8]B. Whiting and C.P. Edwards, ''A Cross-Cultural Analysis of Sex Differences in the Behavior of Children Aged Three Through Eleven,'' *Journal of Social Psychology*, 1973, 91.

[9]Carl R. Ember, ''Female Task Assignment and Social Behavior of Boys,'' *Ethos*, 1973, 1.

[10]See, for example, R.I. Levy, ''Child Management Structure and Its Implications in a Tahitian Family,'' in E. Vogel and N. Bell (eds.), *A Modern Introduction To The Family*, New York, Free Press, 1968 and J.E. Ritchie, *Basic Personality in Rakau*, New Zealand, Victoria University, 1956.

[11]T.S. Weisner and R. Gallimore, ''My Brother's Keeper: Child and Sibling Caretaking,'' *Current Anthropology*, 1977, 18(2).

[12]M. Steward and D. Steward, ''Parents and Siblings As Teachers,'' in E.J. Mash, L.C. Handy and L.A. Hamerlynek (eds.), *Behavior Modification Approaches to Parenting*, New York, Brunner/Mazel, 1976.

[13]J. Jorgensen, ''Indians and the Metropolis,'' in J.O. Waddell and O.M. Watson (eds.), *The American Indian in Urban Society*, Boston, Little Brown, 1971.

[14]R. Gallimore, J.W. Boggs and C.E. Jordan, *Culture, Behavior and Education: A Study of Hawaiian-Americans*, Beverly Hills, Calif., Sage, 1974.

[15]S. MacDonald and R. Gallimore, *Battle in the Classroom: Innovations in Classroom Techniques*, Scranton, Intext, 1971.

[16]P.M. Greenfield, *What We Can Learn from Cultural Variation in Child Care*, paper presented at the American Association for the Advancement of Science, San Francisco, 1974.

[17]A. Weisler and R.B. McCall, ''Exploration and Play: Resume and Redirection,'' *American Psychologist*, 1976, 31(7).

THE JAPANESE BRAIN

Atuhiro Sibatani

Atuhiro Sibatani is a developmental biologist in the Molecular and Cellular Biology Unit of the Commonwealth Scientific and Industrial Research Organization in Sydney, Australia.

The idea that the Japanese live in harmony with their surroundings, turning everyday rituals into art forms, has been ascribed sometimes to genetic inheritance, sometimes to cultural conditioning. But in recent years a clever new technique designed to study speech and hearing defects has yielded one of the most tantalizing —and controversial—theories to date: that Japanese brains function differently from other people's not because of inheritance or conditioning but because of the peculiarities of the Japanese language. If this provocative hypothesis proves to be correct, we may have to revise some of our venerable convictions, for it may turn out that the language we learn alters the physical operation of our brains.

The startling assertion that language shapes the neurophysiological pathways of the brain is the thesis of a dry academic tome that, amazingly, became a best seller in Japan when it appeared in 1978. Written by Tadanobu Tsunoda, *The Japanese Brain: Brain Function and East-West Culture* has yet to be translated into English, but information about its contents has crossed the oceans and is provoking a good deal of discussion in Western circles. What Tsunoda has found, he claims, is that the language one learns as a child influences the way in which the brain's right and left hemispheres develop their special talents.

That the brain's hemispheres specialize in different tasks has been recognized since the 19th century. Neurologists studying men injured by strokes or battle wounds found that damage to the brain's left side often interfered with speech. Since then, scientists have demonstrated hemispheric specialization with techniques such as dichotic listening tests, which present different words simultaneously to both right and left ears to determine which ear "excels" at which tasks.

For most right-handed people, the left hemisphere appears to be the main seat of language, as well as of precise manual manipulations, mathematics, and other analytic functions. The right hemisphere is superior in dealing with spatial concepts, recognition of faces, musical patterns, environmental sounds, and perhaps intuitive and artistic processes. For left-handed people, it is harder to generalize about the brain's organization (see box, pages 26-27).

These are observations made by Western scientists working with Caucasian subjects. Tsunoda's research with Japanese patients, on the other hand, seems to have revealed fundamental differences in the way that Caucasian and Japanese brains divide up the labor of processing sensory data.

A specialist in hearing difficulties at Tokyo Medical and Dental University, Tsunoda was studying patients whose speech was damaged, testing the possibility that he might somehow transfer language processing to the undamaged part of the brain. In the course of his research, Tsunoda devised a unique dichotic listening test designed to be independent of the subject's conscious awareness. The test required subjects to tap simple, regular patterns on a Morse code key. Tones keyed by the tapping process were then fed back to each ear through earphones. One ear received the sound directly; the other ear received the signal delayed by two-tenths of a second. Tsunoda gradually increased the loudness of the delayed signal until it interfered with the subject's key-tapping performance. The purpose was to ascertain whether one ear—and its associated brain hemisphere—predominated in registering this interference.

In addition to the time delay, Tsunoda also supplied each ear with different kinds of sounds—not only pure tones but also such sounds as spoken words, animal noises, Japanese and Western musical instruments, and ocean waves.

One of Tsunoda's findings is that in the brains of right-handed Westerners, Koreans, Chinese, and Ben-

Japanese Pattern

VOWELS
Consonants
Language
Non-verbal human sounds
Japanese instrumental music
Animal sounds
Calculation

Mechanical sounds
Western instrumental music

Left Hemisphere

Right Hemisphere

Western Pattern

VOWELS
Consonants
Language
Calculation

Non-verbal human sounds
Animal sounds
Western and Japanese instrumental music
Mechanical sounds

Left Hemisphere

Right Hemisphere

Tadanobu Tsunoda, a specialist in hearing difficulties, believes that the two halves of the Japanese brain divide up the labor of processing sounds in a way that differs from Western brains. The key to those differences, he postulates, is tnat the Japanese deal with all vowels in the left hemisphere, while Westerners handle isolated vowels in the right hemisphere.

M. E. Challinor

galis, vowel sounds usually get processed in one side of the brain if they occur in isolation, but in the other half if the vowels occur in a spoken context, that is, if they are surrounded by consonants. But right-handed Japanese and Polynesians, Tsunoda discovered, usually process all vowels in the left or dominant half, whether they occur in a spoken context or not.

Mechanical sounds—bells, whistles, and helicopter noises—are among the few sounds that Japanese and Polynesians handle in the right hemisphere as do other ethnic groups. Western instrumental music is also processed in the right hemisphere. By contrast, Japanese subjects handle Japanese music in the left hemisphere, possibly because Japanese music attempts to mimic the human voice.

The Japanese and Polynesians also tend to depend on their left brains for processing nonverbal human utterances that express emotions—sounds such as laughter, crying, or sighing—along with natural sounds such as cricket chirps, cow calls, bird songs, and ocean waves. By contrast, those who speak European languages handle all these sounds in their right hemispheres.

Tsunoda suggests that the Japanese may utilize the left hemisphere more heavily because the Japanese and the Polynesian languages are particularly rich in vowels. One can make up complex sentences in Japanese using vowels only: *"Ue o ui, oi o ooi, ai o ou, aiueo"* means "A love-hungry man who worries about hunger hides his old age and chases love." Polynesians also use lots of vowels, as in the cry of distress, *"Oiaue!"* leading some experts to suggest that the Japanese race has Polynesian roots.

A fundamental discovery of Tsunoda is that the Japanese, in contrast to Westerners, process far more sounds in only one hemisphere. Furthermore, if a sound normally processed in the left hemisphere, whether voices or insect chirps, is buried amidst other background noises, the entire load of processing all the sounds is switched to the left. This switching process, characteristic of all human brains, is not surprising since a first priority of the brain is language.

Curiously enough, certain odors, including perfumes, alcohol, and tobacco smoke, also seem to trigger this switching ability. Because Tsunoda associates the effects of these chemicals with emotional reactions, he infers that the Japanese process emotion in the dominant hemisphere, apparently opposite to Westerners.

Tsunoda is convinced that the switching stimulus is not inherited, but is acquired by every child through the use of the language that he or she speaks. Japanese people brought up in the United States, for example, have the characteristic Caucasian brain lateralization. That is, they process consonants on the left and vowels on the right. Conversely, Americans brought up in Japan and fluent in the language acquire typically Japanese brains. Tsunoda shows that the response of the natively blind, and hence illiterate, Japanese is exactly the same as that of the literate members of the population. From this he concludes that the emergence of the Japanese brain is not triggered simply by learning to read and write Japanese but rather by listening to the language and speaking it.

Tsunoda's results have been heralded by some as beautiful and clear-cut; others have adopted a wait-and-see attitude, since his results have so far not been replicated. Still, brain researchers are intrigued with

1. PERSPECTIVES

Tsunoda's hypothesis that language affects the way the brain's two halves process language.

However, Western scientists are frankly skeptical about some of Tsunoda's sweeping speculations. He conjectures, for example, that in the Japanese brain, logical thinking and emotional responses are not partitioned into separate hemispheres as they seem to be in the West, but are tucked into one and the same verbal hemisphere. This may cause the Japanese to depend more on intuitive and emotional reactions than on logical trains of thought.

Nor do the Japanese distinguish analytic problems and natural sensations in the clear-cut way that Westerners do, according to Tsunoda. The Japanese seem to have a psychological need to live immersed in natural sounds such as bird and insect songs, animal cries, snow thudding off tree branches, ocean waves beating against the shore, and winds whistling through forest pines. That harmony of nature and environment is evident in all aspects of Japanese life, from their calligraphy to the tea ceremony to Noh drama and Ikebana flower arranging. In Japanese landscape and architecture, physical objects melt into the character of space rather than oppose it. For example, the Japanese use few partitions, rooms combine into sweeping space, and the garden may recess into the house so that dwellers can be outdoors while sitting indoors. Tsunoda believes that this blurring of physical barriers and natural elements may help explain the deep sense of harmony within the group that results in the social cohesion of the Japanese nation.

The negative aspect, however, is that the Japanese may be overtaxing the left hemisphere, particularly when forced to learn a variety of foreign languages. After speaking other languages for a few days or months at a time, the Japanese brain seems to switch sounds normally processed on the right to the opposite hemisphere. This undoubtedly places a tremendous burden on the over-utilized verbal hemisphere.

One of the most intriguing spin-offs of Tsunoda's work may be its implications for sociobiology, the science that studies the genetic predisposition of social behavior. Tsunoda argues that some differences in brain function are conditioned by the mother tongue, rather than by genetic factors of ethnic origin. If Tsunoda is right, the patterns of each individual's perceptions, cognitions, mental acts, and social behavior can be dramatically affected during early childhood by one of the most human of activities—language. The debate over strict genetic determinism, so heated on this side of the ocean, has not thus far troubled the Japanese. Whether or not this indifference results from the "inscrutable" nature of the Japanese brain may constitute another absorbing chapter for those interested in Japan's role in modern society.

The sinister hand

Among certain African tribes, children who showed a preference for their left hand were "cured" by having that hand immersed in boiling water and deliberately scalded, ensuring that they would use their right hand while learning important skills. In many other parts of the world, until a generation ago, schoolteachers regularly forced pupils to write with their right hands, regardless of the children's inclinations.

Nevertheless, some people—roughly 10 to 15 percent of the world's population—persisted in remaining left-handed. Recent studies indicate that the trait is inherited, but a century ago no one knew that for sure. In 1865 the French neurosurgeon Paul Broca became fascinated with these people. An early explorer of the human brain, Broca had just discovered that in right-handers, certain speech centers were found only in the left cerebral hemisphere. Among left-handers, he reasoned, the reverse must be true: Their speech center must be in the right hemisphere. Knowing that the left hemisphere controls the right side of the body and vice versa, some scientists pushed this logic even further. They declared that the location of organs such as the heart must be reversed in the bodies of left-handers. Although anatomists soon proved them wrong, Broca's view of left-handedness as a mirror image of right-handedness was dominant for nearly a century.

In the past two decades, however, it has become clear that left-handers vary enormously in the way they process information. While a minority of them do use their right hemisphere for language, as Broca predicted, roughly 60 percent of left-handers process speech in their left hemispheres, just as right-handers do, and another group appears to use both sides; such dual processing makes people less likely to lose their ability to speak if they ever suffer a brain injury or a stroke.

These variations proved so intriguing to brain scientists—especially to those studying the differing specialties of the two cerebral hemispheres—that left-handed people rapidly became their favorite experimental subjects.

The mushrooming studies of left-handers in the past few years have shown that left-handers are less strongly lateralized than right-handers, that is, they show less difference between the two sides of their brain. Thus, while right-handers typically choose their right eye when conditions demand the use of a single eye, among left-handers the chances of being "left-eyed" are only 50-50. Even the right-handed relatives of left-handers are more likely to recover from loss of speech following damage to the left side of their brain than are people whose relatives are all right-handers; this implies a genetic tendency toward weaker laterali-

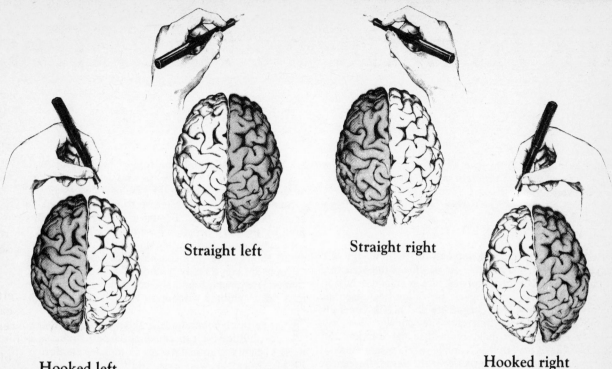

Straight left

Straight right

Hooked left

Hooked right

Scientists are trying to determine if hand and writing posture indicate which brain hemisphere is used for language. Initial results show that left-handers who use the hooked position for writing process language on the left side of their brains, as do the majority of right-handers who use the straight hand position. But left-handers who use a straight hand posture seem to process language in the right hemisphere, as do the small minority of right-handers who use a hooked position.

zation among relatives of left-handers.

But what are the advantages or handicaps of weaker lateralization? This is what researchers are now trying to find out.

In the past, left-handers were all lumped together and accused of being clumsy, retarded, stammerers, stubborn, criminal, neurotic, or homosexual.

"They squint, they stammer, they shuffle and shamble, they flounder about like seals out of water. Awkward in the house, and clumsy in their games, they are fumblers and bunglers at whatever they do," wrote the British psychologist Cyril Burt in 1937. He was particularly scathing about left-handed girls who "often possess a strong, self-willed, and almost masculine disposition: by many little tell-tale symptoms, besides the clumsy management of their hands—by their careless dress, their ungainly walk, their tomboy tricks and mannerisms—they mutely display a private scorn for the canons of feminine grace and elegance."

Today, having shown that none of the above is true, brain scientists are studying left-handers to see whether their particular patterns of brain organization lead to any special talents. Some are looking for clues that might explain why so many artists, including Michelangelo, Leonardo da Vinci, Picasso, and Escher, have been left-handed. Conceivably, since the right hemisphere is superior for imagery and visual abilities, there is some advantage to using the hand that is controlled by the same hemisphere—namely, the left hand—for drawing or painting. This would avoid having to send signals through the corpus callosum which links the two hemispheres.

One recent study has shown that left-handers who process language in both hemispheres do better than other groups on tests of tonal pitch recognition. This talent presumably arises because information regarding pitch is duplicated in the two halves of the brain. Another study has found that a higher than normal percentage of musicians are ambidextrous. Researchers have been unable to show whether this is cause or effect; it is an open question whether or not people who are less lateralized tend to be more gifted musically, or whether playing music develops both hands—or possibly both sides of the brain.

Scientists eventually hope to discover how various patterns of brain organization determine specialized ways of dealing with the world—from spatial, verbal, and mathematical abilities to musical or artistic gifts—not just among those who are left-handed, but among all people. —*Maya Pines*

Washington-based free-lance writer and author of The Brain Changers: Scientists and the New Mind Control.

Just How the Sexes Differ

Captain to Laura: ". . . If it's true we are descended from the ape, it must have been from two different species. There's no likeness between us, is there?"

—"The Father," by August Strindberg

So it has begun to seem, and not only in the musings of a misogynist Swedish playwright. Research on the structure of the brain, on the effects of hormones, and in animal behavior, child psychology and anthropology is providing new scientific underpinnings for what August Strindberg and his ilk viscerally guessed: men and women *are* different. They show obvious dissimilarities, of course, in size, anatomy and sexual function. But scientists now believe that they are unlike in more fundamental ways. Men and women seem to *experience* the world differently, not merely because of the ways they were brought up in it, but because they feel it with a different sensitivity of touch, hear it with different aural responses, puzzle out its problems with different cells in their brains.

Hormones seem to be the key to the difference—and an emerging body of evidence suggests that they do far more than trigger the external sexual characteristics of males and females. They actually "masculinize" or "feminize" the brain itself. By looking closely at the neurochemical processes involved, investigators are finding biological explanations for why women might think intuitively, why men seem better at problem-solving, why boys play rougher than girls.

Whether these physiological differences destine men and women for separate roles in society is a different and far more delicate question. The particular way male brains are organized may orient them toward visual-spatial perception, explaining—perhaps—why they are superior at math. Women's brains may make them more verbally disposed, explaining—possibly—why they seem better at languages. Males of most species appear to be hormonally primed for aggression, pointing—it may be—to the long evolutionary record of male dominance over women.

But few of these presumed differences go unchallenged. And whether they imply anything more—about leadership capacities, for example, or that men are biologically suited for the workplace and women for the hearth—is another part of the thicket. The notion that biology is destiny is anathema to feminists and to many male researchers as well. It is their position that sexual stereotyping, reinforced by a male-dominated culture, has more bearing on gender behavior than do hormones. "As early as you can show me a sex difference, I can show you the culture at work," insists Michael Lewis, of the Institute for the Study of Exceptional Children.

The new research has thus revived, in all its old intensity, the wrangle over whether "nature" or "nurture" plays the greater part in behavior. At the same time, it has become a fresh battle ground for feminism, a continuation of the sex war by other means. Spurred by the women's movement, large numbers of female scientists have moved into an area of inquiry once largely populated by men and by male ideas of gender roles. Both male and female investigators have been challenging male-fostered notions of female passivity and submissiveness. But because some are also acknowledging the role of biology, they are catching flak from hard-core feminists, who fear such findings will be used—as they have been in the past—to deny women equal rights.

Some researchers now refuse to be interviewed, or carefully hedge their assertions. "I found myself being screamed at—this time by the very people whose cause I had supported," wrote sociologist Alice Rossi, after she landed in hot water for talking about the "innate predisposition" of women for child-rearing. "People are being really hounded," agrees anthropologist Sarah Blaffer Hrdy, who found she could not even hypothesize about men's math abilities without provoking feminist wrath.

The research comes under indictment on another count: since possibilities for experimentation with humans are limited, it leans heavily on animal studies. Complains Stanford psychologist Eleanor Maccoby, who reviewed the literature on sex differences: "People look at this and say it is all biological. They generalize wildly from a little monkey research." But most researchers are cautious about making the leap from lower primates to Homo sapiens. Human evolution involved a huge increase in brain flexibility that gave rise to human culture. And over the long course of that evolution, humans have become much less the creatures of their hormones than are rats or rhesus monkeys.

Even so, the researchers are providing some fascinating new glimpses into the biology of behavior. Among their odd assortment of laboratory subjects are male canaries whose song repertoire is imprinted, like a player-piano roll, in a cluster of brain cells; virginal female rats that go through the motions of nursing when confronted with rat pups, and young girls who turn "tomboyish" because they were exposed to male hormones before birth.

There are also enough anomalies—male marmosets that tenderly nurture their young, female langurs that fight ferociously for turf—to cast doubts on some firmly entrenched beliefs about gender behavior. But these contradictions, too, are providing new insights into the essential nature of the sexes.

The everyday perception of sexual differences is a mélange of fact and assumption. That men, for the most part, are larger and stronger than women is something that anyone can see and physiologists can verify. There are also clear-cut differences in primary sexual functions: menstruation, gestation and lactation in women, ejaculation in men. Beyond that, observes Harvard biologist Richard Lewontin, "we just don't know any differences except the plumbing features that unambiguously separate men from women." Other presumed distinctions provide a continuing source of strife for both sexes. For example, the proposition that men are naturally more competitive than women seems increasingly debatable as women move into male jobs and sports, where they often prove as combative as men (A/E p. 31). In any case, the average differences that exist between men and women leave plenty of room for individual variations in the sexes. Not all males in a given group are more aggressive or better at math than all females; not all women are more adept at learning languages. "Women and men both fall along the whole continuum of test results," notes neuropsychologist Eran Zaidel of the University of California, Los Angeles.

Like most behavioral traits, competitiveness is hard to measure objectively, harder yet to attribute to innate causes. Scientists have been trying to zero in on those traits that *can* be measured, by way of psychological tests and brain and hormone studies. A few years ago Diane McGuinness, of the University of California at Santa Cruz, made a study of the vast body of technical literature that has sprung up in the field. She concluded that from infancy on, males and females respond in ways that provide significant clues to their later differences in behavior.

McGuinness believes that girl infants are more alert to "social" cues. They respond more to people, read facial expressions better and seem better able to interpret the emotional content of speech even before they can understand words—a clue to the proverbial "women's intuition." Boy infants are more curious about objects and like to take them apart—the beginning, perhaps, of their superior mechanical aptitude. As infants, they are awake longer and more active and exploratory.

Girls, notes McGuinness, have a "superior tactile sensitivity," even in infancy. They excel in fine-motor coordination and manual dexterity, suggesting why they are better at such tasks as typing and needlework—or neurosurgery. This same affinity for precision and detail seems to account for girls' greater verbal ability. They speak earlier and more fluently and, perhaps aided by superior auditory memory, carry a tune better. (McGuinness doubts a connection, but it has been shown in some studies that mothers tend to carry on more "conversations"—talking and singing—with girl infants than with boy babies.)

Boys stutter more than girls, spell worse and are classified far more often as "learning disabled" or "hyperactive"—quite possibly, McGuinness argues, because the early stress on reading and writing favors girls. But boys have a clear advantage in visual-spatial orientation, marked by a lively interest in geometric forms and in manipulating objects. At an early age boys and girls are about equal in arithmetic, but boys pull ahead in higher mathematics. Their faster reaction time and better visual-spatial ability appear to give them an edge in some sports.

Are these differences real and could they be biologically based? Stanford's Maccoby is skeptical. She and an associate, Carol Jacklin, reviewed more than 1,400 studies of sex differences and concluded that only four of them were well established: verbal ability for girls and visual-spatial ability, mathematical excellence and aggression in boys. Maccoby also contends, as many researchers do, that sex typing and the different set of expectations that society thrusts on men and women have far more to do with any differences that exist—with their divergent abilities—than do genes or blood chemistry. Diane McGuinness, on the other hand, believes there is "compelling" evidence that sex differences are a result of biological determinism, and her colleague, Karl Pribram, agrees. Pribram suspects that men and women may be "programed" differently from the beginning. "We don't know why these things are true," says Pribram. "But it's very difficult to say culture is predisposing males to fail at English and females at math."

To get some inkling how this programing might come about, anthropologists turn to evolutionary scenarios derived from the study of primitive cultures. The most familiar accounts center on hunter-gatherer societies, the prototype of human social organization. There was a clear division of labor along sexual lines in these societies. The risky business of hunting and fighting fell to the males, presumably because they were more expendable: it required only one male to impregnate many females. Thus, males may have evolved the larger musculature, faster reactions and greater visual-spatial acuity for combat, for hurling spears, for spotting distant prey. As an evolutionary result, even today they are more competitive, more at risk. They experience more stress, die younger than women in maturity and suffer more accidental deaths in childhood.

Females were the gatherers and the first agriculturalists. Limited to less venturesome roles by successive pregnancies and the need to care for infants, they may have developed close-to-the-nest faculties of touch and hearing, perhaps even a greater facility for speech as they interacted with their offspring and other females. According to Darwinian theory, these adaptively advantageous traits would then be passed on in the gene pool. Anthropologist Donald Symons carried this thinking a controversial step further in a 1979 book contending that males are predisposed to sexual promiscuity, while females are prone to constancy. His reasoning: a male's reproductive success was determined by the number of females he could impregnate. Females, he said, feared the wrath of jealous stronger mates and found monogamy more conducive to raising offspring—their own measure of reproductive efficacy.

Symons's thesis, wrapped in all its sociobiological trappings, impressed some scientists and infuriated others. In addition, the subject is an explosive one for feminists. If men are driven by testicular fortitude, women reject the notion that they are limited by a kind of ovarian docility. Anne Petersen, director of the adolescent laboratory at Chicago's Michael Reese Hospital, supports this view: "When women really needed to keep making babies because so many died and women themselves didn't live very long, their work was of a different nature and needed to be related to reproduction. That's not true anymore."

Evolutionary evidence does suggest that with the advent of weapons, some physical distinctions between males and females became less necessary. Differences in size and strength have greatly diminished over the millennia, perhaps, according to University of Michigan paleoanthropologist Milford Wolpoff, because "the physical requirements of the male and female roles have become more similar." The main anatomical trend, Wolpoff says, "has been for males to become more feminized."

Harvard's Sarah Blaffer Hrdy is one of the new generation of social scientists who are trying to debunk the concept of female passivity. In a forthcoming book, "The Woman That Never Evolved," Hrdy argues that female territoriality—the aggressive protection of turf—has been overlooked by most anthropologists. "The central organizing principle of primate social life is competition between females, and especially female lineages," she writes. In such matrilinear societies, she argues, "the basic dynamics of the mating system depend not so much on male predilection as on the degree to which one female tolerates another."

Hrdy's work has encouraged other researchers to look beyond the evolutionary stereotypes. Until now, she contends, most such studies have focused on males arranging themselves in order to take advantage of females. "We've really been ass-backwards in trying to understand the primate social organization by looking only at males," asserts Hrdy.

By shifting the anthropological focus, the Hrdy breed of researchers hopes to show that gender roles are not unalterably determined by biology; instead, they may be the product of particular cultures. Even so, the evidence for an inborn masculine "aggression factor" seems inescapable. It is widely agreed that in the majority of animal species, males are more prone to fighting than are females. Biologists trace this to the hormone testosterone, secreted in the testes of the male fetus during a critical period in its development. Although the sex of a fetus is basically determined by its genetic coding (XX chromosomes for females, XY for males), any fetus has the chance of developing either male or female characteristics, depending on the hormones it is exposed to. Testosterone and other male hormones "masculinize" a fetus, differentiating its genitalia from the female's. At the same time, the male hormones prevent the development of ovaries, which secrete female hormones that would stimulate the growth of feminine characteristics.

Scientists first got on the trail of testosterone in 1849. Experiments showed that roosters became less aggressive and lost their sexual drive after they were castrated, then regained their "roosterhood" when extracts from the testes were implanted in them. A century later, in 1959, physiologists Robert Goy and William Young conducted a study still considered a landmark in the field. First they injected pregnant female guinea pigs with massive doses of testosterone. The result: the genetically female offspring in the brood had both male genitalia and ovaries. When the ovaries were removed and the aberrant females were given a fresh dose of testosterone, they behaved like males, even "mounting" other females—the gesture of male dominance in many species.

Goy, now with the University of Wisconsin's Primate Research Center, has confirmed the effects of testosterone in experiments with rhesus monkeys over the past decade. Not only is female behavior partly masculinized by prenatal testosterone, he says, but the robustness and vigor of males depend on how long they have been exposed to the hormone. "The different kinds of behavior that you see young male monkeys display," Goy asserts, "are completely, scientifically and uniquely determined by the endocrine conditions that exist before birth."

To see if hormones play a similar role in human behavior, John Money of Johns Hopkins University and Anke Ehrhardt of Columbia studied one of nature's own experiments—children exposed to abnormally high levels of androgens (male hormones) before birth because of adrenal-gland malfunctions. Among other effects discovered, the researchers at Johns Hopkins found that girls born with this disorder exhibited distinctly "tomboyish" behavior, seldom played with dolls and began dating at a later age than other girls.

The much-cited Money-Ehrhardt research has provided a classic context for the nature-nurture debate. Some scientists maintain that the tomboyism was a clear result of the hormone exposure, and they bolster their argument by noting the scores of animal experiments that demonstrate similar effects. But others criticize the study for failing to emphasize that girls with congenital adrenal hyperplasia do not *look* like normal girls at birth; they often require corrective surgery to restore normal female genitals. Thus, the argument goes, they may be treated differently as they grow up, and their behavior could be more the result of an abnormal environment than of abnormal blood chemistry.

The debate rages back and forth. But at least one scientist who has been on both sides, Rutgers psycho-endocrinologist June Reinisch, recently found evidence to buttress the hormonal argument. Over a period of five years Reinisch studied 25 boys and girls born to women who had taken synthetic progestin (a type of androgen) to prevent miscarriages. When the scientist compared them with their unexposed siblings by giving each child a standard aggression test, she found significant differences between the groups. Progestin-exposed males scored twice as high in physical aggression as their normal brothers; twelve of seventeen females scored higher than their unexposed sisters. "This result was so striking," says Reinisch, "that I sat on the data for a year before publishing."

Reinisch has by no means renounced her belief in the importance of environment. Like many of her colleagues, she suspects that hormones act to "flavor" an individual for one kind of gender behavior or another. But how the individual is brought up is still an important factor. As Robert Goy explains, "It looks as though what the hormone is doing is predisposing the animal to learn a particular social role. It isn't insisting that it learn that role; it's just making it easier. The hormone doesn't prevent behavior from being modified by environmental and social conditions."

As to how the initial "flavor" comes about, researchers now believe that hormones change the very structure of the brain. Some variations in the brains of males and females have been observed in animals. They were found mainly in the hypothalamus and preoptic regions, which are closely connected to the reproductive functions. In those areas males are generally found to have more and larger "neurons"—nerve cells and their connecting processes. Experiments conducted by Dominique Toran-Allerand of Columbia University using cultures of brain cells from newborn mice have shown that neuronal development can be stimulated by hormones, and this suggests a key to the sexual mystery. Says animal

physiologist Bruce McEwen of Rockefeller University: "Growth, as the primary event caused by hormones, could account for the observed differences in brains."

The clearest evidence to date of the brain-hormone link is in songbirds. Several years ago Rockefeller researchers Fernando Nottebohm and Arthur Arnold discovered sex differences in certain clusters of brain nuclei that control the singing function in canaries. The nuclei, they found, are almost four times as large in males as in females—apparently explaining why male songbirds sing and females don't. Singing is part of the mating ritual for the birds, and Nottebohm demonstrated that the size of the nuclei waxed and waned with the coming and going of the mating season. When he treated female songbirds with testosterone, the singing nuclei doubled in size and the females produced malelike songs. "This was the first observation of a gross sexual dimorphism in the brain of a vertebrate," Nottebohm told a scientific meeting last November.

Many scientists are now convinced that hormones "imprint" sexuality on the brains of a large number of animal species by changing the nerve-cell structure. "Even the way dogs urinate—that's a function that is sex different and is determined by hormones," says Roger Gorski, a UCLA neuro-endocrinologist who has done important experiments with animal brains.

But what about humans? So far, no one has observed structural differences between the brains of males and females in any species more sophisticated than rats. In humans, the best evidence is indirect. For years researchers have known that men's and women's mental functions are organized somewhat differently. Men appear to have more "laterality"—that is, their functions are separately controlled by the left or right hemisphere of the brain, while women's seem diffused through both hemispheres. The first clues to this intriguing disparity came from victims of brain damage. Doctors noticed that male patients were much likelier than females to suffer speech impairment after damage to the left hemisphere and loss of such nonverbal functions as visual-spatial ability when the right hemisphere was damaged. Women showed less functional loss, regardless of the hemisphere involved. Some researchers believe this is because women's brain activity is duplicated in both hemispheres. Women usually mature earlier than men, which means that their hemisphere processes may have less time to draw apart. They retain more nerve-transmission mechanisms in the connective tissue between the two hemispheres (the corpus callosum) and can thus call either or both sides of the brain into play on a given task.

On the whole, women appear to be more dominated by the left, or verbal, hemisphere and men by the right, or visual, side. Researchers McGuinness and Pribram speculate that men generally do better in activities where the two hemispheres don't compete with, and thus hamper, each other, while women may be better able to coordinate the efforts of both hemispheres. This might explain why women seem to think "globally," or intuitively and men concentrate more effectively on specific problem-solving.

A few enterprising researchers have tried to find a direct connection between hormones and human-brain organization. UCLA's Melissa Hines studied 16 pairs of sisters, of whom one in each pair had been prenatally exposed to DES (diethylstilbestrol), a synthetic hormone widely administered to pregnant women during the 1950s to prevent miscarriages. Using audiovisual tests, Hines found what appeared to be striking differences between the exposed and unexposed sisters.

First, Hines played separate nonsense syllables into the women's right and left ears. Normally, the researcher explains, most people—but especially males—report more accurately what they hear with the right ear. In her tests, the hormone-exposed women picked the correct syllable heard with the right ear 20 per cent more often than their unexposed sisters. A test of their right and left visual fields produced comparable results. The implication was that the women's brains had been masculinized. "It is compelling evidence that prenatal hormones influence human behavior due to changes in brain organization," says Hines.

Differences in brain organization may have practical implications for education and medicine. Some researchers believe that teaching methods should take note of right-left brain differences, though past attempts at such specialized teaching have been ineffective. Other scientists predict clinical benefits. It is useful to know, for example, that females who are brain-damaged at birth will cope with the defects better than males. Columbia's Toran-Allerand suggests that certain types of infertility might be corrected once scientists understand how hormones mold the reproductive structures of the brain. "I'm interested in clinical applications," she says, "but all these questions get lost in the furor over behavioral differences."

The furor may be inevitable. The very mention of differences in ability between men and women seems to imply superiority and inferiority. Women researchers in the field have had the toughest going at times. Some have found themselves under Lysenkoist pressure to hew to women's-liberation orthodoxy, whatever their data show. University of Chicago psychologist Jerre Levy, a pioneer in studies of brain lateralization, withdrew from public discussion of her work after she was bombarded with hostile letters and phone calls. Harvard's Hrdy recalls sitting on a panel that was cautiously examining the "hypothesis" of male math

superiority when a feminist seated next to her whispered, "Don't you know it's evil to do studies like that?" Says Hrdy: "I was just stunned. Of course it's not evil to do studies like that. It's evil to make pronouncements to say they're fact."

From the time women began moving—rather aggressively—into male-female studies, many researchers have grown wary of making such pronouncements. It has become increasingly difficult to find any statement that is not assiduously qualified. One reason is that differences among members of the same sex are far greater than average differences *between* sexes. Monte Buchsbaum of the National Institutes of Health conducted tests of electrical activity in the brain showing that women tend to have a larger "evoked potential" than men—an indication of greater sensitivity to certain stimuli. But, he cautions, "individuals can vary over about a fivefold range. The variation between the sexes is only about 20 to 40 per cent." Harvard's Richard Lewontin notes that the average male-female differential in math scores is only "half a standard deviation. That's rather small." The math dispute is "just silly," scoffs Lewontin, and assertions about "who's most aggressive or who's most analytical are just the garbage can of barroom speculation presented as science."

Many researchers contend that a child's awareness of gender is more decisive than biology in shaping sexual differences. "The

---

In Sports, 'Lions Vs. Tigers'

When Don Schollander swam the 400-meter freestyle in 4 minutes, 12.2 seconds at the 1964 Olympics, he set a world record and took home a gold medal. Had he clocked the same time against the women racing at the 1980 Moscow Games, he would have come in fifth. In the pool and on the track, women have closed to within 10 per cent or less of the best male times, and their impressive gains raise an intriguing question: will men and women ever compete as equals?

Athletics is one area of sex-difference research that generates little scientific controversy. Physiologists, coaches and trainers generally agree that while women will continue to improve their performances, they will never fully overcome inherent disadvantages in size and strength. In sports where power is a key ingredient of success, the best women will remain a stroke behind or a stride slower than the best men.

Muscle Vs. Fat: A man's biggest advantage is his muscle mass. Puberty stokes male bodies with the hormone testosterone, which adds bulk to muscles. A girl's puberty brings her an increase in fat, which shapes her figure but makes for excess baggage on an athlete. When growing ends, an average man is 40 per cent muscle and 15 per cent fat; a woman, 23 per cent muscle, 25 per cent fat. Training reduces fat, but no amount of working out will give a woman the physique of a man. Male and female athletes sometimes try to build bigger muscles by taking anabolic steroids—artificial male hormones that stimulate muscle growth—even though physicians consider them dangerous and all major sports have outlawed them.

Bulging muscles alone can't make a woman as strong as a man. Men have larger hearts and lungs and more hemoglobin in their blood, which enables them to pump oxygen to their muscles more efficiently than women can. A man's wider shoulders and longer arms also increase his leverage, and his longer legs move him farther with each step. "A female gymnast who puts her hands on a balance beam and raises herself up is showing a lot of strength," says Barbara Drinkwater, a physiologist with the University of California at Santa Barbara, Calif. "But a woman won't throw a discus as far as a man." Although highly conditioned women can achieve pound-for-pound parity with men in leg strength,

their upper-body power is usually only one-half to two-thirds that of an equally well-conditioned male athlete.

A few sports make a virtue of anatomy for women. Extra body fat gives a female English Channel swimmer better buoyancy and more insulation from the cold, and narrow shoulders reduce her resistance in the water. As a result, women have beaten the fastest male's round-trip Channel crossing by a full three hours. In long-distance running contests, women may also be on equal footing with men. Grete Waitz's time of 2 hours, 25 minutes and 41 seconds in last year's New York City Marathon was good enough to bring her in ahead of all but 73 of the 11,000 men who finished the race. "Women tend to do better relative to men the longer the distance gets," says Joan Ullyot, author of "Running Free." "On races 100 kilometers and up, it may turn out that women are more suited to endurance than men." Under the body-draining demands of extended exertion, a woman's fat may provide her with deeper energy reserves. Satisfied that women can take the strain, the International Olympic Committee has authorized a women's marathon for the 1984 Games. In previous years the longest Olympic race for females was 1,500 meters—less than a mile.

Tough: Women athletes have dispelled the myths about their susceptibility to injury. The uterus and ovaries are surrounded by shock-absorbing fluids—far better protected than a male's exposed reproductive equipment. And the bouncing of breasts doesn't make them more prone to cancer, or even to sagging. As for psychological toughness, Penn State physiologist and sports psychologist Dorothy Harris says that "if you give a woman a shot at a $100,000 prize, you discover that she can be every bit as aggressive as a man."

Going one-on-one with a man is not the goal of most women in sports. "It's like pitting lions against tigers," declares Ullyot. "Women's achievements should not be downgraded by comparing them to men's." But as organized women's sports grow up, they will have to face up to at least one serious masculine challenge. According to Ann Uhlir, executive director of the Association for Intercollegiate Athletics for Women, when a college starts taking its women's sports program seriously, it tends to put a man in charge.

ERIC GELMAN

real problem for determining what influences development in men and women is that they are called boys and girls from the day they are born," says biologist Lewontin. He cites the classic "blue, pink, yellow" experiments. When a group of observers was asked to describe newborn infants dressed in blue diapers, they were characterized as "very active." The same babies dressed in pink diapers evoked descriptions of gentleness. When the babies were wearing yellow, says Lewontin, observers "really got upset. They started to peek inside their diapers to see their sex."

It is clear that sex differences are not set in stone. The relationship between hormones and behavior, in fact, is far more intricate than was suspected until recently. There is growing evidence that it is part of a two-way system of cause-and-effect—what Lewontin calls "a complicated feedback loop between thought and action." Studies show that testosterone levels drop in male rhesus monkeys after they suffer a social setback and surge up when they experience a triumph. Other experiments indicate that emotional stress can

change hormonal patterns in pregnant females, which in turn may affect the structure of the fetal brain.

By processes still not understood, biology seems susceptible to social stimuli. Ethel Tobach of New York's American Museum of Natural History cites experiments in which a virgin female rat is presented with a five-day-old rat pup. At first, her response is vague, says Tobach. "But by continuing to present the pup, you can get her to start huddling over it and assuming the nursing posture. How did that come about? There's obviously some biochemical factor that changes . . . When you have the olfactory, visual, auditory, tactile input of a five-day-old pup all those days, it can change the blood chemistry."

A more enigmatic example, says Tobach, is found in coral-reef fish: "About six species typically form a group of female

Sex Research—On the Bias

As they probe deeper into the role of hormones and brain structure, scientists are breaking important new ground in understanding sex differences. But from its mid-nineteenth century beginnings, male-female research has been one of the less glorious chapters in the history of science, a lesson in how wrongheaded scientists can be when they succumb to prevailing biases.

There were no real villains in the drama, only proper Victorians who felt it was in society's interest to show that women were designed for lesser tasks. Scientists argued that if women used their brains excessively, they would impair their fertility by draining off blood cells needed to support the menstrual cycle. Many genuinely believed that the dawning feminist movement threatened the survival of the race. "Women were seen as crazy ideologues, going against nature," says Elizabeth Fee of Johns Hopkins University. "It was the duty of science to prove their inferiority."

Brain Size: Until the turn of the century, researchers sought evidence to support what everyone assumed to be true: that men were smarter than women. Paul Broca, a professor of clinical surgery in Paris, measured hundreds of skulls during the 1870s and found that men's brains were, on the average, 14 per cent larger than women's—clear proof, he said, of men's greater intelligence. Brain size has since been shown to be related to body size, not intelligence.

Broca's measurements at least were accurate, which is more than could be said for the work of some of his zealous contemporaries. When craniologists thought intellect resided in the frontal lobes, autopsies kept finding that frontal lobes were larger in men than in women. When scientists decided the parietal lobes were the true seat of intelligence, male cadavers obligingly turned up with larger parietal lobes. Because the differences were a matter of millimeters, the "objective" measurements had a subjective flaw: researchers usually knew the sex of the cadaver they were examining and invariably found the brain differences they expected.

Craniology fell into disfavor early in this century, a victim of its own convolutions. By the 1920s endocrinologists had discovered what they took to be a real difference: separate

male and female hormones. Later it was found that men had significant amounts of estrogen, the female hormone. "That discovery showed how precarious sex differences really were," says Boston University science historian Diana Long Hall.

Even respected scientists produced flawed research. Psychoanalyst Erik Erikson contended that the nature of human genitalia influences the way males and females think. Boys' thoughts are directed outward, he wrote, while girls look to a nurturing "inner space." In a classic experiment in the 1950s, Erikson asked 11- to 13-year-old children to construct scenes from blocks and other toys. He reported striking sex differences. Male spaces were dominated by thrusting height and downfall, female spaces by static interiors. In short, boys built towers and girls built enclosures.

'Sex Roles': Erikson's notion of inner and outer space is still widely used today. "It fit with traditional sex roles, so people assumed that it was true," says University of Toronto psychologist Paula Caplan. But when Caplan looked at Erikson's data, the evidence seemed flimsy. Fewer than one-tenth of his scenes contained a tower, and boys actually built many more enclosures than towers. The differences that Erikson did find, suggests Caplan, may have occurred because the children chose sex-stereotyped toys. Doll furniture, she points out, does not stack as well as blocks. Caplan repeated the experiment using younger children to reduce the effects of socialization, and found no difference in the number, size or type of towers built by boys or girls.

The problem of bias has not vanished from male-female studies. Some scientists say it has tilted the other way. They complain that feminists seem bent on ascribing all gender differences to male-dominated culture. And many women think the research is still skewed by male views. The continuing controversy illustrates a point that both male and female scientists believe cannot be overlooked: the researchers investigating sex differences are no more than the products of the enigma they are exploring.

DAVID GELMAN with JOHN CAREY

fish with a male on the outside. If something happens to remove the male, the largest female becomes a functional male, able to produce sperm and impregnate females. It has been done in the lab as well as observed in the natural habitat."

The human parallels are limited. No one expects men or women to undergo spontaneous sex changes, and millennia of biological evolution aren't going to be undone by a century of social change. But it is now widely recognized that, for people as well as animals, biology and culture continually interact. The differences between men and women have been narrowing over evolutionary time, and in recent decades the gap has closed further.

Perhaps the most arresting implication of the research up to now is not that there are undeniable differences between males and females, but that their differences are so small, relative to the possibilities open to them. Human behavior exhibits a plasticity that has enabled men and women to cope with cultural and environmental extremes and has made them—by some measures—the most successful species in history. Unlike canaries, they can sing when the spirit, rather than testosterone, moves them. "Human beings," says Roger Gorski, "have learned to intervene with their hormones"—which is to say that their behavioral differences are what make them less, not more, like animals.

DAVID GELMAN with JOHN CAREY and ERIC GELMAN in New York, PHYLLIS MALAMUD in Boston, DONNA FOOTE in Chicago, GERALD C. LUBENOW in San Francisco and JOE CONTRERAS in Los Angeles

NEW GENETIC FINDINGS:
WHY WOMEN LIVE LONGER

PAULA DRANOV

Paula Dranov is a New York-based science writer who specializes in medical reporting.

Picture the female body in all its voluptuous glory: soft curves artfully arranged on a delicate frame, a truly mysterious vessel of passion, pleasure . . . and secret strength.

Forget for a moment the erotic attractions and compare it with the larger, more muscular male physique. Being bigger and stronger, it's small wonder that men can run faster, hit tennis balls harder and heft over their heads weights women can barely budge. No question about it, when it comes to pure physical strength, man is the superior biological specimen.

But that's about it.

Beneath the curves and softness women conceal a biological superiority they share with the female of just about every animal species: a far-reaching capacity for survival males simply do not possess. Not only do women live longer but their survival rate exceeds men's, in the womb and throughout all of life.

Anthropologist Ashley Montagu first brought this female biological advantage to public attention in 1952 with his book *The Natural Superiority of Women.* He candidly conceded that much of the evidence he assembled wasn't exactly new. Indeed, Dr. Estelle Ramey, a professor of physiology at Georgetown University Medical School, notes that "even in the eighteenth century it was observed that women tended to outlive men during their entire life span."

BIOLOGICAL ADVANTAGE

Today, far more is known on the subject than the simple fact that women live longer. Evidence of the female biological advantage continues to build. Consider:

• It's estimated that about 140 males are conceived for every 100 females, but at birth the ratio is down to about 106 to 100. Clearly, males are more vulnerable even before they're born.

• In the first precarious month of life, more boys succumb than girls. The infant death rate is 15 per 1,000 for boys compared with 11 per 1,000 for girls.

• Of 190 birth defects observed, more than 71 percent occur mainly in males, while only 25 percent are found chiefly in females.

• Women in every age group have significantly lower incidence of heart disease than men.

• Even when women have high blood pressure and high levels of cholesterol and other fats (all factors that increase the risks of heart disease and stroke), they have a lower mortality rate than men.

• Even underprivileged and economically deprived women outlive their male counterparts.

• Women are more resistant than men to certain bacterial infections and have persistently higher levels of immune globulin M, a protective blood protein that helps the body resist disease.

• The female sex hormone estrogen appears to enhance immune globulin in experimental animals.

Behind this overwhelming female health and survival advantage are the evolutionary requirements of the human species. We reproduce rather slowly—as a rule, a woman gives birth to only one child at a time over a relatively long (nine month) gestational period. As a result, there is not only a greater need for females than for males but an urgent requirement that females be endowed with the strength and stamina to produce one child after another throughout their reproductive years.

To accomplish all this nature has dealt the female the stronger hand genetically, hormonally and immunologically. The genetic advantage comes down to the striking differences between the X and Y chromosomes. Sperm bearing the Y chromosome, which determines that a child will be male, are smaller than those bearing X chromosomes, which combine with the X contributed by the mother to produce a female. The smaller, lighter sperm carrying the Y (male) chromosomes seem to swim faster than the larger X-bearing (female) ones, get to the eggs faster and thus account for the male-to-female sex ratio at conception.

But that early male numerical advantage is only temporary. More males than females are spontaneously aborted, which seems to indicate that two X chromosomes produce a hardier fetus than an X and a Y.

Clearly, two X chromosomes provide a genetic strength males lack. A defective gene capable of producing a life-threatening abnormality has less chance of expressing itself if it is matched—and as a result, dominated—by a normal gene on the other X chromosome. Males have no such backup system; a recessive gene on the X chromosome doesn't always have a healthy counterpart on the Y to compensate for the weakness. Hence, the more frequent appearance of birth defects in males.

OVERRIDING INFLUENCE

Probably the best-known sex-linked genetic disease is hemophilia, which passes from mother to son but not to daughter. Here, the defective gene on the X chromosome contributed by the mother has no counterpart on the Y to prevent it from expressing itself in males. But the fact that two X chromosomes are needed to produce a female means that a normal gene on the X from the father overrides the influence of the weak maternal gene.

Male sex hormones also exert a deleterious effect in terms of survival. They have been implicated in higher rates of heart disease among men. True, male hormones do confer important benefits. They're responsible for the larger male muscle mass and for the fact that men, unlike women, have almost no age limit on their reproductive capacity.

On the minus side, however, is evidence that the male sex hormone testosterone facilitates development of blockages in the coronary arteries that impede the free flow of blood and often result in heart attacks.

A little medical background: two vital substances produced by the liver are involved in the transport of cholesterol in the blood. One of these, high-density lipoprotein (HDL), appears to facilitate the passage of cholesterol through the coronary arteries. It is believed that the other, low-density lipoprotein (LDL), ties up cholesterol so that instead of flowing through these crucial arteries, it slips through gaps in the linings, gets caught and begins to accumulate in the form of fatty deposits called plaque. The more plaque lining the arteries, the less freely blood can flow to the heart. Should the blood supply be cut off, a heart attack occurs.

Both the male and female sex hormones are involved in this vital cholesterol transport process. In women, estrogen acts on the liver to produce HDL, the beneficial ingredient. Testosterone affects the liver, too, but by way of increasing the LDL-to-HDL ratio—and the higher this ratio gets, the greater the risk of heart disease.

The possibility that testosterone decreases male life expectancy was suggested by a 1969 study by James Hamilton and Gordon E. Mestler, of the Downstate Medical Center in New York, which showed that castrated men outlive normal men and that the earlier the age of castration, the longer a male could expect to live. Indeed, they found that for every year of delay before castration, 0.28 years of life expectancy are lost. In other words, relieving a man of his sex hormones via castration helps prolong his life.

The third female survival mechanism is a superior resistance to most forms of bacterial infections.

All this adds up to a female survival advantage that has in fact been improving over the years. Dr. Louis M. Hellman, professor emeritus at Downstate Medical Center, explains that as life expectancy has increased in the twentieth century "the male-female differential also has increased. The current eight-year gap contrasts with a two-year gap at the beginning of the century."

LOWER MORTALITY
And, contrary to dire predictions, changing social roles have not increased female vulnerability to heart disease and other life-threatening ailments that primarily beset men. Indeed, a recent woman's health survey found that working women are healthier than unemployed women, and a study by the Metropolitan Life Insurance Company determined that women in *Who's Who in America* had a 29 percent lower mortality over the period studied than women in general. (The same results were found when *Who's Who* men were studied, proving, comments Dr. Ramey, that "it's a lot better for your health to be successful and rich than poor and a failure.") Then, too, it's been shown that the only major health risk confronting women today that is greater than in the past is the higher incidence of smoking—but here, too, men's risks are greater.

The popular assumption that women are healthier and live longer because they're more protected never was scientifically accepted. As Dr. Hellman notes, "If this ever was valid, recent changes in social custom should provide an experiment of nature that will prove or disprove the point. There is, however, very little evidence either in the animal world or in human beings that lack of stress plays much of a role in the longevity of the female."

If so, the most pressing question for the future seems to be not whether new women's roles will erode their health advantage but, as Dr. Ramey has asked, whether the secrets of female survival can be tapped to preserve, protect and prolong the lives of men.

THE MORE SORROWFUL SEX

Anywhere from two to five times as many women as men are likely to be diagnosed as depressed. The reasons are largely cultural—having to do with what "being feminine" requires, says a researcher who has investigated the sexually lopsided numbers. The problem may be reaching epidemic proportions.

Maggie Scarf

Maggie Scarf is a science writer who specializes in psychology and psychiatry. She is the author of *Body, Mind, Behavior* (New Republic Books, 1976) and the recipient of three national awards from the American Psychology Foundation. With the aid of an Alicia Patterson Foundation Fellowship, she has been studying depression in women over the various stages of the life cycle.

Do numbers lie? If not, the evidence is clear and overwhelming: females from adolescence onward—and throughout every phase of the life cycle—are far more vulnerable to depression than are males.

It turns up in virtually every study, carried out anywhere and everywhere. And while the figures may vary from one investigation to the next, the general trend is always the same. More women are in treatment for depression, in every institution—inpatient and outpatient—across the country, in state and county facilities, in community mental-health centers. And when the figures are adjusted for age, or phase of life, or socioeconomic circumstances, the outcome is still the same.

For every male diagnosed as suffering from depression, the head count is anywhere from two to six times as many females. The statistics show variation according to who is doing the counting, what the criteria are, the geographic location, and so forth. But the numbers are never equal.

As one might expect, the consistent disparity—those sexually lopsided statistics—has presented itself as a puzzle to experts of every theoretical hue and stripe. And there have been numerous efforts, on the part of psychiatrists, psychologists, epidemiologists (those who study patterns of illness in the population at large), sociologists, and others, to explain what remains a most peculiar phenomenon. Why, after all, should one sex be more vulnerable to depressive disorder than the other? It is, in a way, as strange as the idea of one sex getting flu, or measles, or appendicitis, or some other illness, far more frequently than the other. Some people insist, in fact, that it's not true: the statistical findings are erroneous, they say, because of certain biases in the counting.

That is "the-numbers-do-lie" point of view. According to that view, the unequal figures—and the fact that so many more women are in treatment for some form of depression—bears witness only to the eagerness of doctors and psychotherapists to *label* women. That is, the distressed person who happens to be female will get one particular dog tag—"depression"—hung around her neck with undue alacrity. But women are not in actuality, some experts have insisted, one whit more depressed than males.

So runs the argument made by Phyllis Chesler in her book *Women and Madness*. Psychologist Chesler contends that women tend to get diagnosed—that is, *called*—"depressed," or "disturbed," or "crazy," or "mad" with somewhat sinister readiness. Such psychiatric putdowns are, she maintains, nothing more than a covert societal mechanism, a means for punishing women who do not adjust to and accept their "femininity" under terms laid down by the male-dominated mental-health Establishment (which represents the community at large). The woman who fails to accept her female role, along with its attendant inferior social status, isn't "behaving," and she gets socked with a psychiatric diagnosis. The diagnosis itself, in the view of Chesler and others, is a handy "medical" and scientifically respectable device for keeping women in their place.

Could this explain those strange statistics? Might the high rates of depression among women be just a social phantasmagoria—a "scientific finding" related to nothing more than the biases of psychotherapists and clinicians? Are females actually no more melancholy, afflicted, anxious, troubled, depressed than males—but simply seen as being so?

Chesler's views have a certain compelling quality—especially since those depression statistics are disturbing, and one would like to explain (since one can't simply wish) them away. But I am among those who believe that her argument, as well as the empirical evidence she amasses in support of it, is probably largely false. I have not, after four years of studying the problem of women and depression in a wide variety of clinical settings, actually encountered anyone whom I suspected to be suffering from a disorder that could be called "psychiatric labeling."

The women whom I have come to know well or only slightly, whom I've talked to at length or just a little, whom I've seen just once or several times over periods of months—were all suffering. They were in pain, and in need of help; there just was no question about it.

Debra Thierry, for example, at age 22, had an image of herself as something superfluous in the human world. She told me during one of our interviews that she felt as if she were "litter." She was, she said, like a piece of drifting newspaper, "something that's just floating around, being blown around the sidewalk, underfoot, you know, being kicked aside. . . ." She was excess matter in the universe, unwanted and without value. Such feelings, and the diagnosis of the state she was in, have nothing at all to do with "psychiatric stigmatization."

Neither did the diagnosis ("depressive neurosis") of Kay Ellenberger, the slender, dark-haired wife of a successful Pittsburgh lawyer. I interviewed Kay at Western Psychiatric Institute and Clinic of the University of Pittsburgh Medical School, where she was an inpatient. She told me that she'd come home after successfully playing in a tennis tournament, and suddenly felt "as if the bottom were dropping out of my life, and that I was a nothing. That I'd promised to do too many things I didn't care about for too many people I didn't give a damn about. I was on all these committees, and running like crazy; but it was stupid and meaningless. And I wanted out—to quit trying—to be dead."

"The women whom I've come to know were in pain. They weren't suffering from 'psychiatric labeling.'"

While her children were at school, she had emptied the medicine cabinet and swallowed everything in sight. Could one imagine, even for a moment, that her suffering wasn't "real"—that her extreme misery, her psychological hurting, weren't actually there?

Or that a patient like 43-year-old Muriel Clough, who kept saying that she was "struggling to liberate" herself, was the victim of antifeminine bias? I came to know Muriel well during the six weeks that she spent at the Dartmouth-Hitchcock Mental Health Center. She was struggling to liberate herself, assuredly, but she kept saying that her marriage was much better and her relationship with her husband was "much more intimate" during those times when she was completely *down* and at the nadir of the depression. I could go on and on, but make the same point: these women, none of them, were victims of a "conspiracy" of the male-dominated mental-health Establishment.

One may argue that the experiences of a single investigator who is studying the problem in some of this nation's most advanced psychiatric facilities can't pretend to anything like Chesler's systematic and data-supported overview of the treatment women receive in the entire mental-health-care system. And I will acknowledge at once that mine is a close-up look, the account of where I've gone and what I've seen, of the kinds of things that happened in the types of places that I happened to go to—which is surely open to charges of particularities, quirkiness, biases of every kind.

But so are Chesler's statistics—or, at least, her interpretation of them. Most epidemiologists consider *Women and Madness* a political and polemical tract that bears no resemblance to anything that could be called "scientific" or "objective." The late Marcia Guttentag, who directed a nationwide study of what many consider to be an "epidemic" of depression among women, made a careful study of the same statistical data that Chesler used. Chesler's figures were found to be based on *absolutes*—that is, on simple head counts of female and male patients in county and state psychiatric facilities. Because she found so many more women of age 65 and over in those mental institutions, Chesler concluded that antifeminine bias was rampant among the (mostly male) psychiatric Establishment; and that women were being railroaded, via the "diagnosing" track, into geriatric careers as mental patients.

But what Chesler did not do was adjust those figures for age. If there are more women of age 65 and over in mental institutions, it is because there are more women of that age in the population as a whole. Women live longer. In fact, given the larger number of women of that age in the overall population, there were years, so the Guttentag study group found, when women were underrepresented in psychiatric hospitals. Chesler's claim, that old women are tossed into mental institutions that are actually custodial homes, was simply not found to be substantiated.

Indeed, as one worker on the Guttentag project (officially entitled "The Women in Mental Health Project at Harvard") told me, there were years when *men*, in proportion to their numbers in the population, were overrepresented in mental institutions, as a group. Men are, furthermore, far more likely to be committed to institutions against their will. Male diagnoses very frequently involve problems with other people—alcoholism, aggressive acting-out of psychological problems, and so on—and they are often brought in by the police and sent via the courts to institutions.

Women, on the other hand, tend to receive the more "passive" diagnoses: they become depressed, or perhaps schizophrenic, but they don't usually hurt anyone. For that reason, women are more commonly brought in for treatment by their friends or families, and they are committed *voluntarily*.

Each sex does, intriguingly, appear to take the lead in specific types of psychiatric disturbance. Men, as a group, show far higher rates of alcoholism, drug disorders, and behavior disorders of childhood and adolescence—in short, the more action-oriented, disruptive-to-others kinds of difficulty. Where women are concerned, there is one single category in which they hog the diagnostic stage to an almost preposterous degree. And that is, of course, depression.

The Guttentag group came away from their analysis of the mental-health statistics with a sense that the finding of so much more depression among women was discomfiting but real. Not only was there the excess of treated depressions; moreover, there also seemed to be vast numbers of women who were depressed for various reasons, and who had many of the clear-cut symptoms of depression, but who were walking around, not realizing that they "had" anything, and therefore not seeking treatment.

1. PERSPECTIVES

There could, nevertheless, be a very different sort of a kink in those statistics on women and depression. The fact is that in order to be diagnosed as "suffering from depression," you have to go to a doctor in the first place, and women just plain do go to see their doctors more often. Not only do they consult doctors with a demonstrably greater frequency, but they do so about many more minor kinds of problems and of disabilities.

The difference in what researchers call "health-care-seeking behaviors" springs into being, apparently, sometime just around puberty. Before puberty, as Mitchell Balter, a psychologist at the National Institute of Mental Health, told me, girls and boys see physicians with roughly the same frequency. The younger males may, in fact, take a slight lead in number of visits. But *after* puberty, the picture shifts rapidly: there is a sharp increase on the part of the girls, and a decrease for boys. And the changed pattern—women seeing their doctors more often—will persist throughout adult life.

Going to the doctor seems to be, in essence, a particularly "feminine" way of coping with stress, and of dealing with a variety of difficulties and distresses. Sixty percent of all patient

---

"Vast numbers of women had clear-cut symptoms of depression, but were walking around without realizing it."

---

visits are female-patient visits.

If, furthermore, the patient is not only freely expressive about her sorrows, sadnesses, and life disappointments, but has a few physical symptoms to boot, she may readily be diagnosed as suffering from a depression. She may be given medications—either tranquilizers or antidepressants—to help her weather her current difficulties. The use of medically prescribed mood-altering drugs is a very widespread phenomenon, and the majority of those using them (70 percent in toto) are women. According to NIMH's Balter, a study completed in 1972 indicated that some 23 percent of women between the ages of 18 and 29 (nearly one-fourth the entire age group) had taken some psychotropic medication during the preceding year. Among men of the same age, only 6 percent had done so, and they had used less potent types of medication.

After age 30, the males' use of mood-changing drugs increased: it doubled to 12 percent in the years between 30 and 44. But among females, there was also a steep rise: from 23 percent to a hefty 32 percent. Roughly one-third of all women in this age-group (30 to 44) were using prescription drugs to treat their moods. The psychotropic medications were, for the most part, being prescribed by internists and family practitioners (85 percent of those who said they used mood-altering drugs reported that they had never seen a psychiatrist).

The classic chicken-and-the-egg phenomenon could certainly account for varying rates of male and female patients "in treatment for depression." For, when the numbers of such patients are counted, women—if they go to doctors more often—would obviously be diagnosed and medicated with greater frequency, and therefore have a larger representation.

Many men with the same minor symptoms (as well as the same degree of distress and unhappiness) might never go to their doctors at all. And if a man *did* consult his physician with, for example, a digestive complaint, he would be far less inclined to discuss any emotional components of what would be viewed as a physical problem. Undoubtedly, the same symptoms in a male and female may be seen, by a clinician, as reflecting different underlying problems. Is that antifeminine bias? Or an awareness, on the part of doctors, that women show a much higher rate of depression? Again, the questions and answers are circular.

In any event, the male patient's sex-appropriate behavior will dictate more stoicism, less free expression of sadnesses and weaknesses, and less owning-up to physical and mental difficulties. Men are far more reluctant to take on, even on a transitory basis, the "sick role." It is too inconsistent with standard cultural ideals of independence, autonomy, masculinity.

Men and women may, in other words, be *feeling* in much the same dysphoric, unhappy, anxious, depressed ways but *behaving* very differently. Women may be assuming the culturally "available" sick-and-dependent role when they are stressed, while men fail to seek medical attention. The exorbitant number of depressed women could be a statistical red herring: men and women might be equally depressed, but men may not be counted as "cases" because they have never gone in for treatment.

But that just isn't the case. For, powerful evidence from so-called community studies indicates that women really are more depressed than men—irrespective of who does, or who doesn't, go to the doctor.

What are community studies? Just that: studies of the community at large. What is involved is the selection of a random sample of respondents that is representative of the larger population; and the careful and systematic interviewing of those people in their homes.

If the study happens to be an assessment of psychiatric symptoms in the members of the community, then the research interviewer will fill out a comprehensive questionnaire that covers many aspects of psychological functioning. For instance, to elicit information about the possible existence of a depressive disorder, the interviewer might initially ask: "Did you ever have a period that lasted at least one week when you were bothered by feeling depressed, sad, blue, hopeless, down in the dumps, that you just didn't care anymore, or worried about a lot of things that could happen? What about feeling irritable or easily annoyed?"

If the respondent's replies happen to be in the affirmative, then other questions will follow. Those would touch upon sudden changes in appetite, in sleeping patterns, in energy levels, in interest in customary activities, in sexual functioning. And there would be queries about "feeling guilty," "worthless," or "down on oneself," as well as about problems in concentrating and making decisions, and even in thinking. There would be questions about thoughts of dying and/or suicide.

Myrna M. Weissman, director of the Yale Depression Unit and an epidemiologist by training, has just completed a community study of representative samples of the population in New Haven, Connecticut. Weissman told me that her survey, like all others that have been done, has uncovered the same phenomenon: an almost frightening amount of depression among women respondents. "Regardless of who is or who isn't going to the doctor—and a number of depressed women who were interviewed were *not* in treatment—women are far more depressed than are men."

Much of the female depression discovered in the community study did, she continued, include milder forms of disturbance. "When you get out in the community, and talk to people in their homes, you draw in the somewhat less severe cases, too. Because, you see, for a person to get into psychiatric treatment he or she has to be in pain and be really hurting. Getting help takes energy, and it can be expensive; and some feel it's still a bit of a stigma. And so what our nets take in are people who are doing some suffering, but who aren't in such torment that they can't say, 'Oh well, I can live with it, and besides, what's the

---

"Being female means (frequently) never being encouraged to become a self-sufficient person."

---

use.' They're mildly symptomatic, maybe more transiently symptomatic. But they don't feel bad enough—or maybe don't know enough—to go for help."

In terms of overall volume—sheer numbers of women who are depressed—Weissman is one among a number of experts who have spoken of what seems to be a steady rise in the course of the past decade. Her own research on suicide attempts among females indicates a similar surge upward, primarily among younger women. "This has been well

documented in several countries over the past 10 years," she said. "And, while all suicide attempters may not be depressed, I would say that most of them are."

As rates of depression among women have increased, the age at which they come to clinics for treatment has been inching downward. "Right now," observed Weissman, "the typical person who comes to the Yale Depression Unit is a woman, and she is under the age of 35." This age-shift downward (in the earlier years of this century, the typical depressed patient was described as someone in his 40s or older) may be due to the fact that women who might have come in later in their lives are now appearing earlier, and with far less serious symptoms. "Or," she continued, "it may be due to the fact that help is more available and getting treatment is more acceptable." "Or," Weissman smiled, shrugging slightly, "it may be because the treatments—both drug and psychotherapeutic—are themselves so radically improved."

Weissman, in collaboration with Gerald Klerman of the Alcohol, Drug Abuse and Mental Health Administration (ADAMHA), has written what is considered a classic paper on the topic of women and depression. Reviewing all possible "explanations" for the sexually tilted statistics, they concluded that the figures represent the real situation, and that the findings are true.

Women simply *are* more depressed, in the aggregate, than are men, in the aggregate. The next question is: why?

As one might imagine, theoretical "answers" to that question abound. It would be impossible to review every one of them here. Let me say only that the most obvious sort of solution to the riddle, "female hormones," does not really stand up under close and systematic scrutiny. Endocrines may affect mood, in various ways and at particular times; but they don't really provide reasons for the huge differential in overall male and female rates of depressive illness. Let us therefore turn instead to some of the likelier explanations that have thus far been offered.

One has to do with a trait considered "normally feminine"—meaning, more prominent among females—and that is dependency. Girls and

women are expected in our culture to show higher dependency needs than boys and men do. That is simply one aspect of the designated sex role, and both males and females are being taught their lines from the moment of birth onward. Psychological experiments, for instance, have demonstrated that parents hold new babies differently: an infant in a pink blanket, *believed* to be a girl, will be held and cuddled much more than a newborn who is wrapped in blue.

As the psychologist Judith Bardwick has written: "The dependency, passivity, tears, and affection-seeking normal to both sexes in younger children are defined as feminine in older children, and girls can remain dependent and infantile longer. . . . This has a very pervasive and significant effect: unless something intervenes, the girl will continue to have throughout womanhood a great need for approval from others. Her behavior will be guided by the fear of rejection or loss of love. An independent sense of self with a resulting sense of self-esteem can only evolve when the individual, alone, sets out to attain goals and, with reasonable frequency, achieves them."

This independent sense of self, notes Bardwick (in *Psychology of Women: A Study of Biocultural Conflicts*), is rarely achieved among females. More often, the young girl learns to appraise her worth as a function of the appraisals of others, to value herself insofar as she *is* valued. Being female means (frequently) never being encouraged to become a self-sufficient individual. And girls receive many instructions on that aspect of femininity throughout the long apprenticeship of childhood.

Now it may well be, theorizes Bardwick, that the greater dependency shown by girls and women is not merely due to the powerful acculturating forces that move them in this direction: there may be inborn, biologically based behavioral tendencies of that kind that come into play as well. But in any case, the lack of an *inner* impulse to break away from the dependent relations of early childhood—and the manifest assumption, on everyone's part, that she will not do so—fosters a situation in which the girl, and later, woman, gives her highest priorities to pleasing others, to being attractive to others, to being cared for, and to caring for others.

"An independent sense of self," Bardwick writes, "can only occur when one has many experiences in which he is responsible while he cannot completely depend upon his original sources of love and support." (By switching to the pronoun "he" in this sentence, Bardwick seems to underscore the notion that a person who *does* develop this sort of independence is unlikely to be a "she.") The overall point, however, is that women receive ferocious training in a direction that leads away from thinking, "What do I want?" and toward "What do *they* want or need of me?"

Being successful then translates to being successful in meeting the expectations of others: pleasure has much to do with the act of pleasing. This readily leads to a situation in which good feelings about the self— that is, self-esteem—become dependent upon the esteem of those around one. Feelings of emotional well-being, a sense of one's worthfulness as a person, are hostage to the moods, attitudes, and approval of others (or, maybe, to one critically important other person). One is likable/lov-

> ## "The depressed woman has lost something she vitally depended on. More often than not, it is a love bond."

able/significant only to the extent that one is liked/loved/sigificant to someone else. It follows, then, that in times of interpersonal drought— when sources of emotional supply are unusually low, or not there—the "normally feminine, normally dependent" woman may experience her inner world as emptied of what is good and meaningful to her. The props of her self-regard, if they've been held in place primarily by feedback from the environment, may simply begin to crumble and fall down. Under the circumstances, a woman may become far too harsh in her assessment of herself and of her

worth and usefulness as a human being. She may feel helpless about her life circumstances, and hopelessly ineffectual in terms of her capacity for mastering or changing them. She may, in a word, become depressed.

This is, I suspect, the point at which "normal feminine dependency" becomes transformed into something that can be viewed as pathological—into a clinical depressive illness, involving a very well-recognized cluster of psychological and biological symptoms. The depressed woman is someone who has lost. She has lost "something" upon which she vitally depended. The tone is of something profoundly significant having been taken away, of some crucial life's territory having been surrendered. And what I have seen emerge, with an almost amazing regularity, is that the "loss" in question is the loss of a crucially important and often self-defining emotional relationship.

Despite the complex, varying, dizzyingly diverse matters and difficulties that any person *could* potentially become depressed about, there appeared to be one kind of a loss—more than any other—that had triggered a depressive episode in the women that I interviewed. And that was the loss of a love bond.

Depressions, when they "happen" in women, happen in one kind of context with the greatest of frequency. That context is the loss of emotional relatedness: "attachments" are the critical variable. Some writers have speculated that a much greater investment in one's love attachments might simply go along with culturally transmitted sex-role expectations correlated with "being feminine." Women are not only supposed to be more dependent, but also to be warmer, more expressive, more eager to relate on a personal level, and so forth. Because of intensive and early "femininity training," they might show a greater propensity to put more of themselves into—and therefore to risk more in—a few powerfully important relationships.

To fail in those relationships is, then, equated with "failing in everything." To slip there can mean a headlong slide downward into desperation and misery. It is around such "losses of love" that the depressive clouds tend to gather and to darken: impor-

tant figures leaving or dying; the inability to establish a meaningful bond; being forced, by a natural transition in life, to relinquish an important love tie; a marriage that is ruptured, threatening to rupture, or simply growing progressively distant; the splintering of a love affair or recognition that it is not going to endure. These are among the most common "causes" of female depression.

Men do not appear to become depressed, with this near-predictable regularity, over the rupture, or threatened rupture, of emotional bonds. Again, this could have much to do with role expectations about what "being masculine" involves: independence, action, aggressiveness, and a high motivation toward competing, winning, achieving. For men, the more usual depressive motifs involve work issues, status, and success problems—difficulties in "making it" out there, in the world at large.

Such issues can, of course, be upsetting to women as well, but they don't touch upon the same tender, raw nerve ending of concern. Likewise, this isn't meant to suggest that men don't ever become severely depressed over the loss of a love attachment. But certain core concerns do seem to set the stage for a serious depression in the male and in the female—and in the two sexes, those core concerns are often somewhat different.

And the experience of *being depressed* may be qualitatively different in some important respects for men and for women. One recent survey of college students carried out at Yale University by psychologists Eve Chevron, Donald M. Quinlan, and Sidney J. Blatt found that depressed females suffered "significantly higher levels of . . . experiences of loneliness, helplessness, dependency, and the need for external sources of security." But for the males questioned in the study, the depressive issues had much more to do with "self-criticism and the failure to live up to expectations."

Some experts have gone so far as to hypothesize that a greater sensitivity to fluctuations in a woman's love attachments is bred into the human female genotype because, in the primitive environment in which our species evolved, "loving and protective mother" could be equated with "helpless infant's survival." The mother who fell in love with her baby, and remained emotionally "bound," would stay close by the infant, protecting him or her from being taken as prey. This view has it that the female's high valuation of her emotional attachments was a behavioral tendency that was "selected for" during evolutionary prehistory—because it is a human survival mechanism. And, obviously, those girl babies whose mothers behaved in those ways would have lived to reproduce and pass the same tendencies along to their own female children!

Males and females may, in fact, be marching to somewhat different kinds of inner music. (Or to notes that are similar, but arrangements that are not at all the same.) It is possible that, as Harvard psychologist Jerome Kagan has suggested, the two sexes are "sensitized to different aspects of experience and gratified by different profiles of events."

Everyone does, clearly, keep his or her interior scorecard, a running self-assessment and estimate that continuously monitors the question "How am I doing?" The suggestion raised here is that, in terms of what is "most satisfying" and "most disturbing," males and females do that self-grading somewhat differently.

And for women, whose tendency may be to do their self-rating in terms of the health of their emotional attachments, much more—a far greater amount of "self"—will be at risk in their loving connections. The enhanced vulnerability to depression might be traceable, in fact, to the higher psychic investments being made in that vitally important and crucial business of loving. It could have much to do, as well, with the sort of social climate that we all inhabit nowadays—one not noted, I fear, for the permanence of security-giving emotional attachments.

To be depressed is to be in a state of emotional paralysis, to dare nothing, to try nothing, to freeze. The painfulness of it, that noxious state of helplessness and inferiority, is real—as real as any form of physical pain. Like physical pain, it will often demand a person's complete attention. How often I've spoken with women who seemed preoccupied and distracted; whose energies were being totally absorbed in the effort to retain some equilibrium in their lives. That hurting, like physical hurting, could dominate consciousness with its urgency, its demands. Some women, when I talked with them, were finding it impossible to see beyond the suffering of the moment. Others, less seriously depressed, found it possible to "get out of it for awhile" when life presented some pleasant diversion; but only in the knowledge that they'd be returning home to where the pain was, to the wounded self.

The depressed person *is* wounded, though the injury can't be seen or lo-cated. The person is "changed," diminished; usually he or she has experienced a morbid drop in confidence. Guilt, anxiety, irritability, hostility: those are common ingredients, found in variable proportions, in the depressive stew. It is, experientially, a fairly horrible mixture. And the person suffering from depression, filled with this feast of dysphoric feeling, usually becomes far less able to cope in the outer world, less able to be flexible and to negotiate.

The depressed person must, moreover, operate with less-than-normal energies and capacities. And because there is that true diminution in adaptability, the person's dealings with the environment tend to become off-the-mark, erratic, and often genuinely self-defeating. "Feeling bad," then, has plenty of consequences. The person suffering from depression is, at best, either pedaling-in-place in life, or losing valuable ground.

And the "depressed person" is very liable to be a woman. According to Robert M. A. Hirschfeld, chief of the depression section at the National Institute of Mental Health, survey data indicates that one in every five Americans has at least moderate depressive symptomatology. We are talking, then, about a group of people on the order of 40 million—and two-thirds of that group are women! It is, as Hirschfeld acknowledged, a public-health problem of almost staggering proportions.

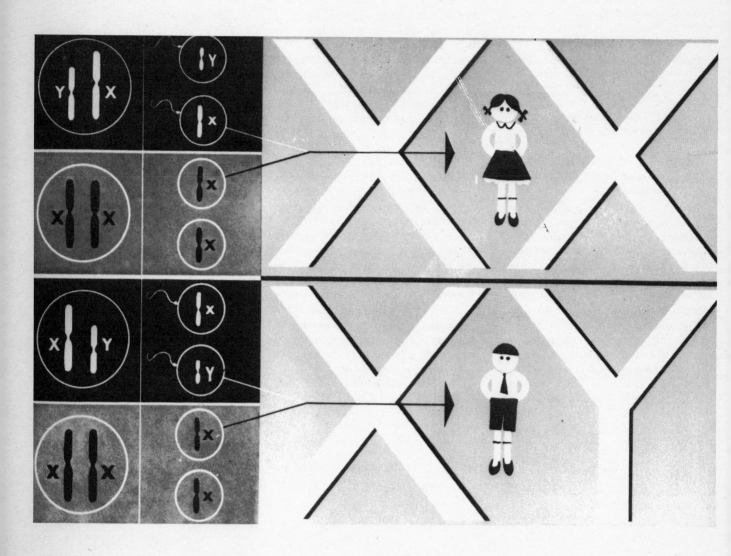

Development During the Prenatal Period

The months of prenatal development are subdivided into three major periods: zygote (conception to two weeks), embryo (two to eight weeks), and fetus (eight weeks to term). Major events of the period of the zygote include reduplication of cells (mitotic cell division) and implantation of the zygote (blastocyst) into the uterine lining. The period of the embryo is characterized by differentiation of the physical structures from three embryonic layers. For example, the nervous system differentiates from the ectoderm, the circulatory system from the mesoderm, and the gastrointestinal tract from endoderm. During the period of the fetus the organism gains weight and undergoes rapid changes in brain growth and development.

During each of the periods of prenatal development the conceptus is vulnerable to a variety of environmental stresses. Factors such as infectious disease, malnutrition, blood incompatibility, drugs, radiation, maternal age, maternal emotional state, and environmental toxins can compromise the intrauterine environment and interfere with normal pathways of development. In addition, a variety of hereditary and non-hereditary genetic factors can place the fetus at risk for death or disability.

Technological advances of the past decades have led to the development of a variety of techniques for assessing the developmental status of the prenatal organism. Whereas each of these techniques contributes to a more complete assessment of the structural and biological viability of the conceptus, each technique also contributes to ethical and moral problems concerning decisions to retain or abort the conceptus. If one knows by the second or third prenatal month that the conceptus has acquired or is at risk for a handicapping condition, does one choose to terminate pregnancy or allow it to continue to term? For some individuals the question of abortion is easily resolved by appeal to some higher ethical or theological dictum. For many other individuals, however, the question has no easy resolution. It is certain, however, that as biomedical technology continues to accelerate, still newer and more precise methods will be developed to evaluate the status of the conceptus. And the ethical issues will become even more complex.

The articles in this section introduce the reader to many of the discoveries that have been made regarding ways to evaluate the biological status of the preterm organism and the psychological influences on mother and fetus during pregnancy.

Looking Ahead: Challenge Questions

Who should have the final word on whether the fetus should be aborted or brought to term? Mother? Father? Priest? Legislator? Physician?

How might the delivery process itself impose risk factors on the fetus? When you sign papers at the hospital authorizing the use of obstetrical medications during delivery, do you really understand what you are signing?

What guidelines would you develop to assist decision making regarding the question of abortion? What are the legal implications of your guidelines?

In what ways does the father contribute to optimal management of pregnancy and delivery?

PREGNANCY: THE CLOSEST HUMAN RELATIONSHIP

Niles Newton and Charlotte Modahl

Niles Newton *is professor of psychology at Northwestern University Medical School in Chicago. She has been studying childbirth and reproduction for many years. With Margaret Mead, she investigated the way different cultures treat reproduction. She conducted some of the first research in this country on human lactation. In one series of experiments she served as a subject, nursing her seven-month-old baby in a study that established oxytocin as the hormone responsible for the reflex that lets down milk in human beings. Newton also has studied the effects of stress during labor, and has surveyed women to discover their reactions to their hysterectomies. Her latest research explores the role of the hormone oxytocin in coitus, childbirth, and lactation. Newton is associate editor of* Birth and Family Journal *and a member of the executive board of the International Society of Psychosomatic Obstetrics and Gynecology.*

Charlotte Modahl *is a doctoral student in psychology at Northern Illinois University. She is collaborating with Newton on a study of hormonal similarities in coitus, childbirth, and lactation. Modahl is also working on a longitudinal study of sex-related legal cases, exploring sexual symbolism in psychic trauma. With Newton, she recently wrote an article on mood differences between mothers who nurse their babies and those who bottle-feed them.*

During its life within the uterus, a baby develops from a single fertilized cell and becomes a human being who, though still immature and dependent, can survive in the outside world. In the course of those 38 weeks, the baby depends on its mother for all its physical needs. She, in turn, is aware that she carries within her a living being. As her baby grows, the mother undergoes profound physical and emotional changes. From conception to birth, the pair deeply affect each other, and their relationship may establish attitudes and ways of interacting that persist for years.

Life begins when sperm and egg unite, and a woman's sexual enjoyment may make it more likely that the sperm will reach the egg. After a woman experiences orgasm, her cervix descends and enlarges, increasing the size of the passageway into the uterus and making it easier for the sperm to ascend and meet the egg. Orgasm also makes her vaginal secretion more alkaline, and according to obstetrician Landrum Shettles, sperm travel more easily in an alkaline environment than in the normally acid vaginal secretion.

At the same time, prostaglandins—substances that are found throughout the body but that are concentrated in the fluid that surrounds the sperm—enter the vaginal wall and are absorbed into the bloodstream. Oxytocin, a hormone, is produced by the woman's own body. In response to these substances the uterus contracts, then relaxes, and this sequence may help the sperm move into the uterus. By remaining relaxed and recumbent after intercourse a woman also helps the sperm make their way into the uterus.

The feeling of closeness that most couples have after intercourse may be the direct result of uterine responses to such substances as prostaglandins and oxytocin, which is released by the pituitary gland during sexual stimulation. Our own pilot studies, in which we measured the moods of both sexes immediately after intercourse, suggest that there is a postcoital decrease in anxiety and depression. The drop was sharper in men than in women. Learning by gradual conditioning also contributes to the closeness of a couple. Because the intense emotions involved in orgasm are extremely pleasurable, the experience of intercourse is reinforcing. Repeated intercourse tends to condition the two people to each other, binding them into a reproductive partnership and providing a foundation for family life and the nurture of the baby.

Of course, orgasm is not necessary for conception, nor is any affection for one's partner. However, there is a strong relationship between a woman's feelings toward her mate and the course of her pregnancy.

Frances K. Grossman and her colleagues at Boston University studied 98 pregnant women in an attempt to assess the effects of their emotions. They found that women with good marital relationships were less likely to be depressed and anxious during pregnancy than women with unhappy marriages.

Whether the good marital relationships led to low levels of anxiety and depression or whether women who are anxious and depressed generally have unhappy marriages cannot be determined. But Grossman did find that women who said, early in their pregnancies, that they had wanted to become pregnant were much less likely to have complications during labor and delivery than women who had not consciously wanted to become pregnant.

Conception does not occur for some time after intercourse and takes place without the awareness of the woman. Sperm travel at only 0.5 cm per minute, but with the aid of muscular contractions and natural chemical and hormonal

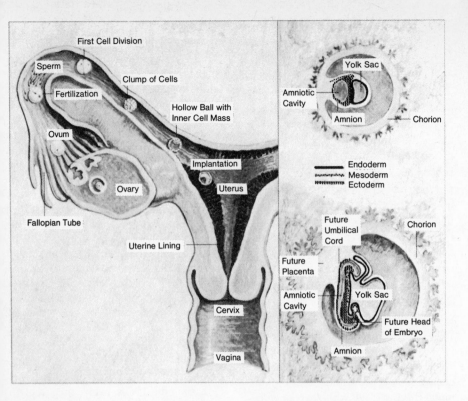

First Cell Division
Sperm
Clump of Cells
Fertilization
Hollow Ball with Inner Cell Mass
Ovum
Implantation
Ovary
Uterus
Fallopian Tube
Uterine Lining
Cervix
Vagina

Yolk Sac
Amniotic Cavity
Amnion
Chorion

Endoderm
Mesoderm
Ectoderm

Future Umbilical Cord
Chorion
Future Placenta
Amniotic Cavity
Yolk Sac
Future Head of Embryo
Amnion

The fertilized egg divides many times on its way to the uterus. In the small drawings of the zygote on the left, cells in the black band will develop into the digestive system; those in the dotted band, into the heart, muscles, and skeleton; those in the lined band, into the skin and nervous system.

assistance, they may reach the Fallopian tubes within a few minutes after they are deposited in the vagina.

Although the egg may survive for approximately 72 hours after it is released from the ovary, it probably can be fertilized only during the first 24 hours. Some of the 400,000,000 sperm that are ejaculated by the man may survive within the woman for as long as seven days, although it is unlikely that they can penetrate and fertilize the egg after the first two days.

Planning the date of intercourse may help determine the baby's sex. About half the sperm released during intercourse carry an X chromosome; they will produce a girl. The other half carry a Y chromosome and they will produce a boy. The male-producing sperm are lighter and move faster, but they die sooner than the sperm that produce females. Because male-producing sperm tend to reach the fertilization site first, intercourse at the time the woman ovulates favors the conception of a boy. Because male-producing sperm die first, intercourse a day or two before ovulation increases the chances of the couple's conceiving a girl.

The first two weeks of a baby's life are called the germinal phase, and the fertilized egg is called a zygote. After the

sperm penetrates the egg, the zygote spends three or four days traveling down the length of the Fallopian tube and then another three to four days floating free in the uterus.

By the time the zygote is about nine days old, it has developed two sacs that surround it completely. The inner one is a fluid-filled sac that protects the zygote from injury, and the outer sac is the one from which the tendrils will grow that attach the zygote to the mother. Throughout this period, the cells of the zygote are dividing rapidly.

By the end of the second week, the zygote is firmly implanted in the uterine wall and has developed three layers of cells. The outer layer will produce the baby's skin, sense organs, and nervous system; the middle layer will develop into the baby's heart and blood vessels, muscles, and skeleton; the inner layer will become the digestive system and related organs such as the liver, the pancreas, and the thyroid gland.

The placenta, which transmits nourishment from mother to baby and takes away all waste products, is also developing at this stage.

Although the mother is unaware of the spectacular growth that is taking place within her, her body is responding to the implanted zygote. Instead of

sloughing off its lining in a menstrual period as it normally does when an egg remains unfertilized, her uterus accepts the zygote in its thickened wall, and the dense network of blood vessels that has developed since ovulation begins to join the placenta.

The end of the baby's germinal phase coincides with the mother's expected menstrual period. At first she may think her period is only delayed, but soon her breasts begin to feel full, heavy, and tender, and she may start each day with nausea and vomiting.

Not all women become sick at the same time of day and many never feel nausea at all. Recently, Marilyn Theotokatos and Niles Newton collected information from over 500 women who were breast-feeding their babies. Sixty-eight percent of them reported experiencing nausea during their pregnancies, although only 16 percent said it had been severe. Among those who became nauseated, 40 percent also vomited.

By now the zygote has become an embryo, a term that describes the baby for the next six weeks. The placenta becomes more developed and from it the umbilical cord runs to the baby's navel. During this phase, the embryo develops its major organs.

At eight weeks, the baby is not much more than an inch long. It has a recognizable brain, a heart that pumps blood through tiny veins and arteries, a stomach that produces digestive juices, a liver that manufactures blood cells, kidneys that function, and an endocrine system. In the male embryo the testes produce androgens. The baby now has limbs and an enormous head with ears, nose, eyes, and mouth. Its eyelids have not yet developed and it has a definite tail, which will recede and become the tip of the spinal column. Nevertheless, it looks human.

Although the mother does not yet look

2. PRENATAL PERIOD

pregnant, her baby has begun to react to its environment. It holds its hands close to its face; should they touch its mouth, the embryo turns its head and opens its mouth wide.

The behavior of the mother has also begun to change in response to changes in hormone production, to her bodily growth, and to her expectations that her way of life will soon be different. She may be unusually tired and sleep a good deal. As the growing embryo presses on her bladder, she may find herself urinating frequently.

Her eating habits are likely to change, perhaps in response to local custom. Many women have cravings for strange food at this time. S. M. Tobin asked 1,000 Canadian mothers "Did you have any peculiar food craving in pregnancy?" and 640 of them said "Yes." Sometimes these cravings may be intense but usually they involve milder yearnings. Craving for cornstarch or clay is common among poor groups in the United States, whereas affluent women may crave ice cream or strawberries in winter. The reasons for these cravings are unknown, though they have been attributed to dietary deficiencies, anxiety, or conformity to cultural expectations.

With the appearance of bone cells at about the ninth week of development, the embryo is called a fetus, which will remain its technical name until birth. For the rest of its gestational period, the fetus is protected from the outside world by the amniotic fluid that fills the space between the inner and outer sacs that developed eight weeks earlier. The amniotic fluid provides a stable, buffered environment, and the fetus floats in a state of relative weightlessness. The fetus urinates directly into the fluid, and its waste products travel from the fluid through the placenta, from which they enter the mother's bloodstream. Exchange between the fetus and the amniotic fluid is slow, while exchange between the mother and the fluid is rapid. The fluid is completely replaced every two or three hours.

By the end of the third month, counting from the mother's last menstrual period, the fetus has grown to a length of three inches and weighs about half an ounce. It shows one of the signs of humanity—its thumb and forefinger are apposed so that, theoretically, it could grasp objects. It bends its finger when its palm is touched; it swallows. It has taste buds, sweat glands, and a prominent nose. By now it has eyelids, but they are sealed shut.

It is during the first trimester that the developing baby may be most sensitive to such influences as drugs, x-rays, disease, and the lack of essential nutrients. Mark Safra and Godfrey Oakley, Jr., of the Georgia Center for Disease Control found that women who take diazepam (a tranquilizer marketed as Valium) during the first trimester are four times as likely to have babies with cleft lips or cleft palate as mothers who do not take the tranquilizer.

During the 1960s, before the sedative was banned, mothers who took thalidomide during the first trimester sometimes gave birth to babies whose

The embryo pictured is eight weeks old. It is about an inch long and has a recognizable brain, heart, stomach, liver, kidneys, and endocrine system. Its tail will recede and become the tip of its spine. Although its head is enormous, it is obviously a developing human being.

arms and legs were nothing more than rudimentary flippers. Other drugs can cause abnormalities ranging from yellowed nails and teeth, which may follow the use of an antibiotic like tetracycline, to blindness, deafness, and gross malformations.

Only within the past six years has the existence of "fetal alcohol syndrome" been established. Kenneth L. Jones and his colleagues at the University of California at San Diego, have followed the pregnancies of alcoholic mothers and report that their babies are likely to be smaller and lighter than most babies, and that they are more likely to have slight facial, limb, and cardiovascular malformations. Some babies who are born to alcoholic mothers have conical heads and are mentally retarded. Animal research indicates that even the regular consumption of moderate doses of alcohol can affect the physical condition of offspring, but such a connection has not yet been established in human beings.

Heroin also affects the fetus. Babies whose mothers take heroin regularly will be born addicted to the drug and soon after birth must endure acute symptoms of heroin withdrawal.

Some minor illnesses may have profound effects. Some mothers who contract rubella (German measles) during the first trimester produce babies who are blind, deaf, mentally retarded, or have diseased hearts.

The fetus depends on its mother for all vitamins, minerals, and nutrients. In experiments with animals, severe protein shortages in early pregnancy have been associated with fetal brain damage. Stephen Zamenhof and his colleagues at the University of California School of Medicine, Los Angeles, found significantly fewer brain cells and cells with lower protein content in rats whose mothers had been placed on a protein-restricted diet before mating. They suggest that protein deprivation leads directly to mental retardation in children.

Research that severely restricts prenatal diets cannot be done on human beings, but records kept during the 1940s, when Germany occupied the Netherlands and Dutch diets were reduced below the minimum requirements for good health, showed an in-

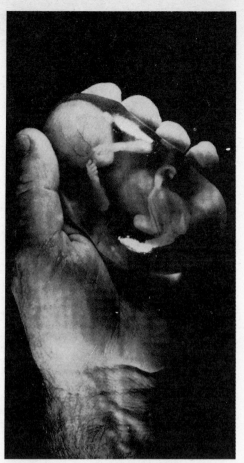

The features of the developing fetus show plainly in these photographs by radiologist Roberts Rugh. The three-month-old fetus can bend its finger and swallow.

crease in stillborn babies and premature births. As soon as diets returned to normal, the rates dropped. In depressed areas of the United States and in countries where the customary diet is deficient, infant mortality rates are high. In such places, dietary supplements have reduced mortality rates.

During the second trimester, the mother first feels her baby's movements. It is this quickening that makes many women acutely aware that they are carrying a living human being. Fetal movement patterns often give rise to maternal fantasies and expectations for the new baby. Women who reject the idea of motherhood when they first discover they have conceived generally come to accept the idea during this period of pregnancy.

The second trimester is generally a time of physical and emotional well-being. The mother no longer finds it necessary to sleep so much, her nausea has

gone, and her appetite has returned. She has lost the continual urge to urinate, and she often feels better than she did when not pregnant. Her condition is now apparent, and each time she looks at herself in the mirror, she has visible evidence of her baby's existence.

Her major worry at this stage may be in regard to her weight, for this is the period of most rapid weight gain. A few years ago, obstetricians urged their patients to restrict weight gain during pregnancy, and many mothers found themselves put on strict diets at a time when they both wanted and needed more nutrients.

Today it is realized that mothers who fail to gain adequately during pregnancy may have less healthy babies. According to the Committee on Maternal Nutrition of the National Research Council, the desirable weight gain during pregnancy is 24 pounds. The extra weight includes, in addition to the baby and the placenta, amniotic fluid, water that enlarges breast tissue, and extra blood needed for circulation. Some of the 24 pounds — on the average, about three and a half pounds — is stored as fat and protein to act as a buffer against the stresses of the postnatal period.

By the time the fetus is five months old, it appears in many ways to be a fully developed human being, but if it were taken from its mother, it could not survive. Only 10 inches long and weighing about half a pound, it has well-formed lungs that are not ready to function and a digestive system that cannot handle food.

This miniature being is often quite active within the uterus and seems to squirm or writhe slowly. Sometimes, by placing a hand on the mother's abdomen, it is possible to identify an elbow, a knee, or a tiny bottom. At this time, if one presses gently on the fetus, it is likely to respond with movements.

At the end of the second trimester, the six-month-old fetus wakes, sleeps, and has sluggish periods. It is likely to nap in a favorite position. It will open and close its eyes, look in all directions, and even hiccup. The grasp reflex is developed, and the fetus is capable of supporting its own weight with one hand. It is about 13 inches long and weighs about one and a

2. PRENATAL PERIOD

The 17-week-old fetus is so active within the uterus that its mother can now feel the separate life within her.

half pounds. With intensive care—regulated temperature, intravenous feeding, and oxygen supplementation—the fetus might now be able to survive outside its mother's body.

At Northwestern University, Marcia Jiminez studied 120 women who were pregnant for the first time, and examined their attitudes toward their jobs and their moods during the last three months of pregnancy. She found that women who expressed satisfaction with their work also felt less anxious, depressed, tired, and guilty and had less difficulty coping with their situation than did women who were dissatisfied.

Although it has been suggested that women who have strong career interests find it difficult to adapt to motherhood, Jiminez' work suggests that a woman's enjoyment of her occupation and her enjoyment of her pregnancy may both be part of a generally positive attitude toward life.

During the last trimester of pregnancy, the mother may respond with pride to the bulge of her enlarging uterus as a sign of fertility. In a society that advocates strict birth control measures, a woman's feelings about her pregnant body may be mixed. A mother who has seen few pregnant women in her life may feel that her body is uncomfortably different. The difference is accentuated if she wears shoes with high heels and stands in a poor posture. Her baby now may kick or punch sharply and sometimes uncomfortably, which can be an annoyance in the middle of the night.

The demand for nutrients is especially heavy at this time, and the baby will absorb 84 percent of all the calcium the mother eats, and 85 percent of the iron. The rapidly developing fetal brain also requires extra protein.

By the last trimester, the baby is aware of events outside its own body. The fetus lives in a noisy environment. Small microphones inserted through the cervix into the uterus of pregnant

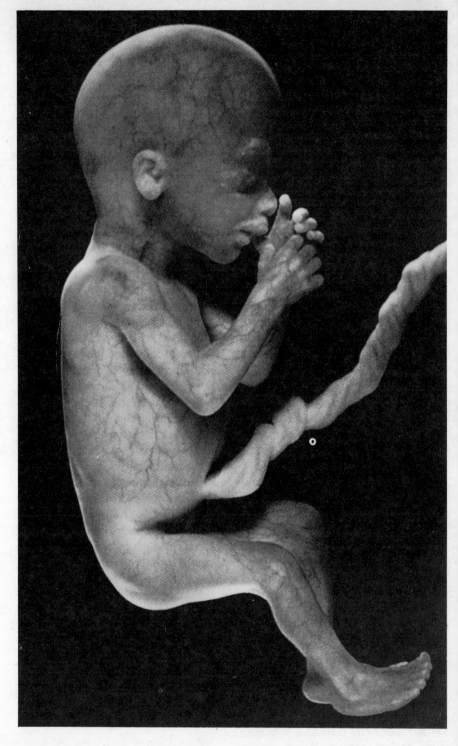

women whose babies were due any day have detected a loud whooshing sound in rhythm with the mother's heartbeat. The sound is produced by blood pulsing through the uterus and is the constant companion of the fetus, which also hears the occasional rumbling of gas in its mother's intestines.

The unborn baby may also respond to outside sounds. Some mothers have complained that they cannot attend

symphony concerts because their babies respond to the music and applause with violent movement. Lester Sontag of the Fels Research Institute placed a small block of wood over the abdomens of women whose babies were due in about five weeks. When the board was struck with a doorbell clapper, about 90 percent of the fetuses immediately began kicking and moving violently. Their heart rates also increased. Besides responding to loud sounds, the fetus responds to the reaction of its mother's heart to such sounds.

Fetuses show individual differences in activity level and heart rate. Like people, some have heart rates that fluctuate greatly while others have hearts that tend to beat more regularly. Sontag has followed a dozen people from their fetal existence to adulthood and discovered that those whose heart rates showed wide variation within the uterus generally show the same wide variation as adults.

On the basis of three-hour interviews with a large group of children and adults, Sontag's colleagues have found that men with highly fluctuating heart rates tend to be reluctant to depend on people they love, to be in conflict over dependency, to be more compulsive, introspective, and indecisive than men with more stable heart rates. They have no explanation for this, but speculate whether they have discovered a genetic component of personality or an example of the influence of the uterine environment.

The relationship between a mother and her unborn baby is so close that the mother's emotions affect her baby's behavior, and the effect may persist after the baby is born. There is no way to ascertain when maternal emotions begin to influence a fetus, because there is no way to measure the effect in early pregnancy. But when a mother is emotionally upset, her body responds with physical changes.

Sontag has observed women in the last trimester of pregnancy who were suddenly faced with grief, fear, or anxiety. One young woman, whose fetus he was checking weekly for activity level and heart rate, came to him terrified after her husband suffered a psychotic breakdown and threatened to kill her. Her baby began kicking so violently that she was in pain, and recordings showed that its activity was 10 times greater than it had been in earlier recordings. The unborn baby of a woman whose husband was killed in an accident showed the same sharp rises in activity. Six other pregnant women who took part in Sontag's studies suffered similar emotional crises. In every case, their babies responded with violent activity. Sontag

followed these babies and found that, although they were physically and mentally normal, they were irritable and hyperactive, and three had severe feeding problems.

The link between mother and child has other postnatal effects. During the last three months of pregnancy, the mother confers on her baby immunity to a number of diseases that she may have contracted in the past. The antibodies manufactured by her immune system cross the placenta and circulate in the baby's bloodstream as well. If the mother has developed such immunities, the baby will have some resistance to measles, mumps, whooping cough, scarlet fever, colds, or influenza for the first few months of its life.

Other substances also cross the placental bridge. Women who smoke on a regular basis give birth to babies who are smaller and lighter than average. Recent studies by Gerhard Gennser and his colleagues showed that when such women smoke a single cigarette, the nicotine concentration in their blood quadruples and their heart rates increase. Within 30 minutes their unborn babies, all at eight months of development, showed a decrease in chest movements and short periods during which they did not "breathe" at all.

When the fetus has completed its 38 weeks in the uterus, the birth process begins. The exact sequence of the physiological changes that initiate labor is not clearly understood, but several substances—oxytocin, vasopressin, progesterone, and prostaglandins—are believed to be involved. Both the mother and the baby produce hormones, and some investigators believe that the fetus initiates the process, or at least gives the signal that tips the balance of factors that start labor.

The experience of childbirth varies from culture to culture. In societies that look upon birth as a fearful and secret experience, women often have long, difficult labors. In societies that are open about childbirth and expect it to be simple, women usually have short, uncomplicated labors. In normal deliveries, a relaxed, undisturbed environment, the presence of trusted helpers, and a minimum of medical interference

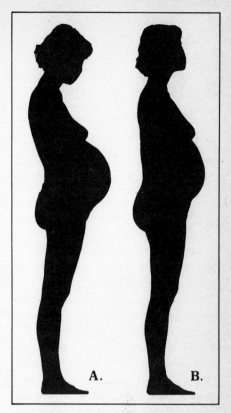

Women who stand improperly may be extremely uncomfortable during their pregnancies. The woman on the left, who is eight months pregnant, complained of pain in her abdomen and lower back. When shown how to stand correctly, she experienced immediate relief, and as the silhouette on the right shows, appeared less pregnant.

encourage the mother and infant to interact successfully.

Fear and disturbance may increase the mother's discomfort, and the excessive amounts of adrenaline that fear can place in a woman's bloodstream may counteract the work of hormones like oxytocin that help labor progress. Anxiety may cause the mother's muscles to become tense, converting simple contractions into painful cramps. Studies have shown that women who have a high level of anxiety during pregnancy are also likely to have complications during delivery.

In the first stage of labor, which may last from two to 16 hours or more, the uterus contracts at regular intervals while the cervix slowly dilates to allow the baby to pass into the birth canal. During this stage, the uterine contractions

Most obstetricians recommend that their patients gain about 24 pounds during pregnancy. Extra interstitial fluid, distributed throughout body tissues, contributes to weight gain. The weight that is not accounted for in the chart on the right is stored as fat and protein and helps the new mother withstand the stresses of the postnatal period.

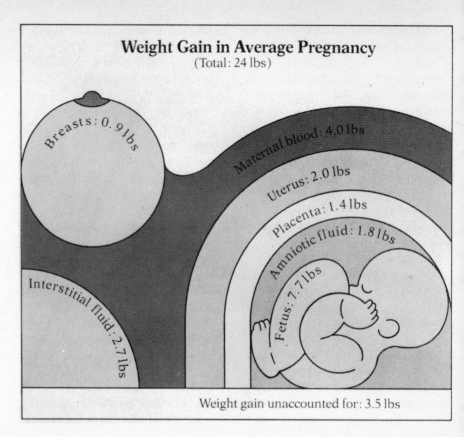

Weight Gain in Average Pregnancy
(Total: 24 lbs)

Breasts: 0.9 lbs

Maternal blood: 4.0 lbs

Uterus: 2.0 lbs

Placenta: 1.4 lbs

Amniotic fluid: 1.8 lbs

Fetus: 7.7 lbs

Interstitial fluid: 2.7 lbs

Weight gain unaccounted for: 3.5 lbs

are faint at first and gradually increase in strength. The mother may not realize that she is in labor until she notices the regular rhythm of the contractions.

In the second stage of labor, the baby usually passes headfirst down the vagina and into the world. This stage may last only a few minutes in women who have previously given birth. With first babies, it generally lasts more than an hour and varies greatly from woman to woman. This is the expulsive stage of labor, and the mother generally helps by bearing down during contractions. General or regional anesthesia may interfere with her efforts to push. The presence of the father in the labor and delivery room often helps the mother relax between contractions and enables the parents to be involved as a couple in the baby's arrival.

In the third stage of labor, the placenta is expelled. This stage can be described as a minilabor that ends with the delivery of the placenta. It usually lasts only a few minutes.

Drugs, even spinal injections, given to relieve a mother's pain or discomfort pass through the placenta to the baby and may interfere with the early bonding process. The baby may receive such a heavy dose that it spends its first few days in a drugged condition. Pediatrician T. Berry Brazelton of Harvard University found that drugged babies had trouble learning to suck and began to gain weight 24 hours later than a control group of unmedicated babies.

Heavily drugged babies may have slow heart and circulatory rates, bluish extremities, and an impaired ability to clear mucus from air passageways. The stimulation of the birth process may

wake them, but only for a short time. Brazelton reports that some drugged babies seem alert in the delivery room, only to lapse into a drugged state in the nursery.

Obviously, a drugged mother and a drugged baby have a hard time relating to each other during the first few days. When Ester Conway and Yvonne Brackbill followed a group of babies, they found that, a year after delivery, babies whose mothers had received medication during labor were still performing much more poorly on standard tests of behavior than babies of undrugged mothers.

And the more heavily drugged the mother had been, the worse the infant performed. Mothers who had required heavy medication differed from mothers requiring less medication, but Conway and Brackbill also discovered that the performance of the baby was related to the type of medication administered, and that the poorest performance levels followed the use of inhalant drugs. Whether the effects are permanent is still unknown.

In the United States, childbirth is often expected to be long and painful, even repellent. Our obstetrical pro-

cedures seem based on the premise that labor and birth are not natural experiences but a serious disease. Until recently, women have responded with deliveries that bore out the premise. Now research and practice have begun to dispel the image of suffering.

Women who have been prepared for childbirth, and who are allowed to go through the birth with a minimum of disturbance and with the support of their husbands, often describe their experience in terms that bear little resemblance to deliveries based on the pain and suffering model. Even women who report a good deal of pain also report feelings of joy, bliss, rapture, and ecstasy.

My own studies have noted a similarity between a woman's behavior during orgasm and at the birth of a baby. In both conditions there is fast, deep breathing in the early stages, the holding of breath at the climax, a tendency to make gasping noises, facial contortions, upper uterine contractions, contraction of abdominal muscles, the loss of sensory perception as delivery or orgasm approaches, and a sense of euphoria at the end.

A mother's emotional attachment to

her baby appears to intensify in the few hours after birth if she is awake and allowed to hold the baby. Breast-feeding is a powerful process that conditions both the mother and the child to a mutually pleasant and healthful interaction. When a baby sucks at the breast, sensory impulses go to the mother's pituitary gland, causing the release of oxytocin. Oxytocin causes the mother's uterus to contract, helping it return to normal size. The hormone also causes the grapelike alveoli in the breast, which hold milk, to contract, letting down milk for the baby.

Research on the nature of pregnancy is beginning to demonstrate what sensitive parents have long known. When pregnancy begins with a rich emotional and sexual relationship, mother and growing fetus continue the pattern. Many psychological and physiological patterns learned in the parents' relationship with each other are carried over to create a bonding between parents and child.

The nutrients, heartbeat, and sleeping patterns that are shared by the mother and the fetus during pregnancy continue after the baby has been born. Hormones such as oxytocin that are present during intercourse, childbirth, and lactation provide a similar physiological basis for these processes, and the emotional reactions appear to be parallel in all. The result of these emotional patterns is the bonding of parents and baby into patterns of mutual pleasure and caring that promote the survival of the species.

For further information:

Brackbill, Yvonne. "Obstetrical Medication and Infant Behavior." *Handbook of Infant Development*, ed. J. D. Osofsky. John Wiley & Sons, 1978.

Gennser, Gerhard, Karel Marshal, and Bo Brantmark. "Maternal Smoking and Fetal Breathing Movements." *American Journal of Obstetrics and Gynecology*, Vol. 123, No. 8, 1975.

Jacobson, H. N. "Nutrition." *Scientific Foundations of Obstetrics and Gynecology*, ed. Elliot E. Philipp, Josephine Barnes, and Michael Newton. William Heineman Ltd., 1977.

Jones, Kenneth L., David W. Smith, Christy N. Ulleland, and Ann Pytkowicz Streissguth. "Pattern of Malformation in Offspring of Chronic Alcoholic Mothers." *The Lancet*, June 9, 1973, pp. 1267-1271.

Mead, Margaret, and Niles Newton. "Cultural Patterning of Perinatal Behavior." *Childbearing: Its Social and Psychological Aspects*, ed. S. A. Richardson and A. F Guttmacher. The Williams & Wilkins Co., 1967.

Newton, Niles. "Emotions of Pregnancy." *Clinical Obstetrics and Gynecology*, Vol. 6, 1963, pp. 639-668.

Newton, Niles. "On Parenthood." *Handbook of Sexology*, ed. John Money and Herman Musaph. Elsevier/North Holland Biomedical Press, 1977.

Sontag, Lester W. "Implications of Fetal Behavior and Environment for Adult Personalities." *Annals New York Academy of Sciences*, Vol. 134, 1966, pp. 782-786.

Walker, David, James Grimwade, and Carl Wood. "Intrauterine Noise: A Component of the Fetal Environment." *American Journal of Obstetrics and Gynecology*, Vol. 109, No. 1, 1971.

Embryo Technology

Lori B. Andrews

When British scientists Dr. Robert Edwards and Dr. Patrick Steptoe announced to the world their creation of a "test-tube baby," Louise Brown, in 1978, they appeared to many people to be solitary pioneers in a revolutionary new field. But actually, work in reproductive technology had been quietly going on all over the world. Today nearly 1,300 doctors and scientists are working on artificial insemination, in vitro fertilization (fertilizing an egg outside a woman's body), embryo transfer (inserting into a woman's uterus an egg that has been fertilized somewhere else, and far, far more bizarre techniques. Last fall, about 500 of these researchers gathered in Kiel, West Germany, for the "World Conference on Instrumental Insemination, In Vitro Fertilization & Embryo Transfer."

Over the three days of the conference, doctors and scientists took the podium at the University of Kiel 162 times to discuss what was possible in reproductive technology. Some also mused about whether what seemed possible—"test-tube babies" or crossing humans and animals—should nevertheless be prohibited for legal, ethical, or social reasons.

To set the tone for the conference, Dr. Kurt Semm, chief of the Department of Gynecology and Obstetrics at the University of Kiel, gave his personal opinion of the field in the keynote address. "There is an old German expression," he said, "which, roughly translated, means, 'One shouldn't poke one's nose in God's affairs.' Let's hope we're not doing that."

Why are scientists like those at Kiel interested in moving procreation from the marriage bed into the scientific laboratory?

Part of the answer is that infertility is becoming a problem for an increasing number of people. According to a study by Dr. Ralph Dougherty, professor of chemistry at Florida State University in Tallahassee, the sperm count of American males has fallen 30 percent in the past 50 years. While in 1929 the median count was 90 million sperm per cubic centimeter, by 1974 it had fallen to 65 million, and in 1979, to 60 million. Possibly because of environmental pollutants, nearly one-quarter of all men now have counts so low that some scientists consider them to be "functionally sterile."

Women, too, are experiencing more problems conceiving. "Because more young women are having sex with a variety of partners and sustaining low-level gynecological infections that may go untreated and damage the reproductive system," explains Dr. Robert T. Francoeur, professor of human sexuality and embryology at Fairleigh Dickinson University in Rutherford, New Jersey, "as many as one in four women between the ages of 20 and 35 are now infertile."

But battling sterility is not the only interest of scientists experimenting with conception. By moving reproduction into the lab, scientists face an unprecedented opportunity to shape the way human beings develop. They can make "improvements" on nature by choosing eggs or sperm from donors with desirable traits—or, potentially, can even rewrite a child's genetic code to develop traits that may be completely new.

It all began with an experiment in artificial insemination.

Reproductive technology is quite a bit older than we may think. In 1884, Dr. William Pancoast, a medical-school professor, was approached by a wealthy Philadelphia couple who had been trying unsuccessfully to have a child. The cause of the problem seemed to be with the husband, so Pancoast looked for someone to donate semen to be injected into the wife's womb. He asked the best-looking member of his class to volunteer, and since the injection was done under anesthesia, Pancoast performed the AID (artificial insemination by donor) without the knowledge of either the husband or the wife.

The baby's birth, though, gave the doctor some pause. The infant so resembled the student that Pancoast felt obliged to explain to the husband what he had done. The rich Philadelphian, happy to have a child, was delighted by the good doctor's creativity. He asked only that his wife never be told the origin of the child.

In the century since then, both the psychological and legal implications of letting a third party into the marriage relationship have proved to be considerably more complex than those encountered by Dr. Pancoast in his critical experiment. At Kiel, doctors reported that their countries differed in their viewpoints on both donor anonymity and liability. In Germany, for example, because of a law that applies to the birth of any child born to German married couples, even a husband who consents to AID could disclaim his paternity anytime during the first two years of the child's life. If the court rules in his favor, the donor of the sperm would then be responsible for the financial support of the child.

In America, fifteen states have laws making a husband who consents to artificial insemination the legal father of the child; in the other states, where no legal precedents exist, it is conceivable that a husband who consents to AID and later changes his mind would be able to charge his wife with adultery or refuse to support the AID child after a divorce.

This uncertain legal status leads some doctors to what some may consider to be unusual practices.

"I purposely inseminate a woman with a number of different sperm samples," said Dr. A. H. Ansari, an Atlanta, Georgia, gynecologist. "Even in the same cycle, I may use four different donors for that individual. I do this so that if the case comes to court and they ask who the father is, it might give the lawyer a hard time to determine which of the four donors should be sued. As for the patient, she is just receiving biological material. She never meets the guy; she doesn't care whose semen you use."

Many scientists and doctors at Kiel disagreed with Ansari. They felt that women *do* care about the donor—and that doctors should, too. If, for example, an AID child developed a genetically transmitted disease, it would be important for the doctor to know who the donor had been to make sure his semen wasn't used again.

Reproduction in a dish.

Of course, AID is not a viable answer to female infertility, two main sources of which are tubal blockage and scarring, which prevent a woman's egg from traveling to the appointed meeting place with the sperm. In the newly developed in vitro fertilization technique, however, a doctor surgically removes an egg from the ovary of a woman with either of these problems and then puts it in a petri dish (a small shallow dish of thin glass with a loose cover) filled with a special medium. He then adds some of the husband's sperm. Once the egg is fertilized, the doctor implants it in the woman's uterus so that it can begin to grow like any other embryo. Although such offspring are generally referred to as "test-tube babies," a test tube is never involved, so it would actually be more accurate to call them "petri-dish progeny."

Currently there are clinics offering this service in Norfolk, Virginia, and Houston, Texas, as well as in Australia, England, France, Germany, and Japan.

Unlike the problems that may occur with AID, the legal and moral problems with in vitro fertilization arise not because there's a third party in the relationship but because the procedure itself is still controversial. In fact, some scientists and ethicists at Kiel wondered whether in vitro fertilization should continue to be used on women at all, since it has not yet been perfected with animals.

"I estimate," said Dr. E. S. E. Hafez, professor of gynecology at Wayne State University School of Medicine in Detroit, Michigan, "that in Sweden, Germany, France, the United Kingdom, and Australia, 20,000 women have been treated, yet we have only three proven births."

Disagreements at the Kiel conference about precisely how to perform in vitro fertilization served to underline that it is still a highly experimental procedure. In fact, there are differences of opinion about everything from how to extract the eggs from a woman to when to put the embryo back inside her.

With debate surrounding so many aspects of in vitro fertilization, it is not surprising that the women who undergo repeated (and often painful) surgery on their ovaries to extract an egg actually have a very slim chance of getting pregnant. One doctor tells his patients the chances are less than 10 percent, but if Hafez's figures of 3 in 20,000 are accurate, the percentage is .015 percent, or 1 in 6,666.

Moreover, although both in vitro fertilization and AID have the potential to give a child to an infertile couple, the enormous differences between the procedures for extracting an egg and collecting semen would seem to require greater protection and more rigid consent requirements than now exist for the women involved in petri-dish procreation.

While semen is usually collected in "masturbatoriums"—softly lit rooms filled with *Playboy* magazines—collecting an egg necessitates surgery and thus presents greater danger. Professor Semm reported that "there has already been one death in Germany during a laparoscopy" (a procedure that is often used in the collection of an egg).

Does an embryo have rights not to be experimented with?

Some ethicists feel that the egg donor is not the only party whose life should be considered; they point out

that the embryo is being manipulated without even the safeguard of the consent process.

For those doctors who don't view the embryo as a human being, the issue of consent is, of course, academic: for them the embryo is just biological material. But for those doctors like Dr. R. F. Harrison of Dublin, Ireland, the whole area of embryo technology is fraught with questions. "Surely there is no problem with embryo transfer when it actually works," Dr. Harrison said. "But what worries me is, if I am experimenting, trying embryo transfer, and I get to the six- to eight-cell stage and then I deliberately stop that, I'm killing a human life."

Some ethicists and lawyers also fear the possibility of creating a grossly defective life with such techniques. Babies Louise, Candice, and Alastair (all "test-tube babies") are healthy, but not enough children have been born through this and similar technology to determine whether or not it is more damaging to the fetus than normal procreation.

Talking about the effects of in vitro fertilization on offspring, Dr. Martin M. Quigley, from the University of Texas Health Science Center at Houston, who established one of America's two in vitro fertilization clinics, commented: "Despite the fact that many animal studies have been done with this technique, we know very little about the abnormalities it can produce. This is because most studies are concerned with achieving fertilization and do not go on to transplant the fertilized embryo in a host mother to see what type of offspring might be created."

However, defects in offspring, including deformities of the pelvis and eyes, *have* shown up in some of the research where animal embryos actually did progress to birth, leading the federal Department of Health and Human Services to request further testing to see if these deformities are a frequent by-product of the process.

Womb for rent.

Because in vitro fertilization is still a relatively new technique, doctors have limited its use to women with blocked fallopian tubes who will ultimately have their own fertilized eggs implanted in their uteri. But some visionaries suggest that as the technique is perfected, it could be used on women who cannot—or do not want to—carry a baby to term.

A busy career woman, for example, could have one of her eggs fertilized with her husband's sperm in a petri dish and then implanted in another woman. The "genetic" mother could pursue her professional interests while her baby grew to term inside a woman who had been hired as a sort of human incubator. After nine months, the egg donor and husband could claim the child, who would be 100 percent theirs genetically, even though he had been nurtured in the womb of a woman whose identity might be unknown to either of them.

Will the baby's real mother please stand up?

Thus far, "mercenary mothers," as Dr. Robert T. Francoeur calls women who rent their wombs, have not yet carried a baby that had developed from another woman's egg. But another type of surrogate mother has come on the scene—one who agrees to be inseminated with the sperm of a man whose own wife is not capable of conceiving or carrying a child to term. (This situation is the reverse of AID, since the child has the husband's genes, but not the wife's.)

Dr. Richard Levin, a 35-year-old Louisville, Kentucky, physician, has set up Surrogate Parenting Center Associates, Inc., the first clinic offering to match couples and surrogates. Last November his first mercenary mother gave birth to a baby boy whom she turned over to a couple who had been trying to have children for ten years. Her reported fee: $10,000.

Her husband, who had had a vasectomy four years earlier, was supportive but other people were horrified. Her parents wouldn't speak to her, a friend called her an adulteress, and her husband lost his job—all because of her surrogate mothering, she says.

The law is no more sympathetic to surrogate mothers than their neighbors are. It views the woman who bears a child as the natural mother, and for that reason a baby created in this way must be adopted by the couple who contracted for it. In at least 26 states, however, there are laws requiring adoptions to go through a public agency and *prohibiting* a mother from accepting money for her child.

Of course, the surrogate-mother situation differs in some very basic ways from paying a pregnant woman for her child. First, the contract is made *before* the woman becomes pregnant. Second, since the adopting father is the natural father, it would seem that he, too, should have rights to the child.

But as a result of the legal complexity surrounding the surrogate-mother situation, it seems that if a surrogate mother wanted to keep the child she delivered and not give him to the couple who contracted for him, she would have that right. "We may see this tested in the courts soon," reported Noel Keane, a Dearborn, Michigan, attorney who has arranged at least ten surrogate gestations. "One of my surrogates who is very close to delivering is beginning to have second thoughts."

Since so many legal questions are raised by a surrogate mother's carrying a child for nine months and then giving him up, some doctors are looking at ways to make the transfer *before* rather than after birth. This has resulted in plans for such new reproductive techniques as artificial embryonation and embryo adoption.

Embryos for sale.

Water Tower Place, a shopping center on Chicago's Magnificent Mile, houses some of the most exclusive shops in the world. There you can buy everything from eighteenth-century Chinese screens to remote-control robots. But now, in the commercial office section of Water Tower Place, in a suite marked "Reproduction and Fertility Clinic," a new item is going up for sale: human embryos.

Since so little is known about the best way to fertilize an egg in vitro, the heads of the clinic, Dr. Randolph Seed, a physician, and his brother, Dr. Richard Seed, a veterinarian, thought that they would rely more closely on nature—they use a volunteer woman as a human petri dish.

Dr. Richard Seed was on hand in Kiel to explain the process. "Artificial embryonation is the transfer, four to five days after fertilization, of a human embryo from the uterus of a fertile donor to the uterus of an infertile recipient, who will then carry the embryo to term," said Seed.

In artificial embryonation (AE), a childless husband and wife pay a fee to the Seed clinic, which, in turn, pays a fee of up to $250 to a fertile woman who agrees to be inseminated with the husband's sperm. Then, four to five days after fertilization, the Seeds try to flush out the embryo so that it can be implanted in the wife, as in in vitro fertilization.

In addition to artificial embryonation, the Seeds plan to offer embryo adoption (EA). "The technology is the same as an artificial embryonation, but donor semen is used instead of the semen of the recipient's husband," said Dr. Richard Seed. "In this case, the embryo has genes from neither of these 'adoptive' parents, even though the 'adoptive' mother will bear the child; thus the situation is analogous to the adoption of a child."

Seed told the Kiel conference how the idea came about. "It's amusing I didn't think of this myself," he recalled. "A patient called up and said he had been trying to adopt a child for five years, and asked, 'Why can't I adopt an embryo?' A few months later I called him back, and that started the whole thing."

Seed believes that neither AE nor EA would involve any legal procedures. "We feel the child delivered to the previously infertile mother will be her child legally," he explained, "and her previously infertile husband will be the legal father, with no necessity for actual adoption proceedings." This view is in keeping with most of the artificial-insemination cases, which say that the woman who bears the child is the natural mother. But considering that in AID cases a child is not only borne by the mother but also conceived of her egg, who knows how a judge would feel if one of Seed's volunteers, who donated the egg—and thus the genes—for the child, sued to have visitation rights?

Illegitimate or just plain weird?

The idea is not as farfetched as it may sound. In England, a British doctor had planned to help an infertile woman have children by doing an ovary transplant. The surgery was stopped, however, when British authorities informed the doctor that although his patient could then conceive and carry a child, any offspring that might result would be considered illegitimate. In such a case the legal mother might actually turn out to be the woman who gave up the ovary and thus donated her eggs.

The big question surrounding the Seeds' program, however, is whether or not it will work. Thus far, they have recovered only one fertilized egg, for which they did not have a ready recipient. Hence the procedure has produced a conception, but not a child. But Richard Seed confidently cited seven years of experience doing embryo transplants in cattle to justify his optimism that eventually they will succeed. "Ten thousand transfers have been done with cattle," he said, "and there are no excess abnormalities." He later added, "I see no reason why it shouldn't work with humans."

Several doctors attacked Seed's attempts to draw conclusions about human beings from experiments done with animals. Indeed, Dr. Quigley had pointed out earlier that it is not always possible even to generalize from one animal species to another. He had come to this conclusion, Dr. Quigley explained, after finding that rat eggs did not fertilize in vitro under the same circumstances in which mouse eggs were routinely fertilized. If one could not generalize in animals so closely related, how could one hope "to establish standards for human in vitro fertilization from laboratory experiments with animals?" he asked.

New possibilities for single parents, including homosexuals.

Although participants in the Kiel conference differed in their medical views and in their moral views, they did share one common perspective: they all referred to their clients as married couples. But married couples who are infertile are not the only people who want to take advantage of the new reproductive technologies. It is estimated that overall 1,500 single women a year receive AID and that 150 lesbian women will conceive children this year via this technique.

Although some doctors who disapprove of single parenthood—especially when the parent-to-be is homosexual—feel they have the right *not* to provide help in such cases, unmarried people maintain that they have legal rights, too. And they are increasingly able to exercise these rights successfully: for example, when a single woman sued Dr. Hafez's clinic at Wayne State University because the clinic would not allow her to apply for AID, the suit was settled only when the clinic agreed to drop its restriction against single women.

Children with four parents?

Now that a larger segment of our society is growing to accept the idea of single parenthood, the question is, How will we respond to other familial combinations that may become possible through technology?

Mice with four parents have already been successfully created by Dr. Beatrice Mintz of the Institute for Cancer Research in Philadelphia, Pennsylva-

nia. Dr. Mintz takes the embryo of a white-mice couple and puts it in a petri dish with the embryo of a black-mice couple. The embryos attach to each other—and then begin dividing until they develop into what looks like a normal 32- to 64-cell embryo. This embryo is put into a surrogate mouse mother, who later gives birth not to a black mouse and a white mouse but to a single black-and-white-colored mouse, which has four genetic parents. The patterned color and crazy quilt of other traits is due to the fact that instead of each cell having chromosomes from each set of parents, some cells have the black mice as "parents" while other cells contain the gene combination of the white-mice parents.

Eliminating fathers altogether.

While Dr. Mintz is trying to breed offspring from multiple parents of both sexes, other researchers are experimenting with producing offspring from females alone. In a radical departure from the traditional concept of family, current research by Clement Markert at Yale University and Pierre Soupart at Vanderbilt University in Nashville, Tennessee, actually has the potential to make males unnecessary to parenthood. Their experiments with mice indicate that two *eggs* can be fused so that they start to divide without the help of any sperm.

"Sperm appear to produce a specific chemical that fuses them to eggs to start the fertilization process," Soupart explained to the audience in Kiel. What Soupart did was to mix such a chemical, an inactive virus, with two mouse eggs. The eggs fused together in what Soupart calls an oocyte fusion product (OFP) and within six days had developed into an embryo with the same structural characteristics as eggs that had been fertilized by sperm.

Soupart sees the main application of his work as being in the animal kingdom: "This method, which would create only female offspring, has considerable nutritional and economic impact, especially in developing countries, like India, where the consumption of meat products is prohibited by local traditions, while the consumption of dairy products is not."

Others, including Lucia Valeska, co-director of the National Gay Task Force, sees its potential for women, particularly lesbians, who might want to have children together. "It's very

understandable that women who want to raise children together should want the children to be part of their biological makeup," she says. "The question that will be raised is, Is this natural? which sort of amuses me because 'naturally' we can do this."

In addition to two-female couples, a single woman who is considering artificial insemination with the sperm of an anonymous donor might be even more pleased to have the opportunity for a daughter who was totally hers genetically. Since the ovaries can produce more than one egg at a time, this could be accomplished by fusing together two eggs from one woman.

Artificial wombs—and animal mothers for human babies.

It's hard enough to imagine that a child could have one, two, or more women as her genetic parents, but it's even more difficult to adjust to the concept of a human child's being "mothered" by an artificial womb. Yet an Italian embryologist, Daniele Petrucci, claims to have developed an artificial womb (thought to be a silicone container filled with amniotic fluid withdrawn by syringe from the wombs of pregnant women) that can serve just such a purpose. According to Petrucci, he let an embryo grow in this substitute home for 29 days, then destroyed it because it looked deformed and enlarged. He later repeated the experiment, he reports, and says he was able to keep a female embryo alive for 59 days.

Work on artificial wombs is in progress all over the world, but creating a machine to replicate the give-and-take equilibrium of mother and child is difficult. A possible alternative would be letting a human baby develop to term not inside a machine—but inside an animal. Recently, cows have been suggested as host mothers. But not enough is known about the mother-fetus interaction to determine exactly what a child would lack from developing within a nonhuman surrogate, whether animal or machine.

Current research points out that the fetus is aware of much about the environment in which it develops; for example, a newborn infant is able to distinguish his mother's voice from that of another woman, presumably from having heard it while he was in the womb. What would it be like for a fetus to develop in a steel or bovine womb—an environment that would expose it to a great many distinctly nonhuman stimuli?

Embryologist Robert T. Francoeur, in examining the possible effects on a child of being implanted in a cow, asked, "Would he look with pride on his nine-month prenatal life grazing in the backyard as being a mark of distinction, sophistication, and aristocracy? Or might he come to resent his mother who put him out to pasture while she played gourmet cook and cultured hostess, lover, and intellectual companion to his father?"

Chlorophyll man: half-human, half-plant?

Some of today's scientists, not bothered by the same considerations as Dr. Francoeur, believe that animals can provide even more than a quiet nine-month resting place for the baby of the future. These researchers are actually looking for ways to combine human genes with those of animals—and even plants—in order to expand our capabilities.

"The test-tube baby and gene-splicing technologies lock into each other," says biologist Dr. Seymour Lederberg of Brown University, Providence, Rhode Island. "Because the fertilized embryo is now accessible in a petri dish, it will be possible to introduce into the cells of an embryo the DNA needed for a genetically defective child to survive."

If this is possible, some scientists ask, why not replace genes not with just a better human gene but with the best gene that you can get for a certain trait from any life form on earth? This is what biologists Martin J. Cline and Winston Salser were doing in their controversial experiments at the University of California at Los Angeles, when they used a herpes-virus gene to transform bone-marrow cells from a patient with a rare blood disease known as thalassemia. After treating the cells with the herpes-virus gene for thymidine kinase—an enzyme used in the synthesis of hemoglobin, the virus form of which Drs. Cline and Salser believe is more efficient than the human variety—the team of biologists reinjected the transformed cells back into the patient, believing that with their increased efficiency, these cells would proliferate and help the patient overcome the disease.

Taking the concept of "gene therapy," as it is called, one step further has led others to think about creating new, improved beings by using other

than human genes to give people traits they *never* had before. These hypothetical combinations of humans and other species are referred to by scientists as "chimeras," an ancient Greek name for a monster who was part lion, part goat, and part serpent.

The interest in such experimentation is explained this way: in the past, when the environment changed, species had to adapt through nature in order to survive, but now the world is changing so rapidly that scientists are searching for ways to give us a helping hand. It is rumored that one American scientist is trying to develop humans with patches of chlorophyll so they can make food as plants do. And another researcher would like to create human beings who, like cows, can live on hay rather than the scarce food we now must eat.

In a less complex experiment, in the late sixties, Chinese surgeon Dr. Ji Yongxiang successfully fertilized a chimp with human sperm to create a "near-human ape" to perform simple tasks. The pregnancy was terminated when rioters destroyed his lab, but Dr. Yongxiang, who believes that the world is ready for animal-human hybrids, advocates trying again.

New forms of life may be here sooner than we think.

As science fiction–like as this may sound, the dawning of the age of chimeras may arrive sooner than we think. Because of the U.S. Supreme Court's 1980 decision that man-made life forms created by gene splicing are patentable, researchers are now allowed to reap the same type of financial rewards for creating life as they would for building a better mousetrap, and discoveries in this area are being sped along by private investment.

Firms that specialize in genetic engineering are doing so well in the marketplace that last October when one such firm, Genentech, Inc., went public, its stock rose from $35 a share to $86 in the first 90 minutes of trading. In fact, the money-making potential of this type of research has led Wall Streeters to dub genetic engineering "the boom business of the 1980s."

Indeed, there are money transactions involved with every aspect of this new technology. The In Vitro Fertilization Clinic of the Eastern Virginia Medical School in Norfolk, Virginia, charges $3,500 to $4,000 for in vitro

fertilization; surrogate mothers may earn $5,000 to $30,000 for renting their wombs; an egg donor in the Seed lab receives $50 per flush, with a bonus of $200 if a fertilized egg is recovered; and a sperm donor at the Tyler Medical Clinic in Los Angeles can contribute two or three times a week for $20 per "donation."

What are we doing to our species?

But for-profit procreation may have effects on our society that are significantly more far-reaching than transferring money from one person's pocketbook into another's. We have already, to use Dr. Semm's phrase, begun to "poke our noses in God's affairs," and some prominent people at Kiel and elsewhere question the advisability of our choosing to do that.

In a recent address, Pope John Paul II criticized artificial insemination, embryo transfers, and genetic engineering, saying they are "risking the survival and integrity of the human person." However, the view that we should set limits on reproductive technology is not only a religious one. As Amitai Etzioni, the George Washington University sociologist who is also director of the Center for Policy Research in Washington, D.C., has pointed out: "What may start as the biological control of illnesses could become an attempt to breed supermen."

Already, embryo technology is being used in animals to "upgrade" species. In Kiel, Dr. R. F. Seamark, professor of obstetrics and gynecology at the University of Adelaide, Australia, reported, "We have transferred embryos from superior angora goats into wild goats." This allows the highly sought after angora goats to have many more children than they themselves could carry to term and stops the wild goats from having their own scruffy offspring. A human analogy would be transferring into the womb of a woman with certain inheritable physical characteristics, like eye color or body type, an embryo from the petri-dish fertilization of the egg and sperm of a couple with physical characteristics that the society deems more attractive or valuable.

Is it right to breed people like animals?

In addition to providing the potential to create supermen, today's reproductive technologies also may make it possible to create a subhuman species to carry on society's menial tasks. One frightening aspect of this process is that, in the words of psychologist Dr. Carl Rogers of La Jolla, California, "We can choose to use our growing knowledge to enslave people in ways never dreamed of before, depersonalizing them, controlling them by means so carefully selected that they will perhaps never be aware of their loss of personhood."

Even if laboratory conception does not actually go so far as to create people like the superintelligent Alphas or the menial Deltas and Epsilons of Aldous Huxley's *Brave New World,* it has the potential to totally refashion the notion of parenthood.

At Kiel, Abbyann Lynch, professor of philosophy at the University of Toronto in Canada, raised the question "Ought man to apply technological skill to his own shaping?

"The more we lose the spontaneity and intimacy of marital relationships," she continued, giving her own view on the subject, "the closer we come to the danger of reducing human reproduction to the processes of a biological laboratory."

Where do we go from here?

Clearly, artificial insemination, petri-dish procreation, egg fusion, and gene splicing affect all of us, even those who do not make use of them, because the mere existence of these techniques changes the concept of the family and creates a potential for new forms of life to share our earth. In that sense the child of reproductive technology is not just the infant of the scientists who create him nor of the particular set of parents who contract for him: he is society's child, and all of us must live with the consequences of his birth.

Test-Tube Babies: Solution or Problem?

Ruth Hubbard

Ruth Hubbard is professor of biology at Harvard University. She is coeditor of Women Look at Biology Looking at Women *(Schenkman, 1979) and* Genes and Gender II: Pitfalls in Research on Sex and Gender *(Gordian Press, 1979). This essay is adapted from a talk she gave at the annual meeting of the American Association for the Advancement of Science in January, 1980 reporting work supported by the National Science Foundation.*

In vitro fertilization of human eggs and the implantation of early embryos into women's wombs are new biotechnologies that may enable some women to bear children who have hitherto been unable to do so. In that sense, it may solve their particular infertility problems. On the other hand, this technology poses unpredictable hazards since it intervenes in the process of fertilization, in the first cell divisions of the fertilized egg, and in the implantation of the embryo into the uterus. At present we have no way to assess in what ways and to what extent these interventions may affect the women or the babies they acquire by this procedure. Since the use of the technology is only just beginning, the financial and technical investments it represents are still modest. It is therefore important that we, as a society, seriously consider the wisdom of implementing and developing it further.

According to present estimates, about 10 million Americans are infertile by the definition that they have tried for at least a year to achieve pregnancy without conceiving or carrying a pregnancy to a live birth. In about a third of infertile couples, the incapacity rests with the woman only, and for about a third of these women the problem is localized in the fallopian tubes

(the organs that normally propel an egg from the ovary to the uterus or womb). These short, delicate tubes are easily blocked by infection or disease. Nowadays the most common causes of blocked tubes are inflammations of the uterine lining brought on by IUDs, pelvic inflammatory disease, or gonorrhea. Once blocked, the tubes are difficult to reopen or replace, and doctors presently claim only a one-in-three success rate in correcting the problem. Thus, of the 10 million infertile people in the country, about 600 thousand (or 6 per cent) could perhaps be helped to pregnancy by in vitro fertilization. (These numbers are from Barbara Eck Menning's *Infertility: A Guide for the Childless Couple*, Prentice-Hall, 1977. Ms. Menning is executive director of Resolve, a national, nonprofit counseling service for infertile couples located in Belmont, Mass.)

Louise Brown, born in England in July, 1978, is the first person claimed to have been conceived in vitro. Since then, two other babies conceived outside the mother are said to have been born — one in England, the other in India. In none of these cases have the procedures by which the eggs were obtained from the woman's ovary, fertilized, stored until implantation, and finally implanted in her uterus been described in any detail. However, we can deduce the procedures from animal experimentation and the brief published accounts about the three babies.

The woman who is a candidate for in vitro fertilization has her hormone levels monitored to determine when she is about to ovulate. She is then admitted to the hospital and the egg is collected in the following way: a small cut is made in her abdomen; a metal tube containing an optical arrangement that allows the surgeon to see the ovaries and a narrow-bore tube (called a micropipette) are inserted through the cut; and the egg is removed shortly before it would normally be shed from the ovary. The woman is ready to go home within a day, at most.

When the procedure was first developed, women were sometimes given

hormones to make them "superovulate" — produce more than one egg (the usual number for most women). But we do not know whether this happened with the mothers of the three "test-tube" babies that have been born. Incidentally, this superovulation reportedly is no longer induced, partly because some people believe it is too risky.

After the egg has been isolated, it is put into a solution that keeps it alive and nourishes it, and is mixed with sperm. Once fertilized, it is allowed to go through a few cell divisions and so begin its embryonic development — the still-mysterious process by which a fertilized egg becomes a baby. The embryo is then picked up with another fine tube, inserted through the woman's cervix, and flushed into the uterus.

If the uterus is not at the proper stage to allow for implantation (approximately 17 to 23 days after the onset of each menstruation) when the embryo is ready to be implanted, the embryo must be frozen and stored until the time is right in a subsequent menstrual cycle. Again, we do not know whether the embryos were frozen and stored prior to implantation with the two British babies; we are told that the Indian one was.

In sum, then, there is a need, and there is a technology said to meet that need. But as a woman, a feminist, and a biologist, I am opposed to using it and developing it further.

Health Risks

As a society, we do not have a very good track record in anticipating the problems that can arise from technological interventions in complicated biological systems. Our physical models are too simpleminded and have led to many unforeseen problems in the areas of pest control, waste disposal, and other aspects of what is usually referred to as the ecological crisis.

In reproductive biology, the nature of the many interacting processes is poorly understood. We are in no position to enumerate or describe the many reactions that must occur at just the right times during the early stages of embryonic development when the fertilized egg begins to divide into increasing numbers of cells, implants itself in the uterus, and establishes the pattern for the different organ systems that will allow it to develop into a normal fetus and baby.

The safety of this in vitro procedure cannot be established in animal experiments because the details and requirements of normal embryonic development are different for different kinds of animals. Nor are the criteria of "normalcy" the same for animals and for people. The guinea pigs of the research and implementation of in vitro fertilization will be:
☐ The women who donate their eggs.
☐ The women who lend their wombs (who, of course, need not be the same as the egg-donors; rent-a-wombs clearly are an option), and
☐ The children who are produced.

The greatest ethical and practical questions arise with regard to the children. They cannot consent to be produced, and we cannot know what hazards their production entails until enough have lived out their lives to allow for statistical analysis of their medical histories.

This example shows the inadequacy of our scientific models because it is not obvious how to provide "controls," in the usual scientific sense of the term, for the first generation of "test-tube" individuals; they will be viewed as "special" at every critical juncture in their lives. When I ask myself whether I would want to be a "test-tube person," I know that I would not like to have to add *those* self-doubts to my more ordinary repertory of insecurities.

A concrete example of a misjudgment with an unfortunate outcome that could not be predicted was the administration of the chemical thalidomide, a "harmless tranquilizer" touted as a godsend and prescribed to pregnant women, which resulted in the births of thousands of armless and legless babies. Yet there the damage was visible at birth and the practice could be stopped, though not until after it had caused great misery. But take the case of the hormone DES (diethyl stilbesterol), which was prescribed for pregnant women in the mistaken (though at the time honest) belief that it could prevent miscarriages. Some 15 years passed before many of the daughters of these women developed an unusual form of vaginal cancer. Both these chemicals produced otherwise rare diseases, so the damage was easy to detect and its causes could be sought. Had the chemicals produced more common symptoms, it would have been much more difficult to detect the damage and to pinpoint which drugs were harmful.

The important point is that both thalidomide and DES changed the environment in which these babies developed — in ways that could not have been foreseen and that we still do not understand. This happened because we know very little about how embryos develop. How then can we claim to know that the many chemical and mechanical manipulations of eggs, sperms, and embryos that take place during in vitro fertilization and implantation are harmless?

A Woman's Right?

The push toward this technology reinforces the view, all too prevalent in our society, that women's lives are unfulfilled, or indeed worthless, unless we bear children. I understand the wish to have children, though I also know many people — women and men — who lead happy and fulfilled lives without them. But even if one urgently wants a child, why must it be biologically one's own? It is not worth opening the hornet's nest of reproductive technology for the privilege of having one's child derive from one's own egg or sperm. Foster and adoptive parents are much needed for the world's homeless children. Why not try to change the American and international practices that make it difficult for people who want children to be brought together with children who need parents?

Advocates of this new technology argue that every woman has a right to bear a child and that the technology will extend this right to a group previously denied it. It is important to examine this argument and to ask in what sense women have a "right" to bear children. In our culture, many women are taught from childhood that we must do without lots of things we

---

Sowing the Seeds of Genius

by Robert Cooke

There is an ultraselect sperm bank in Southern California, the brainchild, as it were, of a California businessman named Robert K. Graham. And its donors are winners of the Nobel Prize in science.

Despite some hilarity at this end of the country, the new sperm bank is apparently serious business for Graham. According to recently published reports, the sperm bank has already supplied sperm for three women, and Graham has tapped at least four Nobel Prize-winning scientists for sperm donations.

The only scientist who admits having donated to Graham's sperm bank is Stanford University's controversial William B. Shockley, 70, who said in a *Los Angeles Times* interview he's disappointed that more of his fellow Nobel winners haven't agreed to donate.

Graham, 74, said he is merely carrying on the dream of 1946 Nobel Prize-winning geneticist Hermann Muller. Muller advocated establishing sperm banks in which donations from brilliant men would be stored until after their deaths. Later, carefully selected women who wished to increase their chances of producing exceptionally bright children would receive the sperm. Muller's ideas were bitterly attacked at the time, and he died a disappointed man.

Much of the opposition to the idea, scientists explain, stems from the fact that it probably won't work, because not enough is yet known about how geniuses are produced. Indeed, a fundamental, often bitter argument is now underway among scientists over whether heredity has much to do with intelligence. "There's no reason to think that individuals born as a result of use of this sperm bank will have any greater chance of being geniuses than anyone else in the world. I think it's ridiculous," said Dr. Jonathan Beckwith, professor of microbiology and molecular genetics at the Harvard School of Medicine.

Graham, who lives in Escondido, Calif., said the women who get the sperm need pay only shipping costs, plus a refundable deposit on the container used to keep the sperm frozen.

An M.I.T. scientist (who asked not to be identified) commented, however, that "it could be a clever confidence game. If people believe they're getting real genius seeds, perhaps the price could go up beyond $10,000."

Biologist George Wald of Harvard, who won the Nobel Prize in 1967, commented: "Oh, this is a crushing blow, to be left out of this sperm bank. I felt bad enough when I only made it into President Nixon's second enemies list." Wald added that "I hope Graham has checked out the sperms' motility if he's starting at Shockley's age level," because the sperm from older men are often unable to "swim" as vigorously as the sperm from young men.

Robert Cooke is the science editor of the Boston Globe.

want — electric trains, baseball mitts, perhaps later an expensive education or a well-paying job. We are also taught to submit to all sorts of social restrictions and physical dangers — we cannot go out alone at night, we allow ourselves to be made self-conscious at the corner drugstore and to be molested by strangers or bosses or family members without punching them as our brothers might do. We are led to believe that we must put up with all this — and without grousing — because as women we have something beside which everything else pales, something that will make up for everything: we can have babies! To grow up paying all the way and then to be denied that child *is* a promise unfulfilled; that's cheating.

But I would argue that to promise children to women by means of an untested technology — that is being tested only as it is used on them and their babies — is adding yet another wrong to the burdens of our socialization. Take the women whose fallopian tubes have been damaged by an infection provoked by faulty IUDs. They are now led to believe that the problems caused by one risky, though medically approved and administered, technology can be relieved by another,

much more invasive and hazardous technology.

I am also concerned about the extremely complicated nature of the technology. It involves many steps, is hard to demystify, and requires highly skilled professionals. There is no way to put control over this technology into the hands of the women who are going to be exposed to it. On the contrary, it will make women and their babies more dependent than ever upon a high-technology, super-professionalized medical system. The women and their babies must be monitored from before conception until birth, and the children will have to be observed all their lives. Furthermore, the pregnancy-monitoring technologies themselves involve hazard. From the start, women are locked into subservience to the medical establishment in a way that I find impossible to interpret as an increase in reproductive freedom, rights, or choices.

Health Priorities

The final issue — and a major one — is that this technology is expensive. It requires prolonged experimentation,

sophisticated professionals, and costly equipment. It will distort our health priorities and funnel scarce resources into a questionable effort. The case of the Indian baby is a stark illustration, for in that country, where many children are dying from the effects of malnutrition and poor people have been forcibly sterilized, expensive technologies are being pioneered to enable a relatively small number of well-to-do people to have their own babies.

In the United States as well, many people have less-than-adequate access to such essential health resources as decent jobs, food and housing, and medical care when they need it. And here, too, poor women have been and are still being forcibly sterilized and otherwise coerced into *not* having babies, while women who can pay high prices will become guinea pigs in the risky technology of in vitro fertilization.

In vitro fertilization is expensive and unnecessary in comparison with many pressing social needs, including those of children who need homes. We must find better and less risky solutions for women who want to parent but cannot bear children of their own.

PRENATAL PSYCHOLOGY:
Pregnant with Questions

CHARLES SPEZZANO

Charles Spezzano practices psychotherapy in Denver, where he is also chief psychologist in the Child Psychiatry Clinic at the University of Colorado Health Sciences Center. He is on the faculties of the Department of Psychiatry at the University of Colorado Medical School and at the School of Professional Psychology at the University of Denver. While he was writing this article, his wife gave birth to their second daughter.

If a pregnant woman is under heavy stress—experiencing grief from a loss, perhaps, or anxiety over her job or her marriage—is she exposing her unborn child to damage? Such is the claim made by several recent books and articles. The belief in the power of the intrauterine environment to shape the future mental and emotional development of the child is an age-old one. Lately, it has been carried to extremes by primal therapists and others who claim that their patients are able to recall experiences in the womb, and that these experiences have influenced subsequent behavior and development.

A new book entitled *The World of the Unborn* by Leni Schwartz states on the cover that "the most important time in our lives may well be the time *before* we are born." Because the author "reexperienced" her own birth in LSD therapy, she believes that "we are aware beings in the womb and that our unconscious retains and stores

the memory of that period." In support of her hypothesis, Schwartz, an environmental psychologist, offers some bits and pieces of research suggesting that a fetus is capable of reacting to movement, light, and sound. The implication is that these intrauterine experiences shape future perceptions and preferences. A similar idea was expressed as far back as 1974 in an article in the *Journal of Primal Therapy* entitled "From Womb to World," suggesting that unhealthy mother-child relationships may begin prenatally. The writer claimed that the fetus can feel frustration, fear, or other painful emotions both before and during birth; if such pain is repressed, a neurosis develops and remains until the conditions of birth are reexperienced in primal therapy.

Another book carries the argument still further. In *The Secret World of the Unborn Child,* to be published next month, primal therapist Thomas Verney of York University argues that, in a number of ways, a woman's psychological well-being during pregnancy is vital to the future healthy development of her child. Writers like Verney feel free to draw upon a rather extensive research literature on prenatal development to support their argument, ignoring everything that might tend to refute it. They have much to choose from; there are a number of scientists who are intrigued by these issues and have been investigating

and writing about them in professional journals for years. Psychiatrist Lester Sontag, for one, has spent more than 30 years exploring human development, from fetus to adulthood, at the Fels Research Institute in Yellow Springs, Ohio. A few comments he made in a 1964 article about maternal emotions during pregnancy are still being quoted by researchers today. He observed that when some of the women he studied who were in the last four months of pregnancy—a small number, only eight in 10 years—became frightened or grief-stricken, their fetuses moved more frequently and forcefully than usual. After birth, these children were more irritable or overactive than others in the study and more likely to have eating problems and very frequent stools.

When I spoke to Sontag a few months ago, he still found his early experiments persuasive. "I definitely feel maternal emotions during pregnancy have both an immediate and a long-term impact on the child," he said. "But it would take multiple measures and long-term observations to really prove anything."

Sontag's comment sums up the current state of what is known as prenatal psychology. Whether stress can have any long-term effects on the fetus—in the first month of pregnancy, the last month, or any time in between—is still uncertain. The research so far can be called suggestive, provocative, even inspiring. What it cannot be called is conclusive, if you want proof of the causal connection between prenatal stress and individual differences in psychological development. Yet the questions raised and their possible implications are too important to be dismissed lightly. For people who expect to be parents, the question of whether such a connection exists is far from academic. The findings of prenatal psychology may also have implications for the current debate over abortion, particularly the issue of when human life begins.

An extreme example from the pediatric literature emphasizes some of the other issues at stake in the question of possible prenatal influences. A healthy 17-year-old girl gave birth to an apparently normal baby boy after a medically uncomplicated pregnancy. Twenty hours of normal infant care followed, with mother and child side by side. Then the baby vomited fresh blood. He was examined and still appeared healthy and vigorous, but his vomiting continued, and one hour later, the baby died. Postmortem examination revealed three peptic ulcers. Since peptic ulcers usually develop in adults who are chronically tense and anxious, the physicians wondered whether the mother could have been under enough stress during pregnancy to pass ulcer-causing hormones across the placenta. In fact, her pregnancy—especially in the last trimester—was extraordinarily stressful. Coerced by her parents into marrying the father of her child, she found herself living with an alcoholic wife-batterer. She went back to live with her parents, but her husband paid her frequent and distressing visits in an effort to persuade her to return to him. Not long before the day of birth he threw a brick through her window, and the police were called in.

A single case certainly does not prove that maternal stress can cause gastric ulcers in a newborn, but the possibility would seem to warrant some kind of counseling for pregnant women who are exceptionally tense, anxious, or unhappy. It would also seem to justify research on prenatal stress, even if the studies are doomed to be less than perfect methodologically.

Several investigators have challenged the old notion of the placenta as a passive and relatively impermeable filter capable of protecting fetuses from all dangerous substances in the mother's blood. In addition to bringing in oxygen and nourishment while eliminating carbon dioxide and waste products, the placenta, according to British physician Peter Beaconfield, "is both active and selective in its transfer of substances essential to the development of the fetus." For example, it doesn't simply select the proteins needed by the fetus from those supplied by the mother. Since these are usually not the exact ones needed, the placenta synthesizes specific fetal proteins from the raw maternal materials.

As investigators were learning that the placenta is not passive, they also found that it is not impermeable to harmful substances in the maternal blood flow. Psychologist Donald Hutchings of The New York State Psychiatric Institute concludes that "with the exception of a few chemical agents that are of large molecular size, biotransformed into inactive metabolites by the placenta, or firmly bound to maternal tissue, most drugs in the maternal circulation readily cross the placenta to enter the embryonic and fetal milieu." Going even further, researchers now believe that the placenta can turn some neutral substances toxic, or it can detoxify a chemical, or it can allow a substance to pass unchanged. Alcohol and narcotics, for example, are apparently transferred intact. Valium crosses so fast that it can put a fetus to sleep in minutes.

The dangers of the permeable placenta were illustrated by what happened with DES (diethylstilbestrol) in the 1960s. During that time, 25,000 women in the United States took the drug to prevent miscarriages and other complications of pregnancy. DES had been in general use since the 1940s, but it wasn't until 1971 that the *New England Journal of Medicine* reported an unusual increase in one form of vaginal cancer among young women born in New England between 1946 and 1951. The disease was eventually linked to the DES given to their mothers. Moreover, according to Yvonne Brackbill, professor of psychology and obstetrics and gynecol-

ogy at the University of Florida, the drug may have also affected the women's personalities: they appear to be "much less aggressive," she said, than women whose mothers have not taken DES.

Teratology is the study of developmental deviations induced by environmental factors. In discussing the effect of prenatal events on later psychological development, we inevitably enter the field of behavioral teratology, which is still in its infancy and therefore hedged with more uncertainties than general teratology. Some good work, however, has been done. For example, Hutchings has found that giving rat mothers large doses of Vitamin A leads to defects in learning and behavior in their offspring. It appears, from this and other studies done in the 1970s, that the areas of the fetal brain that are developing most rapidly at the time mothers are exposed to the damaging agent do not function normally after birth and that behavior associated with those brain areas is impaired.

The animal research was accompanied by new findings in human teratology. The one that drew most attention was fetal alcohol syndrome. The evidence strongly suggests that heavy alcohol consumption during pregnancy leads to damage to the fetus that shows up in the form of mental retardation, poor coordination, defective limbs, and abnormalities of the head and face.

Once it had been demonstrated that substances like thalidomide and alcohol could produce gross physical and mental damage, researchers began to wonder whether more subtle differences between children—such as behavior problems, academic underachievement, depression, or clumsiness—might be attributable to events during pregnancy. The classical method for investigating such a hypothesis, derived from animal studies, would call for taking a group of pregnant women and selecting some to take amphetamines, for example, while the others take nothing. Next, all the children would be removed from the mothers who bore them and raised by a new set of mothers, to distinguish clearly between the mothers who affected them before birth and after. No one could countenance such a study with humans, so researchers have had to rely on animals.

Psychiatrist Myron Hofer of Montefiore Hospital in New York City has summed up these investigations in a chapter of his new book, *The Roots of Human Behavior*: "In animal experiments . . . prenatal exposure to alcohol, methadone, heroin, major tranquilizers, barbiturates, and amphetamines, all have been shown to produce long-term effects on the behavior of offspring when juvenile or adult, weeks and months after the intrauterine period of exposure to these substances."

Despite the enormous and at times seemingly insurmountable obstacles to exploring similar questions with human subjects, many researchers have been attracted to the field of prenatal psychology. Perhaps no investigator has pursued the issue as vigorously as has psychologist Denis H. Stott. For 20 years, from 1957 to 1976, Stott conducted study after study on the prenatal determinants of individual differences, all the while admitting freely that it is a "gray area" of modern psychology. "Knowledge of these determinants," he strongly believes, "is central to our understanding of human development" and is needed not just to plan good prenatal care programs but, more broadly, to create "social conditions conducive to healthy prenatal development."

Stott's best research was done in the mid-1960s in Scotland and Canada. In the Scottish study, about 150 randomly selected mothers were interviewed about their pregnancies during the first month after delivery. The mothers were then visited periodically until their children were four-years-old to obtain information about child health, development, and behavior. When he began the study in 1965, Stott was looking for relationships between a specific type of prenatal stress and a wide range of problems during the first four years of life. The results were unexpected. He wrote: "No relationship was found between the children's health and physical illness, accidents, work stresses or dental operations in the

mothers. Situational stresses such as the death or severe illness of a family member, or shocks and frights experienced by the expectant mother were also not significant. On the other hand, stresses involving severe, continuing personal tensions (in particular, marital discord) were closely associated with child morbidity in the form of ill health, neurological dysfunction, developmental lag, and behavior disturbance."

While Stott cautions that his results are only suggestive, they obviously are provocative. It is tempting, for example, to leap to the conclusion, as some recent authors have done, that marital discord during pregnancy causes a variety of psychological and health problems in the developing child. Then, consistent with the American tendency to overdramatize the hazards of pregnancy, many expectant couples might begin to fear that any significant tension in their relationship will damage the unborn child, thus creating a whole new source of anxiety. To draw such conclusions from Stott's research would be totally unscientific. A closer look at the data shows why.

Most of the mothers in the high-personal-tension group were poor. Half shared bathrooms with other tenants. Almost two-thirds lived in one- or two-room apartments. Some lived in huts. Obviously they remained poor after they delivered. Poverty may well have been the problem, not stress.

Recognizing that possibility, Stott spent considerable time analyzing the relationship of poverty and interpersonal stress to postnatal health and behavioral problems. He found that when high levels of stress were absent, poverty conditions during pregnancy were not clearly related to problems after birth. On the other hand, even among women who were not poor, chronic tension in the mothers during the pregnancy produced more physical and psychological aberrations in their children. Thus, it is fair to say that children whose mothers are both tense *and* poor during pregnancy are most likely to show developmental problems.

Psychologist Arnold Sameroff of the University of Rochester reached exactly the same conclusion in his

comprehensive 1975 review of the research into early influences on development, appropriately subtitled "Fact or Fancy?" In middle- and upper-class families, he found, the negative effects of even severe early physiological trauma, such as oxygen deprivation at birth, tend to disappear with age; in lower-class families, children with the same complications are likely to show severe deficits as they get older. Various prenatal and perinatal traumas have received much attention as possible causes of later hyperactivity, learning problems, and psychiatric disorders. But it is crucial, Sameroff warns, to remember that in every case "it is possible to identify significant numbers of children who, although subjected to these negative influences, nevertheless develop normally." The most potent predictors of how a particular child will develop, if one wants to single out one or two factors, are the quality of the caretaking the child receives from the parents and the family's socioeconomic level.

That doesn't mean prenatal stress is irrelevant. Stott's indictment of chronic marital discord during pregnancy still merits consideration. However, it is essential to treat that form of prenatal stress, and each of the others we will mention, as just one of many interacting factors that determine the behavioral and personality development of any child.

One important word—"father"—has been missing from this analysis of pregnancy and its stresses. There is much speculation about the expectant father's influence on his child's future development through his role in creating a happy or tense prenatal environment. An obvious approach is to compare intact families with those in which the father is absent during pregnancy. The hitch is that if a mother is alone before birth, she is usually alone afterward. Thus, the father's pre- and postnatal influences cannot be teased apart.

Two Finnish researchers, Matti Huttunen and Pekka Niskanen, solved this dilemma by identifying an "index group" of people whose fathers had died before they were born and a "control group" whose fathers had died during the first year of their lives. (Most of the fathers had been killed in World War II.) When the groups were compared, psychiatric and behavioral disorders were more than twice as common in the index group. But here again, more than one explanation is possible. The index group may have suffered the direct and negative biological effect of maternal stress during pregnancy. Or perhaps the mothers who did not lose their husbands until after delivery got off to a better start with their infants at birth and established relationships that were less vulnerable to disruption from their later grief.

Huttunen and Niskanen, psychiatrists at the University of Helsinki, favor the hypothesis that maternal stress acts directly and adversely on the fetus and offer as support an interesting trend in their data. More than two-thirds of the psychiatric patients in the index group had lost their fathers in either the third-to-fifth months of pregnancy or the ninth and tenth months; both periods are thought to be critical for fetal brain development. This is highly speculative, and many experts would favor a psychological rather than a biological explanation. For example, Henry Coppolillo, head of the division of child psychiatry at the University of Colorado Health Sciences Center, emphasizes the father's role in nurturing his pregnant wife. Through his care and attention, says Coppolillo, the father helps ready her for the tremendous demands the infant will make on her caretaking capacities. In either case, it is important to remember Sameroff's warning and note that less than 20 percent of the children who lost their fathers before birth showed any psychiatric disorder. Prenatal loss of father is simply one of the many potential roots of disordered human personality development.

Any theory of prenatal psychology must take into account recent discoveries of how adult sexual behavior is influenced by events between conception and birth. Although the evidence is still fragmented, psychiatrist Myron Hofer claims that this research constitutes "the best documented evidence of long-term effects exerted by the behavioral state of the mother during pregnancy

upon the behavior of her offspring in adulthood." In his new book, mentioned earlier, Hofer points out that once investigators found that everything done to pregnant rats and mice affected their pups, the investigators began to experiment more systematically. Their efforts bore fruit quickly.

In the early 1970s, Ingeborg Ward of Villanova University's psychology department observed that male rats born to mothers who had been subject to restraint and bright light showed both feminized and demasculinized behavior. They made fewer attempts to copulate with females than did other males, and when other males approached them, they often assumed the position female rats take preparatory to being mounted by males. According to Hofer, "All the evidence to date favors the theory that maternal hormones, changed in amounts and pattern during stress, act directly on the fetal brain and/or the fetal endocrine glands to modify the characteristics of the neuronal networks being laid down."

Female rat pups, however, show no such deviation in sexual behavior. It appears that males need a large dose of testosterone at a specific time during the intrauterine period to develop a masculine brain and hormone system. (Richard Pillard, a psychiatric researcher in sex and gender development at Boston University School of Medicine, points out a parallel phenomenon that has been observed in human adults: during combat, men burn up testosterone.) If stress on the mother uses up the testosterone her fetus needs, the basic substrates of male sexual behavior will simply not exist. Female offspring are not vulnerable in the same way, since they don't require testosterone to develop feminine sexual behavior. As for humans, Hofer says, "if someone wants to be sensational, it is possible to claim that homosexuality is due to prenatal maternal stress." He prefers not to be that sensational and sees this kind of speculation as potentially dangerous if presented as fact.

This sex-development research is important in two ways. It is the best evidence available of the importance of prenatal experience in fostering

normal postnatal behavior. And, as Hofer points out, it demonstrates that while the *potential* for adult behavior is influenced by prenatal events, the *expression* of this behavior requires a long series of social interactions after birth. For example, it has been shown that the stress-related sexual deviations of male rats in Ward's experiment can be reversed if they spend time in a cage with females their own age before they reach puberty. If rats are that flexible, it seems likely that human beings are even more so.

Despite these research problems, fetal psychology and the related disciplines of behavioral embryology and behavioral teratology will continue to expand. There is now an International Society for the Study of Prenatal Psychology, based in Austria, as well as a new international journal called *Early Human Development*, which deals with the continuity of fetal and postnatal life. One man who is surprised that more research is not currently being done with human beings is Gilbert Gottlieb of Dorothea Dix Hospital in Raleigh, North Carolina, who has edited four volumes of studies on the development of behavior and the nervous system. As investigators from embryology, teratology, and psychology find out about one another's work, he feels, more research will occur and will be likely to yield new insights. Denis Stott has recently learned that some of his work is being successfully replicated in Europe, with similar results suggesting that prenatal stress can alter postnatal development.

Psychologist Yvonne Brackbill is especially concerned about various medications taken by pregnant women that she says can impair a newborn's development. The 40 or more studies that have been done suggest that analgesics (Demerol and Nisentil, for example) and anesthetics (Cycloprane and Penthrane) can affect infants' motor skills such as their ability to push up with their arms or kick their legs. They also affect cognitive development, such as the tendency to perseverate—that is, to continue paying attention to a repeated stimulus. Since most infants soon shift their attention elsewhere in this situation, perseveration is considered a sign of cognitive deficit; the ability to shut out irrelevant stimuli is an important part of normal development. During our discussions, however, Brackbill drew my attention to the fact that research has not yet been done to show whether these developmental delays and deviations persist beyond the first year.

This kind of integrity is common to scholars working in the prenatal field. They are willing to tolerate ambiguities and uncertainties in their research and are extremely careful about offering advice to parents on how much stress during pregnancy is too much and may cause harm. All the evidence to date encourages parents to avoid "pregnancy paranoia" in which they begin to fear that every wrong move they make will doom their child to an abnormal future. Stress on a pregnant woman may well alter the fetus in some as-yet-unproven ways and create a risk of behavioral deficits later on. But how much of a risk? There is still no consensus. Hampered by the enormous expense of more definitive long-term studies with large numbers of parents and children, and by the lack of reliable, valid measures of either stress or deviant outcomes, current research is open to widely divergent interpretations.

It appears more productive, therefore, to turn our attention to the larger number of children who appear to develop normally despite early problems, whatever the cause. The prenatal period is merely one stage in the total life span of an individual, and infants show great plasticity. Each new developmental stage brings reorganization of past experiences and the chance for neutralizing or aggravating the effects of previous traumas. Some parents create an atmosphere of hopefulness, acceptance, and flexibility within which a child's early problems tend to right themselves as he or she advances to maturity. Other infants who have physical or mental problems meet with parental anger, pessimism, and rigidity, making it more likely that early deviances will become permanent. Exploring the most successful parenting practices seems a more useful way to prevent future suffering than getting bogged down in the perhaps futile attempt to identify prenatal stress as a necessary or sufficient cause of a child's later problems.

For more information, read:

Gottlieb, Gilbert. *Studies on the Development of Behavior and the Nervous System.* Volume 1 through 4 Academic Press, 1973, 1974, 1976, and 1978 respectively $39.50, $37.50, $43.00, $28.50

Hofer Myron. *Roots of Human Behavior.* W.H. Freeman 1981. $22.50 paper $11.50

Sameroff Arnold J. "Early Influences on Development. Fact or Fancy?" *Merrill-Palmer Quarterly of Behavior Development.* Vol. 21. (1975) No. 4

Sontag, Lester W. "Implications of Fetal Behavior and Environment for Adult Personalities." *Annals of the New York Academy of Science* Vol. 134 (1966): 782-786.

Ward, Ingeborg. "Prenatal Stress Feminizes and Demasculinizes the Behavior of Males." *Science,* Vol. 175 (1972): 82-84

A PERFECT BABY

VIRGINIA APGAR, M.D., M.P.H.,
and JOAN BECK

Much more is known today than ever before about the diseases
and abnormalities, inherited and acquired, that cause
birth defects. There are still great gaps in our understanding, of course. But
if all the knowledge now available could be put into use by physicians,
by public health officials and especially by men and
women who are still to become parents, probably more than half of the
birth defects which now occur could be prevented.

The prevention of birth defects should begin long before a baby is born—ideally, even before he is conceived. For example, a man or woman who thinks that a close relative has a disorder which might be hereditary should take advantage of genetic counseling. This consultation should be obtained before marriage, but certainly before the conception of a child. It is particularly important for parents who have already had an infant with a genetic disorder.

Many people worry for years about the possibility of passing on to children a defect which has affected a parent or grandparent, an uncle or cousin, niece or nephew. Yet, very often, genetic counseling can prove such fears groundless or greatly exaggerated.

Often, the defect that a person fears transmitting is not hereditary at all, or is caused by a combination of hereditary factors with something that goes wrong during prenatal life. Even if a family disorder is hereditary, a genetic counselor may be able to determine from a family medical history that a particular individual could not possibly be a carrier of the abnormality. Or, if the prospective parent could be a carrier, the counselor may be able to suggest specific laboratory tests which will confirm or deny this possibility. For carriers, a genetic counselor can spell out in percentages how much risk each child would face.

A genetic counselor is also a good source of information about ways in which some birth defects can be diagnosed early in prenatal life, while the pregnancy can still be terminated if the prospective parents so decide. The counselor can also report on new treatments that may make the risk of having a baby with a specific birth defect seem less devastating to potential parents.

A considerable number of couples faced with the risk of having a baby with a major birth defect decide not to bear children but to form their family by adoption.

As more studies are done, researchers note a definite connection between the timing of pregnancy in the lives of parents and the occurrence of birth defects, prematurity, and stillbirths. Recent findings indicate that the ideal age for a woman to bear children is between twenty and 35. If possible, it is best not to begin having babies before the age of eighteen and to complete childbearing before 40. Countless mothers younger than eighteen and older than 40 have given birth to perfectly normal healthy babies. But young girls who become pregnant often do so in circumstances which make it unlikely they will have good prenatal care from the beginning of pregnancy. Many such girls are not adequately nourished, and many are exposed to other conditions which are associated with prematurity, high infant mortality, and an increase in certain birth defects.

Mothers older than 40 run greatly increased risks of having a child with a chromosomal abnormality, particularly Down's syndrome.

Many physicians now suggest that for women who become pregnant after 40, amniocentesis—an examination of the fluid in the amniotic sac—be used to discover if there are any chromosomal errors in the unborn infant's cells. New techniques make it possible to diagnose these conditions in time to terminate the pregnancy if the mother wishes.

It's desirable for a man to father his children before he reaches the age of 45. The chances are somewhat greater that a baby will be stillborn or have a congenital malformation if the father is older than 45, regardless of the age of the infant's mother. However, the risks to the baby are not as great with an older father as with an older mother.

Ideally, there should be an interval of at least two years between the end of one pregnancy and the beginning of another. The shorter the time period between pregnancies, the greater the likelihood of birth defects and obstetrical difficulties. The younger the mother is, the greater the risks to which she exposes her offspring by having them too close together.

The more children a mother has, beginning with the third, the fewer the chances that each will be born healthy and normal. In part, these risks are related to factors such as the increasing age of the mother, short spacing between pregnancies, and poor living conditions.

When a couple plans to conceive a child, intercourse should take place at intervals of no more than twenty-four hours for several days just preceding and about the time of ovulation. Some birth defects occur because the ovum is fertilized late in the monthly cycle, just as it is beginning to disintegrate. This delayed fertilization greatly increases the likelihood of chromosomal abnormalities and miscarriage.

The chances that conception will occur and that the resulting infant will be normal and healthy are greatest when the uniting sperm and ovum are both fresh. Frequent intercourse during the time when conception is possible is thought to increase the odds that neither sperm nor egg will be "overripe" at the time they join.

It can be difficult to determine precisely when ovulation takes place during a particular menstrual cycle. It is usually estimated that ovulation occurs approximately fourteen days before the beginning of a menstrual period, but this varies from one woman to another and, sometimes, even from one month to another in the same woman.

One method of approximating the day of ovulation is for a woman to take her temperature every morning when she wakes up and to keep a chart of the readings. In many women, the temperature increases by a small percentage of a degree at the time of ovulation and remains higher until the beginning of the next period. If conception occurs, the temperature stays at the higher level. This method is not completely reliable, however.

A few women are aware when ovulation occurs because they feel a pain in the lower abdomen when the tiny follicle containing the maturing ovum bursts to release the egg cell. This pain or discomfort, about midpoint in the menstrual cycle, is called "mittelschmerz."

Because a mother provides the total prenatal environment for an unborn baby during the first, most critical period of his existence, her health and wellbeing are inextricably linked with his. There is much a woman can do—not only during this crucial nine months, but before—to make this environment healthy, nourishing, and free of the hazards that can produce defects in her unborn baby.

Ideally, the kind of medical care that helps a woman provide a healthy, nourishing environment for her unborn child begins long before pregnancy. Before she marries, a woman—and her future husband—should be given adequate information about family planning in accordance with their needs and beliefs. Their family histories may suggest the need for genetic counseling, not so much because it might influence the decision to marry but for its relevance in planning ahead for children.

When she is ready to have a baby, a woman should have a medical checkup before conceiving, to make sure that she has no infections, nutritional deficiency, or physical abnormality that might interfere with the baby's development or safe birth. At this pre-pregnancy checkup, a woman should be warned against the dangers of drugs, radiation, and infections during the earliest weeks of pregnancy. Her doctor should make sure that her immunizations are up to date, to protect her and the baby she hopes to conceive against as many viral infections as possible. Medical conditions which might harm an unborn infant, such as thyroid deficiency, venereal disease, tuberculosis, and diabetes should be checked for and, if present, treated before pregnancy begins. Disorders like sickle cell anemia and heart disease, which are more hazardous to a pregnant woman than to her unborn infant, should also be evaluated before pregnancy starts.

If either the husband or wife has a history of occupational exposure to radiation, or if either has been taking drugs, the doctor may decide that an examination of their chromosomes is desirable to detect any possible abnormalities. A few studies link the father's exposure to radar equipment within a few weeks prior to conception to an increase in mongolism in their babies.

Regular checkups are essential during pregnancy, too, even if they seem routine and unnecessary. This medical monitoring provides an early warning system to detect the possibility of conditions such as toxemia, Rh incompatibility, premature separation of the placenta, and premature birth. It can also alert the physician to the presence of twins, which place extra strain on the mother's body and complicate delivery.

To encourage prenatal care, most obstetricians and family doctors arrange a fee in advance to cover prenatal supervision, delivery, hospital visits, and postnatal checkups. The fee remains the same, even if the mother requires more of the doctor's time than usual because of a complication during pregnancy or at the time of birth. Often, medical insurance plans cover most of these costs, or the physician may work out an installment type of payment program if a couple wishes.

City, county, and state health departments often provide free or low-cost prenatal care in neighborhood clinics, along with health and nutrition information. Many hospitals, especially those affiliated with a medical school, often give free prenatal services and obstetrical care. The U.S. Children's Bureau has made grants of many millions of dollars to help finance these municipal

2. PRENATAL PERIOD

and hospital services. In many states, some prospective parents are also eligible for Medicaid funds for prenatal and obstetrical care.

There are several sources prospective parents seeking free or low-cost prenatal care can contact for information. These include: local, county, or state health departments; the nearest office of the U.S. Department of Health, Education, and Welfare; the closest clinic or chapter of The National Foundation—March of Dimes; the nearest hospital, particularly if it is connected with a medical school; the local community referral agency or Community Fund headquarters.

No woman should become pregnant unless she is sure she has had rubella (German measles) or has been immunized against it. This cause of serious birth defects can now be eliminated completely by means of a vaccine, but such protection must be assured before pregnancy begins.

The rubella vaccine contains live virus, greatly weakened so that it is not strong enough to produce a full-blown infection but will still trigger the body to manufacture antibodies as protection against the disease. No one knows whether these attenuated viruses could harm an unborn baby if a prospective mother were vaccinated early in pregnancy. Physicians, therefore, won't give rubella vaccine to any woman who might possibly be pregnant, or who might become pregnant within two or three months. To make sure that no woman in early pregnancy is inadvertently vaccinated, most doctors will not immunize any adolescent girl unless she is known to them as a regular patient. Women are immunized only if a physician can be sure they are using a reliable method of birth control and understand the necessity for avoiding conception for at least two months following vaccination.

Before the rubella vaccine was developed, it was estimated that about twenty percent of women reached adulthood without ever having had the disease. Their unborn infants were thus in danger should they be exposed to the virus during the early months of pregnancy.

With the development of a simple blood test for rubella antibodies in 1966, doctors discovered, however, that a large percentage of women were mistaken about whether or not they had had the disease. Many women who assumed they were immune to rubella had apparently had a brief illness with a mild rash caused by another virus. Antibodies were found in the blood of another large group of women who were sure they had never had the disease, indicating that they must have had such a light case that there were no noticeable symptoms at all.

Because it is so easy to be mistaken about rubella, a woman should either have the vaccination or a blood test proving antibodies are present before she becomes pregnant.

From the very beginning of pregnancy, a woman should do everything possible to keep herself in good health and to avoid exposure to contagious diseases. This is particularly important during the first three months of pregnancy, when all of the new baby's organs and body structures are being formed and when the hazards of malformations are greatest.

An unborn infant, particularly during the first eight to thirteen weeks of pregnancy, is very vulnerable to certain viral infections. Destruction of just a few cells in his tiny body so early in his development can cause major malformations or disorders, while a similar loss in an adult would not even be noticeable. Even viruses which produce only brief, mild symptoms in an adult, can result in severe and lifelong handicaps for an unborn child.

To protect an unborn infant against the hazards of viral infections, a woman should make sure she is immunized, not only against rubella, but also against whooping cough, measles, mumps, polio, diphtheria, and smallpox before she becomes pregnant. During pregnancy, she should avoid exposure to any persons who might have a viral infection. Should she become ill during pregnancy, she should consult her physician immediately. Avoiding illness also lessens the likelihood that she might need medication during pregnancy—another hazard to the unborn child.

All during pregnancy, a woman should avoid eating undercooked red meat or contact with any cat which might be the source of a toxoplasmosis infection. Toxoplasmosis is usually a mild disease in adults. Many individuals may have it without even being aware of it, although it may produce a brief rash, cough, swollen glands, and other symptoms much like the common cold. But if a woman has toxoplasmosis during pregnancy, the organism may also attack her unborn baby. The mother recovers quickly from the disease. The unborn infant may not, but may continue to have active infection all during the months before birth and afterward.

Of the unborn infants who have toxoplasmosis during pregnancy, about twenty percent are born with major defects, including mental retardation, hydrocephalus, epilepsy, eye damage, and hearing loss. Some may also be premature.

A pregnant woman should not take any drugs whatever unless absolutely essential—and then only when prescribed by a physician who is aware of her pregnancy. This prohibition is particularly important during the first two or three months of pregnancy when the unborn infant's body and organs are developing. The reason this ruling has to be so strict and inclusive is that scientists now think many birth defects are caused by the action of a drug taken by a pregnant woman on the vulnerable tissues of her unborn child whose particular, individual, genetic make-up makes these tissues susceptible to damage.

Most drugs—over the counter as well as those sold by prescription—are tested on experimental animals under laboratory conditions before they are given to human beings. But the unborn offspring of laboratory species do not consistently react the same way to medications as unborn human children. So this kind of research cannot provide enough answers to protect all infants.

Drugs, in this sense, include not only medicines, such as aspirin, sleeping pills, tranquilizers, but also such things as nose drops and sprays, laxatives, mineral oil, douches, reducing aids, and even baking soda, vitamin supplements, and other common remedies for various conditions. Spray-can insecticides are also forbidden, along with other potent substances which can be inhaled.

X-ray examinations or radiation treatments should not be given to a pregnant woman; this warning applies particularly to the abdominal area during the first three months of pregnancy.

So hazardous is radiation to an unborn infant during the earliest weeks of his life that many physicians and hospitals make it a rule not to X-ray the abdominal area of any woman of childbearing age except in serious emergencies or during the first ten days following the start of a menstrual period. There is much less danger to an unborn child during the last part of pregnancy, after all of the infant's organs and bodily structures have been formed. X-rays at this stage may be essential to diagnose the condition of the unborn infant, or to help ensure his safe birth.

Another hazard of radiation during pregnancy is the possibility of damage to genes and chromosomes in the egg cells already formed within the tiny body of an unborn baby girl. This kind of abnormality would not show up for at least another generation, until this girl herself began to have children or perhaps for several generations.

It's best to avoid cigarettes during pregnancy. The babies born to mothers who smoke average about half a pound less in weight than the infants of women who don't smoke.

Researchers aren't sure precisely how cigarette smoking affects an unborn infant. They theorize that the growth retardation may be caused by the nicotine, which is known to pass through the placenta into the body of the developing infant, or by the high level of carbon monoxide in the mother's blood which reduces the amount of oxygen the blood brings to the unborn infant.

A nourishing diet, rich in proteins, vitamins, and minerals, and adequate in total calories, is essential during pregnancy. It is also important all of the years of a girl's life before she has children.

The body of an unborn infant must be built out of the nutrients in his mother's body. If she is gravely undernourished, her child will be, too.

In the United States, there are still many prospective mothers who do not have an adequate diet. Poor nutrition in pregnancy usually reflects a lifetime of undernourishment due primarily to poverty, but also to lack of information about good nutrition and sometimes to cultural customs.

Poor women are not the only ones who are inadequately nourished. A woman who diets too strictly during pregnancy for the sake of her appearance may be exposing her unborn infant to unnecessary risk, too. According to a three-year study by the National Research Council's committee on maternal nutrition, severe caloric restrictions during pregnancy may have harmful effects on the unborn baby's neurological development and may make his birth weight hazardously low. Ideal weight gain during pregnancy should be twenty-four pounds, or a range of twenty to 25 pounds, the Council concluded.

A prospective mother who is Rh-negative should make sure her physician takes the necessary steps to protect her unborn baby and subsequent children from Rh disease.

An Rh-negative woman who has never been pregnant with an Rh-positive baby and who has never received a transfusion of Rh-positive blood can now almost always be protected against the danger of having a child with Rh disease. All that is necessary is to have an injection of a special gamma globulin containing antibodies against Rh-positive blood within 72 hours after she's given birth to an Rh-positive infant. The Rh antibodies in the vaccine attack any Rh-positive red blood cells which might have entered the mother's circulation at the time of birth and destroy them before they can trigger the mother's own immune system to produce antibodies. Should she become pregnant again with an Rh-positive infant, she will not have any of the dangerous antibodies which could destroy the baby's red blood cells before birth.

It is essential, however, that every Rh-negative mother receive the vaccine following the birth of every Rh-positive infant, every miscarriage, or every abortion, spontaneous or induced.

Unfortunately, the Rh vaccine cannot help an Rh-negative mother who has already begun to produce antibodies because she had an Rh-positive baby or an abortion or a miscarriage before the vaccine was developed. If she becomes pregnant, her physician should monitor the well-being of her unborn infant carefully by checking the level of antibodies in the mother's blood and, if necessary, by amniocentesis. If the baby is in danger, he can often be helped by an intrauterine blood transfusion.

Every precaution should be taken to prevent a baby from being born prematurely. This caution applies to all weighing less than five-and-a-half pounds at birth.

The handicaps of prematurity extend from subtle forms of learning difficulties and behavior problems in the almost-normal ranges of birth weight to severe retardation, blindness, hearing loss, and even death in the tiniest newborns. Prematurity is often linked, too, with cerebral palsy and retarded physical development.

Prematurity is related to many direct and indirect causes: poor nutrition, illness of the mother during pregnancy, poverty, lack of good prenatal care, too short an interval following a previous pregnancy, cigarette smoking, anemia, the mother's age, unfavorable living conditions, and twins. Most of the recommendations already made will help to reduce the possibility of prematurity.

Good, regular, prenatal care is probably the most important of these recommendations, for there is much a physician can do to prevent prematurity. He can prescribe supplementary proteins, irons, and vitamins, if necessary. He can treat toxemia, bleeding, anemia, infections and other disorders in the mother before they become a serious threat to her unborn child. Sometimes, he can stop a premature labor so that the pregnancy can continue until close to full term. He can diagnose the

2. PRENATAL PERIOD

presence of twins and advise the mother to get extra rest, especially during the last three months of pregnancy, to help postpone labor as long as possible.

A good physician also knows how to balance any need the mother may have for pain-relieving drugs with the safety of her child. For all of the medications given to the mother also affect her infant as long as blood is circulating through the umbilical cord. Normally, the processes of labor and birth decrease the amount of oxygen that reaches the infant; too much anesthetic or anesthetic given at the wrong time can cause brain injury or even death from the drastically lowered oxygen supply. Medications such as muscle relaxants and depressants given to the mother during labor also affect her child. Because they don't wear off as quickly in the baby as they do in the mother, their effect can be an extra handicap for the newborn who is struggling to survive on his own immediately after birth, especially if he has other handicaps.

Education-for-childbirth classes which teach a prospective mother what to expect during labor and delivery are beneficial to many women. By easing tensions and fears about the unknown, this instruction can help to reduce the amount of anesthetic a woman needs, show her ways to cooperate with her physician, and teach her how to work with the powerful muscular forces of her body—all important to the safety of her baby.

An infant changes faster during these first nine months than he ever will again. He is more vulnerable to injury before birth than he ever will be after.

Following these recommendations may seem tedious and unnecessary, especially to women who have already given birth to healthy children. But each pregnancy is different. Each child has a unique make-up. No effort is too great to increase the chances that a baby will be born without handicaps.

Information about family planning, prenatal care and genetic counseling can usually be found by contacting a large medical center or university-affiliated medical school, or from The National Foundation—March of Dimes, 1275 Mamaroneck Avenue, White Plains, New York 10605.

Premature Birth: Consequences for the Parent-Infant Relationship

The normal pattern of interaction in which both infant and parent initiate and respond to mutually complementary behavior is difficult to establish when the infant is premature

Susan Goldberg

Susan Goldberg is Assistant Professor of Psychology at Brandeis University. She has been doing research in infant development since 1967, including studies of infant cognitive development and parent-infant interaction. Her research on the effects of prematurity on early social relations has been carried out over the past four years with the cooperation of the Boston Hospital for Women. Address: Psychology Department, Brandeis University, Waltham, MA 02154.

Imagine, if you will, the sound of a young infant crying. For most adults it is a disturbing and compelling sound. If it is made by your own infant or one in your care, you are likely to feel impelled to do something about it. Most likely, when the crying has reached a particular intensity and has lasted for some (usually short) period of time, someone will pick the baby up for a bit of cuddling and walking. Usually, this terminates the crying and will bring the infant to a state of visual alertness. If the baby makes eye contact with the adult while in this alert state, the caregiver is likely to begin head-nodding and talking to the baby with the exaggerated expressions and inflections that are used only for talking to babies. Babies are usually very attentive to this kind of display and will often smile and coo. Most adults find this rapt attention and smiling exceedingly attractive in young infants and will do quite ridiculous things for these seemingly small rewards.

I have used this example to illustrate that normal infant behaviors and the behaviors adults direct toward infants seem to be mutually complementary in a way that leads to repeated social interactions enjoyed by both infants and adults. Consider now the experiences of a baby whose cry is weak and fails to compel adult attention, or the baby (or adult) who is blind and cannot make the eye contacts that normally lead to social play. When the behavior of either the infant or the adult is not within the range of normal competence, the pair is likely to have difficulties establishing rewarding social interactions. Premature birth is one particular situation in which the interactive skills of both parents and infants are hampered.

Recent studies comparing interactions of preterm and full-term parent-infant pairs have found consistently different patterns of behavior in the two groups. Before we turn to these studies, it will be useful to introduce a conceptual framework for understanding parent-infant interaction and a model within which the findings can be interpreted.

A conceptual framework

In most mammalian species, the care of an adult is necessary for the survival, growth, and development of the young. One would therefore expect that such species have evolved an adaptive system of parent-infant interaction which guarantees that newborns will be capable of soliciting care from adults and that adults will respond appropriately to infant signals for care. Where immaturity is prolonged and the young require the care of parents even after they are capable of moving about and feeding without assistance, one would also expect the interactive system to be organized in a way that guarantees the occurrence of social (as opposed to caregiving) interactions that can form the basis for a prolonged parent-child relationship. It is not surprising to find that when these conditions are met, the parent-infant interaction system appears to be one of finely tuned reciprocal behaviors that are mutually complementary and that appear to be preadapted to facilitate social interaction. Furthermore, as the example given earlier illustrates, both parents and infants are initiators and responders in bouts of interaction.

This view is quite different from that taken by psychologists in most studies of child development. For most of the relatively short history of developmental psychology it was commonly assumed that the infant was a passive, helpless organism who was acted upon by parents (and others) in a process that resulted in the "socialization" of the child into mature forms of behavior. In popular psychology this emphasis appeared as the belief that parents (especially mothers) were *responsible* for their child's development. They were to take the credit for successes as well as the blame for failures.

In the last fifteen years, the study of infant development has shown that the young infant is by no means passive, inert, or helpless when we consider the environment for which he or she is adapted—that is, an environment which includes a responsive caregiver. Indeed, we have discovered that infants are far more skilled and competent than we originally thought. First, the sensory systems of human infants are well developed at birth, and their initial perceptual capacities are well matched to the kind of stimulation that adults normally present to them. Infants see and discriminate visual patterns from birth, although their visual acuity is not up to adult standards. Young infants are especially attentive to visual movement, to borders of high contrast, and to relatively complex stimuli. When face to face with infants, adults will normally present their faces at the distance where newborns are best able to focus (17–22 cm) and exaggerate their facial expressions and movements. The result is a visual display with constant movement of high contrast borders.

A similar phenomenon is observed in the auditory domain. Young infants are most sensitive to sound frequencies within the human vocal range, especially the higher pitches, and they can discriminate many initial consonants and vocal inflections. When adults talk to infants, they spontaneously raise the pitch of their voices, slow their speech, repeat frequently, and exaggerate articulation and inflection. Small wonder that young infants are fascinated by the games adults play with them!

In addition, researchers have found that when adults are engaged in this type of face-to-face play they pace their behavior according to the infant's pattern of waxing and waning attention. Thus the infant is able to "control" the display by the use of selective attention. At the same time, studies have found that babies are most likely to smile and coo first to events over which they have control. Thus, infants are highly likely to smile and gurgle during face-to-face play with adults, thus providing experiences which lead the adult to feel that he or she is "controlling" an interesting display. We will return to the notion of control and the sense of being effective as an important ingredient in parent-infant relationships.

A second respect in which infants are more skilled and competent than we might think is their ability to initiate and continue both caregiving and social interactions. Although the repertoire of the young infant is very limited, it includes behaviors such as crying, visual attention, and (after the first few weeks) smiling, which have compelling and powerful effects on adult behavior. Almost all parents will tell you that in the first weeks at home with a new baby they spent an inordinate amount of time trying to stop the baby's crying.

Crying is, at first, the most effective behavior in the infant's repertoire for getting adult attention. When social smiling and eye contact begin, they too become extremely effective in maintaining adult attention. In one study, by Moss and Robson (1968), about 80 percent of the parent-infant interactions in the early months were initiated by the infant. Thus, the normally competent infant plays a role in establishing contacts and interactions with adults that provide the conditions necessary for growth and development.

Competence motivation: a model

The actual process by which this relationship develops is not clearly understood, but we can outline a plausible model that is consistent with most of the available data. A central concept in this model is that of competence motivation, as defined by White (1959). In a now-classic review of research on learning and motivation in many species, White concluded that behaviors that are selective, directed, and persistent occur with high frequency in the absence of extrinsic rewards. He therefore proposed an intrinsic motive, which he called competence motivation, arising from a need to cope effectively with the environment, to account for behavioral phenomena such as play, exploration, and curiosity. Behavior that enables the organism to control or influence the environment gives rise to feelings of efficacy that strengthen competence motivation. White pointed out that much of the behavior of young infants appears to be motivated in this manner. Why else, for example, would infants persist in learning to walk when they are repeatedly punished by falls and bruises?

At the other extreme, Seligman (1975) has demonstrated that animals, including humans, can quickly learn to be helpless when placed in an unpleasant situation over which they have no control. This learned helplessness prevents effective behavior in subsequent situations where control is possible. It has been suggested that an important part of typical parent-infant interaction in the early

months is the prompt and appropriate responses of the parent to the infant's behavior, which enable the infant to feel effective. The retarded development often seen in institutionalized infants may arise from learned helplessness in a situation where, though apparent needs are met, this occurs on schedule rather than in response to the infant's expression of needs and signals for attention. There is a general consensus among researchers in infant development that the infant's early experiences of being effective support competence motivation, which in turn leads to the exploration, practice of skills, and "discovery" of new behaviors important for normal development.

I have suggested elsewhere (1977) that competence motivation is important to parents as well. Parents bring to their experiences with an infant some history that determines their level of competence motivation. However, their experiences with a particular infant will enhance, maintain, or depress feelings of competence in the parental role. Unlike infants, parents have some goals by which they evaluate their effectiveness. Parents monitor infant behavior, make decisions about caregiving or social interaction, and evaluate their own effectiveness in terms of the infant's subsequent behavior.

When parents are able to make decisions quickly and easily and when subsequent infant behavior is more enjoyable or less noxious than that which prompted them to act, they will consider themselves successful in that episode. When parents cannot make decisions quickly and easily and when subsequent infant behavior is more aversive or less enjoyable than that which led them to intervene, they will evaluate that episode as a failure. Figure 1 illustrates this process, and the following discussion is intended to clarify the model depicted.

The normally competent infant helps adults to be effective parents by being readable, predictable, and responsive. Readability refers to the clarity of the infant's signaling—that is, how easily the adult can observe the infant and conclude that he or she is tired, hun-

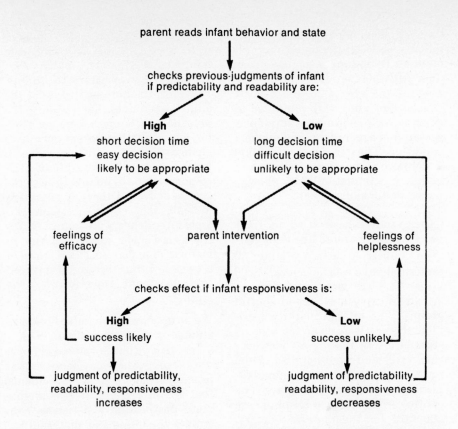

parent reads infant behavior and state

↓

checks previous judgments of infant
if predictability and readability are:

High
short decision time
easy decision
likely to be appropriate

Low
long decision time
difficult decision
unlikely to be appropriate

feelings of
efficacy

parent intervention

feelings of
helplessness

checks effect if infant responsiveness is:

High
success likely

Low
success unlikely

judgment of predictability,
readability, responsiveness
increases

judgment of predictability,
readability, responsiveness
decreases

Figure 1. An adult who has experienced a successful interaction with an infant (*left*) perceives the infant as "readable" and predictable and acquires a feeling of competence in further interactions. The good and sensitive care that results causes the infant to feel more competent in turn at eliciting the appropriate responses, and thus a cycle of successful interaction is established. The reverse of this pattern is illustrated in the right side of the figure.

gry, eager to play, etc. Although there may be some infants who are easier for everyone to read than others, readability within the parent-infant pair is a joint function of infant behavior and the adult's skill in recognizing behavior patterns.

Predictability refers to the regularity of the infant's behavior—whether sleeping, waking, feeding, and elimination follow a recognizable pattern and whether the infant repeatedly responds to similar situations in a similar fashion. Again, both infant behavior and adult behavior and sensitivity to the infant determine predictability within a given pair. Responsiveness is the infant's ability to react to external stimulation, whether animate or inanimate. To the extent that an infant responds promptly and reliably to adult behavior he or she contributes directly to the adult's feelings of effectiveness as a caregiver.

The left side of Figure 1 shows that when an infant is readable and pre-dictable the adult is able to make caregiving decisions quickly and easily and is highly likely to make decisions that lead to successful or desirable outcomes. When an adult has interacted with an infant in ways that have led to an evaluation of success, the adult is likely to perceive the infant as more readable and predictable than before. Thus, the infant who is readable, predictable, and responsive can capture an initially disinterested adult in cycles of mutually rewarding and effective interaction. Notice also that, in this part of the figure, the adult is able to respond promptly and appropriately to the infant's behavior, providing the infant with what we would describe as good or sensitive care that enhances the infant's feelings of competence. In addition, since these successes make the adult feel efficacious, he or she now has more confidence and is better able to make judgments about infant behavior and caregiving in the future. The right side of the figure illustrates the situation in which the infant is

unreadable, unpredictable, and un-responsive as a joint function of poorly organized infant behavior and/or poorly developed adult skills.

Problems of preterm pairs

Under normal conditions, the natural reciprocity of adult and infant behavior guarantees that each member of the pair is provided with frequent opportunities to feel effective. A review of what is known about preterm infants and their parents will indicate that such pairs have a greater probability of falling into the patterns on the right side of the figure than do their full-term counterparts. Most preterm pairs eventually do develop successful relationships. However, the available data also indicate that they must make compensatory adjustments to enable them to overcome initial disadvantages.

Premature infants are those who are born after fewer than 37 weeks of gestation and weigh under 2,500 g. Infants who were born small for their age or with known congenital defects are not included in the samples of the studies I am describing. The most obvious fact of premature birth is that parents are confronted with an infant who is relatively immature and may not have developed the care-soliciting or social behaviors available to the full-term infant.

Several studies, including my own (Goldberg et al., in press), which have systematically evaluated the behavior of preterm infants (close to term or hospital discharge), have reported that they spent less time alert, were more difficult to keep in alert states, and were less responsive to sights and sounds (a ball, a rattle, a moving face, and a voice) than the full-term comparison group. Furthermore, preterm infants who had experienced respiratory problems following birth rarely cried during the newborn examination, even though some of the procedures (e.g. undressing, testing reflexes) are mildly stressful and provoke crying from full-term infants. This suggests that these preterm infants are not likely to give adults clear distress signals.

The effectiveness of the preterm infant's cry in compelling adult attention has not been studied extensively. However, at the University of Wisconsin (Frodi et al. 1978), mothers and fathers were shown videotapes of full-term and preterm infants in crying and quiescent states. A dubbed sound track made it possible to pair the sound and the picture independently. Physiological recordings taken from the viewing parents indicated that the cry of the premature infant was physiologically more arousing than that of the full-term infant, and particularly so when paired with the picture of the preterm baby. Furthermore, ratings of the cries indicated that parents found that of the premature baby more aversive than that of the full-term infant. Thus, although the preterm infant may cry less often, this can be somewhat compensated for by the more urgent and aversive sound of these cries. If a parent is able to quiet these cries promptly, they clearly serve an adaptive function. If, however, the infant is difficult to pacify or frequently irritable, it is possible that the aversive experience will exceed the parent's level of tolerance or that he or she will experience repeated feelings of helplessness that can be damaging to the interactive relationship.

Thus far, we have assumed that the less competent behavior of the preterm infant is primarily attributable to immaturity. Often prematurity is associated with other medical problems that depress behavioral competence. In addition, the early extrauterine experiences of preterm infants in intensive-care nurseries probably do little to foster interactive competence and may, in fact, hinder its occurrence. Procedures such as tube feedings, repeated drawing of blood samples, temperature taking, and instrument monitoring often constitute a large proportion of the preterm infant's first encounters with adults. There are few data on the effects of specific medical procedures, and since these procedures cannot ethically be withheld on a random schedule, this is a difficult area to study. However, numerous studies have attempted to foster early growth and development of preterm infants by adding specific kinds of experiences.

An example of a study from the first category is one in which 31 preterm infants were gently rocked for 30 minutes, three times each day, from their fifth postnatal day until they reached the age of 36 weeks postconception. They were compared to 31 unrocked preterm babies of similar gestational age, weight, and medical condition. The experimental infants were more responsive to visual and auditory stimulation and showed better motor skills as well.

Other studies have tried to treat preterm infants more like their full-term counterparts. In one study 30 preterm infants weighing 1,300–1,800 grams at birth were randomly assigned to experimental and control groups. The infants in the experimental group were given extra visual stimulation by placing mobiles over their cribs and were handled in the same manner as full-term infants for feeding, changing, and play. The control group received hospital care standard for the preterm nursery, which meant that handling was kept to a minimum. Although initial weights and behavioral assessments had favored the control group, at 4 weeks postnatal age, the experimental group had gained more weight and showed better performance on the behavioral assessment than the controls.

Like these two examples, all of the other studies which provided extra stimulation to preterm infants showed gains in growth and/or development for the babies in the experimental group beyond those of the control group. Thus, although we do not know whether routines of intensive care interfere with early development, we do know that the behavioral competence of preterm infants can be enhanced by appropriate supplemental experiences.

On the parents' side, premature birth means that parenthood is unexpectedly thrust upon individuals who may not yet be fully prepared for it. Beyond the facts of not having finished childbirth preparation classes or having bought the baby's crib, it may be that more fundamental biological and psychological processes are disrupted. A beautiful series of studies by Rosenblatt (1969) has explored the development of maternal behavior in rats. As in humans, both male and female adult rats are capable of responding appropriately to infants. However, the hormonal state of the adult determines how readily the presence of infants elicits such behaviors. Furthermore, hormonal changes during pregnancy serve to bring female rats to a state of peak responsiveness to infants close to the time of delivery. Other animal studies indicate that experiences immediately after delivery are important for the initiation of maternal behavior. In many mammalian species removal of the young during this period may lead to subsequent rejection by the mother.

We do not have comparable hormonal studies of human mothers, but it seems likely that hormonal changes during pregnancy may serve similar functions. There is some evidence that among full-term births, immediate postpartum experiences contribute to subsequent maternal behavior. A series of studies by Klaus and Kennell (1976) and their colleagues provided some mothers with extra contact with their infants soon after birth. In comparison with control groups, these mothers were observed to stay closer to their babies, to touch and cuddle them more, and to express more reluctance to letting others care for their babies after leaving the hospital. Klaus and Kennell have summarized these studies and interpreted them as indicating that there is an optimal or "sensitive" period for initiating maternal behavior in humans. As further evidence they cite studies in which preterm infants are found to be overrepresented among reported cases of child abuse, neglect, and failure to thrive. These disturbing statistics, they suggest, reflect the effects of parent-infant separation during the sensitive period.

Even if one does not accept the idea of the sensitive period as described by Klaus and Kennell (and many developmental psychologists do not), it is clear that parents whose preterm infants must undergo prolonged hospitalization have few opportunities to interact with them. Even in the many hospitals that encourage parents to visit, handle, and care for their babies in intensive care, the experiences of parents with preterm infants are in no way comparable to those of parents with full-term infants.

If you have ever visited a friend under intensive care in the hospital, you will have some idea of the circumstances under which these parents must become acquainted with their infants. Neither parents nor infants in this situation have much opportunity to practice or develop interactive skills or to experience the feelings of competence that normally accompany them. Parents also have little opportunity to learn to read, predict, or recognize salient infant behaviors. In a study conducted at Stanford University, Seashore and her colleagues (1973) asked mothers to choose themselves or one of five other caregivers (e.g. nurse, grandmother) as best able to meet their infants' needs in numerous caregiving and social situations. Mothers who had not been able to handle their first-born preterm infants chose themselves less often than mothers in any other group sampled.

Thus, both infants and parents in preterm pairs are likely to be less skilled social partners than their full-term counterparts, because the development of interactive capacities has been disrupted and because they have had only limited opportunities to get acquainted and to practice. In addition, during the hospital stay, parents of preterm infants already have little self-confidence and lack the feeling of competence. Ordinarily, an interactive pair in which one member has limited competence can continue to function effectively if the partner is able to compensate for the inadequate or missing skills. In the case of parent-infant pairs, because the infant's repertoire and flexibility are limited, the burden of such compensatory adjustments necessarily falls upon the parent.

Observations of interactions

Six studies to date have compared parent-infant interaction in full-term and preterm pairs. They were carried out in different parts of the country with different populations and different research methodologies. Yet there seems to be some consistency in findings that is related to the age of the infant at the time of observation (which also reflects the duration of the parent-infant relationship). Each study involved repeated observation of the same parent-infant pairs, though the number of observations and the length of the studies vary.

Those studies which observed parents and infants in the newborn period typically report that parents of preterm infants are less actively involved with their babies than parents of full-term infants. Relative to full-term infants, preterm infants were held farther from the parent's body (Fig. 2), touched less, talked to less, and placed in the face-to-face position less often. Subsequent observations of the same pairs usually indicated that the differences between preterm and full-term pairs diminished with time, as parents in the preterm group became more active. Thus, it appears that the initiation of interaction patterns considered "normal" for full-term pairs is delayed in preterm pairs. In my own study (Di Vitto and Goldberg, in press) I found that for one kind of parental behavior—cuddling the baby—the preterm infants never received as much attention as the full-term infants in spite of increases over time. Over the first four months, parents cuddled preterm infants more at later feeding observations, but they were never cuddled as much as the full-term infants at the very first (hospital) observation. Thus, the development of some kinds of interactions in the preterm group can be both delayed and depressed.

In contrast with these observations of very young infants, studies of older infants reported a very different pattern. Regardless of the observation situation (feeding, social play, or object play), preterm infants were less actively engaged in the task than were full-term infants, and their parents were more active than those of full-term infants. In one study of this type, Field (1977) placed each mother face to face with her baby, who sat in an infant seat, and asked her to "talk to your baby the way you would at home." Infant attention and parent activity were coded. Infants in the preterm group squirmed, fussed, and turned away from their mothers more than those in the full-term group, and preterm mothers were more active than full-term mothers. Instructions that decreased maternal activity ("Try to imitate everything your baby does") increased infant attention in both groups, while those that increased maternal activity ("Try to get your infant to look at you as much as possible") decreased infant attention in both groups.

Field's interpretation of these findings assumed that infants used gaze aversion to maintain their exposure to stimulation within a range that would not overtax their capacities for processing information. Thus, when mothers' activity decreased, infants were able to process the information provided without the need to reduce stimulation. Field also suggested that since the *imitation* condition provided stimulation that was matched to the infant's behavior, it might be more familiar and thus easier for the infant to process. It is possible that the greater initial fussing and gaze aversion reflected information-processing skills that were less developed than those of full-term infants.

Brown and Bakeman (in press) observed feedings in the hospital, one month after discharge, and three months after discharge. Their findings are somewhat different from the overall trend because they were similar at all observations. Behavior segments were assigned to four categories: *mother acts alone, baby acts alone, both act,* and *neither acts.* In comparing preterm and full-term pairs, they reported that preterm infants acted alone less frequently than full-term infants, while mothers of preterm infants acted alone more often than those of full-term infants. Furthermore, in preterm pairs, the *neither acts* state was most likely to be followed by *mother acts alone,* while in the full-term pairs, it was equally likely to be followed by activity by the baby or the mother.

In my own research (DiVitto and Goldberg, in press; Brachfeld and Goldberg 1978) there are two sets of data consistent with these findings. First, we found that parent behavior during feedings in the hospital, and at the 4-month home and laboratory visits, was related to infant behavior in the newborn period. Regardless of their condition at birth, infants who had been difficult to rouse as newborns received a high level of functional stimulation from parents (e.g. position changes, jiggling the nipple). Infants who had been unresponsive to auditory stimulation as newborns received high levels of vocal and tactile stimulation during feedings. Thus, the parents of infants who were unresponsive as newborns appeared to work harder during feedings in the first four months than did the parents whose newborns were more responsive.

We also observed the same pairs at 8 months of age in a free-play situation. Four toys were provided and parents were asked to "do what you usually do when [name] is playing." In this situation, both at home and in the laboratory, preterm infants (particularly those who had been very young and small at birth and had respiratory problems) played with toys less and fussed more than the full-term group. Parents in the preterm group stayed closer to their babies, touched them more, demonstrated and offered toys more, and smiled less than those in the full-term group.

Another study with somewhat younger infants also fits this pattern. Beckwith and Cohen (1978) observed one wake–sleep cycle at home one month after discharge. Since babies were born and discharged at different ages, the age of the infants varied: some were relatively young, while others were closer in age to the older groups in other studies. All were born prematurely. However, Beckwith and Cohen found that mothers whose babies had experienced many early complications devoted more time to caregiving than those who had babies with fewer problems.

All these studies concur in indicating that parents with preterm infants or preterm infants with more serious early problems devote more effort to interacting with their babies than do their full-term counterparts. In most

Figure 2. Parents of preterm infants seem, in general, to be less actively involved with their babies in the early postnatal period than parents of full-term infants. For example, this tendency is sometimes revealed in the distances at which mothers hold their preterm babies (*left*) relative to mothers of full-term babies (*right*), as shown in these posed pictures. (Photographs by B. Di Vitto.)

of the studies this was coupled with a less responsive or less active baby in the preterm pairs. Thus, it appears that parents adapt to the less responsive preterm baby by investing more effort in the interaction themselves. As Brown and Bakeman put it, the mother of the preterm infant "carries more of the interactive burden" than her full-term counterpart. From our own laboratory, there is some evidence that other adults have a similar experience with preterm infants. At our regular developmental assessments at 4, 8, and 12 months, staff members rated the preterm group as being less attentive to the tasks, less persistent in solving them, and less interested in manipulating objects than the full-term group. In addition, staff members found it necessary to spend more time with the preterm group to complete the required tasks.

The consistency of these findings suggests that in pairs with a preterm infant, adults use a common strategy of investing extra time and effort to compensate for their less responsive social partner. It is important to note that while this seems to be a widely adopted strategy, it is not necessarily the most successful one. In Field's study (1977) a decrease in maternal activity evoked infant attention more effectively than an increase. In our own observations of 8-month-old infants, increased parent involvement did not reduce the unhappiness of the sick preterm group, and some play sessions had to be terminated to alleviate infant distress. Hence, while there seem to be some consistent strategies by which parents compensate for the limited skills of their preterm infants, these pairs may continue to experience interactive stress in spite of or even because of these efforts. Continuation of such unrewarding interactions, as Figure 1 indicates, is a threat to continued effective functioning of the interactive system.

The data reviewed above provide little evidence on the duration of interactive differences between full-term and preterm pairs. Among researchers in the field, there seems to be an informal consensus which holds that such differences gradually disappear, probably by the end of the first year. In my own research, a repetition of the play sessions at 12 months revealed no group differences. At 11–15 months, Leiderman and Seashore (1975) report only one difference: mothers of preterm babies smile less frequently than those of full-term babies. However, in Brown and Bakeman's study, group differences were observed as late as preschool age in rated competence in social interactions with adults (teachers) and peers. These data are too meager and scattered to support a firm conclusion on the duration of group differences.

This review has focused only upon the ways in which premature birth may stress the parent-infant interaction system. Preterm infants are generally considered to be at higher risk for subsequent developmental and medical problems than their full-term counterparts. In order to understand the reasons for less than optimal developmental outcomes, it is important to bear in mind that premature births occur with high frequency among population subgroups where family stress is already high (e.g. young, single, black, lower-class mothers). Most of the research designs which would allow us to disentangle the independent contributions of each medical and social variable to long-term development are unethical, impractical, or impossible to carry out with human subjects.

The early approach to studying the consequences of prematurity was to consider each of these medical and social variables as "causes" and the physical and intellectual development of the child as the "effect." The data reviewed here indicate that we cannot think in such simple terms. Prematurity (or any other event which stresses the infant) stresses the parent-infant interaction system and indeed the entire family. The way in which the family is able to cope with these stresses then has important consequences for the child's development. A major finding of the UCLA study was that for preterm infants, as for full-term infants, a harmoniously functioning parent-infant relationship has beneficial effects on development in other areas, such as language, cognition, motor skills, and general health. Prematurity, like many other developmental phenomena, can best be understood as a complex biosocial event with multiple consequences for the child and the family.

Furthermore, in the absence of sophisticated medical technology, the vast majority of the births we have been discussing would not have produced live offspring. In evolutionary history, though it would have been adaptive for infants' initial social skills to be functional some time before birth was imminent, there was no reason for these preadapted social skills to be functioning at 6 or 7 months gestation. Premature births include only a small proportion of the population, but our ability to make such infants viable at younger and younger gestational ages by means of artificial support systems may be creating new pressures for differential selection. The fact that the majority of preterm pairs do make relatively successful adaptations indicates that the capacity to compensate for early interactive stress is one of the features of the parent-infant interaction system.

References

Beckwith, L., and S. E. Cohen. 1978. Preterm birth: Hazardous obstetrical and postnatal events as related to caregiver-infant behavior. *Infant Behav. and Devel.* 1.

Brachfeld, S., and S. Goldberg. Parent-infant interaction: Effects of newborn medical status on free play at 8 and 12 months. Presented at Southeastern Conference on Human Development, Atlanta, GA, April 1978.

Brown, J. V., and R. Bakeman. In press. Relationships of human mothers with their infants during the first year of life. In *Maternal Influences and Early Behavior*, ed. R. W. Bell and W. P. Smotherman. Spectrum.

DiVitto, B., and S. Goldberg. In press. The development of early parent-infant interaction as a function of newborn medical status. In *Infants Born at Risk*, ed. T. Field, A. Sostek, S. Goldberg, and H. H. Shuman. Spectrum.

Field, T. M. 1977. Effects of early separation, interactive deficits, and experimental manipulations on mother-infant interaction. *Child Development* 48:763–71.

Frodi, A., M. Lamb, L. Leavitt, W. L. Donovan, C. Wolff, and C. Neff. 1978. Fathers' and mothers' responses to the faces and cries of normal and premature infants. *Devel. Psych.* 14.

Goldberg, S. 1977. Social competence in infancy: A model of parent-infant interaction. *Merrill-Palmer Quarterly* 23:163–77.

Goldberg, S., S. Brachfeld, and B. DiVitto. In press. Feeding, fussing and play: Parent-infant interaction in the first year as a function of newborn medical status. In *Interactions of High Risk Infants and Children*, ed. T. Field, S. Goldberg, D. Stern and A. Sostek, Academic Press.

Kennell, J. H., and M. H. Klaus. 1976. Caring for parents of a premature or sick infant. In *Maternal-Infant Bonding*, ed. M. H. Klaus and J. H. Kennell. Mosby.

Klaus, M. H., and J. H. Kennell. 1976. *Maternal-Infant Bonding.* Mosby.

Leiderman, P. H., and M. J. Seashore. 1975. Mother-infant separation: Some delayed consequences. In *Parent-Infant Interaction.* CIBA Foundation Symp. 33. Elsevier.

Moss, H. A., and K. S. Robson. The role of protest behavior in the development of parent-infant attachment. Symposium on attachment behavior in humans and animals. Am. Psych. Assoc. Sept. 1968.

Rosenblatt, J. S. 1969. The development of maternal responsiveness in the rat. *Am. J. Orthopsychiatry* 39:36–56.

Seashore, M. J., A. D. Leifer, C. R. Barnett, and P. H. Leiderman. 1973. The effects of denial of early mother-infant interaction on maternal self-confidence. *J. Pers. and Soc. Psych.* 26:369–78.

Seligman, M. R. 1975. *Helplessness: On Development, Depression and Death.* W. H. Freeman.

White, R. 1959. Motivation reconsidered: The concept of competence. *Psych. Review* 66: 297–333.

Development During Infancy

3

No age period in human development has received more attention during the past 20 years than that of infancy. To be sure, we are more certain than ever before that the first several years of life are of fundamental importance for subsequent human development. However, it is equally clear that events of infancy do not predetermine developmental outcome. For example, providing optimal care for infants lays the foundation for successful negotiation of the preschool and childhood years, but optimal care of the infant does not make the preschool or school-aged child any less vulnerable to inadequate caregiving. Child abuse first experienced during the school-age years can be as devastating to subsequent development as child abuse first experienced during infancy.

Despite massive research efforts, or perhaps because of them, many problems of early development remain to be solved. What are the long term effects of infant day care? What are the effects of fostercare placement on the infant's social and emotional development? Much developmental research is directed toward determining the kinds of experiences children should have, at what times these experiences should be provided, and by whom they should be provided. To what extent do critical periods set the course of development during infancy? Critical periods refer to times when the rate of organization occurs most rapidly. Presumably, it is during these times that the individual is most susceptible to environmental stress. For example, the critical period for structural differentiation of the organism is the period of the embryo, and it is during this period that the fetus is most susceptible to environmental insults. Similarly, it has been suggested that the critical period for the organization of social attachments is from the third to the twelfth postnatal month. Studies of the effects of maternal deprivation and poor institutional care suggest that failure to establish attachments during the first year of life are detrimental to the infant's subsequent social-emotional development. The events of a critical period are not irreversible as they once were thought to be. However, the more one moves beyond a critical period the more difficult it is to develop the phenomenon unique to that period.

Part of the explosion in infant research involves the study of caregivers and infants in cultures other than Western ones. Cross-cultural research has broadened our concept of individual differences and has helped to clarify the relative influences of culture and species on behavioral organization. Not only are there cultural differences in rearing patterns, but there are differences in infant behavior as well. And, irrespective of culture, each infant's unique characteristics will play an important role in structuring the quality of interaction with his or her caregiver(s) and caregiving environment.

Looking Ahead: Challenge Questions

Are there reasons why premature infants especially ought to be breast fed? What affect might it have on maternal bonding?

What sort of evidence would support the notion that infants are competent, active cognizers of the external world?

Why has it been so difficult to demonstrate behavioral continuity from the period of infancy to subsequent periods of development?

If the two-year-old shows empathy and concern for others, can this be taken as evidence to support the notion that the newborn enters the world intrinsically good, and then is corrupted by society?

Biology is one key to the bonding of mothers and babies

Hara Estroff Marano

Hara Estroff Marano is a staff editor for Medical
World News *and the mother of two young children.*

Pink and plump, the infant shrieks his discomfort to
the world he entered less than 48 hours before. Pediatrician T. Berry Brazelton materializes at one side of
the wailing, flailing infant, and lowers his head to
its level. "Hi there," he croons. "Can you turn over
here if I talk to you?" The baby turns his head to the
voice, seems to search for its source, and stops crying.

With this simple demonstration (although this
youngster may be exceptionally responsive), Dr.
Brazelton contradicts the old idea that babies can
neither see nor hear at birth. He also dramatizes a
major discovery about human development. It is a
discovery only in the sense that science is beginning
to recognize what sensitive mothers have suspected all
along—that newborns come highly equipped for their
first intense meetings with their parents, and in particular their mothers.

At birth, Dr. Brazelton and others have found,
not only do many babies orient themselves to sound,
they prefer the sound of the human voice, especially
the high pitch of the female voice. They choose to look
at faces over other objects and will follow with their
eyes the turning of a nearby human face. They may
imitate facial gestures such as sticking out a tongue,
and they synchronize their body movements to the
rhythm of the human voice.

Unable to cling to the fur of a highly mobile mother
(as do monkeys), the human infant depends for all of
its early needs on the strength of its mother's emotional attachment. How mother and infant become
emotionally attached—some call it bonded—to each
other is a subject of serious investigation by growing
numbers of psychologists, psychiatrists and pediatricians in the United States. The current wave of interest began building about 15 years ago, as technological gains dramatically improved premature babies'
chances of survival.

Drs. Marshall Klaus and John Kennell at Rainbow
Babies' and Children's Hospital in Cleveland, and
others who run the intensive-care nurseries that sprang
up to treat premature infants, made a startling observation. They found that in spite of their near-heroic
measures to save tiny premature infants, a dispropor-
tionate number of them wound up back in hospital
emergency rooms, battered and abused by their parents, even though they had been sent home intact and
thriving. One of the contributing factors seemed to
them to be the prolonged separation of mothers and
infants which was then routine following premature
births. Such separations can sometimes interfere with
the bond that usually forms under more natural circumstances, Drs. Klaus and Kennell believe.

Getting a handle on this bond was not easy, Dr.
Klaus now admits. He first had to overcome the limitations of his training as a pediatrician. "It took me
some time to turn my eyes from the baby to the
mother; only then did I fully realize that the mother
is the instrument through which the baby can thrive."

Nor is the problem limited to premature babies.
Reports of child abuse and neglect are increasingly
common. According to the National Center for Child
Abuse, three of every hundred infants need medical
attention for injuries suffered at the hands of parents.
Then there are the subtler forms of abuse—neglect and
growth failure without organic cause—that do not always find their way to hospitals or headlines. In addition, there is the growing sense that the ability to cope
and handle stress is acquired through the parent-
infant relationship.

Among those who are probing mother-infant bonds,
some researchers adopt for their models monkeys and
other animals, both in the wild and in the laboratory;
but while mother-infant attachment occurs in many
animals as well as humans, bonding in humans occurs
in a very specific and complex way. Other investigators comb through the family histories of children for
whom such bonds failed to develop. Still others chron-

icle the actions of normal mothers and children going about their daily routines at home, or observe them in special laboratories set up to film them during brief separations and reunions.

But all are aimed at gathering information to help improve the relations between parents and children in the United States. Taken together, their work constitutes the first major attempt to analyze scientifically what is often seen only through rose-colored glasses, if at all—how human relationships develop.

Dr. Brazelton is one of the numerous investigators parsing the relationship between mothers and infants and between fathers and infants as well. His observations at Harvard and at Boston Children's Hospital lead him to believe that a mother's emotional attachment to her infant does not descend on her by some obstetrical deus ex machina in the delivery room. It develops gradually, beginning in the feelings and dreams many women experience about themselves and their unborn babies during pregnancy.

Anxiety and disruption of old concepts through dreams, Dr. Brazelton thinks, may become part of a normal process, an "unwiring" of all the old connections to be ready for the new role. Prenatal anxiety whips up the energy a woman needs for greeting the individuality of the new baby. Evidence is amassing that as she does so, she is responding to the hormonal signals guiding her pregnancy to a close.

Evolutionary biologists, who take a long-range view of human behavior, see parenting as much too important to be left to whimsy. They hold that behavior that is so critical to the survival of both individuals and the species must have some biologic base, must somehow be wired into the nerve circuits of the brain. This notion strikes a strong chord with one of the emerging themes of modern neuroscience—that behavior has more innate components than most of us have been led to believe.

Most of the evidence supporting the idea of a "biologic base" comes from experiments with rats and sheep. As science proceeds by the careful building of inferences, scientists look to such animals for clues to what might be going on in humans. Not that we, like the experimental rat, are held in hormonal thrall. But the rigorous experiments that are possible with animals can lead to new insights, confirmed by close observation of human behavior. In looking at the earliest stages of maternal attachment in rats, Drs. Jay Rosenblatt and Harold Siegel of Rutgers University have discovered two distinct phases, a hormonally triggered burst of interest followed by a more psychologically based need for continued contact.

Hormones may trigger mothering in animals

About a day before mother rats give birth (ten days before birth in sheep and three weeks before birth in humans) their hormonal status changes dramatically. The blood level of progesterone, the hormone that sustains pregnancy by keeping the laden uterus quiescent, declines sharply, and the level of estrogen rises. The main point of his work, Dr. Rosenblatt says, is to show that "the onset of maternal behavior before delivery is based on this rise in estrogen." An animal facing a litter of pups, as young rats are called, before the change, stays as far away from them as possible. So do virgin rats. If, however, the virgins are caged with pups, left with them day in and out, then they become maternal toward them after four to seven days, and the smaller the cage, the faster it happens. That can be contrived in a laboratory, where it is possible to remove the litter each day and give them needed care while substituting a fresh batch of age-mates. But in real life, such animals would be dead within 48 hours. Yet when a virgin rat is given a transfusion of blood from a new mother, she becomes maternal within ten to 14 hours. She will preside over the litter as if it were her own. Although she cannot lactate, she assumes the nursing position, crouching over her pups. She will do all in her power to keep the pups close, retrieve them when they stray. Blood taken from animals before the hormonal shift, or more than 24 hours after birth, has no effect on her.

"We know," says Dr. Rosenblatt, "that without the hormones, the rats' initial aversion to pups is a fear of them. It appears to be part of a basic fearfulness of anything new. And little pups are certainly something new and novel for them."

The hormones, he says, reduce the fearfulness. "They tip the balance of responding to the young, making the mother ignore or overcome the unattractive features." In essence, he says, the hormones change the animals' motivation. "Of all the stimuli a mother faces, she selects certain ones to attend to. Of all the things she can do, she does only certain things. If, before a rat gives birth, we offer her a baby and a piece of food, she chooses the food. Right after birth, she chooses the baby. That is a motivational shift."

Because all rats—virgins and even males, as well as new mothers—become maternal if given enough time with pups, Dr. Rosenblatt concludes that all rats share the neural system that underlies maternal behavior. He thinks the same is probably true of primates. Among the colony of rhesus monkeys at the Rutgers Institute of Animal Behavior, which he heads, it is not uncommon for a never-pregnant animal to adopt a baby. "Like the rat, rhesus monkeys have some inherent capacity to be maternal. But that doesn't mean there's no physiologic basis for it." Unfortunately, he notes, no one is testing hormones in monkeys.

Closeness keeps mothers maternal

Rosenblatt's rat studies show that hormones trigger

a maternal response, but also that hormones don't keep the animals responsive. Only the pups themselves do that. A certain amount of close contact with the pups shortly after birth is needed for the mother to maintain the responsiveness that the hormones activated. If a rat gives birth but isn't allowed any contact with her litter, but is given her pups back later, she won't be maternal toward them. It will take her about a week to overcome her resistance to them, as if she had never been pregnant. But if a new mother is allowed several hours of contact with her new litter before they are removed, she will still be highly maternal if they are given back days or even weeks later.

Do Rosenblatt's studies throw any light on observations of human mothers and infants? Shortly after Klaus and Kennell suspected that mother-infant separation might be partly responsible for the unusually high rate of battering among premature babies, they opened the doors of the premature nursery to mothers, although in taking such a radical step they had no proof that what they were doing was safe or even helpful. Indeed, the medical canons of the era held that the mothers could contaminate their babies. (Dr. Klaus, with Drs. Cliff Barnet, Herbert Leiderman and Rose Grobstein of Stanford University, demonstrated that the infection rate didn't change in preemie nurseries after mothers were let in.)

Invariably, the mothers approached their infants with a substantial amount of residual fear, as if precipitous birth had cut short their hormonal and psychological ripening. At first, they stalked the incubators, circling them for one to two visits; eventually, they would make an approach, poking their babies with their fingertips. Finally, they stroked their babies' arms and legs, moving, after four to eight visits, from the extremities to the babies' trunk—something mothers of full-term infants do exuberantly within minutes of greeting their babies, provided they are afforded the privacy and emotional support for a "rendezvous."

This period of rendezvous after normal births, which Rosenblatt's studies suggest is important, has been commonly overlooked in American obstetrical practice. Seen with the cool eye of clinical observation, what develops between mothers and babies who are permitted close contact after birth is a sort of "mating dance," an elaborate courtship in which, ideally, they link their behavior through every available channel of communication into a smooth pattern of signal and response.

A high state of alertness in both mother and infant is helpful in getting things rolling. Not only does the infant come headlong into the world unfurling such a repertoire of social skills that those who notice them are apt to refer to "the amazing newborn," but babies born to unmedicated mothers have an incredible capacity for alertness in the first hours after birth. Newborns, says Dr. Klaus, spend nearly 40 minutes of their first hour on Earth in a state of rapt attention. During this time, their eyes are shiny-bright, wide open, and capable of focusing and fixating on objects. During this state of high alertness newborns will follow with their eyes and head a slowly moving object, especially a face. This prolonged alertness, says the Cleveland pediatrician, "is especially suited for meeting parents. It often fosters parental feeling and a sense of ecstasy in parents."

In humans, it's love at first sight

It is the visual attentiveness that· seems to matter most to parents. When mothers talk to their babies just after birth, 80 percent of what they say is related to eyes. "Please open your eyes. If you open your eyes I'll know you're alive," is a phrase Dr. Klaus has heard many times. Most striking, when mothers are put together with their babies, they shift around to arrange themselves *en face* with their infants—that is, they align themselves so that their eyes and the babies' eyes are in the same plane of rotation. In doing so, mothers seem to be searching for a signal. "A mother can't easily become bonded to her infant unless the baby responds to her in some manner," explains Dr. Klaus, who restates his observation as possibly one of the basic principles of bonding: "You can't love a dishrag."

Giving flesh to Yeats' declaration that "love comes in at the eye," mother and infant often move through a behavioral sequence which, Drs. Klaus and Kennell find, is not very different from what happens when a man and woman fall in love: mutual gazing, touching, fondling, nuzzling, kissing. Together, they fall into step with each other, their synchrony epitomized sometimes by a baby moving its head in rhythm to its mother's words.

Nothing better illustrates the interdependence of needs than the act of a mother breast-feeding her baby. There are both emotional and physical benefits, for mother and infant. Breast-fed infants, for example, have measurably fewer infections and develop fewer allergies than bottle-fed babies. "The most powerful way to forge a strong bond between mother and infant is through breast-feeding," says Dr. Kennell. "It continues the process begun early after birth. As a social exchange, it is rewarding for mother and baby over and over." His observations have convinced him that the more mothers and babies are together shortly after birth, the stronger is the attachment between them and the more natural mothering becomes. (At the same time, Dr. Klaus points out that it does not always happen instantly. He cites studies showing that about 25 percent of mothers "fall in love" with their babies before birth, from 25 to 30 percent in the first hours. For about 40 percent, it takes a week or even longer.) So powerful is the early attachment that its effects on both mothers and babies are discernible years later.

Drs. Klaus and Kennell have demonstrated this experimentally, in their own pediatric bailiwick. "The hospitals served as a natural laboratory," says Klaus. "In America, where hospital separation is routine, most of us look separately at mothers and babies. Being together is biologic and natural. Separation is abnormal and frightening." They studied a group of mothers and babies who got routine care. For comparison, they followed another group of mothers and babies who were put in a warm room where they could lie together unclothed for an hour after birth and who were again together five more hours on each of the first three days after delivery.

Starting a month after birth and throughout the next five years—the study is still under way—the Cleveland pediatricians found important differences between the two groups of mothers and babies. A month after birth, when they came back for a special office visit, the early-contact mothers stood closer to their infants, picked up their crying babies more. During feedings (which were filmed), they more often stayed in the *en face* position and fondled their babies. Interviews revealed they were more reluctant to leave their infants with someone else. And they reported that when they did go out, they found themselves constantly thinking about the baby.

A year later, they still were more attentive to their babies during an uncomfortable office visit. They helped the pediatrician approach the reluctant babies. After two years, the early-contact mothers talked differently to their children from the control mothers, although there were no differences among the two groups of mothers that could account for such an observation—all were from the same socioeconomic class and scored about the same on I.Q. tests. Yet the early-contact mothers used richer language constructions and more words, especially descriptive adjectives. They issued fewer commands to their children but asked more questions. What is more, the early-contact mothers continued to speak to their children when other adults came into the room, while the control mothers talked more to the adult interviewer.

"The early-contact mothers presented a model of better learning," says Dr. Kennell. Their speech had more variety, more opportunity for response. "They gave more stimulation to the child's thinking. They seemed to be more aware of the needs of growing children. They could interpret the environment better to them.

"To think all that is the result of just 16 hours of being together in the first three days of life," sighs Dr. Kennell. "The hours after birth seem to comprise a sensitive period for maternal-infant attachment."

Early contact may reduce child abuse

At Nashville General Hospital, Dr. Susan O'Connor has found that, by putting mothers together with their babies during the hospital stay after birth, she can help a significant proportion of women to overcome some of the personal and cultural pulls that could lead them to neglect their infants. Dr. O'Connor, assistant professor of pediatrics at Vanderbilt University, feels that "this is an area of research that shows real promise." She cautions, however, that we must not "look upon it as the answer to all our problems." Indeed, none of the researchers feels early bonding solves all the problems of parenting. Nor do they believe that the lack of very early contact must preclude the forming of an affectionate bond. Dr. Klaus points out that adoptive mothers often do "a great job."

If the early falling in love is somewhat magical, coaxed out of a built-in bag of tricks, staying in love over the next months and years takes work. Even so, Harvard pediatrician Brazelton and Cornell University psychiatrist Daniel Stern see it, quite literally, as a form of child's play. They have recorded such play—in separate but similar studies—by videotaping several mothers and babies in face-to-face encounters, at home and in special laboratories, and later analyzing and reanalyzing the film. Both doctors find that mother-infant play is built on tiny advances and retreats, thrusts and parries, that go by too fast for most people to see what is happening but which sensitive mothers nevertheless apprehend. It has taken Dr. Brazelton as long as 28 hours to catch all there is to see in three minutes of film!

In the earliest stages, mother leads the game, reaching over to draw her baby out. Gradually, the baby essays a few smiles, throws out a few vocalizations, and starts up the game himself. He reaches out to get her, she brightens to his overtures, feeding him talk and facial expressions—the elements of communication—until, overloaded with stimulation, he averts his head to recover himself, before beginning again.

Throughout, the mother serves as a protective "envelope" for her baby, screening out the background noise of the world, letting through only as much stimulation and information as she learns to expect he can handle, enlarging the envelope as her baby signals his readiness for more before he turns away. (Fathers, Dr. Brazelton finds, do something different with their babies. They excite them more, even when they assume a primary care-taking role. Where mothers soothe, fathers tend to whisk their babies off onto gleeful excursions—and babies seem to expect it of them.)

As the mother forms a protective envelope for her baby, so does baby form one for her. In ways that are physiologically measurable, the attachment of mother and infant buffers them both from the effects of stress. The most striking evidence of this comes from studies

3. INFANCY

with monkeys, showing that mother-infant contact lowers the levels of the "stress hormone" cortisol.

This buffering system, says Stanford University psychologist Seymour Levine, "permits a mother to focus on her infant and the baby to focus on his mother. Through the mother-infant relationship, he develops a coping response. He finds he can do something about his environment."

Biologically speaking, today's mothers and babies are two to three million years old. Are there behavioral characteristics that come to us as part of our genetic makeup, as do such characteristics as bone size and blood type? "A mother must take care of her baby, must be close to him," says Dr. Klaus. And mothers have been taking care of their babies for millions of years. "We think that when we put the body of a mother close to her baby, something is turned on that is part of her genetic makeup." The human infant, he concludes, seems to have been built to communicate with its mother. The evidence points to a sophisticated signaling system between them. "If this is so," he says, "then babies make sense. Then the puzzle of their unusual talents begins to unravel."

The Importance of Mother's Milk

Graham Carpenter

Graham Carpenter is an assistant professor of biochemistry and medicine in the Department of Biochemistry of Vanderbilt University's School of Medicine in Nashville, Tennessee.

Reproduction of the species is considered to be the driving force in the evolution of biological systems, and nature has a large investment in a newborn organism. This is particularly true of humans and other large mammals that have relatively long gestation periods and usually produce only one or two offspring at each birth. Survival of these newborns is critical to the continuance of each species, and for almost all mammals, milk is the material that provides total nourishment during the initial stages of life. The exception is the human species, which in the last several decades has in large numbers shifted from milk to substitutes.

At the turn of the century approximately 50 percent of all newborn babies in the United States were breast-fed for at least the first twelve months of life. Recent surveys indicate that during the 1970s about one in three babies was nursed during the first month of life. Concurrently, the duration of breast-feeding declined rapidly, to the point that about 5 percent of all infants were nursed until six months of age and 1 percent until twelve months of age. At present there is a shift among new mothers to return to breast-feeding, but this group is small in relation to the total population. In today's era, when information is prized and decision making based on biological instinct is, perhaps unfortunately, frowned upon,

it would seem that data about such an important commodity as milk (and its commercial formula substitutes) should be of concern to both the public and the medical profession.

Too frequently the assumption is made that "milk is milk," and whether a newborn human drinks mother's milk or cow's milk is of no large consequence. A second assumption is that today's technology is so advanced that artificial formulas can duplicate the quality of natural human milk. Both of these assumptions are incorrect. Comparison of the composition of milk from various species shows that there exist significant qualitative and quantitative differences, which seem to reflect the varying needs of each species for proper development in early life. For example, compared to human milk, the milk of the bovine species has twice as much protein, which after digestion provides the newborn with a source of raw materials, that is, amino acids, for muscle growth. This increased protein content reflects the needs of almost all newborn animals (excluding humans) to grow rapidly in order to avoid predators, become independent quickly, and survive. A newborn cow, for instance, will double its birth weight in about 50 days, whereas a newborn human requires 180 days. Because human milk is relatively low in protein does not mean that it is insufficient in that respect. Human infants do not need to add bulk as rapidly as other newborn animals, but have different needs, including maintaining a higher rate of brain growth.

Myelination—the surrounding of nerve axons with a lipoprotein membrane, or sheath, necessary for the proper conduction of nerve impulses—is an important process of human brain growth in the first year of life. Myelination requires substantial quantities of lipid, and human milk has a relatively

high lipid content. Lipid intake is also important for the baby of lower weight who, to maintain a proper body temperature, must produce fat for insulation.

Comparative analyses of milk compositions provide many examples such as these and quite strongly demonstrate that each species has during its evolution devised a milk composition that is optimal for its specific needs. The high degree of sophistication of the biological mechanisms that control the composition of milk is demonstrated by the capacity of mothers of preterm babies to produce a milk with a higher content of protein and lipid than is found in the milk of mothers of full-term babies. Nature apparently provides the preterm, low-weight infant with extra protein to increase its body mass, or weight, and with extra lipid to stabilize body temperature and to provide more fuel for energy production.

Most of our present knowledge of milk composition concerns those components that are present in milk in the largest amounts—macronutrients such as protein, carbohydrate, and lipid. There is also a reasonable body of data describing the levels and different types of minerals and vitamins. Milk, however, is more complex than is currently understood, and there are increasing feelings in the scientific community that, in addition to satisfying nutritional needs, milk may provide other, subtle but important factors for growth and development in the newborn. For example, the antibodies passed through milk from mother to child help the newborn, whose ability to produce antibodies has not fully developed, resist microbial infections. Breast-fed babies do, in fact, suffer fewer infections, particularly of the gastrointestinal tract, than formula-fed babies.

Hormones, the body's chemical messengers, are other kinds of molecules present in milk. Most hormones are either proteins (insulin, for example) composed of many amino acids or have chemical structures related to steroids. A characteristic of all hormones is that they are present in body fluids in exceedingly small quantities compared with other components. Nevertheless, hormones play a vital role in regulating body chemistry and physiology. For instance, an individual can ingest large amounts of carbohydrates but unless insulin is present the carbohydrates will not be utilized properly. Similarly, other hormones control virtually all processes that the body carries out. There are a number of hormones known to be present in human milk, but their significance has not been demonstrated. This does not mean these molecules have no significance; it simply means that none is yet understood.

Also present in milk is an interesting class of hormones—often referred to as growth factors—that controls cell growth and differentiation. Research in this area is relatively new and it is very likely that many more growth factors will be identified in the future. Of those now known, one of the most thoroughly investigated is epidermal growth factor, or EGF. Epidermal growth factor is a small protein molecule composed of fifty-three amino acid residues. Discovered in the early 1960s in mice by Stanley Cohen, a biochemist at Vanderbilt University, this hormone has since been detected in many mammalian species. It is present in almost all human body fluids, including milk, where its concentration (approximately thirty nanograms per milliliter) is quite significant for a hormone.

The classic procedure for determining the function of a hormone has been to surgically remove the hormone-producing organ and record the resultant effects of the hormone deficit. The observed effects can be attributed to the hormone deficit if the effects are prevented or reversed by injection of the hormone into the animal. This experimental strategy is not always possible, however. In the case of EGF, it is not known where in the body the hormone is produced, and the available evidence suggests that there are probably several distinct sites. Therefore, it has not been possible to determine unequivocally what functions EGF serves. There are experiments, however, that demon-

strate quite interesting biological effects when EGF is administered to animals. These results probably provide good clues to the natural function of this hormone.

Cohen's pioneering studies demonstrated that EGF had a pronounced stimulatory effect on the growth and differentiation of the outermost layer of the body, the epidermis of the skin—hence the name epidermal growth factor. In those initial experiments, newborn mice were injected daily with small quantities of EGF, and an intriguing result was observed. Normally, a newborn mouse is born with its eyelids shut and, without exception, its eyelids open at thirteen to fourteen days after birth. But baby mice injected with the epidermal growth factor opened their eyelids in seven days. The proliferation of skin cells in the eyelid area and their differentiation, a process called keratinization, had been accelerated. Later studies have demonstrated that the effect was not limited to the eyelid area; all areas of the epidermis were similarly stimulated to proliferate and keratinize in newborn animals treated with EGF.

More recent experiments have shown that EGF stimulates epithelial tissues other than the skin. The entire gastrointestinal tract is lined with epithelial tissue, and several studies have shown that EGF stimulates cell division and accelerates certain differentiation events in this tissue. During development in the perinatal period, the cells of the gastrointestinal tract begin to produce hydrolytic enzymes—protein molecules required for the digestion of food. The rates of activity of these enzymes are generally low at birth and increase as the infant matures. If these digestive enzymes do not develop at the correct time, the infant is less able to utilize food to sustain its nutritional requirements, and obviously, a long delay in the maturation of the digestive enzymes has serious consequences. A recent study by a Canadian group at the University of Sherbrooke in Quebec reports that when newborn mice are injected with EGF the maturation of several digestive enzymes is accelerated. Thus the hormone may have a significant function in the growth and development of this important system.

A second critical transition for the newborn infant occurs at birth, when

maternal oxygen, which was delivered *in utero,* is no longer available; suddenly oxygen must be taken in from the air and, via the lungs and circulatory system, made available to the tissues. Failure to adjust successfully to atmospheric oxygen at birth creates a respiratory distress syndrome called hyaline membrane disease. This is, of course, a serious condition, and one that is seen not infrequently in the baby born prematurely. The premature infant is especially vulnerable to this problem because the final stages of lung maturation in the fetus normally occur just prior or to full-term birth. Hyaline membrane disease cannot be predicted before birth, and the mortality rate is low but significant.

This respiratory syndrome has been studied in newborn lambs, and the results show that in cases where pregnant ewes are subjected to stress and their lambs delivered prematurely, many of the lambs (75 percent) die soon after birth of respiratory distress similar to the hyaline membrane disease seen in human infants. When the fetal lambs are infused *in utero* with EGF for a few days prior to premature delivery, however, development of the lung epitheliums is accelerated, respiratory distress is infrequent, and almost all of the lambs survive. Moreover, in these same experiments, accelerated growth of epithelial tissue is noted not only in the lungs but also in other areas of the pulmonary epithelium, such as the trachea and esophagus.

An epithelial tissue in adults that appears to be very sensitive to EGF is the corneal epithelium in the eye. This tissue is important for sight and can be easily injured. Research has shown that in both experimental animals and in humans, EGF accelerates the proliferation of cells in a wounded area of the corneal epithelium, thereby hastening the wound-healing process. Because EGF is present in milk, a natural fluid, clinical trials are being conducted to determine whether milk might be a feasible treatment for corneal lesions. In fact, the use of EGF as a general wound-healing agent is being considered. In this regard, it is interesting to note that folk medicine often prescribes the application of urine to wounds. And in war situations this practice has been employed when other treatments were not immediately available. Urine is usually a sterile solution and contains high amounts of EGF.

Nature is often remarkably ingenious in devising strategies to maximize the utility of its products. In the case of protein molecules, it is not unusual for one protein to be capable of carrying out more than one distinct function. At the level of molecular architecture this is not a trivial engineering task. In addition to its capacity to stimulate cell growth and differentiation, the EGF molecule has a second biological activity. When administered to experimental animals and to humans, this hormone blocks the release of excess gastric acid induced by a variety of chemicals, such as histamine or pentagastrin.

The ability of EGF to control acid secretion in the stomach was discovered unexpectedly in 1975 when a British research group at Imperial Chemical Industries isolated and determined the structure of urogastrone—a protein hormone known since the 1920s to be present in high quantities in the urine of pregnant women and to be an inhibitor of acid secretion. Computer comparisons of urogastrone with other known proteins revealed that the primary structures (that is, the sequences of amino acids) of urogastrone and EGF are so nearly identical (within the limits of experimental technology) that one can conclude they are the same molecule. Thus milk-fed infants receive a hormone that may help to prevent discomforts and injuries to the stomach lining (ulcers, for example) resulting from excess gastric acid.

Scientists have been increasingly successful at removing selected cells from the intact animal and maintaining these cells in a growing state for long periods of time in the laboratory. This technique, called cell culture, involves placing the cells in a plastic dish with a defined medium of glucose, minerals, vitamins, and amino acids. Under these conditions the cells remain viable but do not grow. For growth and cell division to occur, serum—a blood fraction—is added to the medium. Serum apparently contains hormones, or growth factors, necessary to regulate cell growth and division and for many years has been considered to be indispensable for this purpose. Recent experiments, however, have shown that milk is also able to stimulate cell proliferation in cell culture and can make up for approximately 95 percent of the serum requirement. Biologically, this suggests that milk is a fluid rich in the factors that control cell growth, and EGF appears to be a key ingredient in milk for the stimulation of cell division. This has been demonstrated by treating milk with antibodies to EGF.

Antibodies are highly specific reagents that are able to recognize one particular protein molecule in a mixture of thousands of slightly different protein molecules. Antibodies bind to the protein they recognize and often inactivate it. When milk is treated with antibodies to EGF, the capacity of the milk to stimulate cell growth is reduced by 90 percent. This has been demonstrated in cell culture with human fibroblasts (connective tissue cells) and human glia cells (brain-derived cells). Different types of cells are known to vary in their responsiveness to any particular hormone; therefore, it is possible that milk may contain additional growth factors other than EGF that stimulate different cell types. For example, erythropoietin, a protein growth factor that controls the development of red cells in the blood, is known to be present in milk.

In summary, we know that EGF is able to exert a significant influence on the proliferation and differentiation of various types of epithelial cells in the intact animal. We also know that this hormone is present in physiologically significant quantities in milk. Is there a connection? Is milk a physiological vector for the delivery of important hormones to the infant? These intriguing questions will have to wait for further research before answers can be provided. To resolve some obvious questions, however, a few additional comments can be made at this time. If the hormone is ingested in milk and affects the growth and development of tissue in the gastrointestinal tract, it must be able to resist conditions in the gastrointestinal tract that cause the digestion of most protein molecules. Studies of the chemistry of EGF have shown that the hormone's activity is not destroyed by exposure to strong acid or proteases (enzymes that digest protein molecules). Therefore, it is not unreasonable to suggest that this growth factor may remain biologically active in the normally adverse conditions, for a protein molecule, of the gastrointestinal tract.

If EGF also affects other epithelial tissues, such as that of the lungs, then there is a second problem: the growth factor would have to be absorbed into the circulatory system by passing through the lining of the gastrointestinal tract. That this apparently can occur has been demonstrated in the experiments conducted with newborn mice. The results showed that oral administration, as well as subcutaneous injection, of EGF resulted in precocious eyelid opening, indicating that the hormone was absorbed from the gastrointestinal tract into the systemic circulation. This has not, however, been demonstrated with humans, and animal species differ in terms of which proteins will be absorbed.

As already mentioned, large numbers of babies in the United States are not fed with mother's milk but receive commercial formulas with either a soybean or bovine milk base. Does this make a difference as far as growth factors are concerned? Cells grown in culture can be stimulated to grow and divide by either human or bovine milk. When various commercial formulas were tested, however, none were able to stimulate cell proliferation. This is not to be taken as an indictment of the commercial formulas—certainly many children have been raised on formula without apparent adverse consequences. These results indicate that human milk is not duplicated by the industrial products and that our scientific technology is not always able to better or even equal nature.

WHAT A BABY KNOWS

A baby comes into the world loving sweets,
feeling pain,
and preferring the sight of human beings.

AIDAN MACFARLANE

Aidan Macfarlane, *a pediatrician who has conducted postdoctoral research in collaboration with developmental psychologists, approaches his research both as physician and psychologist. He is senior medical officer in Community Pediatrics at Radcliffe Infirmary, Oxford.*

Macfarlane, who was born in London, graduated from Cambridge University and took his general medical training at St. Mary's Hospital in London. He came to Children's Hospital in Boston for a residency in pediatrics and has been a teaching fellow at Harvard Medical School. For several years he was a research officer in the Department of Experimental Psychology at Oxford and a member of Wolfson College. He remains an honorary research officer at Oxford and he is also an editor for Spastics International Medical Publications.

Macfarlane has conducted research with T. Berry Brazelton and Jerome Bruner. Most of his recent work has centered on newborn babies and their mothers, and he is especially interested in the bond that develops between parent and child. The Psychology of Childbirth, from which this article is drawn, is so sympathetic to both mother and baby that one reviewer presumed it had been written by a woman.

People used to think that the world of a newborn baby was a confusion of sounds and smells, of shifting light and shadow. Psychologists believed that the infant automatically received all sensations from the environment without exerting any selectivity or discrimination. This is not so. Although we are still in the process of discovering just how competent the newborn baby is, we know that just by being born human the infant fresh from the mother's womb is especially attracted to the features of another human face. This is remarkable because he or she has had no previous experience of the external sight or sound of human beings.

There are many ways of finding out what tiny babies can see, hear, or feel. Undoubtedly the pleasantest is that used by mothers to find out what their babies are thinking—watching them very carefully. One master at baby watching is Peter Wolff of Harvard University, who studied a group of 10 babies in their own homes, observing each for 30 hours a week. He was particularly interested in learning the amount of time infants spend in an attentive state—fully awake, breathing evenly, with eyes wide open, and quiet. Wolff found that the time a baby spends attentively increases from 11 percent in the first week to 21 percent by the fourth. The rest of the time he is either drowsy, asleep, or crying. Wolff also observed that during the first six hours after birth the amount of alert inactivity varied greatly from infant to infant. Some stayed awake for an hour and a half before the first sleep and did not have another alert period of equal length until the end of the first month; others fell asleep within 15 minutes of delivery and did not wake up fully until the second or third day.

I mention this because a newborn baby's reactions to his changing environment largely depend on whether he is drowsy or alert, quiet or crying. The best time to observe a baby's reactions to sights, sounds, and smells is when he is alert but quiet. Unfortunately, as Wolff's study suggests, the time a baby spends like this in the first few days after birth is very short.

Many factors affect a baby's state of alertness. Heinz Prechtl of the State University in Groningen, Holland, gathered from the brain-wave patterns of very small babies that infants under two weeks were never fully awake when flat on their backs; they were much more alert when they were lying at an angle, with their heads higher than their feet, or when they were upright. Alertness may be one reason crying babies often become quiet as soon as they are held up to the shoulder. Once babies are fully alert, the sights and sounds around them take their attention from crying.

A few years ago, many psychologists tested babies when they were flat on their backs, and got negative results. In some cases, they repeated the tests with the babies sitting up, and the babies responded. Today, we frequently use small supportive chairs, inclined at an angle, for babies to sit in. Other factors affecting response include the temperature of the surroundings—too hot and the baby goes off to sleep, too cold and he starts to cry. Touch also matters—if the baby is tightly swaddled, he tends to go to sleep, but he may become distressed if totally naked. Light is an influence—too much and he shuts his eyes with a grimace, too dim and he may go to sleep again. Sound also affects response—many babies go to sleep more quickly when background noise is present.

If the baby is to respond to certain things in the outside world, he needs to encounter certain features that attract

"What a Baby Knows," by Aidan MacFarlane, *Human Nature*, February 1978. From THE PSYCHOLOGY OF CHILDBIRTH
© 1977 by Aidan MacFarlane. Reprinted by permission of Harvard University Press.

Smells help the newborn baby learn about the world. Researchers measured this two-day-old infant's reactions to smells like acetic acid by monitoring his activity level, his heart rate, and his breathing patterns. When he stopped reacting to a smell, a change in odors caused him to respond again.

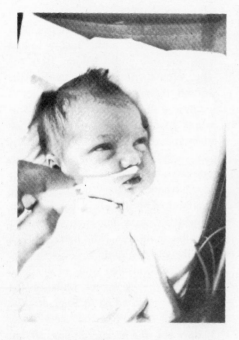
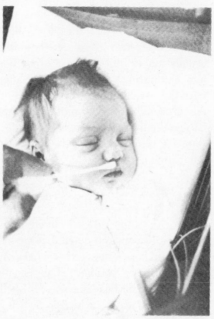
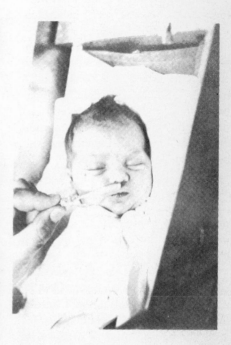

his attention, and he has to link them to his previous experiences. If the stimulus is too intense, too bright, too noisy, or has too many novel features, he may defend himself by turning away or crying. If the intensity of the stimulus is too low or too familiar, he may ignore it. These changes in behavior seem to be accompanied by changes in the heart rate. If the baby is attentive to an event he finds interesting, his heartbeat tends to slow down. If a stimulus is too intense or frightening, his heart speeds up, perhaps in preparation for some sort of defense.

Even when a newborn baby is asleep, he will screw up his eyes, frown, and tense his muscles if a bright light suddenly shines on his face. If, however, he is awake and is brought near a window but not into direct sunlight, he will often turn toward the light, indicating that he knows where it is and is attracted by it. There is evidence that a baby responds to light even inside the uterus, and it seems possible that if a bright light were shone on the pregnant mother's belly, the baby might turn toward it. Perhaps some of

the baby's movements before birth stem from this attraction to light.

After he is born, a baby is much more discriminating; he looks at some things and not at others. If you move your head slowly from side to side nine inches away from his face, he may follow it for short distances with jerky eye movements. To discover which features of the human face a baby prefers to look at, Robert Fantz of Western Reserve University showed a series of babies three flat objects the size and shape of a head. On the first a stylized face was painted in black on a pink background; the second had the same features as the first, but scrambled; and the third had a solid black patch at one end, equal in area to the features of the first two. The babies tested were from four days to six months old, but regardless of age all of them looked most at the real face, somewhat less at the scrambled face, and least at the object with the black patch.

Most studies on what babies prefer to look at involve showing them two things together and observing their eye move-

ments to see which one they look at more. This method shows that babies are attracted by contrast. They like complex patterns with sharp demarcations, and it is not surprising that they watch the eyes more than other features of the human face. With their whites, darker irises, and black pupils, the eyes present definite contrasts. Obviously it is significant that infants should be attracted to eyes and faces as they are.

It was not realized until recently that very young babies could see at all, because it had not been discovered that they have a fairly rigid distance of focus, approximately nine inches. If you want a baby to look at something, it is best to show it to him at this distance—which is incidentally just about the same distance the mother's face is from her baby's when she is breast-feeding. More recent work shows that this distance may not be completely fixed, for by examining the movements of each eye individually and the distance between the center of the pupils, we can observe that the eyes converge and diverge when babies look at objects 10 to 20 inches away but not when the objects are nearer or farther away. This suggests that babies see objects only within this range.

The baby expects to feel what he can see. Tom Bower of the University of Edinburgh filmed the reactions of a series of babies less than two weeks old while a large object moved at different speeds toward them. Examining the films, he found that as the object approached, the babies pulled back their heads and put their hands between themselves and the object. Apparently they already have a reaction that would defend them against being hit by the object, and they expect it to be solid. The baby's reaction to an object approaching at a certain speed on a "hit path" was specific, but if the object moved away from or to one side of the baby, or if the speed of approach changed, he did not react.

In another experiment, Bower created the optical illusion of an object by using polarized light. If the baby was sitting up in a supportive chair so that his arms and hands were free to move, he appeared to try to grasp the object if it seemed within reach. When he tried to grasp an illusory object only to find that it was not there, he was greatly disturbed. This experiment seems to indicate that at least one aspect of the coordination between eye and hand is present at birth and that the newborn baby expects to be able to touch the objects he sees.

The most common complicated sight a baby sees after birth is his mother's face, not as a still object in one plane, like a photograph, but as a dynamic and continually moving object with varied expressions and different associated contexts such as food and warmth. How soon does a baby come to distinguish his mother's face from others? Genevieve Carpenter of the Behavior Development Research Unit at St. Mary's Hospital in London sat two-week-old babies in supportive chairs and, when they were alert, presented each with either his mother's face or the face of a woman the baby had never seen before in a framed opening in front of him. She observed the babies' general behavior, where they looked,

and for how long. At two weeks the babies spent more time looking at their mothers' faces than they did at the strangers' faces. In fact, when they were presented with the other women's faces, they frequently showed strong gaze aversion, looking right away, almost over their shoulders. This kind of withdrawal suggests that the babies found the stimulus too intense or too novel.

It is obvious to most mothers that their babies can hear. If the baby is alert, a loud noise, such as a door slamming, usually makes him tense up or startles him. But during the first few days after birth the middle part of the ear behind the eardrum is full of amniotic fluid that only gradually is absorbed or evaporates. Until it is, sounds reaching the baby's ear are dampened.

Michael Wertheimer of the University of Colorado sounded clicks at either of a newborn's ears and noted that the infant turned his eyes toward the clicks. At four days or even younger, a baby can be taught to turn his head one way for a bell and the other way for a buzzer, by being rewarded when he is right. But we know a baby hears even before he is born. Researchers put a very small microphone on the end of a flexible narrow tube and inserted it into the mother's uterus after her membranes broke — generally a sign that birth will take place within a few hours. The tube was placed close to the baby's ear and recorded the noise actually reaching him. The recordings showed that, before birth, the baby is continuously exposed to the loud rhythmical sounds of the mother's blood flowing through the uterine wall, and to the intermittent sounds of gas rumbling in the mother's intestine. Since it is possible to record an infant's heart rate at this stage, researchers produced a loud noise near the mother's abdominal wall, recorded it from inside the uterus, and noted the effects it had on the baby's heart rate. As might be expected, the baby's heart speeded up.

It seems from this, and from work done with babies right after birth, that patterned, rhythmic sounds produce more response than pure, continuous tones, and that the most effective sound

of all is one that includes the fundamental frequencies found in the human voice. There is also evidence from research done by John Hutt at Oxford University that babies respond better to high than to low frequencies. In light of this, it is interesting that both mothers and fathers often pitch their voices higher when talking to their babies than when talking to adults.

It is also true that a newborn baby held upright, with a man talking to him from one side and a woman from the other, often turns more frequently to the woman. If the woman is the mother, it seems as if the baby recognizes her. This may be so, for another study shows that, if a mother consistently calls her baby by name every time she picks him up, feeds him, or comes near him, by the third day he will frequently turn toward her if she stands out of his line of sight and calls his name. If a woman who is not his mother does the same thing, the baby is much less likely to respond.

A more controlled study of this kind was done by Margaret Mills of the University of Reading in England. She tested babies who were between 20 and 30 days old, and arranged things so that when an infant sucked on a nipple, he heard a recording of either his own mother's or a woman's voice he had not heard before, matched for volume. She found that a baby would suck significantly more to conjure up the sound of his own mother. This is another subtle difference in babies' behavior toward mothers and strangers.

Turning the eyes toward a sound is sophisticated, but five-day-old babies seem to do even better than this. If we disregard his eyes and watch him turn his head instead, we find that he seems able to tell from what angle a sound is coming. Babies will turn their heads more toward a sound coming from an angle of 80 degrees from the midline than they will toward a sound coming from 15 degrees from the midline on the same side.

Also, if you turn your face from side to side in front of the baby's face while talking, his head and eyes follow you — as if he were trying to keep the sound coming from a position between his ears. I have observed babies who under these conditions do not move their eyes at all, but track the face by holding the eyes straight ahead and moving the head only.

Finally, there are the intriguing findings of William Condon of Boston University, who for many years analyzed films of adults talking to one another. He found that people talking seemed to move in synchrony with speech. This "dance" can be detected only by careful film analysis, when even the very smallest movements can be examined. Louis Sander, also of Boston University, later joined Condon in a project of videotaping a series of infants who were between 12 hours and two days old. They showed that even at this incredibly early age the babies seemed to move in precise time to human speech. They did so whether the speech was English or Chinese.

Smell is another way to gather information about people and the world around. Human beings are capable of considerable discrimination among smells, and smells can be evocative. Apparently, smell is a fairly old sense in terms of evolutionary development and is associated with parts of the brain that are themselves ancient in terms of evolution.

That very young babies can smell has been shown in a study conducted at Brown University by Trygg Engen, Lewis Lipsitt, and Herbert Kay. They observed the activity, heart rate, and breathing patterns of 20 two-day-old babies. Each baby was presented with two of four smells: anise oil, asafetida, acetic acid, and phenyl alcohol. When a smell was first presented to the baby, his activity, heart rate, and breathing patterns changed. If the smell continued, the baby gradually learned to take no notice of it (habituation), but when the smell changed, up went the activity and the heart rate, and breathing patterns changed again—the infant recognized the smell as being different from the one he had become used to.

Recent studies I did with newborn babies were designed to test their sense of smell. I had noticed that a baby may turn his face to

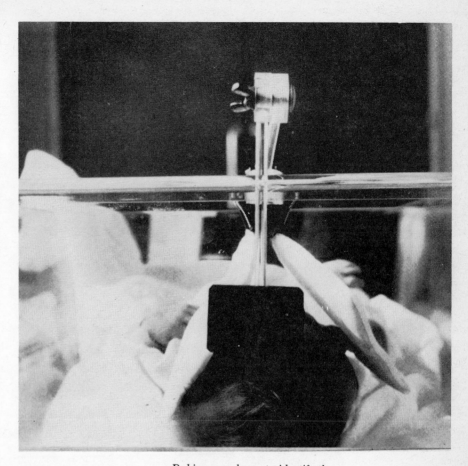

Babies soon learn to identify the scent of their own mothers. Placed between two gauze breast pads, a six-day-old infant will turn his head toward his own mother's pad. Tests show that the baby responds to the smell of her body and not to the milk.

his mother's breast even before he has looked at it or before his face has been touched by the nipple. This might be because he senses the heat of the breast, and indeed infrared photographs show that the breasts and lips of a lactating woman are the warmest areas of her skin. It might also be that the baby learns very rapidly that when one side of his body is held against his mother he has to turn his head to that side to get fed. But the baby might also be smelling the breast; each time the baby is fed his nose is in such close contact that the food and the smell may become associated.

Initially, I took a four-inch-square gauze breast pad that had been placed inside the mother's bra between feedings, and put it at one side of the baby's head,

touching his cheek. At the same time, I put a clean breast pad against his other cheek and for one minute filmed all the movements of the baby's head; at the end of the minute I reversed the pads for another minute. Previous studies had shown that babies tend spontaneously to turn their heads more to the right than to the left, perhaps because mothers tend to hold their babies more often on their own left. Analysis of the films showed that, five days after birth, babies spent more time with their heads turned toward their mothers' pads than toward the clean pads.

To get subtler results I repeated the test using another mother's breast pad instead of a clean one. At two days the babies spent an equal amount of time

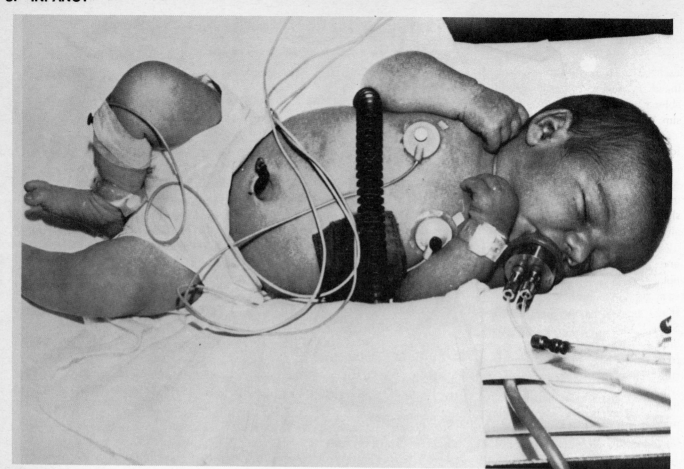

with both pads, but at six days the babies generally spent more time turned to their own mothers' pads than to the other mothers' pads. At 10 days this effect was even more striking. All the babies were breast-fed and were tested when hungriest, just before a feeding.

This experiment still left open the question of whether it was the breast milk itself, the mother's smell, or a combination of the two that attracted the baby. I did one further experiment, in which the mothers expressed milk from their breasts onto gauze pads and then used those pads as in the previous experiment. This revealed that babies do not turn toward breast milk alone — perhaps the smell of the milk is not strong enough. Here again we witness the ability of the baby to distinguish his own mother from other mothers when he is only six days old.

Perhaps mothers can also recognize their own babies' smells. In a pilot study, we asked a group of mothers to sniff their babies. We blindfolded the mothers and put earphones on them so they could neither see nor hear their babies. When

A baby is born with a taste for sweets. The more sugar Lewis Lipsitt added to a liquid, the more slowly the baby sucked and the faster his heart beat Lipsitt suggests the baby sucks slowly to savor the sweet taste.

they were given two babies to smell, a significant number managed to identify their own, though I suspect they may have been using other clues.

Taste is a relatively simple sense in human beings, and the fine discriminations that we think we can make by taste we actually make by smell. Two observations indicate that the baby inside the uterus is able to taste. The baby, surrounded by amniotic fluid, swallows it continually and then eliminates it. In certain pregnancies too much fluid accumulates too quickly. Forty years ago a physician developed a novel way of treating this. By sweetening the fluid surrounding the baby with an injection of saccharine, he discovered that he could reduce the amount of fluid, possibly because the baby was encouraged to swallow more.

More recently another doctor found

that injecting a substance opaque to x-rays into the amniotic fluid caused a decrease in swallowing; this was clear from the x-ray pictures of the substance being swallowed by the child. The opaque substance he had injected had an extremely unpleasant taste. These observations suggest that a fetus can taste.

Lewis Lipsitt and his colleagues at Brown have been looking at the effect of sweetness on a baby's sucking patterns and on his heart rate. Working with babies two to three days old, they arranged a system by which, each time the baby sucked, he got a tiny fixed amount of fluid containing varying amounts of sugar. They found that the more sugar, the slower the baby sucked and the more his heart rate increased. This is in some ways not what we might expect: Certainly I would have thought that if the baby tasted a sweet substance he would have sucked faster to try to get more. But in this case, where the baby was getting only a very small, fixed amount with each suck, he sucked more slowly. This should mean that less effort was used and the heart should have slowed down, but it did not. One explanation that Lipsitt put forward is that the baby sucks more slowly in order to savor the sweeter fluid, and perhaps the excitement of tasting it increases the heart rate. It is probable, though, that the answer is not this simple.

The findings of David Salisbury of Oxford University are also remarkable. He looked at the effects of different fluids on babies' swallowing, sucking, and breathing patterns. Previous work showed that water, cow's milk, and glucose, when introduced into the back of a calf's throat, interrupted its breathing, while a salt-water solution did not; this suggested that a calf has specialized taste receptors to prevent it from breathing when swallowing. Salisbury studied the effects of feeding different solutions to human babies. He found that, when given salt water, they suppressed breathing very poorly, but when he fed them sterile water, none inhaled it—which suggests that a baby may also have these specialized taste receptors.

Salisbury also examined the swallowing, breathing, and sucking patterns of the infants when fed breast milk or a substitute. Again the patterns were different, indicating that a young baby's taste is sensitive enough to distinguish breast milk from a substitute.

The composition of breast milk alters during the day and during each feeding. At the beginning of a feeding, breast milk is somewhat diluted; it becomes more concentrated toward the end. So the baby gets his fluids at the beginning and his food at the end. Shifting from one breast to another has been likened to having a drink in the middle of a meal.

We still accept the idea that newborn babies are not as susceptible to pain as they are later. This is demonstrated by the practice in the United States of circumcising newborn baby boys without anesthetics or analgesics. If it were necessary to remove an equivalent area of highly sensitive skin from some other part of the body, say the little finger, most people would be horrified if it were done

without an anesthetic; yet few demand anesthetics for circumcision.

Little systematic research has been done on a baby's perceptions of pain, since no one wants to hurt babies simply to discover the nature of their reactions. However, watching any baby who has to have a blood test shows us that babies are aware of pain. Blood samples are commonly obtained by pricking a baby's heel with a small stylette and collecting the drops in a container. The normal reaction of the baby to this procedure is immediately to try to withdraw his foot and to wail with anguish. He may also tense all his muscles and turn bright red. His heartbeat and breathing patterns change rapidly as well. But the reaction varies greatly from baby to baby, depending on his temperament and his alertness at the time. And a baby's reactions to pain can be modified: The increases in the heart rate after circumcision can be reduced by having the baby suck at the same time, though the slowed heartbeat is no evidence that the baby feels any less pain.

After a newborn baby has been circumcised, his sleep patterns are disturbed, just as they are after blood has been taken from his heel. In one study the disturbance followed very soon after circumcision; in a second study there was a prolonged period of wakefulness and fussing, which was followed by an increase in quiet sleep. Circumcision has also been implicated in sex differences. Most studies of newborn babies have been done in the United States, where the majority of males are circumcised at birth. It is possible that some reported differences in behavior between male and female babies are results of this surgical procedure instead of any innate differences.

One ability that more than any other demonstrates the sophistication of newborn babies is imitation. In a series of very well controlled studies conducted in England and the United States, Andrew Meltzov demonstrated that at two weeks of age children can stick out

their tongues and clench and unclench their hands when they watch someone else do these things. These actions are not immediately obvious to the eye and can be detected only by very close analysis of video tape and film. The baby has to watch another person stick out a tongue, and then he has to realize that his own tongue is equivalent to that person's even though he has never seen his own tongue. Without being able to see what he is doing himself, he must match his movements to the ones he sees. These studies are very recent and are likely to have considerable impact on theories of social development in children.

All this research points in the same direction. It shows that when a baby is born, he already has the capacity to respond selectively and socially to human beings instead of to the things in his new environment. By the age of two weeks, he has learned to recognize the features of individual human beings—especially those of his usual caretaker. At the same time, the baby's appearance and behavior help ensure that his physical and emotional needs are looked after. These systems of mutual attraction and needs did not develop independently, but evolved over millions of years as an interdependent series of factors that ensure the survival of the newborn baby and thus of the human species.

For further information:
Bower, T. G. R. *Development in Infancy.* W. H. Freeman, 1974.
Carpenter, G. "Mother's Face and the Newborn." *New Scientist,* March, 1974.
Crook, K. and Lipsitt, L. P. "Neonatal Nutritive Sucking: Effects of Taste Stimulation upon Sucking Rhythm and Heart Rate." *Child Development,* No. 47, 1976.
Macfarlane, J. A. *The Psychology of Childbirth.* Harvard University Press, 1977.
Macfarlane, J. A. "Olfaction in the Development of Social Preferences in the Human Neonate." *Parent-Infant Interaction.* CIBA Foundation Symposium, No. 33, new series, ASP, 1975.
Richards, M. P. M., J. F. Bernal, and Y. Brackbill. "Early Behavioral Differences: Gender or Circumcision." *Developmental Psychobiology,* No. 9, 1976.

ATTACHMENT AND THE ROOTS OF COMPETENCE

The attachment between mother and baby
can support or retard the later development
of healthy independence and personal competence.

L. ALAN SROUFE

L. Alan Sroufe *is professor of child psychology at the Institute of Child Development, University of Minnesota, and adjunct professor in the department of psychiatry at the University of Minnesota School of Medicine. He served four years on the Personality and Cognition Research Review Committee of the National Institute of Mental Health. Sroufe has written more than a dozen articles on early emotional development, and his book,* Knowing and Enjoying Your Baby, *emphasizes the infant's social, emotional, and intellectual development.*

Some two-year-olds approach problems with great enthusiasm. They are eager, persistent, and flexible. They show obvious joy in the mastery of problems and are not easily frustrated. If they do get stuck, they seek help from their caregivers.

Other children present a different picture. They give up quickly or spend a great deal of time in activities that bear no relation to the problem. They may whine, pout, fuss, or become frustrated—stamping their feet, banging objects, or becoming petulant. Even in the absence of frustration some children show no joy in mastery, no eagerness in the face of challenge. They either ignore their caregivers' suggestions, refuse them loudly, or do exactly the opposite of whatever is suggested—even when it keeps them from solving the problem. Some throw tantrums or hit or kick their caregivers.

Do such wide differences among young children reflect the emergence of individual personalities? If we are observing characteristic behavior in these children, can we predict later developmental problems and later competence from what we have seen? Finally, have these children simply been different from birth, or has their early experience—including the quality of the relationship with their caregivers—played an important role in the development of their personalities?

The issue of whether a baby's early experiences affect the development of attitudes, fears, and expectations that will later influence the way he or she behaves as a child or an adult is important. Although some researchers profess to find little continuity in human development, the question is not settled. If it is possible to show strong links in the behavior of individual children over time and across situations, we clearly have captured something enduring in the child. And if we can predict which children are likely to show later developmental problems, it may be possible to intervene before the problems develop. At the same time, establishing continuity in behavior is important for understanding the roots of competence and the nature of human development itself.

But attempts to predict individual development are doomed if researchers simply measure the same behavior throughout a baby's growth. Infants and young children develop rapidly—they learn new behavior, and old behavior takes on new meaning. All behavior becomes increasingly complex. Two-year-olds, for example, are learning language and solving problems, running and climbing, and showing complex emotions like shame and defiance. None of these activities are characteristic of young infants. Babies babble, crawl, uncover objects, and push away the hand that wipes their noses, but we cannot assume that these actions are identical to later reactions, even when they appear similar. Attempts to show that particular behavior or capacities are stable have consistently failed.

Early attempts to show consistency in characteristics like dependency or aggression also failed—and for good reason. Dependency, for example, is a natural, universal state in infancy. Seeking physical closeness and contact is both normal and adaptive. Therefore such behavior in a baby cannot indicate whether he or she will be overly dependent as a four-year-old. Aggression, on the other hand, is not within the capacity of the infant, so preschool aggression cannot be predicted from an infant's vigor in nursing, an activity which might be used by researchers as an early measure of aggressiveness. Vigorous babies may or may not be vigorous toddlers, but if they are, they may be vigorous in climbing and in play without necessarily being aggressive. In general, all approaches that have assumed some simple identity between earlier and later behavior have failed.

In our work at the University of Minnesota we assumed that despite rapid advances in development, and despite dramatic changes in a child's behavior, there is continuity in the quality of a child's adaptation, in his or her personal style or orientation. The child who functions well emotionally and socially at one age will be likely to function well at the next, even though different areas of behavior might be examined. Although toddlers are much more developed, they are fundamentally the same people they were as infants, and four-year-olds are the same people they were as toddlers.

Thus by assessing how well a child functioned with respect to the important issues in one developmental period it should be possible to forecast the quality of the child's functioning in the next period. As complex and subtle as the task was, we sought to capture something basic about the nature of the individual personality as it emerged in infancy and as it was manifest in the preschool years.

The developmental issue we decided to study in infancy was the effectiveness of the attachment relationship between the infant and his or her primary caregiver. (In order to avoid confusing caregivers and babies, I will refer to the caregiver as "she," since in most cases the caregiver is a mother, and to the baby as "he.") Attachment refers to the special intimacy and closeness that develop toward the end of the first year between the infant and his caregiver. The infant feels comfortable and secure with her and derives comfort and security from her presence. The baby who is doing well is the baby who has a healthy attachment to his caregiver, a relationship that promotes exploration and mastery of the environment. In studying the way infants used the relationship with their mothers to support their exploration and play, we felt we could capture the quality of their adaptation.

We predicted that the one-year-old infant who showed a secure, effective attachment would also be competent and successfully autonomous as a toddler, in terms of both his style in approaching problems and his emotional involvement in them. Such a link was entirely reasonable. The infant who uses the caregiver as a base for moving out into the world, and as a haven when threatened or distressed, develops motor skills and a sense of himself as effective. In sharing his play with his caregiver at a distance, the infant evolves a new way of maintaining contact while operating independently. The infant is free to invest himself in challenging the environment because he is confident that he can maintain his tie with his caregiver even while he is widening his world.

By the preschool years, children's capacity for independent functioning has vastly increased. They are concerned with developing peer relations, establishing their sense of themselves as boys or girls, expanding their skills at symbolic play and fantasy, and learning to manage their impulses. Children who have adapted well will look at challenging situations positively, and their expectations concerning people will also be positive. Such a child should be competent with peers and well liked by teachers, but this competence should be predicted not from measures of play between infants but from the quality of the attachment relationship between infant and caregiver.

If later personality can be predicted from studies made during infancy, then attachment is a likely starting point, for it is the focus of social, emotional, and mental development in infancy. A great deal of the infant's experience occurs with, or is controlled by, the caregiver. The tendency to become attached to an available figure has been built into the infant through the course of evolution, and virtually all infants become attached. But the quality of the attachment relationship varies widely, and these differences must have a strong impact on the emerging person. We are all defined by our relationships with others. Within this first relationship, the child also learns a great deal about his impact on the world—about who he is.

In our studies we used Mary Ainsworth's system for assessing attachment. In this system the quality of a baby's attachment is defined in terms of its balance with exploration—the use of the caregiver as a base for exploration and as a source of comfort when distressed—and in terms of the baby's effectiveness in reestablishing contact following separation. The emphasis is on the organization of the baby's behavior across contexts that vary in their degree of stress. Separation and reunion experiences are especially critical for the one-year-old infant and are therefore central in assessing attachment. In Ainsworth's system there are three major patterns of attachment: securely attached infants; anxiously attached, avoidant infants; and anxiously attached, ambivalent infants. Obviously, only the first group shows good adaptation.

If a securely attached baby comes into an unfamiliar playroom, he leaves his mother and becomes absorbed in the toys, though he may share his play with her. He rarely becomes wary in response to a stranger who chats with his mother before engaging him in play. Although he may hesitate at first, he soon warms up and begins to play with the stranger.

When his mother departs, leaving him with the stranger, stress is presumed to increase. Even so, some securely attached babies do not become upset. Distress at separation is influenced by a host of factors, including the baby's mood, hunger or fatigue level, health, age, recent experience, degree of involvement with the stranger, and understanding of the situation. Some babies do become upset. On reunion they go directly to their mothers, actively seeking contact and maintaining it until they are settled. They often cling, sink in, mold, or otherwise clearly show their desire for contact and its effectiveness in providing comfort. Securely attached infants who are not distressed by separation commonly do not seek physical contact on reunion; instead they smile, bounce, vocalize, or show a toy, actively seeking interaction instead of contact. They are happy to see their mothers.

When a securely attached infant is left alone, his stress is greatly increased. Many 12- to 18-month-old infants are obviously distressed, and almost all reveal some degree of upset by subdued or otherwise altered play. Although such an infant may become more settled when a stranger enters, and may even promote contact with the stranger, he clearly shows that this contact will not do. He may continue to be distressed or he may mix contact seeking with squirming and pushing. When his mother returns, he immediately leaves the stranger and goes to her.

It is impossible to reduce the quality of attachment to a single kind of behavior. Securely attached infants may or may not seek physical contact with their mothers. They may cry a lot or a little. Sometimes they are friendly to strangers, at other times they are not. But securely attached infants do actively reestablish some kind of contact on reunion, whether physical or social, and they are able to use their mothers as a base for exploration and play.

Anxiously attached babies show quite different patterns of behavior. The ambivalent group does little exploring in an unfamiliar playroom, even when the mother is present. They may cry and seek physical contact before they are separated. They are obviously wary of the stranger and acutely distressed when the mother leaves—even when they are not left alone. Most significant, they cannot settle down when the mother returns. Some clearly reveal their ambivalence by mixing contact seeking with contact resistance, behaving much as distressed infants do with strangers. They may push the mother away, or hit, kick, or bat away offered toys. Others simply continue crying, fussing, or pouting. Such an attachment relationship obviously fails to support the infant in his exploration and mastery of new surroundings.

The other major group of anxiously attached infants is not so obvious. In an unfamiliar playroom, infants in this avoidant group separate readily from their mothers and do not seek contact. A baby from this group is not wary of a stranger and does not cry when his mother departs, unless perhaps he is left alone. But when his mother returns, he greets her casually, if at all, and commonly shows frank signs of avoidance. He may look, turn, or move away, or start to approach his mother and then turn or back off. He typically settles down when a stranger enters the room. His avoidance of his mother becomes ex-

treme during a second reunion when he is under increased stress. The more these infants are distressed, the less likely they are to seek contact. The system is turned around and has become maladaptive.

In our research Everett Waters, Leah Matas, and I sought to determine whether the quality of an infant's attachment would indicate the nature of his or her emerging personality. First we examined 108 infants, some at 12 and then at 18 months, others at only one of these ages. Each time independent coders rated them according to Ainsworth's system. Ninety percent of these suburban infants fit one of Ainsworth's three patterns closely. Of the 50 infants seen at both ages, 48 were placed in the same group at 18 months as they had been at 12 months. A baby may have become less (or more) upset the second time, may have smiled and vocalized more (or less), or may have sought less (or more) physical contact, but the quality of attachment had remained the same.

Demonstrations of reliability and stability are essential starting points for any attempts to show long-range continuity in development. In addition to our own results we had Ainsworth's data showing that a baby's attachment pattern at 12 months could be predicted from an assessment of his mother's behavior from his sixth to his 15th week of life, and that the baby's behavior in the laboratory was related to his behavior at home.

If these measures of attachment were capturing something essential about the emerging personality of the child, then the differences in the quality of attachment should have clear consequences for autonomous functioning in the next developmental period. To assess the quality of these babies' adaptations as two-year-olds, we selected a situation in which each toddler had to solve increasingly difficult problems. The early problems, such as pushing a toy from a tube with a stick, were easy for two-year-olds to solve. But the final problem—which required the child to weight down a lever with a block in order to raise candy through a hole in the top of a plexiglass box—was well beyond the capacity of a two-year-old. When the child got stuck, however, his mother was

nearby to help him. As when the infants had been subjected to separation and stress, this procedure taxed their capacities for maintaining organized behavior and for drawing on their own resources and those of their mothers.

How quickly a child solved the problem was unimportant, just as at 12 or 18 months an infant's level of cognitive development was not our central interest. We were interested in three aspects of competence: (1) emotional dimensions such as enthusiasm, positive feelings, and the ability to face challenges without becoming frustrated; (2) motivation, reflected by the time spent away from the task (the inverse of persistence); and (3) ability to use maternal assistance effectively, as shown by the child's compliance or attempted compliance, or by his ignoring suggestions, saying "no," acting aggressively, or throwing tantrums. We also assessed the quality of the mother's assistance (the timing, pacing, and clarity of her hints) and her degree of support (whether she encouraged, attended to, or was otherwise available to her child).

Relationships between the quality of a baby's attachment at 18 months and his adaptive functioning at two years of age were dramatic and powerful. Children who were assessed as secure in their attachment as infants were much more enthusiastic and persistent and showed more positive feelings than did children who had earlier been assessed as anxiously attached. The children who had been securely attached showed fewer negative emotions, and in a situation where assistance was essential, they were more compliant, threw fewer tantrums, and ignored or opposed their mothers less often. In contrast to the anxiously attached groups, they showed no aggression.

The relationships with the quality of maternal behavior impressed us even more. Mothers whose infants had earlier been assessed as securely attached were strikingly higher on both "supportive presence" and "quality of assistance" than were mothers of anxiously attached infants. The infant's attachment reflects the history of his interaction with his mother; and it is reasonable to expect that the caregiver of a securely attached infant will be similarly responsive in the child's second year. In our assessment of the infant's behavior we had apparently captured the quality of the mother-child relationship.

We had tested the children with the Bayley scales and knew that the three groups were similar in their general level of development, but we needed to make sure that we were not simply measuring intellectual capacity or some simple in-born trait of the infant. We also had to determine whether or not the continuity we observed was entirely a result of the mother's presence. Our final study was done in collaboration with Wanda Bronson of the University of California. As part of her studies of young children she video-taped 34 mothers and their 15-month-old infants in an unfamiliar setting, using a procedure similar to Ainsworth's. Bronson also had descriptive codings of these same children, which were made in a nursery-school setting when the children were three and a half years old. In the nursery school, two independent observers watched the children for five weeks and then arranged a series of statements according to how well they described each child.

We viewed only the infant-mother tapes and had no knowledge of the children's later behavior. But when our judgments of the infants' attachments were compared with codings of their behavior made more than two years later, relationships again were clear and powerful. Infants we assessed as secure in their attachment were distinguished by 11 of the 12 statements relating to competence with peers, and by five of the 12 statements relating to ego strength from the nursery-school study. As preschoolers, the securely attached infants were more purposeful, involved, and effective with peers than were the insecurely attached infants. There were no differences in the groups on a developmental level at 15 months or in IQ scores at three and a half years. Because mothers were not present when the children were assessed at nursery school, it was clear that the quality of infant-mother attachment indicated the child's emerging personality.

Why did we succeed in demonstrating continuity when others have failed?

First, we assessed the child's overall quality of adaptation rather than particular capacities or behavior. Second, we emphasized emotions and motivation rather than intellectual abilities. Third, we linked our assessment to issues that are important in various periods of development. Finally, we began with measures having demonstrated power.

It should be pointed out that our studies and others demonstrating continuity have been the result of research with children in stable environments. In our study, infants came from two-parent families characterized by stable employment, a low divorce rate, and low mobility. Much greater change, toward both better and poorer adaptation, would be expected in studies of children from rapidly changing environments. Work by Byron Egeland and Amos Deinard, which is now in progress, shows this to be true.

Within our model of continuity, however, the issue is not stability but the coherence of individual development. We would not expect a child to be permanently scarred by early experience or permanently protected from environmental assaults. Early experience cannot be more important than later experience, and life in a changing environment should alter the quality of a child's adaptation.

Still, early experience may be of special importance in two ways. First, the child is engaged in active transactions with the environment. The child not only interprets experience, the child creates experience. As Alfred Adler has suggested, the child is both the artist and the painting. If because of early experience the preschooler isolates himself from the peer group, he removes himself further from positive social experiences. Second, if self-esteem and trust are established early, children may be more resilient in the face of environmental stress. They may show poor adaptation during an overwhelming crisis, but when the crisis has passed and the environment is again positive, they may respond more quickly. Even when floundering, some children may not lose their sense that they can affect the en-

vironment and that, ultimately, they will be all right.

In these first studies we did not explore temperament. Children also differ in their reactivity, their tempo, and their particular likes and dislikes. Children who are secure in their attachment may be highly active or placid, cuddly or stiff. Some toddlers are more interested in one kind of activity, some prefer another. Well-functioning infants and children do not fit one particular mode.

Moreover, every child will at times be difficult, irritable, or have tantrums. Even securely attached infants squirm, kick, and push away from strangers. It is not the capacity that distinguishes the children but when and how they show it. The securely attached infants who cooperated with their mothers in solving a problem were not so cooperative when they were asked to stop playing with attractive toys. In this situation they showed as much opposition as did anxiously attached infants.

The personality differences we stress cannot be reduced to temperament (though the influence of temperament is part of our current research). We emphasize the child's self-concept and his or her approach to problems and opportunities. A child who has a rapid tempo may be seething with anger, hostile to other children, unable to control his or her impulses, and filled with feelings of worthlessness. But a child who has a rapid tempo also may be eager, spirited, effective, and a pleasure to others, and may like him- or herself.

In our work we emphasize the quality of maternal care. This is not because care from fathers and others is unimportant. If a father provides consistent, high-quality care, we would expect the baby to become securely attached to that father. All of the people in the child's environment help to shape the development of his personality. Nonsocial experiences are clearly important as well. A baby needs to play with objects if he is to develop a sense of mastery. And inborn differences in babies certainly affect the quality of adaptation, especially in the case of the ambivalent, difficult-to-settle babies. But our emphasis on maternal care is based on clear and substantial evidence concerning the impact of

THE EMERGENCE OF PERSONALITY

PEER COMPETENCE	Average scores Securely attached babies	Insecurely attached babies	Level of significance*				EGO STRENGTH
Sympathetic to peers' distress	16.7	8.3	.01	.04	10.2	13.2	Forcefully goes after what he wants
Spectator in social activities	8.2	12.2	.02	.05	10.0	11.8	Likes to learn new intellectual skills
Hesitant with other children	6.8	10.8	.03	.25	6.2	5.3	Does not persevere when goals are blocked
Characteristically unoccupied	7.3	9.4	.14	.36	9.6	8.8	Suggestible
Hesitates to engage	7.7	12.4	.007	.23	13.0	14.2	Becomes involved in whatever he does
Peer leader	10.4	6.5	.01	.15	11.4	13.2	Confident of own ability
Other children seek his company	12.4	7.5	.001	.01	7.4	4.8	Uncurious about the new
Attracts attention	11.2	6.8	.02	.01	8.3	12.8	Self-directed
Suggests activities	12.3	6.9	.005	.03	8.4	5.6	Unaware, turned off, "spaced out"
Socially withdrawn	6.7	12.1	.002	.19	11.0	12.5	Sets goals that stretch abilities
Withdraws from excitement and commotion	5.3	8.2	.03	.25	7.5	6.3	Samples activities aimlessly, lacks goals
A listener (not full participant in group activities)	8.3	11.8	.05	.38	7.0	6.4	Indirect in asking for help

*Any number lower than .05 reflects a statistically significant relationship. Numbers lower than .01 reflect highly significant relationships.

When observations of nursery school children were compared with ratings of their attachment to their mothers (measured when the children were 15 months old), strong relationships appeared between the quality of the early relationship and later personality.

maternal care on the quality of an infant's attachment.

Certain behavior patterns among caregivers appear to be linked with later secure attachment and competence among babies. At least half a dozen studies point to the same key characteristic, which is best described as responsiveness or sensitivity. Ainsworth's research has shown that highly sensitive mothers who are neither interfering nor rejecting consistently have infants who are secure in their attachment at 12 months.

Good maternal care involves responding to the infant's signals promptly and effectively. When during face-to-face interaction the infant turns his head away, signaling that he needs less stimulation, the sensitive caregiver relaxes and waits. Not until the baby signals his readiness does she reengage him. When the infant cries, the sensitive caregiver responds promptly, and effectively puts an end to the infant's distress. When the baby seeks contact, the sensitive caregiver responds warmly and affectionately, teaching the infant that his signals are effective. The sensitive caregiver provides smooth transitions and meshes her (or his) stimulation or assistance with the infant's behavior. She does not thrust interaction on an

unreceptive infant. Sensitivity requires that the mother respond to the individual needs and nature of her infant: This is why sensitive care generally promotes healthy emotional development in vastly different babies.

Through sensitive interaction the infant learns that he can have an impact on the world and that stimulation in the presence of the caregiver is not threatening. In the presence of the caregiver he can tolerate the excitement of new experiences because he has learned that the caregiver is available when needed. Ultimately the baby comes to believe that such resources lie within himself, and he develops a sense of trust in his caregiver. This trust eventually becomes a belief in his own competence.

As a child develops, the characteristics of sensitive care change, although responsiveness to the child's signals probably remains a central feature. Psychoanalyst Margaret Mahler has described how, during the child's second year, the caregiver must give her infant a gentle nudge to encourage his move toward independent functioning. In our study of two-year-olds the sensitive caregiver was not the one who immediately solved the problem for the child, nor the one who allowed him to become frustrated before assistance was offered. She was the mother who gave

the child space to work, then offered minimal hints to help him solve the task and retain a sense of having solved it by himself.

The process of becoming an autonomous individual continues in different ways throughout one's life. It is reasonable to expect that difficult beginnings could be overcome by good social support during later development. And it is doubtful that good adaptation can be maintained without continued support. Competence in early years does not guarantee competence later in life. But it's a good start.

For further information:

Ainsworth, M. D. S., and S. M. Bell. "Mother-Infant Interaction and the Development of Competence." *The Growth of Competence*, edited by K. J. Connolly and Jerome Bruner. Academic Press, 1974.

Brazelton, T. B., Barbara Koslowski, and Mary Main. "The Origins of Reciprocity: The Early Mother-Infant Interaction." *The Effect of the Infant on Its Caregiver*, edited by Michael Lewis and L. A. Rosenblum. John Wiley & Sons, 1974.

Matas, Leah, R. Arend, and L. A. Sroufe. "Continuity of Adaptation in the Second Year of Life: Quality and Infant-Caregiver Attachment and Later Competence." *Child Development*, (in press).

Sroufe, L. A. *Knowing and Enjoying Your Baby.* Prentice-Hall, 1977.

Sroufe, L. A., and Everett Waters. "Attachment as an Organizational Construct." *Child Development*, Vol. 48, 1977, pp. 1184-1199.

Are Young Children Really Egocentric?

Janet K. Black

Janet K. Black, Ph.D., is Assistant Professor of Education, College of Education, North Texas State University, Denton.

Recently a friend decided it was best if she gave away her dog. After screening responses to an ad she had placed in the newspaper, three-year-old Julie and her parents were invited to visit Zoe. Upon becoming satisfactorily acquainted with Zoe, Julie's mother and father asked her if she would like to take Zoe home. Julie replied in a concerned tone, "But if we take Zoe, she (Zoe's owner) won't have a dog."

Another child, two years of age, accidentally knocked off his grandmother's glasses while she was reading to him. Marc patted her and said, "Sowy Grandma." This behavior was repeated several times throughout the afternoon.

Five-year-old Jonathan asked his mother how many quarters there were in a football game. After internalizing her explanation he said, "Five quarters would make one game and one quarter, and eight quarters would make two games."

The above behaviors are typical of observations that parents, teachers, and caregivers have made while interacting with young children. These behaviors pose some questions about theories of child development which suggest that young children are totally egocentric. According to Piaget, egocentric behavior prevents young children from conserving and from acting altruistically. However, there is a growing body of research evidence that indicates even young children are not always egocentric.

Piagetian and recent research on egocentrism

Yarrow and Zahn-Waxler (1977) conducted a detailed study of children's altruistic behavior between the ages of ten months and two-and-one-half years. After analyzing some 1,500 incidents, Yarrow and Zahn-Waxler conclude that children, from at least the age of one, have a capacity for compassion and various kinds of prosocial behavior. Pines (1979) discusses Flavell's conclusions about young children's egocentric behavior. As a result of his research, Flavell concludes that Piaget overestimated how egocentric young children are. His investigations with three-year-olds indicate that they behave in a consistently nonegocentric manner. Flavell believes that some of the role taking exercises used in earlier research may have been too complicated for young children, while simpler, more relevant tests are more reliable indicators of their true ability. For example, if children are shown a card with a dog on one side and a cat on the other, they will be able to tell you correctly what animal you see and ignore the one that faces them. Flavell concludes, "Though young children may not completely understand, they can pay attention to how other people feel" (Pines 1979, p. 77).

One of Piaget's best known experiments which allegedly documents young children's inability to decenter is the mountain task. Piaget used a three-dimensional model of three mountains (Piaget and Inhelder 1967) which are distinguished from one another by color, as well as the presence of snow on one mountain, a house on another, and a red cross on the third.

A child is seated at one side of the table upon which the above model is placed. The experimenter then stands a doll at varying positions around the mountains and asks the child, "What does the doll see?" Because it is hard for the child to give a verbal description, she or he is given a set of ten pictures and is asked to choose the one that shows what the doll sees. Generally, children up to the age of eight or nine cannot successfully do this task. Piaget (Piaget and Inhelder 1967, p. 212) concluded that these responses indicate that young children are unable to decenter.

Donaldson (1979) states:

We are urged by Piaget to believe that the child's behavior in this situation gives us deep insight into the nature of the world. This world is held to be one that is composed largely of "false absolutes." That is to say the child does not appreciate that what he sees is relative to his own position; he takes it to represent truth or reality—*the world as it really is.* . . . Piaget believes that this is how it is for the young child: that he lives in the state of the moment, not bothering himself with how things were just previously, with the relation of one state to those which came before or after it. . . . The issue for Piaget is how the momentary states are linked or fail to be linked in the child's mind. The issue is how well the child can deal conceptually with the transitions between them. (pp. 12-13).

A task, similar to yet different from Piaget's mountain experiment, designed by Hughes (1975) demonstrates that there are important considerations in the experimenter's, as well as in children's, behavior which Piaget failed to take into account. After acquainting the child with a square board divided by barriers into four equal sectors, a doll was put in each of the sectors one at a time and a policeman was positioned at points of the sectors. The child was asked if the policeman could see the doll. Then another policeman was introduced and the child was told to hide the doll from both policemen. This was repeated three times so that each time a different sector was left as the only hiding place. Thirty children between the ages of three-and-one-half and five were given this task with 90 percent of their responses correct. Even the ten youngest children (average age, three years nine months) achieved a success rate of 88 percent. Thus, it seems that young children are capable of considering and coordinating two different points of view.

Donaldson (1979) explains the differences in children's behaviors on the mountain and policeman tasks from an experiential perspective. In short, 'the mountain task is abstract in a psychologically very important sense: in the sense that it is abstracted from all basic human purposes, feelings, and endeavors" (p. 17). However, "young children know what it is to try to hide. Also they know what it is to be naughty and to want to evade the consequences" (p. 17).

In the policeman task, the motives and intentions of the characters are comprehensible even to three-year-olds. If tasks require children to act in ways which make human sense, that is, are in line with very basic human purposes, interactions, and intentions, children show none of the difficulties in decentering which Piaget maintained. Thus, while adults, as well as children, are egocentric in certain situations throughout life, the extent to which young children are egocentric seems to be much less than Piaget suggested.

Related to the issue of basic human purpose and feeling is the nature and context of the language used by the experimenter. Once again Donaldson raises valid questions about various Piagetian class-inclusion tasks. In one of these tasks a child is presented with four red flowers and two white flowers and asked: Are there more red flowers or flowers? Children of five usually respond that there are more red flowers. Donaldson (1979) states:

> There is not much doubt about what a child *does* when he makes the standard type of error and says there are more red flowers than flowers; he compares one subclass with the other subclass. . . . The question is why does he compare subclass with subclass? Is it because he *cannot* compare subclass with class as Piaget maintains? Or is it because he thinks that this is what he is meant to do? Is there. . . a failure of communication? (p. 39)

In discussing a variety of task-inclusion experiments in which there was perceptual and/or language modification, Donaldson (1979) concludes that the questions the children were answering were very often not the questions the experimenter was asking. In short, the children's interpretations did not correspond to the experimenter's intentions. The children either did not know what the experimenter meant, what the language meant, or had expectations about questions or the experimental material which shaped their interpretation. In other words, when children hear words that refer to a situation which they are at the same time perceiving, their interpretation of the words is influenced by the expectations they bring to the situation. While Piaget was aware of the differences of what language is for the adult and what language is for the child, he failed to keep these differences in perspective in using language for studying children's thinking (Donaldson, 1979).

Gelman (1979) believes that if researchers stopped using incorrect assumptions and inappropriate measures, they would learn that the cognitive skills of younger children are far greater than has been assumed. She developed a magic task which does not ask a child to distinguish between more or less but rather to designate a winner or loser in number conservation activities. "According to the results of the 'magic task,' preschoolers know full well that lengthening or shortening the array does not alter the numerical value of a display" (p. 903).

Conclusions

There is a growing body of evidence which indicates that young children are not as egocentric as Piaget suggested. If children are in contexts where they can use their knowledge of very basic human purposes, intentions, and interactions, they show no difficulty in decentering. Likewise, children's interpretation of the language of others is dependent upon (1) their knowledge of language; (2) their assessment of the experimenter's, caregiver's, teacher's, and parent's nonverbal intentions; and (3) the manner in which children would represent the physical situation if the adults were not present.

Because Julie had had very basic human experiences of relinquishing prized possessions or opportunities, she understood how people might feel if they had their pets taken away. Because Mac had had the very basic human experience of being hurt and comforted, he understood how his grandmother felt and what he could do to comfort her. Because Jonathan had had the very basic human experience of acquiring desired possessions with his allowance frequently given to him in four quarters, he readily indicated his ability to decenter when discussing the quarters in a football game. Contrary to Piaget, these three children's behavior indicates that they are not completely egocentric if the context is meaningful to them. While Piaget's vast contribution to the understanding of young children's cognitive development cannot be denied, no theory is final. Early childhood researchers and practitioners need to add to and clarify the roots of later cognitive and prosocial competence.

Implications

Astute and observant adults can renew their faith in their abilities to observe and know children in ways that researchers often misinterpret. Perhaps there should be more action research on the part of early childhood teachers or more dialogues between teachers and researchers.

Providing situations in which young children can demonstrate their knowledge of very basic human purposes and interactions is necessary to make a more accurate assessment of their competencies. Young children demonstrate varying competencies in different environments. There are contexts in which children indicate they do not know, when, in fact, they do (Black, 1979a, 1979b; Cazden 1975).

A third implication concerns the language adults use with children. Teachers and parents constantly need to be aware of young children's knowledge of language, of our nonverbal behavior, and children's probably understanding of physical situations.

By acknowledging that children in basic human situations and meaningful communicative contexts are more capable than previously thought, teachers, caregivers, and parents can obtain more accurate information about children's abilities. An awareness of these competencies facilitates provision of more appropriate experiences for young children.

Finally, researchers need to become cognizant of inappropriate methodology in studying young children. Conclusions and recommendations emanating from misguided research do not help practitioners provide appropriate experiences for young children. In fact, practices based on misinformation may actually impede children's development. Gelman (1979) cautions professionals about continuing to measure children with tasks that are more appropriate to older children (and even adults): "The time has come for us to turn our attention to what young children can do as well as what they cannot do. We should study preschoolers in their own right" (p. 904).

References

Black, J.K. "Formal and Informal Means of Assessing the Competence of Kindergarten Children." *Research in the Teaching of English* 13, no. 1 (1979a): 49-68.

Black, J.K. "There's More to Language Than Meets the Ear: Implications for Evaluation." *Language Arts* 56, no. 5 (1979b): 526-533.

Cazden, C. "Hypercorrection in Test Responses." *Theory into Practice* 14 (1975): 343-346.

Donaldson, M. *Children's Minds.* New York: Norton, 1979.

Gelman, R. "Preschool Thought." *American Psychologist* 34 (1979): 900-905.

Hughes, M. "Egocentrism in Pre-School Children." Unpublished doctoral dissertation, Edinburgh University, 1975.

Piaget, J., and Inhelder, B. *The Child's Conception of Space,* New York: Norton, 1967.

Pines, M. "Good Samaritans at Age Two?" *Psychology Today* (June 1979): 66-77.

Yarrow, M.R., and Zahn-Waxler, C. "The Emergence and Functions of Prosocial Behaviors in Young Children." In *Readings in Child Development and Relationships,* ed. R. Smart and M. Smart, New York: Macmillan, 1977.

Development During Childhood

So pronounced are the cognitive and social changes occurring during the transition from preschool to childhood that one developmental psychologist refers to this time period as the "5-to-7 shift." During the 5-to-7 shift, significant changes occur in the child's attention span, recall memory, learning and problem solving skills. Articulation improves, vocabulary size increases, and the child achieves a new understanding of syntactical aspects of language. Preference for touch as the primary means of object exploration yields to preference for visual exploration, although there is also a marked increase in the ability to integrate touch and visual information. Most of the changes that occur during the transition to childhood, as well as those that occur during childhood itself, involve social, emotional, cognitive, and language development.

Social-Emotional Development. During the school years the child's social network expands. School extends the child's peer group beyond the confines of the immediate neighborhood and exposes the child to a new set of authority figures. The quality of the child's interaction with each of his or her available role models will influence the structure of his or her sense of social and emotional competence. Erik Erikson suggests that this age period is important for resolving a conflict between industry and inferiority. If the child's attempts to master the environment are encouraged and approved by family, peers, and teachers, a sense of competence or mastery will develop. However, if the child's experiences lead to failure and disapproval, a sense of inferiority will develop. The extent to which conflict affects personality development during childhood is a topic of special interest to contemporary developmentalists.

Cognitive Development. During the past two decades the study of cognitive development has been dominated by the theory of Jean Piaget. Piaget equates cognitive operations with mental structures corresponding to mathematical operations such as transitivity, equivalence, and addition. Piaget's theory attempts to explicate the structural elements of the mind by focusing on the functional abilities of the child at various stages of development. Although it provides a rich description of what the child can and cannot do during a particular stage of development, it is less adequate for explaining how the child acquires various cognitive skills. Thus, many developmentalists are becoming interested in information processing models in an effort to integrate classic cognitive psychology with the phenomena of cognitive development.

Language Development. Early language theorists were interested in determining whether linguistic ability is acquired as a consequence of exposure to a language environment or is somehow "built in" at birth. Research has focused on the role played by imitation, on differences and similarities across languages, and on the relationship between the developing individual's linguistic abilities and more general cognitive and emotional processes. Prior to the 1960s theories of language development stressed environmental determinants. The theory of Noam Chomsky shifted emphasis to genetic explanations of language development. Today, investigators stress organizational models which are based on the assumption that genetic and environmental factors work in concert to determine behavioral outcome, including the behavior we call language.

Looking Ahead: Challenge Questions

What are the long term effects of divorce, abuse and neglect, family death, poverty, or disability on child development? Why are some children so devastated by conflict whereas others seem to take even the worst of conditions in stride?

How do the cognitive competencies of the child differ from those of the infant and preschooler? Why would developmentalists be dissatisfied with stage theories of development?

Language often is cited as a uniquely human characteristic. Yet, non-human primates have demonstrated the ability to learn highly complex sequential behaviors that reflect some of the rudiments of early language development in humans. What evidence would provide unquestionable support for the contention that non-human primates are capable of learning human language?

Is there an innate capacity which can be called intelligence? Are there several kinds of intelligence? Is intellectual capacity fixed at birth, or can it be modified by experience?

If your child doesn't get along with other kids

There are ways for parents to help kids who have chronic difficulties getting along with their peers. And if you think a problem may be in the making, there are things you can do before it gets serious.

THE PROBLEM may announce itself with a call from another parent. The voice is regretful but firm.

"Please don't send your child over to play anymore. He spends the whole time fighting."

Or it may be a baby-sitter, a principal or a teacher who's the first to tell you your child doesn't get along with other children.

No parent likes to get that kind of report, but is it really so important for children to get along well with their peers?

"Children get unique things from each other," says Dr. Willard W. Hartup, director of the University of Minnesota's Institute of Child Development and a leading researcher on children's peer relations. For one thing, he says, "it's better to practice and experience the consequences of aggression with someone your own size."

But if a child doesn't learn how to handle his feelings and communicate his ideas and opinions to his peers, says Chet Brodnicki of Child and Family Services in Hartford, Conn., "he'll probably encounter a great deal of difficulty as an adult along those very lines. And that will affect his ability to hold a job, complete his education and have stable relationships with other adults."

Other professionals agree. Furthermore, they agree that peer problems may go along with other problems, such as learning difficulties. And these troubles may be your best signal that all is not going well with your child.

Whether they are aggressive or shy and withdrawn, the youngsters who can't get along with children their age are isolated, and as various studies suggest, these children are more likely than others to drop out of school, become juvenile delinquents and have mental health problems.

The good news is that you can prepare your children—even very young ones—to get along with their peers and you can get help if you need it.

A healthy pattern

Granted, children need children. Does this mean you should worry about a child who wants to spend his recess reading? Not at all, says Elizabeth Jacob of the Virginia Frank Child Development Center of the Jewish Family Service in Chicago. "Maybe he is by temperament a quiet child. Maybe this year he happens to be fascinated with reading. Many children's development is uneven, and there's a wide range of 'normal unevenness,' as well as a wide range of temperaments."

With most children social development follows a predictable course. At first the baby grows attached to the parents. A secure child will soon start responding in friendly fashion to a strange adult or child. Two-year-olds usually enjoy playing alongside each other, but they're really not ready to play *with* each other. By the time a child is 3, he may be ready to take turns or share toys if an attentive adult is nearby to give occasional help.

Five-year-olds can often manage cooperative play by themselves. By the time a child is 6 to 8, he wants to be off with other children much of the time.

From then until adolescence and beyond, peers are king. If you say there's no Santa Claus and a peer says there is, your son or daughter will probably believe there is a Santa Claus.

What can you do to help along the healthy process of establishing peer friendships?

▶ Make sure your baby has a warm, secure, continuing relationship with the person who takes principal care of him. This is particularly important from five to 18 months. One study found that the security of a child's attachment to his mother at 15 months accurately predicted his competence with his peers at 3½.

▶ Let your child be with other children at a very early age. Infancy isn't too young for informal get-togethers in the mother's presence. Even children too young for

As Dr. Benjamin Spock puts it, "Children encountered for the first time can seem as strange and dangerous to inexperienced children as gorillas would to grownups."

▶ Pay special attention to early social experiences of an only child to make up for the practice he's not getting with a brother or sister. Spock suggests taking a first child to a playground or a friend's backyard several times a week, especially when the child begins to walk. "Let her learn all by herself how to meet push with push or how to hang on to a toy when another child is trying to grab it. . . . She gets her basic feelings about the meaning of aggressiveness of other children from her parents. If her parents consider it dangerous or cruel, she is frightened by it. If they take it casually, she learns to do the same."

▶ When a 3-year-old is ready for cooperative play, it's helpful if a responsible adult is around, not to take over but to see that the children learn, in Jacob's words, to "take turns, negotiate, moderate, wait." At this age children can benefit from a good nursery school or other regularized play situation.

▶ Try to spot developmental problems early. Mary MacCracken, a learning disability specialist who wrote *Lovey: A Very Special Child* and former chairperson of the National Association for Mental Health, favors screening tests in kindergarten or earlier to look at developmental indicators—motor, verbal and paper-and-pencil skills—and behavioral characteristics, such as how the child gets along with other children. If there is trouble with any of these indicators, MacCracken says, the child should be watched carefully and given one-to-one help.

▶ Try pairing a withdrawn preschooler with one other playmate. This worked especially well in a study in which the playmate was 15 months younger than the shy child.

▶ If your school-age child isn't making friends on his own, lend a hand. Spock suggests inviting children over, one by one, for a meal or a special outing. If it takes a bribe to get them to come, so be it. "You can't make an obnoxious child popular with bribes," he says, "but you can ensure that your child's good qualities will be given fair consideration."

▶ Read children's books with a friendship theme with your child. It gives him an opening to air his feelings about other children.

▶ Work with your child's teacher to find clubs and recreation groups that follow the youngster's interests.

Becoming better playmates

New research projects are throwing more light on qualities that make a child attractive to other children. Studies point to four social skills that seem to count in gaining acceptance. Dr. Sherri Oden of the University of Rochester and Dr. Steven Asher of the University of Illinois call these skills participating, cooperating, communicating, and validating or supporting.

Children earn acceptance by *participating* in a game or activity, initially by getting started and paying attention; by *cooperating,* as in taking turns and sharing materials; by *communicating,* through talking and listening; and by *validating or supporting,* as in looking or smiling at playmates and offering help or encouragement.

Oden and Asher set out to see whether coaching in these four skills would make unpopular third- and fourth-graders more acceptable to their classmates. After five coached play sessions over four weeks, the unpopular kids got higher ratings from their classmates. A year later their ratings were higher still.

Harvard researcher Dr. Robert L. Selman has been studying children's ideas about friendship. It is Selman's theory that children move through an overlapping five-stage sequence in understanding friendship and may get stuck at any stage along the way. The process begins at ages 3 to 7 with what Selman labels Momentary Playmateship, in which friends are valued because they're nearby and have nice toys. At ages 4 to 9 comes One-Way Assistance, in which a friend is a friend because he does what you want to do. Between 6 and 12 Two-Way Fair-Weather Cooperation may develop, followed at 9 to 15 by Intimate, Mutually Shared Relationships. Finally, at age 12 or older, children may gain enough perspective for Autonomous Interdependent Friendships to become possible.

Selman, who has used his research to help troubled youngsters make friends, observes that a child may not get along with his peers if his understanding of friendship is either delayed or too advanced. For instance, a youngster who's ready for intimate, possessive friendships may not be accepted by playmates who are still working their way through the one-way, self-interest stages.

Selman has been able to help disturbed children catch up to stages of friendship skills that fit their ages. He's also developing "friendship therapy" for children who have been too aggressive or too shy to get along with children of their age.

Stanford's Dr. Philip G. Zimbardo is another social psychologist working on the problems of shyness. He tells parents and teachers to head off shyness by helping the child discover his attractive qualities, and he recommends games and exercises for combating shyness in both children and adults.

Should you get help?

Dr. Charles R. Shaw, a child psychiatrist and author of *When Your Child Needs Help,* writes that "the time to take your child for professional help is when you feel he is unhappy. . . . If your child is happy, he is probably okay."

Dr. Spock says that "a child who can't make or keep friends by six or who is aggressive or cruel or timid with other children, should have help."

The Hartford Child and Family Services finds that peer problems are most often first identified in the classroom. This agency sees the largest concentration of peer problems in overaggressive 7- to 12-year-old boys, especially "just after the report cards come out and in the spring, when promotions for next year are being planned," says Brodnicki.

The agency may assign such a child to an activity group in which six or eight children with similar difficulties meet for an hour and a half each week for at least ten weeks with one or two therapists. Through group activities, including decision making and discussion, the children learn ways of dealing with anger and frustration.

At the same time a counselor meets with the family. This does not imply that the family is to blame. However, methods that worked with other children in the family may not work with this one, and possibly the counselor can suggest alternatives. Or the whole family, including the child, may be reacting to a traumatic family event—a death, a divorce, a move.

Community mental health centers and family service agencies like the one in Hartford get government or United Way funds as well as other public or private grants, so they can often adjust their fees to the family income. Charges are usually moderate compared with the fees of private practitioners. Look for family agencies in the Yellow Pages under "Marriage and Family." Ask whether they belong to the Family Service Association of America (there are about 270 members in 42 states). Community mental health centers set up with federal funds are required by law to have children's services. Look for them under "Health," "Mental Health," "Clinics" or "Social Services." Help or referral assistance may also be available through the child's school if your school system has social workers and psychologists on the staff, as many do.

Your health insurance may include counseling from private therapists or funded agencies.

It's a good idea to have a youngster who has behavior problems checked by a physician. There may be some medical problem that's causing the trouble—for instance, a vision or hearing defect. Dr. Lee Salk, a family psychologist, points out that even minimal dysfunction of the central nervous system, often easily treated, can bring on aggressive behavior, hyperactivity, temper tantrums—and the concomitant trouble with peers—if untreated.

Whatever is bothering a child who doesn't get along with other kids—fear or insecurity, lack of social skills, trouble in the home, medical problems—it's worth finding out what it is and doing something about it.

Most children respond well to help, says Dr. Rosemary Burns, director for Youth and Family Service at a Virginia community mental health center. "In the length of a school year in treatment, most children can work their way out of peer difficulties." □

BOOKS FOR PARENTS AND KIDS

The following publications are for parents.

• *Childhood and Adolescence: A Psychology of the Growing Person,* by L. Joseph Stone and Joseph Church (Random House; $14.95). Fourth ed.

• *Families: Applications of Social Learning to Family Life,* by Gerald R. Patterson (Research Press, Box 3177, Champaign, Ill. 61820; $4.95). Rev. ed.

• *Help for Your Child: A Parent's Guide to Mental Health Services,* by Sharon S. Brehm (Prentice-Hall; $10.95 hardcover, $4.95 paperback).

• *Raising Children in a Difficult Time,* by Benjamin Spock, M.D. (W. W. Norton; $7.95 hardcover. Pocket Books; $1.95 paperback).

• *Shyness: What It Is, What to Do About It,* by Philip G. Zimbardo (Addison-Wesley; $9.95 hardcover, $5.95 paperback).

These books are for children.

• *Amos and Boris,* by William Steig (Farrar, Straus & Giroux; $6.95 hardcover. Penguin; $1.95 paperback).

• *Frog and Toad Are Friends,* by Arnold Lobel (Harper & Row; $5.95 hardcover, $1.95 paperback).

• *Let's Be Enemies,* by J. M. Udry (Scholastic Book Service, 906 Sylvan Ave., Englewood Cliffs, N.J. 07632; 85 cents). Illustrated by Maurice Sendak.

• *Tales of a Fourth Grade Nothing,* by Judy Blume (Dutton; $6.95 hardcover. Dell; $1.25 paperback). See also other books by Blume.

• *Two Good Friends,* by Judy Delton (Crown; $5.50).

• *Will I Have a Friend?* by Miriam Cohen (Collier Books; $6.95 hardcover, $1.95 paperback).

THE MYTH
OF THE VULNERABLE
CHILD

Arlene Skolnick

Arlene Skolnick is a research psychologist at the Institute of Human Development, University of California, Berkeley. Skolnick, who received her Ph.D from Yale, is chiefly interested in marital relationships and changes in self-concepts in later life. She has written *The Intimate Environment*, coauthored *Family in Transition* (both published by Little, Brown), and is now writing a developmental-psychology textbook for Harcourt Brace Jovanovich.

Anxious parents should relax. Despite what some psychologists have been telling them for years, they do not have make-or-break power over a child's development.

Americans have long been considered the most child-centered people in the world. In the 20th century, this traditional American obsession with children has generated new kinds of child-rearing experts—psychologists and psychiatrists, clothed in the authority of modern science, who issue prescriptions for child-rearing. Most child-care advice assumes that if the parents administer the proper prescriptions, the child will develop as planned. It places exaggerated faith not only in the perfectibility of the children and their parents, but in the infallibility of the particular child-

rearing technique as well. But increasing evidence suggests that parents simply do not have that much control over their children's development; too many other factors are influencing it.

Popular and professional knowledge does not seem to have made parenting easier. On the contrary, the insights and guidelines provided by the experts seem to have made parents more anxious. Since modern child-rearing literature asserts that parents can do irreparable harm to their children's social and emotional development, modern parents must examine their words and actions for a significance that parents in the past had never imagined. Besides, psychological experts disagree among themselves. Not only have they been divided into competing schools, but they also have repeatedly shifted their emphasis from one developmental goal to another, from one technique to another.

Two Models of Parenting

Two basic models of parental influence emerge from all this competition and variety, however. One, loosely based on Freudian ideas, has presented an image of the vulnerable child: children are sensitive beings, easily damaged not only by traumatic events and emotional stress, but also by overdoses of affection. The second model is that of the behaviorists, whose intellectual ancestors, the empiricist philosophers, described the child's mind as a *tabula rasa*, or blank slate. The behaviorist model of child-rearing is based on the view that the child is malleable, and

parents are therefore cast in the role of Pygmalions who can shape their children however they wish. "Give me a dozen healthy infants, well-formed, and my own specified world to bring them up in," wrote J. B. Watson, the father of modern behaviorism, "and I'll guarantee to take any one at random and train him to be any type of specialist I might—doctor, lawyer, artist, merchant, chief, and yes, even beggar man and thief!"

The image of the vulnerable child calls for gentle parents who are sensitive to their child's innermost thoughts and feelings in order to protect him from trauma. The image of the malleable child requires stern parents who coolly follow the dictates of their own explicit training procedures: only the early eradication of bad habits in eating, sleeping, crying, can fend off permanent maladjustments.

Despite their disagreements, both models grant parents an omnipotent role in child development. Both stress that (1) only if parents do the right things at the right time will their children turn out to be happy, successful adults; (2) parents can raise superior beings, free of the mental frailties of previous generations; and (3) if something goes wrong with their child, the parents have only themselves to blame.

Contemporary research increasingly suggests, however, that both models greatly exaggerate the power of the parent and the passivity of the child. In fact, the children's own needs, their developing mental and physical qualities, influence the way they perceive and interpret external events. This is not to say that parents exercise no influence on their children's development. Like all myths, that of parental determinism contains a kernel of truth. But there is an important difference between influence and control. Finally, both models also fail to consider that parent-child relations do not occur in a social vacuum, but in the complex world of daily life.

Traditionally, child-study researchers have assumed that influence in the parent-child relationship flowed only one way, from active parent to passive child. For example, a large number of studies tested the assumption, derived from Freudian theory, that the decisive events of early childhood centered around feeding, weaning, and toilet-training. It is now generally conceded that such practices in themselves have few demonstrable effects on later development. Such studies may have erred because they assumed that children must experience and react to parental behavior in the same ways.

Even when studies do find connections between the behavior of the parents and the child, cause and effect are by no means clear. Psychologist Richard Bell argues that many studies claiming to show the effects of parents on children can just as well be interpreted as showing children's effects on parents. For instance, a study finding a correlation between severe punishment and children's aggressiveness is often taken to show that harsh discipline produces aggressive children; yet it could show instead that aggressive children evoke harsh child-rearing methods in their parents.

A Methodological Flaw

The image of a troubled adult scarred for life by an early trauma such as the loss of a parent, lack of love, or family tension has passed from the clinical literature to become a cliché of the popular media. The idea that childhood stress must inevitably result in psychological damage is a conclusion that rests on a methodological flaw inherent in the clinical literature: instead of studying development through time, these studies start with adult problems and trace them back to possible causes.

It's true that when researchers investigate the backgrounds of delinquents,

mental patients, or psychiatric referrals from military service, they find that a large number come from "broken" or troubled homes, have overpossessive, domineering, or rejecting mothers, or have inadequate or violent fathers. The usual argument is that these circumstances cause maladjustments in the offspring. But most children who experience disorder and early sorrow grow up to be adequate adults. Further, studies sampling "normal" or "superior" people—college students, business executives, professionals, creative artists, and scientists—find such "pathological" conditions in similar or greater proportions. Thus, many studies trying to document the effects of early pathological and traumatic conditions have failed to demonstrate more than a weak link between them and later development.

The striking differences between retrospective studies that start with adult misfits and look back to childhood conditions, and longitudinal studies that start with children and follow them through time, were shown in a study at the University of California's Institute of Human Development, under the direction of Jean Macfarlane. Approximately 200 children were studied intensively from infancy through adolescence, and then were seen again at age 30. The researchers predicted that children from troubled homes would be troubled adults and, conversely, that those who had had happy, successful childhoods would be happy adults. They were wrong in two-thirds of their predictions. Not only had they overestimated the traumatic effects of stressful family situations, but even more surprisingly, they also had not anticipated that many of those who grew up under the best circumstances would turn out to be unhappy, strained, or immature adults (a pattern that seemed especially strong for boys who had been athletic leaders and girls who had been beautiful and popular in high school).

Psychologist Norman Garmezy's work on "invulnerability" offers more recent evidence that children can thrive in spite of genetic disadvantages and environmental deprivations. Garmezy began by studying adult schizophrenics and trying to trace the sources of their problems. Later, he turned to developmental studies of children who were judged high risks to develop

schizophrenia and other disorders at a later age. When such children were studied over time, only 10 or 12 percent of the high-risk group became schizophrenic, while the majority did not.

Other Sources of Love

The term "invulnerables" is misleading. It suggests an imperviousness to pain. Yet, the ability to cope does not mean the child doesn't suffer. One woman, who successfully overcame a childhood marked by the death of her beloved but alcoholic and abusive father, and rejection by her mother and stepmother, put it this way: "We suffer, but we don't let it destroy us."

The term also seems to imply that the ability to cope is a trait, something internal to the child. One often finds in the case histories of those who have coped with their problems successfully that external supports softened the impact of the traumatic event. Often something in the child's environment provides alternative sources of love and gratification—one parent compensating for the inadequacy of the other, a loving sibling or grandparent, an understanding teacher, a hobby or strong interest, a pet, recreational opportunities, and so on.

Indeed, the local community may play an important role in modulating the effects of home environments. Erik Erikson, who worked on the study at the Institute of Human Development, was asked at a seminar, "How is it that so many of the people studied overcame the effects of truly awful homes?" He answered that it might have been the active street life in those days, which enabled children to enjoy the support of peers when parent-child relations got too difficult.

Psychologist Martin Seligman's learned-helplessness theory provides a further clue to what makes a child vulnerable to stress. Summarizing a vast array of data, including animal experiments, clinical studies, and reports from prisoner-of-war camps, Seligman proposes that people give up in despair not because of the actual severity of their situation, but because they feel they can have little or no effect in changing it. The feeling of helplessness is learned by actually experiencing events we cannot control, or by being led to believe that we have no control.

Seligman's theory helps to explain two puzzling phenomena: the biographies of eminent people that often reveal stressful family relations, and Macfarlane's findings that many children who did come from "ideal" homes failed to live up to their seeming potential. The theory of learned helplessness suggests that controllable stress may be better for a child's ego development than good things that happen without any effort on the child's part. Self-esteem and a sense of competence may not depend on whether we experience good or bad events, but rather on whether we perceive some control over what happens to us.

Parents Can't Be Pygmalions

Many of the same reasons that limit the effect of events on children also limit the ability of parents to shape their children according to behavioral prescription. The facts of cognition and environmental complexity get in the way of best-laid parental plans. There is no guarantee, for example, that children will interpret parental behavior accurately. Psychologist Jane Loevinger gives the example of a mother trying to discipline her five-year-old son for hitting his younger sister: if she spanks him, she may discourage the hitting, or she may be demonstrating that hitting is okay; if she reasons with the child, he may accept her view of hitting as bad, or he may conclude that hitting is something you can get away with and not be punished for.

Other factors, interacting with the child's cognitive processes and sense of self, limit the parents' ability to shape their children. Perhaps the most basic is that parents have their own temperamental qualities that may modify the message they convey to their children. One recurrent finding in the research literature, for example, is that parental warmth is important to a child's development. Yet warmth and acceptance cannot be created by following behavioral prescriptions, since they are spontaneous feelings.

Further, the parent-child relationship does not exist apart from other social contexts. A study of child-rearing in six cultures, directed by Harvard anthropologists John Whiting and Beatrice Whiting, found that parents' behavior toward children is based not so much on beliefs and principles as on a "horde of apparently irrelevant considerations": work pressures, the household work load, the availability of other adults to help with household tasks and child care, the design of houses and neighborhoods, the social structure of the community. All these influences, over which parents usually have little control, affect the resources of time, energy, attention, and affection they have for their children.

The effects of social class may also be very hard to overcome, even if the parent tries. Psychiatrist Robert Coles has written about poor and minority children who often come to learn from their families that they are persons of worth—only to have this belief shattered when they encounter the devaluing attitudes of the outside world. Conversely, middle-class children from troubled homes may take psychological nourishment from the social power and esteem that are enjoyed by their families in the community.

Science and the Family: Historical Roots

Given the lack of evidence for the parental-determinism model of child-rearing, why has it been so persistent? Why have we continued to believe that science can provide infallible prescriptions for raising happy, successful people and curing social problems?

As psychologist Sheldon White has recently observed, psychology's existence as a field of scientific research has rested upon "promissory notes" laid down at the turn of the century. The beginnings of modern academic psychology were closely tied to education and the growth of large public expenditures for the socialization of children. The first psychologists moved from philosophy departments to the newly forming education schools, expecting to provide scientific methods of education and child-rearing. The founding fathers of American psychology—J. B. Watson, G. S. Hall, L. M. Terman, and others—accepted the challenge. Thus, learning has always been a central focus of psychologists, even though the rat eventually came to compete with the child as the favored experimental animal.

If the behaviorists' social prescriptions conjure up images of *Brave New World* or *1984*, a more humane promise was implicit in Freudian theory. The earliest generations of Freudians encouraged the belief that if the new knowledge derived from psychoanalysis was applied to the upbringing of children, it would be possible to eliminate anxiety, conflict, and neurosis. The medical miracles achieved in the 19th and early 20th centuries gave the medical experts immense prestige in the eyes of parents. There seemed little reason to doubt that science could have as far-reaching effects on mental health as it had on physical health. Furthermore, as parents were becoming more certain of their children's physical survival, children's social futures were becoming less certain. When the family was no longer an economic unit, it could no longer initiate children directly into work. Middle-class parents had to educate their children to find their way in a complex job market. The coming of urban industrial society also changed women's roles. Women were removed from the world of work, and motherhood came to be defined as a separate task for women, the primary focus of their lives. Psychological ideas became an intrinsic part of the domestic-science movement that arose around the end of the 19th century; this ideology taught that scientific household management would result in perfected human relationships within the home, as well as in the improvement of the larger society.

The Limits of Perfectibility

As we approach the 1980s, Americans are coming to reject the idea that science and technology can guarantee limitless progress and solve all problems. Just as we have come to accept that there are limits to growth and to our natural resources, it is time we lowered our expectations about the perfectibility of family life. Instead of trying to rear perfectly happy, adjusted, creative, and successful children, we should recognize that few, if any, such people exist, and even if they did, it would be impossible to produce such a person by following a behavioral formula. Far from harming family relations, lowered expectations could greatly benefit them.

What is more, the belief in parental determinism has had an unfortunate influence on social policy. It has encouraged the hope that major social

4. CHILDHOOD

problems can be eradicated without major changes in society and its institutions. For example, we have in the past preferred to view the poor as victims of faulty child-rearing rather than of unemployment, inadequate income, or miserable housing. Ironically, while we have been obsessed with producing ideal child-rearing environments in our own homes, we have permitted millions of American children to suffer

basic deprivations. A seemingly endless series of governmental and private commissions has documented the sorry statistics on infant mortality, child malnutrition, unattended health needs, and so on, but the problems persist. In short, the standards of perfection that have been applied to child-rearing and the family in this century have not only created guilt and anxiety in those who try to live up to them, but

have also contributed to the neglect of children on a national scale.

For further information, read:

Clarke, Ann M and A D Clarke, eds *Early Experience Myth and Evidence*. Free Press. 1977. $13.95

Garmezy, Norman "Vulnerable and Invulnerable Children Theory, Research and Intervention." American Psychological Association. MS 1337. 1976

Goertzel, Victor and Mildred G. Goertzel *Cradles of Eminence*. Little. Brown. 1978. paper. $4.95

Macfarlane, Jean "Perspectives on Personality Consistency and Change from the Guidance Study." *Vita Humana*. Vol 7. No 2. 1964

WE WANT YOUR ADVICE

Any anthology can be improved. This one will be—annually. But we need your help.

Annual Editions revisions depend on two major opinion sources: one is the academic advisers who work with us in scanning the thousands of articles published in the public press each year; the other is you—the person actually using the book.

Please help us and the users of the next edition by completing the prepaid article rating form on the last page of this book and returning it to us. Thank you.

American Research on the Family and Socialization

John A. Clausen

John A. Clausen, Ph.D., is chairman, Department of Sociology, University of California, Berkeley, and research sociologist at the university's Institute of Human Development. His article is based on an address delivered at the US-USSR Seminar on Preschool Education in Moscow, U.S.S.R.

The family is a basic unit of social organization in all societies. However, its composition, the functions that it serves, the division of labor within it and the allocation of its resources vary greatly from one society to another. Anthropologists, sociologists, psychologists, educators, psychiatrists, historians and political scientists have all been active in family research in the United States. They seek to answer many questions, but one focal point for almost all is how the family orients a child to the world and how it prepares him or her for full participation in society, what we commonly call *socialization*—the whole process of the child's becoming a competent member of society.

There are several approaches to studying the family as an instrument of socialization. The family is, first of all, a context or matrix for development, set in the larger social environments of the neighborhood, the culture of which it is part and the social structure, including the economic and political systems.

The family has been characterized as a unit of interacting personalities, though it has been peculiarly difficult to conceptualize that unity in operational terms. Studies of the effect of family context on child development tend therefore to focus on structural features—the size and composition of the family, the partial or total absence of a parent (usually the father, since families without mothers tend to be seen as non-families)—or they focus on salient experiences or conditions such as economic deprivation, conflict between parents or the effects of life crises. In such research the investigator may seek (1) to delineate the ways in which socialization practices are influenced by the family's structural features or conditions of life or (2) to show that outcomes in child behavior are directly responsive to these features or conditions.

Related to the study of general features of the family as they affect the development of the child are those studies that ask how placement of the family in the larger social structure or cultural milieu influences the childrearing orientations and behaviors of parents and other agents of socialization. Social class and religious and ethnic background have been most utilized as indices of placement but features of parents' occupational experiences have also been examined recently.

Another type of investigation examines the ways in which particular orientations toward childrearing and the actual practices of parents impinge upon a child. The majority of psychologists' studies of childrearing are of this type. They seek to learn how particular modes of parental control, techniques of reward and punishment, communication patterns and the like influence a child and his cognitive, emotional and social development and competence. For the most part, the technique of analysis is correlational, seeking to relate current parental practices to a child's current behaviors. In a few instances, longitudinal studies make it possible to trace relationships between early parental behaviors and later child outcomes.

Each student of the family is influenced by the perspective and preoccupations of his own discipline. I shall comment on a number of trends in American research that seem to me both interesting and significant; as a sociologist I shall stress studies of contextual and structural features more than studies seeking to delineate relationships between specific parental practices and child behavior.

Interest in the history of the family and of childrearing has increased enormously in the past decade, with younger historians playing a leading role. Although historical studies can tell us little about childrearing in the distant past, they have called into question earlier assumptions about the family. For example, they have demonstrated a distinct difference between the household and the family as social units in pre-industrial Europe and North America. They indicate also the greater prevalence of the nuclear family of parents and their unmarried offspring than had previously been suspected. Moreover, studies using materials available in local archives which reflect life in the past century or two do permit reconstruction of the family life course in ways that illuminate the developmental experiences of children.[1]

Another area of family research that has implications for child socialization, though not directly concerned with it, relates to the correlates and consequences of age at marriage (especially the consequences of early marriage and of pregnancies of teenagers) and to decisions about childbearing and the spacing of pregnancies. Early marriages tend to be less stable and a substantial body of research now attests to the undesirable consequences of teenage pregnancy for both mother and child.

4. CHILDHOOD

Social Change

Social change has continued to impinge sharply on the American family. During the past decade we have witnessed an increase by one-third in the proportion of women who remain single to age 25, along with a marked decrease in the average number of children born in a family.[2] The number of working mothers has more than doubled since the end of World War II—and half of all mothers of school-aged children are now in the labor force.

Working mothers of preschool children turn where they can to a variety of child care arrangements, formal and informal. In 1974, for example, approximately 1.3 million children were in licensed or approved day care centers or in family day care, 1.7 million were in informal out-of-home care and nearly 5 million were in nursery school or pre-kindergarten programs (about three-quarters of them part-time).[3] The care received in full-time day care centers has varied greatly; research on the effects of substitute care in some of the better centers suggests few differences between children reared at home and those raised in group care settings, but many day care centers, especially proprietary centers conducted for profit, give care that can only be rated "fair" or "poor."[4]

As the rate of marriage has declined, the rate of divorce has increased at all class levels. The number of single-parent families (most often a mother and young children) has nearly doubled in a decade, as a consequence both of increasing divorce on the part of parents of young children and a rise in illegitimacy, largely accounted for by adolescent pregnancies. The very strong movement for increased rights and equal treatment for women, both in and out of the home, has undoubtedly played a part in these trends.

Household size has steadily decreased over recent decades, as more housing has become available. Alternate living arrangements as opposed to the traditional family household are on the increase. In the decade of the 1960s there was an eightfold increase in the number of household heads who live apart from relatives but share their living quarters with an adult partner of the opposite sex. Communal living arrangements of groups of adults and children have also increased. In such households children tend to be warmly loved and well cared for in the early years but not to be much supervised as they grow older.[5]

Research on the effects of these changes is still very limited. There have been a number of studies which examined the effects of the presence or absence of a parent on the child's development. Most frequently, of course, it is the father who is absent. A very careful review on the topic of "Children in Fatherless Families" concludes that there is little solid evidence that mere absence of the father produces serious distortion in the child's cognitive and emotional development or gender identity.[6] In general, the way in which a mother functions with her children is far more important than the number of adults in the household. At the same time, there can be no question but that a mother who must cope alone with problems of household and income maintenance faces a much more stressful situation than one who can count on a husband's help. Much depends on the kind of child care arrangements that can be made if the mother works, and on the mother's ability to arrange some time for relaxed enjoyment of her children. Perhaps the most consistent finding relating to the effect of father absence on children's cognitive development is that boys from such families tend to have higher verbal than mathematical skills, a reversal of normal expectation when both parents are present.

More important than his mere presence or absence is the actual role of the father in the socialization of a child. Although fathers were long neglected by students of child development, they are now being studied increasingly. Indeed, several recent books are devoted entirely to the father's role in childrearing, drawing on considerable research evidence.[7] The father is important not only as a role model for the child and a source of emotional support and childrearing partner to the mother but also as a source of orientation to that segment of the culture shared by males. The father also plays an important part in the establishment and maintenance of the set of standards and values that provides the moral climate of a family. This role may be filled by others, as can the emotional support role, but there is increasing evidence that both mothers and children benefit markedly from a father's active participation in child care.

Social Class and Socialization

It is 20 years since Urie Bronfenbrenner published his analysis of "Socialization and Social Class Through Time and Space," showing evidence that in some respects, at least, the social classes were moving closer together in their childrearing practices.[8] Social class has continued to be a major variable in examining not only parental practices but general orientations toward childrearing, contexts of development and transitions such as those to school and work. Two decades ago Daniel Miller and Guy Swanson examined differences among families classified not only on the basis of social class but also on the setting of the father's occupation, whether "entrepreneurial" or within a large bureaucratic organization.[9]

More recently, Melvin Kohn, building upon the insights of Marx and Weber, has attempted to analyze how the conditions of occupational life affect the psyche, values and childrearing orientations of American parents.[10] In his early work, Kohn demonstrated that middle-class parents are more likely than working-class parents to want their children to be considerate of others, intellectually curious, responsible and self-controlled, while working-class parents are more likely to want their children to have good manners, to do well in school and to be obedient. Thus the middle-class parent tends to put emphasis upon self-direction for the child, the working-class parent to place a higher value on conformity. But Kohn does not stop at social class. He is able to demonstrate that fathers whose jobs entail self-direction—who are not closely supervised, who work with ideas rather than things, and who face great complexity on the job—value self-direction in their children, while those whose work requires them to conform to close supervision and a highly structured work situation are more likely to want their children to be conforming.

Categorization by social class, as used in American research, rests largely on educational attainment and occupational status. So measured, social class indexes a whole host of differences in life experiences, among them the type of housing occupied and the neighborhood of residence, the role differentiation between the parents, the fluency of language usage in the family, tendencies toward concrete versus abstract verbal expression and thought and the breadth of social participation of the parents. Under these circumstances, to know that various aspects of child development are associated with social class is merely to reiterate that life style and life chances are markedly influenced by a family's place in the social structure. Many efforts are now being directed to delineating the specific mechanisms by which such effects are mediated, since childrearing practices are associated with, but by no means wholly determined by, class position.[11]

Much recent research focuses on the development of competence in the child. In general, the competent child is characterized as self-reliant, self-controlled and uninhibited in his relationships with others but not overly aggressive or demanding. Competence obviously entails the development of cognitive, physical and social skills as well as emotional control. Parental warmth and encouragement coupled with parental control seem to be essential ingredients in the production of competent children, but it appears that the combinations of parental acceptance and control appropriate for producing competent young children differ for boys and girls. For the induction of competence away from home in children of preschool age, recent research indicates that neither the affectionate and permissive parent nor the cold, authoritarian parent is as effective as the parent who combines affection with strict control and yet encourages joint discussion of family-related issues.[12]

Another topic which has received a great deal of attention in the past decade is the nature and explanation of sex differences in personality and performance. A recent assessment of available research evidence demonstrates that many of the widely held beliefs about sex differences are simply not substantiated. For example, the long-standing belief that girls are more "social," more "suggestible" and have lower achievement motivation than boys, or that girls are better at rote learning and simple repetitive tasks while boys are better at those that require a higher level of cognitive processing—all these have been disproven by systematic research.[13] A few sex differences seem more firmly established: that girls have greater verbal ability than boys, that boys excel in mathematical ability and in visual-spatial ability beyond age 12, and that males are more aggressive.

Differences in parental response to girls and boys seem to increase as children grow older and to be greater in the case of fathers than in that of mothers, but studies bearing directly on sex-differentiated responses of fathers or of either parent toward older children are relatively rare. Recent research on the interactions of

biological and social systems suggests that parenting behaviors are more influenced by biology than some celebrants of unisex would like to believe; in particular, hormonally related responsiveness of mothers to infants seems firmly established.[14]

The fact that boys and girls are responded to quite differently by their parents in all known societies cannot be unimportant in leading to typical patternings in the development of the sexes. Since parents are often unaware of the ways in which their behaviors differ, however, observational studies of the actual behaviors of the parents toward boys and girls are needed to supplement interview materials. Such research is now going on, and, indeed, a major development in recent socialization research has been the turn to painstaking observation in place of or as a supplement to interviewing.

Methodological advances include the systematic assessment of bias and unreliability when retrospective data are obtained on early family experiences and childrearing practices; the development of observational techniques and category systems for use in both naturalistic and laboratory settings; and the utilization of short-term longitudinal studies of over-lapping cohorts in order to differentiate cohort effects—that is, the effects of particular historical events or social changes—from the effects of age.[15]

Studies by anthropologists continue to increase our knowledge of the ways in which very different social structural arrangements and cultural themes influence childrearing. Especially influential here has been the work of Beatrice and John Whiting in their studies of six different cultures.[16] Their publications have focused successively on the cultures and their general patterns of childrearing, on mothers and their behaviors (with some examination of fathers' roles as well) and on the children themselves. It should be noted that attempts to delineate modal personality types in various cultures or nations has been largely abandoned. Efforts are rather concentrated on delineating the specific linkages between features of the basic maintenance systems of societies, household composition, parental practices and child behaviors.

The anthropological perspective leads naturally to a consideration of the family as a whole. The family is an organized unit, one that functions through ongoing transactions among members and between the family and the larger community. What attributes of that unit as a whole best index its effects on the development of children? Past research leaves much to be desired, but the following are surely among those important features of families that are associated with favorable child development in the United States: harmony rather than conflict between parents; equality of authority or at most modest differences in parental power and authority; clearly patterned, mutually acceptable procedures for dealing with important problems and decisions; involvement of the child in decisions affecting him or her, as appropriate to the child's age; and accurate labeling of feelings and intentions in communication within the family.

The study of communication processes in the family has been of particular interest to workers in the field of psychiatric disorders. Parents in families with a schizophrenic child, in particular, have been found to "mislabel" their emotions and actions, to have difficulty in achieving a shared focus of attention, to disqualify their own utterances, and to maintain shared fictions about the family as a unit.[17] There is much evidence to suggest a genetic component in schizophrenia, but since fewer than half of identical twins are concordant for schizophrenia, experiential features are probably involved as well.

We are now beginning to get much more systematic research on communication processes in normal families, usually involving the observation of the family in a laboratory situation but occasionally entailing videotaping in the home.[18] It is not yet clear what the major correlates and consequences of communication deviance are, but the field seems a promising one. For example, observational studies of parent-child interaction in the home suggest that very low levels of communication from mother to child are involved in instances of markedly retarded language development.[19]

The ways in which illness or psychopathology of a parent impinge upon the child in other respects, particularly as threatening behaviors or child neglect are entailed, are also under investigation in many places. Mental illness may go unnoted and untreated, often leading to conflict between parents and feelings of abandonment by the child.[20]

Long-term longitudinal studies continue to add to our knowledge of socialization in the family. Two major contributions have come from the Institute of Human Development at Berkeley, which has for more than 40 years followed several hundred study members born in the 1920s. Jack Block and Norma Haan have examined continuities and discontinuities in personality from the pre-adolescent years to the late thirties[21] Parental warmth, acceptance and stability tend to affect the child's development and performance at all age levels, but there is much personality change well beyond childhood. Some study members who had unhappy childhoods and showed

serious problems in adolescence nevertheless arrived at age 40 competent and self-accepting. Others who gave early promise have had a more problematic time in middle age.

A recent book by Glen Elder traces the effects of economic deprivation during the depression of the 1930s upon the lives of the then pre-adolescent study members.[22] The boys, in particular, appear to have been challenged by the dismal experiences of their families and many rose to the challenge by taking on part-time jobs to help their parents. Girls, on the other hand, were more largely pushed into help with domestic chores. Not accidentally, the birth cohort to which these women belong has been the most home-oriented of any in recent decades.

In conclusion, the family mediates between the larger community and its demands and the developing child. It is responsive to the conditions of life in the society, to economic conditions, wars, and to natural catastrophes that from time to time afflict all societies. A major thrust in interdisciplinary studies now in process is to seek to trace the effects of such family adaptations upon the child.

[1]See, for example, the papers prepared for a workshop, "The Family Life Course in Historical Perspective," published in Volume I of the newly founded *Journal of Family History* (Winter 1977).

[2]See Arthur J. Norton and Paul C. Glick, "Changes in American Family Life," *CHILDREN TODAY,* May-June 1976 and Urie Bronfenbrenner, "The Challenge of Social Change to Public Policy and Developmental Research," paper presented at the President's Symposium, "Child Development and Public Policy," at the Annual Meeting of the Society for Research in Child Development, Denver, April 1975.

[3]See *Toward A National Policy for Children and Families,* Report of the Advisory Committee on Child Development, National Academy of Sciences, Washington, D.C., 1976.

[4]*Ibid,* Appendix: "Research on the Effects of Day Care on Child Development." See also Mary Keyserling, *Windows on Day Care,* National Council of Jewish Women, New York, 1972.

ACYF Family Research

In 1974 the Office of Child Development—now the Administration for Children, Youth and Families—began a 5-part research effort centering on the family as a focal point for research bearing on child development. During the first two phases of the effort, ACYF supported research projects which looked at child development within different types of families—among them, one-parent families, minority families, low-income families and families in which both parents work. Research projects also examined the interaction of families with community services, schools and other institutions. Television was studied both within the family system and as part of the institutional system.

Now, in the third phase, ACYF is supporting pilot studies to test the kinds of changes in services and interventions needed to enhance child development and improve interaction between families and institutions. Full-scale demonstrations of promising programs will be supported in the next phase.

Finally, the knowledge derived from these and other studies will be used to design new programs and policies to

meet the needs of children and families and to provide guidance for existing ACYF programs.

ACYF is also funding a series of projects which will look at ways to communicate scientific knowledge about child development to different types of families, examine the kinds of information that families have and need on child development and its impact on childrearing and study ways to make information on supportive services more accessible.

In addition, the Children's Bureau, ACYF, supports demonstration projects that are attempting to prevent foster placement of children by providing comprehensive services to families under stress. The use of social service contracts between agencies and families to improve case planning is being tested in other projects.

Demonstration projects that provide comprehensive services to abused and neglected children and their families, and research projects that are exploring factors contributing to child abuse and neglect and promising prevention and treatment techniques are supported by the National Center on Child Abuse and Neglect in the Children's Bureau.

[5] See, for example, Bennett M. Berger and Bruce M. Hackett, "On the Decline of Age Grading in Rural Hippie Communes," *Journal of Social Issues,* Volume 30, No. 2, 1974; also J. Rotnchild and S. B. Wolf, *Children of the Counterculture,* New York, Doubleday, 1976.

[6] Elizabeth Herzog and Cecelia E. Sudia, "Children in Fatherless Families," in *Review of Child Development Research,* Volume 3, edited by Bettye M. Caldwell and Henry N. Riccuti, Chicago, University of Chicago Press, 1973.

[7] For example, David B. Lynn, *The Father: His Role in Child Development,* Belmont, California, Wadsworth, 1974; Michael E. Lamb (ed.), *The Role of the Father in Child Development,* New York, Wiley, 1976.

[8] Urie Bronfenbrenner, "Socialization and Social Class Through Time and Space," in *Readings in Social Psychology,* E. E. Maccoby, T. M. Newcomb and E. L. Hartley (eds.), New York, Holt, Rinehart and Winston, 1958.

[9] Daniel R. Miller and Guy E. Swanson, *The Changing American Parent: A Study in the Detroit Area,* New York, Wiley, 1958.

[10] Melvin L. Kohn, *Class and Conformity: A Study in Values,* Second edition, Chicago, University of Chicago Press, 1977.

[11] The evidence is summarized in Alan C. Kerchkoff, *Socialization and Social Class,* Englewood Cliffs, New Jersey, Prentice Hall, 1972.

[12] See Diana Baumrind, "The Development of Instrumental Competence Through Socialization," *Minnesota Symposium on Child Psychology,* Anne Pick (ed.), Minneapolis, University of Minnesota Press, 1973.

[13] Eleanor E. Maccoby and Carol N. Jacklin, *The Psychology of Sex Differences,* Stanford, California, Stanford University Press, 1974. A critical evaluation of the shortcomings of existing evidence is provided by Jeanne Block in "Another Look at Sex Differentiation in the Socialization Behaviors of Mothers and Fathers," to be published in *Psychology of Women: Future Directions for Research,* New York, Psychological Dimensions, Inc., 1978.

[14] See Alice S. Rossi, "A Biosocial Perspective on Parenting," *Daedalus,* Spring 1977.

[15] Marian Radke Yarrow *et al.,* "Recollections of Childhood: A Study of the Retrospective Method," *Monographs of the Society for Research in Child Development,* Volume 35, No. 5, 1970. The work of Baumrind, referred to in note 12, is illustrative of developments in observational methods. Issues of cohort and longitudinal analysis are dealt with in John R. Nesselroade and H. W. Reese, *Life Span Developmental Psychology: Methodological Issues,* New York, Academic Press, 1973.

[16] See Beatrice Whiting (ed.), *Six Cultures: Studies of Child Rearing,* New York, Wiley, 1963; Leigh Minturn and W. W. Lambert, *Mothers of Six Cultures: Antecedents of Child Rearing,* New York, Wiley, 1964; and B. B. and J. M. Whiting, *Children of Six Cultures: A Psychocultural Analysis,* Cambridge, Mass., Harvard University Press, 1975.

[17] Some of the research relating to schizophrenia is reported in David Rosenthal and Seymour Kety (eds.), *The Transmission of Schizophrenia,* London, Pergammon Press, 1968. (See especially Part III, "Social, Cultural and Interpersonal Studies.")

[18] A more general review of studies of communication in the family is given by Theodore Jacob, "Family Interaction in Disturbed and Normal Families: A Methodological and Substantive Review," *Psychological Bulletin,* January 1975.

[19] Margaret Wulbert *et al.,* "Language Delay and Associated Mother-Child Interactions," *Developmental Psychology,* January 1975.

[20] Several studies examining such effects are contained in *The Child in His Family,* edited by E. James Anthony and C. Koupernik, New York, Wiley, 1970. See also J. Clausen and C. Huffine, "The Impact of Parental Mental Illness on the Children," in *Research on Community and Mental Health,* Roberta Simmons (ed.), Greenwich, Conn., JAI Press. (To be published in 1978.)

[21] Jack Block and Norma Haan, *Lives Through Time,* Berkeley, California, Bancroft Press, 1972.

[22] Glen H. Elder, Jr., *Children of the Great Depression,* Chicago, University of Chicago Press, 1974.

MOOD & MEMORY

GORDON H. BOWER

Gordon H. Bower is chairman of the psychology department at Stanford University. An experimental psychologist, he specializes in human learning and memory, and was coauthor, with Ernest Hilgard, of the textbook Theories of Learning *(Prentice-Hall), now in its fifth edition. Bower, a member of the National Academy of Sciences, describes his studies of the impact of emotion on learning as a recent sideline. This article is adapted from his Distinguished Scientific Contributions Award address given last year at a meeting of the American Psychological Association. The full address first appeared in the* American Psychologist.

An American soldier in Vietnam blacked out as he stared at the remains of his Vietnamese girlfriend, killed by Vietcong mortar fire. Vowing revenge, he plunged into the jungle. Five days later an American patrol discovered him wandering aimlessly, dazed, disoriented. His memory of the preceding week was a total blank. He had no idea where he'd been or what he'd been doing for that period. Even after his return to the U.S., he could not recall the blackout period.

Several years later, a psychiatrist treating him for depression put him under hypnosis and encouraged him to reconstruct events from his combat days, both before and during the blackout. He calmly recalled earlier events, but when he neared the traumatic episode, he suddenly became very agitated, and more memories came pouring out. He began to relive the trauma of seeing his girlfriend's body and felt again the revulsion, outrage, and lust for revenge. Then, for the first time, he remembered what had happened after the mortar attack. He had commandeered a jeep, traveled alone for days deep into Vietcong territory, stalked Vietcong in the jungles, and set scores of booby traps with captured weapons before stumbling upon the American patrol. Curiously, after awakening from his hypnotic trance, the patient could remember only a few incidents singled out by the psychiatrist. But further treatments, described in the book *Trance and Treatment* by psychiatrists Herbert and David Spiegel, enabled him to bring more details into consciousness.

This case illustrates an extreme memory dissociation; the blackout events could be recalled in one state (of hypnotic agitation) but not in another (normal consciousness). Hypnosis helped the person return to the psychic state he was in when the blackout started; at that point, the emotional feelings returned, as did memories of the details of the blacked-out events. Psychoanalysts might call this a case of severe repression, which refers to the avoidance of anxiety-provoking memories. I believe such a label equates an observation with an explanation that may or may not be correct. Instead, I believe the soldier's case is an example of state-dependent memory, a more encompassing theory that refers to people's difficulty in recovering during one psychological state any memories acquired in a different state. State-dependency and repression are competing theories of forgetting. Each offers an explanation of why the soldier's blacked-out memories returned as he relived his trauma. But repression could not explain why a happy person can find happy memories easier to recover than sad ones.

The idea of studying the efficiency of memory during different psychological states—for example, while in hypnosis, under the influence of drugs, or after sensory deprivation—has been around for more than 50 years. However, previous investigations have been limited both in method and scope. While many clinical examples of state-dependency occur—for instance, violent "crimes of passion" are often blocked out but hypnotically recoverable by the assailant—such cases are really too rare, inconvenient, and complex for an adequate scientific analysis. In an earlier article in *Psychology Today* ("I Can't Remember What I Said Last Night, But It Must Have Been Good," August 1976), Roland Fischer described several examples and conjectured that memories are bound up with specific levels of physiological arousal. But my research shows that arousal level is not nearly as critical as the type of emotion felt—whether fear, depression, anger, or happiness. The most common laboratory method in previous studies of state-dependency used rats, learning with or without an injection of a drug like Amytal and later tested in either a drugged or nondrugged state.

As an experimentalist, I was challenged to produce state-dependent memory in the laboratory, using normal people and trying to evoke commonly occurring emotions as "states." Two of my students, Steve Gilligan and Ken Monteiro, and I were especially interested in trying to produce such learning using different emotions, such as depression, joy, fear, and anger. This turned into a more ambitious project when we found evidence not only of state-dependent memory but also of related emotional influences on think-

 From *Psychology Today*, June 1981. The full address first appeared in *The American Psychologist*, Vol. 36, No. 2, February 1981. Reprinted by permission.

ing, judging, and perceiving. First, I'll describe our work on state-dependent memory.

The technique we employed for inducing moods used our subjects' imaginations, guided by hypnotic suggestion. College students who were very hypnotizable volunteered for our study. After hypnotizing them, we asked them to get into a happy or sad mood by imagining or remembering a scene in which they had been delightfully happy or grievously sad. Often the happy scene was a moment of personal success or of close intimacy with someone; the sad scenes were often of personal failure or the loss of a loved one. We told them to adjust the intensity of their emotion until it was strong but not unbearable—it was important for them to function well enough to learn. After getting into a mood state, the subjects performed a learning experiment for 20 or 30 minutes, after which they were returned to a pleasantly relaxed state before debriefing. (These procedures are harmless and our subjects have willingly volunteered for further experiments.)

After some pilot work, we found that strong mood state-dependent memory could be produced by teaching people two sets of material (such as word lists)—one while happy, the other while sad—and then asking them to remember one set in a happy or a sad mood. In one study, groups of hypnotized subjects learned List A while happy or sad, then learned List B while happy or sad, and then recalled the List A while happy or sad. The lists were 16 words long; memory was always tested by free recall. The groups can be classified into three conditions. In the first, control subjects learned and recalled both lists in a single mood, happy for half of them and sad for the other half. In the second condition, the subjects learned List A in one mood, learned List B in a different mood, and recalled List A in their original mood; these subjects should have recalled more than the control subjects because their different learning moods helped them to isolate the two lists, thus reducing confusion and interference from List B when they tried to recall List A. The third, interference condition was just the reverse; those students tried to recall List A when they were in their second, List B mood. Their recall of List A should have suffered, because the recall mood evokes memories of the wrong List B rather than the target List A.

When we returned subjects to their original moods, we did so by having them call up scenes different from their original ones. For example, if a woman originally induced happiness by reliving a scene of herself scoring the winning goal in a soccer match, we would instruct her to return to the happy mood by imagining a different scene, such as riding a horse along the beach. We had subjects use a second imagined situation so that any memory advantage obtained for same-mood testing would be due to overlap of moods, not to overlap of imaginary scenes.

A person's retention score was calculated as the percentage of originally learned items that were recalled on the later test. The results are in the chart on page 64; there is an obvious state-dependent effect. People who were sad during recall remembered about 80 per-

cent of the material they had learned when they were sad, compared with 45 percent recall of the material they had learned when they were happy. Conversely, happy recallers remembered 78 percent of their happy list, versus 46 percent of their sad list. The state-dependent memory effect shows up in the crossover of these lines on the chart. A good metaphor for this is to suppose that you have one bulletin board for happy moods and another for sad moods. On each board you post the messages you learn while in that mood. You will be able to read off your messages from the happy bulletin board best if you first get into a happy mood, and the messages on the sad bulletin board best when you get into a sad mood.

Aside from the state-dependent effect, I am often asked whether people learn better when they are happy or when they are sad. Others have found that clinically depressed patients are often poor learners. However, in all of our experiments with word lists, we never have found a difference in overall learning rate or later retention that was due to the subject's mood. I suspect this reflects our control over the hypnotic subjects' motivation to do as well as possible in the learning task despite their happy or sad feelings.

We next addressed the issue of whether state-dependency would occur for recall of actual events drawn from a person's emotional life. We enlisted some volunteers who agreed to record such emotional events in a daily diary for a week. We gave these subjects a book-

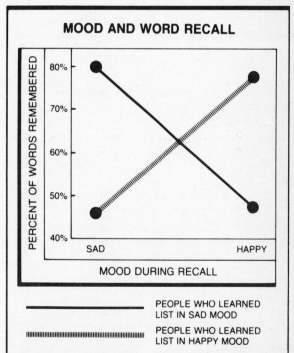

MOOD AND WORD RECALL

PERCENT OF WORDS REMEMBERED

MOOD DURING RECALL

―――――― PEOPLE WHO LEARNED LIST IN SAD MOOD

|||||||||||||||||| PEOPLE WHO LEARNED LIST IN HAPPY MOOD

Results of an experiment in which groups of students learned a list of words while in one mood and later tried to recall as many as they could while in the same mood or a different mood. They were able to remember a much larger percentage when their learning mood matched their recall mood. This "state-dependency" effect is seen in the big difference between scores in the two recall situations, dramatized by the steep incline of the two lines connecting them. (The black dots show average percentages for both groups.)

let for recording emotional incidents and discussed what we meant by an emotional incident. Examples would be the joy they experienced at a friend's wedding or the anger they experienced in an argument at work. For each incident they were to record the time, place, participants, and gist of what happened and to rate the incident as pleasant or unpleasant on a 10-point intensity scale.

Conscientious diary-keeping is demanding, and we dropped nearly half of our subjects because they failed to record enough incidents in the proper manner consistently over the week. We collected usable diaries from 14 subjects and scheduled them to return a week later. At that one-week interval they were hypnotized; half were put into a pleasant mood and the other half into an unpleasant mood, and all were asked to recall every incident they could remember of those recorded in their diaries the week before.

The percentages of recall showed the expected results: people in a happy mood recalled a greater percentage of their recorded pleasant experiences than of their unpleasant experiences; people in a sad mood recalled a greater percentage of their unpleasant experiences than of their pleasant experiences.

Remember that when subjects originally recorded their experiences, they also rated the emotional intensity of each experience. These intensity ratings were somewhat predictive: recall of more intense experiences averaged 37 percent, and of less intense experiences 25 percent. The intensity effect is important, and I will return to it later.

After subjects had finished recalling, we asked them to rate the current emotional intensity of the incidents they recalled. We found that they simply shifted their rating scale toward their current mood: if they were feeling pleasant, the recalled incidents were judged as more pleasant (or less unpleasant); if they were feeling unpleasant, the incidents were judged more unpleasant (or less pleasant) than originally. That should be familiar—here are the rose-colored glasses of the optimist and the somber, gray outlook of the pessimist.

Is it possible that recording incidents in a diary and rating them as pleasant or unpleasant encourages subjects to label their experiences in this manner and in some way gives us the results we want? Perhaps. To avoid such contaminants, in our next experiment we simply asked people to recall childhood incidents. We induced a happy or sad mood in our subjects and asked them to write brief descriptions of many unrelated incidents of any kind from their pre-high school days. Subjects were asked to "hop around" through their memories for 10 minutes, describing an incident in just a sentence or two before moving on to some unrelated incident.

The next day, we had the subjects categorize their incidents as pleasant, unpleasant, or neutral while unhypnotized and in a normal mood (so that their mood would not influence how pleasant or unpleasant they rated an event). The few neutral incidents recalled were discarded, and the chart below shows the main results. Happy subjects retrieved many more pleasant than unpleasant memories (a 92 percent bias); sad sub-

jects retrieved slightly more unpleasant than pleasant memories (a 55 percent bias in the reverse direction).

What the subjects reported was enormously dependent on their mood when recalling. That is state-dependent memory: the subjects presumably felt pleasant or unpleasant at the time the incidents were stored, and their current mood selectively retrieves the pleasant or the unpleasant memories.

What kind of theory can explain these mood-state dependent effects? A simple explanation can be cast within the old theory that memory depends upon associations between ideas. All we need to assume is that an emotion has the same effect as an "active idea unit" in the memory system. Each distinct emotion is presumed to have a distinct unit in memory that can be hooked up into the memory networks. The critical assumption is that an active emotion unit can enter into association with ideas we think about, or events that happened, at the time we are feeling that emotion. For instance, as the ideas recording the facts of a parent's funeral are stored in memory, a powerful association forms between these facts and the sadness one felt at the time.

Retrieval of some contents from memory depends upon activating other units or ideas that are associated with those contents. Thus, returning to the scene of a funeral, the associations activated by that place may cause one to reexperience the sadness felt earlier at the funeral. Conversely, if a person feels sad for some reason, activation of that emotion will bring into consciousness remembrances of associated ideas—most likely other sad events.

This theory easily explains state-dependent retrieval. In the first experiment, for example, the words of List A became associated both with the List A label and with the mood experienced at that time. Later, the words from List A can be retrieved best by reinstating the earlier List A mood, since that mood is a strongly associated cue for activating their memory. On the contrary, if a person had to recall List A while feeling in a different (List B) mood, that different mood would arouse associations that competed with recall of the correct items, thus reducing the memory scores. The same reasoning explains how one's current mood selectively retrieves personal episodes associated originally with pleasant or unpleasant emotions.

Beyond state-dependent memory, the network theory also helps to explain a number of influences of emotion on selective perception, learning, judgment, and thinking. When aroused, an emotion activates relevant concepts, thoughts, and frameworks for categorizing the social world. We have confirmed, for example, that people who are happy, sad, or angry produce free associations that are predominantly happy, sad, or angry, respectively. Similarly, when asked to fantasize or make up an imaginative story to pictures of the Thematic Apperception Test (TAT), they produce happy, sad, or hostile fantasies, depending on their emotional state. If asked for top-of-the-head opinions about their acquaintances, or the performance of their car or TV, they give highly flattering or negative evaluations, according to their mood. Also, their mood causes them to be optimistic or pessimistic in prognosticating future events about themselves and the nation. These influences can be seen as veiled forms of state-dependent retrieval of either the positive or negative memories about the person, event, or object.

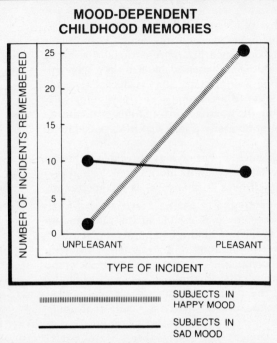

MOOD-DEPENDENT CHILDHOOD MEMORIES

NUMBER OF INCIDENTS REMEMBERED

TYPE OF INCIDENT

|||||||||||||||||||| SUBJECTS IN HAPPY MOOD

————————— SUBJECTS IN SAD MOOD

Another experiment showing the state-dependency effect. Groups of students were put into a sad or a happy mood and then asked to remember incidents from childhood. Later, they labeled the incidents as either pleasant or unpleasant. Happy subjects recalled far more pleasant than unpleasant incidents. Sad subjects retrieved slightly more unpleasant memories. (The black dots show averages for both groups.)

Mood affects the way we "see" other people. Social interactions are often ambiguous, and we have to read the intentions hidden behind people's words and actions. Is that person being steadfast in arguing his position or is he being pigheaded and obstructive? Was his action courageous or reckless? Was that remark assertive or aggressive? In reading others' intentions the emotional premise from which we begin strongly influences what we conclude. Thus the happy person seems ready to give a charitable, benevolent interpretation of social events, whereas the grouch seems determined to find fault, to take offense, or to take the uncharitable view. We find that these effects appear just as strongly when people are judging themselves on competence or attractiveness as well as when they're judging others. For example, when our subjects were in a depressed mood, they tended to judge their actions moment-by-moment in a videotaped interview as inept, unsociable, and awkward; but if they were in a happy mood, they judged their behaviors as confident, competent, and warmly sociable. Thus, social "reality" is constructed in the eye of the beholder, and that eye is connected to the emotions.

The network theory further predicts that an emotion should act as a selective filter in perception, letting in signals of a certain emotional wavelength and filtering out others. The emotional state adjusts the filter so that the person will attend more to stimulus material that agrees with or supports the current emotion. An analogy is that our feelings are like a magnet that selects iron filings from a heap of dust, attracting to itself whatever incoming material it can use.

Emotional effects can de demonstrated in attention and perception as well as learning. Thus, a sad person will look at pictures of sad faces more than happy faces; a happy person will dwell longer on happy faces. People who are happy from having just succeeded at an intelligence task have lower thresholds for seeing "success" words; subjects who've failed have lower thresholds for "failure" words.

The main work we've done on this salience effect concerns selective learning. In one of our experiments, subjects were made happy or sad by a posthypnotic suggestion as they read a brief story about two college men getting together and playing a friendly game of tennis. André is happy—everything is going well for him; Jack is sad—nothing is going well for him. The events of the two men's lives and their emotional reactions are vividly described in the story, which is a balanced, third-person account. When our subjects finished reading the story, we asked them to tell us who they thought the central character was and who they identified with. We found that readers who were happy identified with the happy character, thought the story was about him, and thought the story contained more statements about him; readers who were sad identified with the sad character and thought there were more statements about him.

Our subjects tried to recall the text the next day while in a neutral mood. Eighty percent of the facts remembered by the sad readers were about the sad character; 55 percent of the facts remembered by the happy readers were about the happy character. This is a mood-congruity effect; readers attend more to the character whose mood matches their own. Since all recallers were in a neutral mood, their differing recall results from their selective learning; it is not a state-dependent effect, since that requires varying subjects' mood during recall as well as during learning.

How is the mood-congruity effect explained? Why is mood-congruent material more salient and better learned? Two explanations seem worth considering.

The first hypothesis is that when one is sad, a sad incident in a story is more likely than a happy incident to remind one of a similar incident in one's life; vice versa, when one is happy. (Note that this is simply the state-dependent retrieval hypothesis.) An additional assumption is that the reminding is itself an event that enhances memory of the prompting event. This may occur because the old memory allows one to elaborate on the prompting event or to infuse it with greater emotion. In other studies, we have found that people remember descriptions of events that remind them of a specific incident in their lives far better than they recall descriptions that don't cause such reminiscence. To summarize, this hypothesis states that the mood-congruity effect is produced by selective reminding.

The second hypothesis, which complements the first, is that the mood-congruity effect comes from the influence of emotional intensity on memory. We demonstrated this idea in a study in which subjects were put in a sad or happy mood during hypnosis and then asked to read a story that went from a happy incident to a sad incident to a happy incident, and so on. Although our hypnotized subjects in several experiments tried to maintain steady moods, they reported that a mood's intensity would wane when they read material of the opposite quality. Thus happy subjects would come down from their euphoria when they read about a funeral or about unjust suffering; those topics intensified the sad subjects' feelings.

But why are intense emotional experiences better remembered? At present, there are many explanations. One is that events that evoke strong emotional reactions in real life are typically events involving personally significant goals, such as attaining life ambitions, elevating self-esteem, reducing suffering, receiving love and respect, or avoiding harm to oneself or loved ones. Because of their central importance, those goal-satisfying events are thought about frequently and become connected to other personal plans and to one's self-concept.

Intense experiences may also be remembered better because they tend to be rare. Because they are distinctive, they are not easily confused with more numerous, ordinary experiences; they tend to be insulated from interference.

The explanation of the mood-congruity effect that fits our lab results best is that mood-congruous experiences may be rehearsed more often and elaborated or thought about more deeply than experiences that do not match our mood. Thus sad people may be quickly able to embroider and elaborate upon a sad incident, whereas they don't elaborate on happy incidents. Because their sad incidents are elaborated and processed more deeply, sad people learn their sad incidents better than their happy ones. The same principle explains why happy people learn happy incidents better.

Having reviewed some evidence for mood-congruity and mood-dependency effects, let me speculate a bit about the possible implications for other psychological phenomena.

One obvious phenomenon explained by mood dependency is mood perpetuation—the tendency for a dominant emotion to persist. A person in a depressed mood will tend to recall only unpleasant events and to project a bleak interpretation onto the common events of life. Depressing memories and interpretations feed back to intensify and prolong the depressed mood, encouraging the vicious circle of depression. One class of therapies for depression aims at breaking the circle by restructuring the way depressed people evaluate personal events. Thus patients are taught to attend to and rehearse the positive, competent aspects of their lives and to change their negative evaluations.

State-dependent memory helps us to interpret several other puzzling phenomena. One is the impoverished quality of dream recall shown by most people. Most people forget their dreams, which is surprising considering that such bizarre, emotionally arousing events would be very memorable had they been witnessed in the waking state. But the sleep state (even the REM state of dreaming) seems psychologically distinct from the waking state, and dream memories may thus not be easily transferred from one state to the other.

State-dependent retention may also explain the fact that people have very few memories from the first year or two of their lives. In this view, as infants mature, their brains gradually change state, so that early memories become inaccessible in the more mature state. The problem with this hypothesis is that it leads to no novel predictions to distinguish it from the plethora of competing explanations of infantile amnesia, which generally range from Freud's repression theory to the theory that the infant's and adult's "languages of thought" mismatch so badly that adults can't "translate" records of infant memories.

State-dependent memory has been demonstrated previously with psychoactive drugs like marijuana, alcohol, amphetamines, and barbiturates. For example, after taking amphetamines, subjects remember material they have learned while high on the drug in the past better than when they are not high on it. Since such substances are also mood-altering drugs, a plausible hypothesis is that they achieve their state-dependent effect by virtue of their impact on mood.

To summarize, we have now found powerful influences of emotional states upon selective perception, learning, retrieval, judgments, thought, and imagination. The emotions studied have been quite strong, and their temporary psychological effects have been striking. What is surprising to me is that the emotional effects on thinking uncovered so far seem understandable in terms of relatively simple ideas—the notion that an aroused emotion can be viewed as an active unit in an associative memory and that it stimulates memories, thoughts, perceptual categories, and actions. Perhaps this is as it should be—that theories developed in one field (memory) aid our understanding of phenomena in another field (for example, emotional fantasies in the psychiatric clinic). Certainly that is the goal of all basic science.

Erik Erikson's Eight Ages Of Man
One man in his time plays many psychosocial parts

David Elkind

DAVID ELKIND *is professor of psychology and psychiatry at the University of Rochester.*

At a recent faculty reception I happened to join a small group in which a young mother was talking about her "identity crisis." She and her husband, she said, had decided not to have any more children and she was depressed at the thought of being past the child-bearing stage. It was as if, she continued, she had been robbed of some part of herself and now needed to find a new function to replace the old one.

When I remarked that her story sounded like a case history from a book by Erik Erikson, she replied, "Who's Erikson?" It is a reflection on the intellectual modesty and literary decorum of Erik H. Erikson, psychoanalyst and professor of developmental psychology at Harvard, that so few of the many people who today talk about the "identity crisis" know anthing of the man who pointed out its pervasiveness as a problem in contemporary society two decades ago.

Erikson has, however, contributed more to social science than his delineation of identity problems in modern man. His descriptions of the stages of the life cycle, for example, have advanced psychoanalytic theory to the point where it can now describe the development of the healthy personality on its own terms and not merely as the opposite of a sick one. Likewise, Erikson's emphasis upon the problems unique to adolescents and adults living in today's society has helped to rectify the one-sided emphasis on childhood as the beginning and end of personality development.

Finally, in his biographical studies, such as "Young Man Luther" and "Gandhi's Truth" (which has just won a National Book Award in philosophy and religion), Erikson emphasizes the inherent strengths of the human personality by showing how individuals can use their neurotic symptoms and conflicts for creative and constructive social purposes while healing themselves in the process.

It is important to emphasize that Erikson's contributions are genuine advances in psychoanalysis in the sense that Erikson accepts and builds upon many of the basic tenets of Freudian theory. In this regard, Erikson differs from Freud's early co-workers such as Jung and Adler who, when they broke with Freud, rejected his theories and substituted their own.

Likewise, Erikson also differs from the so-called neo-Freudians such as Horney, Kardiner and Sullivan who (mistakenly, as it turned out) assumed that Freudian theory had nothing to say about man's relation to reality and to his culture. While it is true that Freud emphasized, even mythologized, sexuality, he did so to counteract the rigid sexual taboos of his time, which, at that point in history, were frequently the cause of neuroses. In his later writings, however, Freud began to concern himself with the executive agency of the personality, namely the ego, which is also the repository of the individual's attitudes and concepts about himself and his world.

It is with the psychosocial development of the ego that Erikson's observations and theoretical constructions are primarily concerned. Erikson has thus been able to introduce innovations into psychoanalytic theory without either rejecting or ignoring Freud's monumental contribution.

The man who has accomplished this notable feat is a handsome Dane, whose white hair, mustache, resonant accent and gentle manner are reminiscent of actors like Jean Hersholt and Paul Muni. Although he is warm and outgoing with friends, Erikson is a rather shy man who is uncomfortable in the spotlight of public recognition. This trait, together with his ethical reservations about making public even disguised case material, may help to account for Erikson's reluctance to publish his observations and conceptions (his first book appeared in 1950, when he was 48).

In recent years this reluctance to publish has diminished and he has been appearing in print at an increasing pace. Since 1960 he has published three books, "Insight and Responsibility," "Identity: Youth and Crisis" and "Gandhi's Truth," as well as editing a fourth, "Youth: Change ·and Challenge." Despite the accolades and recognition these books have won for him, both in America and abroad, Erikson is still surprised at the popular interest they have generated and is a little troubled about the possibility of being misunderstood and misinterpreted. While he would prefer that his books spoke for themselves and that he was left out of the picture, he has had to accede to popular demand for more information about himself and his work.

The course of Erikson's professional career has been as diverse as it has been unconventional. He was born in Frankfurt, Germany, in 1902 of Danish parents. Not long after his birth his father died, and his mother later married the pediatrician who had cured her son of a childhood illness. Erikson's stepfather urged him to become a physician, but the boy declined and became an artist instead—an artist who did portraits of children. Erikson says of his post-adolescent years, "I was an artist then, which in Europe is a euphemism for a young man with some talent and nowhere to go." During this period he settled in Vienna and worked as a tutor in a family friendly with Freud's. He met Freud on informal occasions when the families went on outings together.

These encounters may have been the impetus to accept a teaching appointment at an American school in Vienna founded by Dorothy Burlingham and directed by Peter Blos (both now well known on the American psychiatric scene). During these years (the late nineteen-twenties) he also undertook and completed psychoanalytic training with Anna Freud and August Aichhorn. Even at the outset of his career, Erikson gave evidence of the breadth of his interests and activities by being trained and certified as a Montessori teacher. Not surprisingly, in view of that training, Erikson's first articles dealt with psychoanalysis and education.

It was while in Vienna that Erikson met and married Joan Mowat Serson, an American artist of Canadian descent. They came to America in 1933, when Erikson was invited to practice and teach in Boston. Erikson was, in fact, one of the first if not the first child-analyst in the Boston area. During the next two decades he held clinical and academic appointments at Harvard, Yale and Berkeley. In 1951 he joined a group of psychiatrists and psychologists who moved to Stockbridge, Mass., to start a new program at the Austen Riggs Center, a private residential treatment center for disturbed young people. Erikson remained at Riggs until 1961, when he was appointed professor of human development and lecturer on psychiatry at Harvard. Throughout his career he has always held two or three appointments simultaneously and has traveled extensively.

Perhaps because he had been an artist first, Erikson has never been a conventional psychoanalyst. When he was treating children, for example, he always insisted on visiting his young patients' homes and on having dinner with the families. Likewise in the nineteen-thirties, when anthropological investigation was described to him by his friends Scudder McKeel, Alfred Kroeber and Margaret Mead, he decided to do field work on an Indian reservation. "When I realized that Sioux is the name which we [in Europe] pronounced "See us" and which for us was the American Indian, I could not resist." Erikson thus antedated the anthropologists who swept over the Indian reservations in the post-Depression years. (So numerous were the field workers at that time that the stock joke was that an Indian family could be defined as a mother, a father, children and an anthropologist.)

Erikson did field work not only with the Oglala Sioux of Pine Ridge, S. D. (the tribe that slew Custer and was in turn slaughtered at the Battle of Wounded Knee), but also with the salmon-fishing Yurok of Northern California. His reports on these experiences revealed his special gift for sensing and entering into the world views and modes of thinking of cultures other than his own.

It was while he was working with the Indians that Erikson began to note syndromes which he could not explain within the confines of traditional psychoanalytic theory. Central to many an adult Indian's emotional problems seemed to be his sense of uprootedness and lack of continuity between his present life-style and that portrayed in tribal history. Not only did the Indian sense a break with the past, but he could not identify with a future requiring assimilation of the white culture's values. The problems faced by such men, Erikson recognized, had to do with the ego and with culture and only incidentally with sexual drives.

The impressions Erikson gained on the reservations were reinforced during World War II when he worked at a veterans' rehabilitation center at Mount Zion Hospital in San Francisco. Many of the soldiers he and his colleagues saw seemed not to fit the traditional "shell shock" or "malingerer" cases of World War I. Rather, it seemed to Erikson that many of these men had lost the sense of who and what they were. They were having trouble reconciling their activities, attitudes and feelings as soldiers with the activities, attitudes and feelings they had known before the war. Accordingly, while these men may well have had difficulties with repressed or conflicted drives, their main problem seemed to be, as Erikson came to speak of it at the time, "identity confusion."

It was almost a decade before Erikson set forth the implications of his clinical observations in "Childhood and Society." In that book, the summation and integration of 15 years of research, he made three major contributions to the study of the human ego. He posited (1) that, side by side with the stages of psychosexual development described by Freud (the oral, anal, phallic, genital, Oedipal and pubertal), were psychosocial stages of ego development, in which the individual had to establish new basic orientations to himself and his social world; (2) that personality development continued throughout the whole life cycle; and (3) that each stage had a positive as well as a negative component.

Much about these contributions—and about Erikson's way of thinking—can be understood by looking at his scheme of life stages. Erikson identifies eight stages in the human life cycle, in each of which a new dimension of "social interaction" becomes possible—that is, a new dimension in a person's interaction with himself, and with his social environment.

TRUST vs. MISTRUST

The first stage corresponds to the oral stage in classical psychoanalytic theory and usually extends through the first year of life. In Erikson's view, the new dimension of social interaction that emerges during this period involves basic *trust* at the one extreme, and *mistrust* at the other. The degree to which the child comes to trust the world, other people and himself depends to a considerable extent upon the quality of the care that he receives. The infant whose needs are met when they arise, whose discomforts are quickly removed, who is cuddled, fondled, played with and talked to, develops a sense of the world as a safe place to be and of people as helpful and dependable. When, however, the care is inconsistent, inadequate and rejecting, it fosters a basic mistrust, an attitude of fear and suspicion on the part of the infant toward the world in general and people in particular that will carry through to later stages of development.

It should be said at this point that the problem of basic trust-versus-mistrust (as is true for all the later dimensions) is not resolved once and for all during the first year of life; it arises again at each successive stage of development. There is both hope and danger in this. The child who enters school with a sense of mistrust may come to trust a particular teacher who has taken the trouble to make herself trustworthy; with this second chance, he overcomes his early mistrust. On the other hand, the child who comes through infancy with a vital sense of trust can still have his sense of mistrust activated at a later stage if, say, his parents are divorced and separated under acrimonious circumstances.

This point was brought home to me in a very direct way by a 4-year-old patient I saw in a court clinic. He was being seen at the court clinic because his adoptive parents, who had had him for six months, now wanted to give him back to the agency. They claimed that he was cold and unloving, took things and could not be trusted. He was indeed a cold and apathetic boy, but with good reason. About a year after his illegitimate birth, he was taken away from his mother, who had a drinking problem, and was shunted back and forth among several foster homes. Initially he had tried to relate to the persons in the foster homes, but the relationships never had a chance to develop becuase he was moved at just the wrong times. In the end he gave up trying to reach out to others, because the inevitable separations hurt too much.

Like the burned child who dreads the flame, this emotionally burned child shunned the pain of emotional involvement. He had trusted his mother, but now he trusted no one. Only years of devoted care and patience could now undo the damage that had been done to this child's sense of trust.

AUTONOMY vs. DOUBT

Stage Two spans the second and third years of life, the period which Freudian theory calls the anal stage. Erikson sees here the emergence of *autonomy*. This autonomy dimension builds upon the child's new motor and mental abilities. At this stage the child can not only walk but also climb, open and close, drop, push and pull, hold and let go. The child takes pride in these new accomplishments and wants to do everything himself, whether it be pulling the wrapper off a piece of candy, selecting the vitamin out of the bottle or flushing the toilet. If parents recognize the young child's need to do what he is capable of doing at his own pace and in his own time, then he develops a sense that he is able to control his muscles, his impulses, himself and, not insignificantly, his environment—the sense of autonomy.

When, however, his caretakers are impatient and do for him what he is capable of doing himself, they reinforce a sense of shame and doubt. To be sure, every parent has rushed a child at times and children are hardy enough to endure such lapses. It is only when caretaking is consistently overprotective and criticism of "accidents" (whether these be wetting, soiling, spilling or breaking things) is harsh and unthinking that the child develops an excessive sense of shame with respect to other people and an excessive sense of doubt about own abilities to control his world and himself.

If the child leaves this stage with less autonomy than shame or doubt, he will be handicapped in his attempts to achieve autonomy in adolescence and adulthood. Contrariwise, the child who moves through this stage with his sense of autonomy buoyantly outbalancing his feelings of shame and doubt is well prepared to be autonomous at later phases in the life cycle. Again, however, the balance of autonomy to shame and doubt

set up during this period can be changed in either positive or negative directions by later events.

It might be well to note, in addition, that too much autonomy can be as harmful as too little. I have in mind a patient of 7 who had a heart condition. He had learned very quickly how terrified his parents were of any signs in him of cardiac difficulty. With the psychological acuity given to children, he soon ruled the household. The family could not go shopping, or for a drive, or on a holiday if he did not approve. On those rare occasions when the parents had had enough and defied him, he would get angry and his purple hue and gagging would frighten them into submission.

Actually, this boy was frightened of this power (as all children would be) and was really eager to give it up. When the parents and the boy came to realize this, and to recognize that a little shame and doubt were a healthy counterpoise to an inflated sense of autonomy, the three of them could once again assume their normal roles.

INITIATIVE vs. GUILT

In this stage (the genital stage of classical psychoanalysis) the child, age 4 to 5, is pretty much master of his body and can ride a tricycle, run, cut and hit. He can thus initiate motor activities of various sorts on his own and no longer merely responds to or imitates the actions of other children. The same holds true for his language and fantasy activities. Accordingly, Erikson argues that the social dimension that appears at this stage has *initiative* at one of its poles and *guilt* at the other.

Whether the child will leave this stage with his sense of initiative far outbalancing his sense of guilt depends to a considerable extent upon how parents respond to his self-initiated activities. Children who are given much freedom and opportunity to initiate motor play such as running, bike riding, sliding, skating, tussling and wrestling have their sense of initiative reinforced. Initiative is also reinforced when parents answer their children's questions (intellectual initiative) and do not deride or inhibit fantasy or play activity. On the other hand, if the child is made to feel that his motor activity is bad, that his questions are a nuisance and that his play is silly and stupid, then he may develop a sense of guilt over self-initiated activities in general that will persist through later life stages.

INDUSTRY vs. INFERIORITY

Stage Four is the age period from 6 to 11, the elementary school years (described by classical psychoanalysis as the *latency phase*). It is a time during which the child's love for the parent of the opposite sex and rivalry with the same sexed parent (elements in the so-called family romance) are quiescent. It is also a period during which the child becomes capable of deductive reasoning, and of playing and learning by rules. It is not until this period, for example, that children can really play marbles, checkers and other "take turn" games that require obedience to rules. Erikson argues that the psychosocial dimension that emerges during this period has a sense of *industry* at one extreme and a sense of *inferiority* at the other.

The term industry nicely captures a dominant theme of this period during which the concern with how things are made, how they work and what they do predominates. It is the Robinson Crusoe age in the sense that the enthusiasm and minute detail with which Crusoe describes his activities appeals to the child's own budding sense of industry. When children are encouraged in their efforts to make, do, or build practical things (whether it be to construct creepy crawlers, tree houses, or airplane models—or to cook, bake or sew), are allowed to finish their products, and are praised and rewarded for the results, then the sense of industry is enhanced. But parents who see their children's efforts at making and doing as "mischief," and as simply "making a mess," help to encourage in children a sense of inferiority.

During these elementary-school years, however, the child's world includes more than the home. Now social institutions other than the family come to play a central role in the developmental crisis of the individual. (Here Erikson introduced still another advance in psychoanalytic theory, which heretofore concerned itself only with the effects of the parents' behavior upon the child's development.)

A child's school experiences affect his industry-inferiority balance. The child, for example, with an I.Q. of 80 to 90 has a particularly traumatic school experience, even when his sense of industry is rewarded and encouraged at home. He is "too bright" to be in special classes, but "too slow" to compete with children of average ability. Consequently he experiences constant failures in his academic efforts that reinforces a sense of inferiority.

On the other hand, the child who had his sense of industry derogated at home can have it revitalized at school through the offices of a sensitive and committed teacher. Whether the child develops a sense of industry or inferiority, therefore, no longer depends solely on the caretaking efforts of the parents but on the actions and offices of other adults as well.

IDENTITY vs. ROLE CONFUSION

When the child moves into adolescence (Stage Five—roughly the ages 12-18), he encounters, according to traditional psychoanalytic theory, a reawakening of the family-romance problem of early childhood. His means of resolving the problem is to seek and find a romantic partner of his own generation. While Erikson does not deny this aspect of adolescence, he points out that there are other problems as well. The adolescent matures mentally as well as physiologically and, in addition to the new feelings, sensations and desires he experiences as a result of changes in his body, he develops a multitude of new ways of looking at and thinking about the world. Among other things, those in adolescence can now think about other people's thinking and wonder about what other people think of them. They can also conceive of ideal families, religions and societies which they then compare with the imperfect families, religions and societies of their own experience. Finally, adolescents become capable of constructing theories and philosophies designed to bring all the varied and conflicting aspects of society into a working, harmonious and peaceful whole. The adolescent, in a word, is an impatient idealist who believes that it is as easy to realize an ideal as it is to imagine it.

Erikson believes that the new interpersonal dimension which emerges during this period has to do with a sense of *ego identity* at the positive end and a sense of *role confusion* at the negative end. That is to say, given the adolescent's newfound integrative abilities, his task is to bring together all of the things he has learned about himself as a son, student, athlete, friend, Scout, newspaper boy, and so on, and integrate these different images of himself into a whole that makes sense and that shows continuity with the past while preparing for the future. To the extent that the young person succeeds in this endeavor, he arrives at a sense of psychosocial identity, a sense of who he is, where he has been and where he is going.

In contrast to the earlier stages, where parents play a more or less direct role in the determination of the result of the developmental crises, the influence of parents during this stage is much more indirect. If the young person reaches adolescence with, thanks to his parents, a vital sense of trust, autonomy, initiative and industry, then his chances of arriving at a meaningful sense of ego identity are much enhanced. The reverse, of course, holds true for the young person who enters adolescence with considerable mistrust, shame, doubt, guilt and inferiority. Preparation for a successful adolescence, and the attainment of an integrated psychosocial identity must, therefore, begin in the cradle.

Over and above what the individual brings with him from his childhood, the attainment of a sense of personal identity depends upon the social milieu in which he or she grows up. For example, in a society where women are to some extent second-class citizens, it may be harder for females to arrive at a sense of psychosocial identity. Likewise at times, such as the present, when rapid social and technological change breaks down many traditional values, it may be more difficult for young people to find continuity between what they learned and experienced as children and what they learn and experience as adolescents. At such times young people often seek causes that give their lives meaning and direction. The activism of the current generation of young people may well stem, in part at least, from this search.

When the young person cannot attain a sense of personal identity, either because of an unfortunate childhood or difficult social circumstances, he shows a certain amount of *role confusion*—a sense of not knowing what he is, where he belongs or whom he belongs to. Such confusion is a frequent symptom in delinquent young people. Promiscuous adolescent girls, for example, often seem to have a fragmented sense of ego identity. Some young people seek a "negative identity," an identity opposite to the one prescribed for them by their family and friends. Having an identity as a "delinquent," or as a "hippie," or even as an "acid head," may sometimes be preferable to having no identity at all.

In some cases young people do not seek a negative identity so much as they have it thrust upon them. I remember another court case in which the defendant was an attractive 16-year-old girl who had been found "tricking it" in a trailer located just outside the grounds of an Air Force base. From about the age of 12, her mother had encouraged her to dress seductively and to go out with boys. When she returned from dates, her sexually frustrated mother demanded a kiss-by-kiss, caress-by-caress description of the evening's activities. After the mother had vicariously satisfied her sexual needs, she proceeded to call her daughter a "whore" and a "dirty tramp."

As the girl told me, "Hell, I have the name, so I might as well play the role."

Failure to establish a clear sense of personal identity at adolescence does not guarantee perpetual failure. And the person who attains a working sense of ego identity in adolescence will of necessity encounter challenges and threats to that identity as he moves through life. Erikson, perhaps more than any other personality theorist, has emphasized that life is constant change and that confronting problems at one stage in life is not a guarantee against the reappearance of these problems at later stages, or against the finding of new solutions to them.

INTIMACY vs. ISOLATION

Stage Six in the life cycle is young adulthood; roughly the period of courtship and early family life that extends from late adolescence till early middle age. For this stage, and the stages described hereafter, classical psychoanalysis has nothing new or major to say. For Erikson, however, the previous attainment of a sense of personal identity and the engagement in productive work that marks this period gives rise to a new interpersonal dimension of *intimacy* at the one extreme and *isolation* at the other.

When Erikson speaks of intimacy he means much more than love-making alone; he means the ability to share with and care about another person without fear of losing oneself in the process. In the case of intimacy, as in the case of identity, success or failure no longer depends directly upon the parents but only indirectly as they have contributed to the individual's success or failure at the earlier stages. Here, too, as in the case of identity, social conditions may help or hinder the establishment of a sense of intimacy. Likewise, intimacy need not involve sexuality; it includes the relationship between friends. Soldiers who have served together under the most dangerous circumstances often develop a sense of commitment to one another that exemplifies intimacy in its broadest sense. If a sense of intimacy is not established with friends or a marriage partner, the result, in Erikson's view, is a sense of isolation—of being alone without anyone to share with or care for.

GENERATIVITY vs. SELF-ABSORPTION

This stage—middle age—brings with it what Erikson speaks of as either *generativity or self-absorption,* and stagnation. What Erikson means by generativity is that the person begins to be concerned with others beyond his immediate family, with future generations and the nature of the society and world in which those generations will live. Generativity does not reside only in parents; it can be found in any individual who actively concerns himself with the welfare of young people and with making the world a better place for them to live and to work.

Those who fail to establish a sense of generativity fall into a state of self-absorption in which their personal needs and comforts are of predominant concern. A fictional case of self-absorption is Dickens's Scrooge in "A Christmas Carol." In his one-sided concern with money and in his disregard for the interests and welfare of his young employee, Bob Cratchit, Scrooge exemplifies the self-absorbed, embittered (the two often go together) old man. Dickens also illustrated, however, what Erikson points out: namely, that unhappy solutions to life's crises are not irreversible. Scrooge, at the end of the tale, manifested both a sense of generativity and of intimacy which he had not experienced before.

INTEGRITY vs. DESPAIR

Stage Eight in the Eriksonian scheme corresponds roughly to the period when the individual's major efforts are nearing completion and when there is time for reflection—and for the enjoyment of grandchildren, if any. The psychosocial dimension that comes into prominence now has *integrity* on one hand and *despair* on the other.

The sense of integrity arises from the individual's ability to look back on his life with satisfaction. At the other extreme is the individual who looks back upon his life as a series of missed opportunities and missed directions; now in the twilight years he realizes that it is too late to start again. For such a person the inevitable result is a sense of despair at what might have been.

These, then, are the major stages in the life cycle as described by Erikson. Their presentation, for one thing, frees the clinician to treat adult emotional problems as failures (in part at least) to solve genuinely adult personality crises and not, as heretofore, as mere residuals of infantile frustrations and conflicts. This view of personality growth, moreover, takes some of the onus off parents and takes account of the role which society and the person himself play in the formation of an individual personality. Finally, Erikson has offered hope for us all by demonstrating that each phase of growth has its strengths as well as its weaknesses and that failures at one stage of development can be rectified by successes at later stages.

The reason that these ideas, which sound so agreeable to "common sense," are in fact so revolutionary has a lot to do with the state of psychoanalysis in America. As formulated by Freud, psychoanalysis encompassed a theory of personality development, a method of studying the human mind and, finally, procedures for treating troubled and unhappy people. Freud viewed this system as a scientific one, open to revision as new facts and observations accumulated.

The system was, however, so vehemently attacked that Freud's followers were constantly in the position of having to defend Freud's views. Perhaps because of this situation, Freud's system became, in the hands of some of his followers and defenders, a dogma upon which all theoretical innovation, clinical observation and therapeutic practice had to be grounded. That this attitude persists is evidenced in the recent remark by a psychoanalyst that he believed psychotic patients could not be treated by psychoanalysis because "Freud said so." Such attitudes, in which Freud's authority rather than observation and data is the basis of deciding what is true and what is false, has contributed to the disrepute in which psychoanalysis is widely held today.

Erik Erikson has broken out of this scholasticism and has had the courage to say that Freud's discoveries and practices were the start and not the end of the study and treatment of the human personality. In addition to advocating the modifications of psychoanalytic theory outlined above, Erikson has also suggested modifications in therapeutic practice, particularly in the treatment of young patients. "Young people in severe trouble are not fit for the couch," he writes. "They want to face you, and they want you to face them, not a facsimile of a parent, or wearing the mask of a professional helper, but as a kind of over-all individual a young person can live with or despair of."

Erikson has had the boldness to remark on some of the negative effects that distorted notions of psychoanalysis have had on society at large. Psychoanalysis, he says, has contributed to a widespread fatalism—"even as we were trying to devise, with scientific determinism, a therapy for the few, we were led to promote an ethical disease among the many."

Perhaps Erikson's innovations in psychoanalytic theory are best exemplified in his psycho-historical writings, in which he combines psychoanalytic insight with a true historical imagination. After the publication of "Childhood and Society," Erikson undertook the application of his scheme of the human life cycle to the study of historical persons. He wrote a series of brilliant essays on men as varied as Maxim Gorky, George Bernard Shaw and Freud himself. These studies were not narrow case histories but rather reflected Erikson's remarkable grasp of Europe's social and political history, as well as of its literature. (His mastery of American folklore, history and literature is equally remarkable.)

While Erikson's major biographical studies were yet to come, these early essays already revealed his unique psycho-history method. For one thing, Erikson always chose men whose lives fascinated him in one way or another, perhaps because of some conscious or unconscious affinity with them. Erikson thus had a sense of community with his subjects which he adroitly used (he calls it *disciplined subjectivity*) to take his subject's point of view and to experience the world as that person might.

Secondly, Erikson chose to elaborate a particular crisis or episode in the individual's life which seemed to crystallize a life-theme that united the activities of his past and gave direction to his activities for the future. Then, much as an artist might, Erikson proceeded to fill in the background of the episode and add social and historical perspective. In a very real sense Erikson's biographical sketches are like paintings which direct the viewer's gaze from a focal point of attention to background and back again, so that one's appreciation of the focal area is enriched by having pursued the picture in its entirety.

This method was given its first major test in Erikson's study of "Young Man Luther." Originally, Erikson planned only a brief study of Luther, but "Luther proved too bulky a man to be merely a chapter in a book." Erikson's involvement with Luther dated from his youth, when, as a wandering artist, he happened to hear the Lord's Prayer in Luther's German. "Never knowingly having heard it, I had the experience, as seldom before or after, of a wholeness captured in a few simple words, of poetry fusing the esthetic and the moral; those who have suddenly 'heard' the Gettysburg Address will know what I mean."

Erikson's interest in Luther may have had other roots as well. In some ways, Luther's unhappiness with the papal intermediaries of Christianity resembled on a grand scale Erikson's own dissatisfaction with the intermediaries of Freud's system. In both cases some of the intermediaries had so distorted the original teachings that what was being preached in the name of the master came close to being the opposite of what he had himself proclaimed. While it is not possible to describe Erikson's treatment of Luther here, one can get some feeling for Erikson's brand of historical analysis from his sketch of Luther:

"Luther was a very troubled and a very gifted young man who had to create his own cause on which to focus his fidelity in the Roman Catholic world as it was then. . . . He first became a monk and tried to solve his scruples by being an exceptionally good monk. But even his superiors thought that he tried much too hard. He felt himself to be such a sinner that he began to lose faith in the charity of God and his superiors told him, 'Look, God doesn't hate you, you hate God or else you would trust Him to accept your prayers.' But I would like to make it clear that someone like Luther becomes a historical person only because he also has an acute understanding of historical actuality and knows how to 'speak to the condition' of his times. Only then do inner struggles become representative of those of a large number of vigorous and sincere young people—and begin to interest some troublemakers and hangers-on."

After Erikson's study of "Young Man Luther" (1958), he turned his attention to "middle-aged" Gandhi. As did Luther, Gandhi evoked for Erikson childhood memories. Gandhi led his first nonviolent protest in India in 1918 on behalf of some mill workers, and Erikson, then a young man of 16, had read glowing accounts of the event. Almost a half a century later Erikson was invited to Ahmedabad, an industrial city in western India, to give a seminar on the human life cycle. Erikson discovered that Ahmedabad was the city in which Gandhi had led the demonstration about which Erikson had read as a youth. Indeed, Erikson's host was none other than Ambalal Sarabahai, the benevolent industrialist who had been Gandhi's host—as well as antagonist—in the 1918 wage dispute. Throughout his stay in Ahmedabad, Erikson continued to encounter people and places that were related to Gandhi's initial experiments with nonviolent techniques.

The more Erikson learned about the event at Ahmedabad, the more intrigued he became with its pivotal importance in Gandhi's career. It seemed to be the historical moment upon which all the earlier events of Gandhi's life converged and from which diverged all of his later endeavors. So captured was Erikson by the event at Ahmedabad, that he returned the following year to research a book on Gandhi in which the event would serve as a fulcrum.

At least part of Erikson's interest in Gandhi may have stemmed from certain parallels in their lives. The 1918 event marked Gandhi's emergence as a national political leader. He was 48 at the time, and had become involved reluctantly, not so much out of a need for power or fame as out of a genuine conviction that something had to be done about the disintegration of Indian culture. Coincidentally, Erikson's book "Childhood and Society," appeared in 1950 when Erikson was 48, and it is that book which brought him national prominence in the mental health field. Like Gandhi, too, Erikson reluctantly did what he felt he had to do (namely, publish his observations and conclusions) for the benefit of his

Erikson in a seminar at his Stockbridge, Mass., home.

"Young analysts are today proclaiming a 'new freedom' to see Freud in historical perspective, which reflects the Eriksonian view that one can recognize Freud's greatness without bowing to conceptual precedent."

ailing profession and for the patients treated by its practitioners. So while Erikson's affinity with Luther seemed to derive from comparable professional identity crises, his affinity for Gandhi appears to derive from a parallel crisis of generativity. A passage from "Gandhi's Truth" (from a chapter wherein Erikson addresses himself directly to his subject) helps to convey Erikson's feeling for his subject.

"So far, I have followed you through the loneliness of your childhood and through the experiments and the scruples of your youth. I have affirmed my belief in your ceaseless endeavor to perfect yourself as a man who came to feel that he was the only one available to reverse India's fate. You experimented with what to you were debilitating temptations and you did gain vigor and agility from your victories over yourself. Your identity could be no less than that of universal man, although you had to become an Indian—and one close to the masses—first."

The following passage speaks to Erikson's belief in the general significance of Gandhi's efforts:

"We have seen in Gandhi's development the strong attraction of one of those more inclusive identities: that of an enlightened citizen of the British Empire. In proving himself willing neither to abandon vital ties to his native tradition nor to sacrifice lightly a Western education which eventually contributed to his ability to help defeat British hegemony—in

all of these seeming contradictions Gandhi showed himself on intimate terms with the actualities of his era. For in all parts of the world, the struggle now is for *the anticipatory development of more inclusive identities* . . . I submit then, that Gandhi, in his immense intuition for historical actuality and his capacity to assume leadership in 'truth in action,' may have created a ritualization through which men, equipped with both realism and strength, can face each other with mutual confidence."

There is now more and more teaching of Erikson's concepts in psychiatry, psychology, education and social work in America and in other parts of the world. His description of the stages of the life cycle are summarized in major textbooks in all of these fields and clinicians are increasingly looking at their cases in Eriksonian terms.

Research investigators have, however, found Erikson's formulations somewhat difficult to test. This is not surprising, inasmuch as Erikson's conceptions, like Freud's, take into account the infinite complexity of the human personality. Current research methodologies are, by and large, still not able to deal with these complexities at their own level, and distortions are inevitable when such concepts as "identity" come to be defined in terms of responses to a questionnaire.

Likewise, although Erikson's life-stages have an intuitive "rightness" about them, not everyone agrees with his formulations. Douvan and Adelson in their book, "The Adolescent Experience," argue that while his identity theory may hold true for boys, it doesn't for girls. This argument is based on findings which suggest that girls postpone identity consolidation until after marriage (and intimacy) have been established. Such postponement occurs, says Douvan and Adelson, because a woman's identity is partially defined by the identity of the man whom she marries. This view does not really contradict Erikson's, since he recognizes that later events, such as marriage, can help to resolve both current and past developmental crises. For the woman, but not for the man, the problems of identity and intimacy may be solved concurrently.

Objections to Erikson's formulations have come from other directions as well. Robert W. White, Erikson's good friend and colleague at Harvard, has a long standing (and warm-hearted) debate with Erikson over his life-stages. White believes that his own theory of "competence motivation," a theory which has received wide recognition, can account for the phenomena of ego development much more economically than can Erikson's stages. Erikson has, however, little interest in debating the validity of the stages he has described. As an artist he recognizes that there are many different ways to view one and the same phenomenon and that a perspective that is congenial to one person will be repugnant to another. He offers his stage-wise description of the life cycle for those who find such perspectives congenial and not as a world view that everyone should adopt.

It is this lack of dogmatism and sensitivity to the diversity and complexity of the human personality which help to account for the growing recognition of Erikson's contribution within as well as without the helping professions. Indeed, his psycho-historical investigations have originated a whole new field of study which has caught the interest of historians and political scientists alike. (It has also intrigued his wife, Joan, who has published pieces on Eleanor Roosevelt and who has a book on Saint Francis in press.) A recent issue of Daedalus, the journal for the American Academy of Arts and Sciences, was entirely devoted to psycho-historical and psycho-political investigations of creative leaders by authors from diverse disciplines who have been stimulated by Erikson's work.

Now in his 68th year, Erikson maintains the pattern of multiple activities and appointments which has characterized his entire career. He spends the fall in Cambridge, Mass., where he teaches a large course on "the human life cycle" for Harvard seniors. The spring semester is spent at his home in Stockbridge, Mass., where he participates in case conferences and staff seminars at the Austen Riggs Center. His summers are spent on Cape Cod. Although Erikson's major commitment these days is to his psycho-historical investigation, he is embarking on a study of preschool children's play constructions in different settings and countries, a follow-up of some research he conducted with preadolescents more than a quarter-century ago. He is also planning to review other early observations in the light of contemporary change. In his approach to his work, Erikson appears neither drawn nor driven, but rather to be following an inner schedule as natural as the life cycle itself.

Although Erikson, during his decade of college teaching, has not seen any patients or taught at psychoanalytic institutes, he maintains his dedication to psychoanalysis and views his psycho-historical investigations as an applied branch of that discipline. While some older analysts continue to ignore Erikson's work, there is increasing evidence (including a recent poll of psychiatrists and psychoanalysts) that he is having a rejuvenating influence upon a discipline which many regard as dead or dying. Young analysts are today proclaiming a "new freedom" to see Freud in historical perspective—which reflects the Eriksonian view that one can recognize Freud's greatness without bowing to conceptual precedent.

Accordingly, the reports of the demise of psychoanalysis may have been somewhat premature. In the work of Erik Erikson, at any rate, psychoanalysis lives and continues to beget life.

Freud's "Ages of Man"

Erik Erikson's definition of the "eight ages of man" is a work of synthesis and insight by a psychoanalytically trained and worldly mind. Sigmund Freud's description of human phases stems from his epic psychological discoveries and centers almost exclusively on the early years of life. A brief summary of the phases posited by Freud:

Oral stage—roughly the first year of life, the period during which the mouth region provides the greatest sensual satisfaction. Some derivative behavioral traits which may be seen at this time are *incorporativeness* (first six months of life) and *aggressiveness* (second six months of life).

Anal stage—roughly the second and third years of life. During this period the site of greatest sensual pleasure shifts to the anal and urethral areas. Derivative behavioral traits are *retentiveness* and *expulsiveness*.

Phallic stage—roughly the third and fourth years of life. The site of greatest sensual pleasure during this stage is the genital region. Behavior traits derived from this period include *intrusiveness* (male) and *receptiveness* (female).

Oedipal stage—roughly the fourth and fifth years of life. At this stage the young person takes the parent of the opposite sex as the object or provider of sensual satisfaction and regards the same-sexed parent as a rival. (The "family romance.") Behavior traits originating in this period are *seductiveness* and *competitiveness*.

Latency stage—roughly the years from age 6 to 11. The child resolves the Oedipus conflict by identifying with the parent of the opposite sex and by so doing satisfies sensual needs vicariously. Behavior traits developed during this period include *conscience* (or the internalization of parental moral and ethical demands).

Puberty stage—roughly 11 to 14. During this period there is an integration and subordination of oral, anal and phallic sensuality to an overriding and unitary genital *sexuality*. The genital sexuality of puberty has another young person of the opposite sex as its object, and discharge (at least for boys) as its aim. Derivative behavior traits (associated with the control and regulation of genital sexuality) are *intellectualization* and *estheticism*.

—D.E.

PIAGET

"He is advocating a revolutionary doctrine about human knowing that undermines the assumptions of much of contemporary social science."

DAVID ELKIND

David Elkind, Ph.D., is professor of psychology, psychiatry and education at the University of Rochester and the director of graduate training in developmental psychology.

It is probably fair to say that the single most influential psychologist writing today is famed Swiss developmentalist Jean Piaget. His work is cited in every major textbook on psychology, education, linguistics, sociology, psychiatry and other disciplines as well. There is now a Jean Piaget Society that each year draws thousands of members to its meetings. And there are many smaller conferences, both here and abroad, that focus upon one or another aspect of Piaget's work. It is simply a fact that no psychologist, psychiatrist or educator today can claim to be fully educated without some exposure to Piaget's work.

The man who has made this tremendous impact upon social science is now in his 79th year and shows no signs of letting up his prodigious pace of research, writing and lecturing. In the last few years, he has published more than a half-dozen books, has traveled and lectured extensively (he received an honorary degree from the University of Chicago last fall) and continues to

A/E Editor's note: Jean Piaget died in September of 1980 at the age of 84.

lead a year-long seminar attended by interdisciplinary scholars from around the world. The seminar is held in Geneva at Piaget's Center for Genetic Epistemology, which he founded more than 15 years ago. Each year Piaget invites scholars from all over the world to attend the Center for a year.

My first extensive exposure to Piaget occurred when I spent the 1964–65 year at the Center and learned something of Piagetian psychology firsthand. Although I have never been among the small intimate group of colleagues and students who surround Piaget, we have remained good friends over the years. I last visited with him about two years ago in New York when he came to America to receive the first International Kittay Award (of $25,000) for scientific achievement. The ceremony was held at the Harvard Club and was attended by a small group of invited guests, many of whom, like myself, had worked with or been associated with Piaget in some way. He presented a paper in the afternoon and a brief acceptance speech at the formal dinner that evening.

When Piaget arrived, he wore his familiar dark suit and vest with the remarkably illusionary sweater that somehow seems to keep appearing and disappearing as you watch him. He is of average height, solid in build and looks a little like Albert Einstein, a resemblance that is heightened by the fringe of long white hair that surrounds his head and by the scorched meerschaum that is inevitably in his hand or in his mouth. Up close, his most striking feature is his eyes, which somehow give the impression that they possess great depth and insight. My fantasy has always been

that Freud's eyes must have looked something like that. (Piaget's eyes are remarkably keen as well, despite his glasses. A year before the Kittay ceremony, I visited him in Geneva and we took a walk together. As we climbed the small mountain in the back of his home, he pointed out wild pigeons and flora and fauna towards the top that I could not see at all!)

That afternoon, Piaget talked about his research on conscious-awareness. As one has come to expect from him and his coworkers, the studies were most original. In one investigation, he asked children to walk upon all fours and then describe the actions they had taken. "I put my left foot out, then my right hand" and so on. What he found was that young children have great difficulty in describing their actions and that it is not until middle childhood that they can describe their actions with any exactness. Piaget also said (but was most probably joking) that he also asked some psychologists and logicians to perform the same task. The psychologists did very well but the logicians, at least according to Piaget, constructed beautiful models of crawling that had nothing to do with the real patterning of their actions.

At the dinner meeting that evening, Piaget accepted the award. In his talk, he related his fantasy of the committee meeting at which it was decided that the award should go to him. He imagined, he said, that the physicians on the committee were reluctant to give the award to a neurologist or physiologist who in turn were reluctant to see it go to a neurochemist or molecular biologist. Piaget appeared as the compromise candidate because he belonged to no particular discipline (except to the one he himself created, although he did not say this) and was, therefore, the only candidate upon whom everyone could agree. The speaker who gave Piaget the award assured him that while his fantasy was most amusing, it had no basis in fact, and that Piaget was the first person nominated and unanimously chosen by the selection committee.

There was not much chance to talk to Piaget after the dinner, but it was probably just as well. He does not really like to engage in "small talk," and at close quarters it is often difficult to find things to say to him other than discussion of research, which seems rather inappropriate at dinner parties. Yet his difficulty with small talk does not seem to extend to women, and with them he can be most charming in any setting. He is even not above clowning a bit. It should be said, too, that on formal social occasions, when he is officiating or performing some titular function, he is most gracious and appropriate. It is the small interpersonal encounters, such as occur at the dinner table, that seem most awkward for him. Perhaps his total commitment to his work has produced this social hiatus. It is certainly a small price to pay for all that he has accomplished.

Despite the enormity of his collected works, the extent of Piaget's influence is surprising for several reasons. For one thing, it has been phenomenally rapid and recent. Although Piaget began writing in the early decades of this century, his work did not become widely known in this country until the early 1960s. It is only in the past 10 years that Piaget's influence has grown in geometric progression from his previous recognition. Piaget's influence is also surprising because he writes and speaks only in French; so all of his works have had to be translated. Moreover, his naturalistic research methodology and avoidance of statistics are such that many of his studies would not be acceptable for publication in American journals of psychology. But most surprising of all is the fact that Piaget is advocating a revolutionary doctrine regarding the nature of human knowing that, if fully appreciated, effectively undermines the assumptions of much of contemporary social science.

What then is it about Piaget's work and theory that has made him so influential despite his controversial ideas and his somewhat unacceptable (at least to a goodly portion of the academic community) research methodology? The fact is that, theory and method aside, his descriptions of how children come to know and think about the world ring true to everyone's ear. When Piaget says that children believe that the moon follows them when they go for a walk at night, that the name of the sun is in the sun and that dreams come in through the window at night, it sounds strange and is yet somehow in accord with our intuitions. In fact, it was in trying to account for these strange ideas (which are neither innate, because they are given up as children grow older, nor acquired, because they are not taught by adults) that Piaget arrived at his revolutionary theory of knowing.

In the past, two kinds of theories have been proposed to account for the acquisition of knowledge. One theory, which might be called *camera* theory, suggests that the mind operates in much the same way as a camera does when it takes a picture. This theory assumes that there is a reality that exists outside of our heads and that is completely independent of our knowing processes. As does a camera, the child's mind takes pictures of this external reality, which it then stores up in memory. Differences between the world of adults and the world of children can thus be explained by the fact that adults have more pictures stored up than do the children. Individual differences in intelligence can also be explained in terms of the quality of the camera, speed of the film and so on. On this analogy, dull children would have less precise cameras and less sensitive film than brighter children.

A second, less popular theory of knowing asserts that the mind operates not as a camera but rather as a projector. According to this view, infants come into the world with a built-in film library that is part of some natural endowment. Learning about the world amounts to running these films through a projector (the mind), which displays the film on a blank screen—the world. This theory asserts then that we never learn anything new, that nothing really exists outside of our heads and that the whole world is a product of our own mental processes. Differences between the world of adults and the world of children can be explained by arguing that adults have projected a great many more films than have children. And individual differences can be explained in terms of the quality of the projection equipment or the nature and content of the films.

The projector theory of knowing has never been very popular because it seems to defy common sense. Bishop Berkeley, an advocate of this position, was once told that he would be convinced that the world was not all in his head if, when walking about the streets of London, the contents of a slop bucket chanced to hit him on the head. The value of the projector, sometimes called the *idealistic* or *platonic* theory of knowing, has been to challenge the copy theorists and to force them to take account of the part that the human imagination plays in constructing the reality that seems to exist so independently of the operations of the human mind.

In contrast to these ideas, Piaget has offered a nonmechanical, creative or *constructionist* conception of the process of human knowing. According to Piaget, children construct reality out of their experiences with the environment in much the same way that artists paint a picture from their immediate impressions. A painting is never a simple copy of the artist's impressions, and even a portrait is "larger" than life. The artist's construction involves his or her experience but only as it has been transformed by the imagination. Paintings are always unique combinations of what the artists have taken from their experience and what they have added to it from their own scheme of the world.

In the child's construction of reality, the same holds true. What children understand reality to be is never a copy of what was received by their *sense* impressions; it is always transformed by their own ways of knowing. For example, once I happened to observe a friend's child playing at what seemed to be "ice-cream wagon." He dutifully asked customers what flavor ice cream they desired and then scooped it into make-believe cones. When I suggested that he was the ice-cream man, however, he disagreed. And when I asked what he was doing, he replied, "I am going to college." It turned out his father had told him that he had worked his way through college by selling Good Hu-

"When he was 10, his observations on a sparrow were published in a science journal, initiating a publications career equalled by few."

In a few free moments, Piaget studies his own garden handiwork.

mor ice cream from a wagon. The child has recreated his own reality from material offered by the environment. From Piaget's standpoint, we can never really know the environment but only our reconstructions of it. Reality, he believes, is always a reconstruction of the environment and never a copy of it.

Looked at from this standpoint, the discrepancies between child and adult thought appear in a much different light than they do for the camera and projector theories. Those theories assume that there are only quantitative differences between the child and adult views of the world, that children are "miniature adults" in mind as well as in appearance. In fact, of course, children are not even miniature adults physically. And intellectually, the child's reality is qualitatively different from the adult's because the child's means for constructing reality out of environmental experiences are less adequate than those of the adult. For Piaget, the child progressively constructs and reconstructs reality until it approximates that of adults.

To be sure, Piaget recognizes the pragmatic value of the copy theory of knowing and does not insist that we go about asserting the role of our knowing processes in the construction of reality. He does contend that the constructionist theory of knowing has to be taken into account in education. Traditional education is based on a copy theory of knowing that assumes that if given the words, children will acquire the ideas they represent. A constructionist theory of knowing asserts just the reverse—that children must attain the concepts before the words have meaning. Thus Piaget stresses that children must be active in learning, that they have concrete experiences from which to construct reality and that only in consequence of their mental operations on the environment will they have the concepts that will give meaning to the words they hear and read. This approach to education is not new and has been advocated by such workers as Pestalozzi, Froebel, Montessori and Dewey. Piaget has, however, provided an extensive empirical and theoretical basis for an educational program in which children are allowed to construct reality through active engagements with the environment.

Piaget's concern with the educational implications of his work comes naturally because he has, for the whole of his career, been associated with the J. J. Rousseau Institute, which is essentially a training school for teachers. And Rousseau made explicit a theme that has permeated Piaget's work, namely, that child psychology is the science of education. The union of child development and educational practice is thus quite natural in Switzerland, particularly in Geneva where Rousseau once lived and worked. Indeed, Piaget's Swiss heritage, while it does not explain his genius, was certainly an important factor in determining the directions towards which his genius turned.

Besides its beauty, perhaps the most extraordinary thing about Switzerland is the number of outstanding psychologists and psychiatrists it has produced in relation to the modest size of its population (two million people). One thinks of Claparède who preceded Piaget at the Institut de Rousseau in Geneva; of Carl Gustav Jung, the great analytic psychologist; of Herman Rorschach, who created the famed Rorschach inkblot test; and of Frederich Binswanger, the existential psychiatrist. And then, of course, there is Jean Piaget. But it is important to recognize that there appears to be something in the Swiss milieu and gene pool that is conducive to producing more than its share of exceptional social scientists.

Piaget himself was born in a small village outside Lausanne. His father, a professor of history at the University of Lausanne, was particularly well-known for his gracious literary style. Piaget's mother was an ardently religious woman who was often at odds with her husband's free thinking and lack of piety. Growing up in this rather conflictual environment, Piaget turned to intellectual pursuits, in part because of his natural genius, but perhaps also as an escape from a difficult and uncomfortable life situation.

As often happens in the case of true genius, Piaget showed his promise early. When he was 10, he observed an albino sparrow and wrote a note about it that was published in a scientific journal. Thus was launched a career of publications that has had few equals in any science. When Piaget was a young adolescent, he spent a great deal of time in a local museum helping the curator, who had a fine collection of mollusks. This work stimulated Piaget to undertake his own collection and to make systematic observations of mollusks on the shores of lakes and ponds. He began reporting his observations in a series of articles that were published in Swiss journals of biology. As a result, he won an international reputation as a mollusciologist, and, on the basis of his work, he was offered, sight unseen, the curatorship of a museum in Geneva. He had to turn the offer down because he was only 16 and had not yet completed high school.

Although Piaget had a natural bent for biological observation, he was not inclined toward experimental biology. The reason, according to Piaget, was that he was "maladroit" or not well coordinated enough to perform the delicate manipulations required in the laboratory. But Piaget had other intellectual pursuits. He was very interested in philosophy, particularly in Aristotle and Bergson, who speculated about biological and natural science. Piaget was initially much impressed by the Bergsonian dualism between life forces (*élan vital*) and physical forces, but he eventually found this dualism unacceptable. More to his liking was the Aristotelian position that saw logic and reason as the unifying force in both animate and inanimate nature. What living and nonliving things have in common is that they obey rational laws. Not surprisingly, Piaget came to regard human intelligence, man's rational function, as providing the unifying principle of all the sciences—including the social, biological and natural disciplines. It was a point of view that was to guide him during his entire career.

In 1914, Piaget had intended to go to England for a year to learn English as many young Europeans did, but the war intervened. Consequently, despite rumors to the contrary, Piaget does not speak or understand spoken English very well, although he has a fair command of written English.

At the University of Lausanne, Piaget majored in biology and, not surprisingly, conducted his dissertation on mollusks. Early in his college career, Piaget took what Erik Erikson might call a "moratorium"—a period away from his studies and his family. Piaget's moratorium was spent in a Swiss mountain spa. There, he wrote a novel that described the plan of research he intended to pursue during his entire professional career. To a remarkable degree, he has followed the plan he outlined in that book.

After obtaining his doctorate, Piaget explored a number of traditional disciplines seeking one that would allow him to combine his philosophical interest in epistemology (the branch of philosophy concerned with how we know reality) and his interest in biology and natural science. He spent a brief period of time at the Burgholzli in the psychiatric clinic in Zurich where Carl Gustav Jung had once worked. In those years, he was much impressed by Freudian theory and even gave a paper on children's dreams in which Freud showed some interest. But he never had any desire to be a clinician and left the Burgholzli after less than a year.

From Zurich, Piaget traveled to Paris where he worked in the school that had once been used as an experimental laboratory by Alfred Binet. Piaget was given the chore of standardizing some of Sir Cyril Burt's reasoning tests on French children. Although the test administration was

At the age of 79, the eminent psychologist carries a challenging research load.

boring for the most part, one aspect of the work did capture his interest: often when children responded to an item, they came up with unusual or unexpected replies. Although these replies were "wrong" or "errors" for test purposes, they fascinated him. In addition, when children came up with the wrong answer to questions such as "Helen is darker than Rose and Rose is darker than Joyce, who is the fairest of the three?" Piaget was curious about the processes by which the wrong response was arrived at. It seemed to him that the contents of the children's errors and the means by which they arrived at wrong solutions were not fortuitous but systematic and indicative of the underlying mental structures that generated them.

These observations suggested to Piaget that the study of children's thinking might provide some of the answers he sought on the philosophical plane. He planned to investigate them, then move on to other problems. Instead, the study of children's thinking became his lifelong preoccupation. After Paris, Piaget moved permanently to Geneva and began his investigations of children's thinking at the J. J. Rousseau Institute. The publication of his first studies in the field, *The Language and Thought of the Child*, and later, *Judgment and Reasoning in the Child*, *The Child's Conception of the World* and *The Moral Judgment of the Child*, gained worldwide recognition and made Piaget a world-renowned psychologist before he was 30. Unfortunately, these books, which Piaget regarded as

preliminary investigations, were often debated as finished and final works.

When Claparède retired from his post as director of the Institute of Educational Science at the University of Geneva, Piaget was the unanimous choice to succeed him. Piaget retained this post, as well as his professorship at that university, until his recent retirement. As Piaget's work became more well-known, many students came to work with him and collaborate in his research efforts. One of these students was Valentine Châtenay, whom Piaget proceeded to court and to wed. In due course they had three children, Jacqueline, Laurent and Monique. These children, all grown now, have been immortalized by Piaget in three books that are now classics in the child-development literature, *The Origins of Intelligence in the Child*, *The Construction of Reality in the Child* and *Play, Dreams and Imitation in Childhood*.

The books came about in this way. After Piaget's initial studies of children's conceptions of the world, he turned to the question of how these notions came to be given up and how children arrive at veridical notions about the world. What he was groping for was a general theory of mental development that would allow him to explain both the "erroneous" ideas he had discovered in his early works and the obviously valid notions arrived at by older children and adults. It seemed clear to Piaget that the mental abilities by which children reconstruct reality have to be sought in the earliest moments of psychic exis-

"Piaget took a novel tack: he didn't posit an outer reality for the infant; he saw construction of reality as the infant's basic task."

tence; hence, the study of infants.

In his study of infants, Piaget, as had other investigators such as Milicent Shinn and Bronson Alcott, used his own children as subjects. However, Piaget's infant studies were unusual in several respects. Perhaps the most novel aspect had to do with his own perspective. He did not assume that there was an external reality for the infant to simply copy and become acquainted with. Rather, he saw the construction of reality as being the basic task of the infant. This way of looking at infant behavior allowed him to observe and study aspects of the infants' reactions that had previously been ignored or the significance of which had not been fully appreciated. Piaget noted, for example, that infants do not search after desired objects that disappear from view until about the end of the first year of life. To him, this meant that young infants had not yet constructed a notion of objects that continue to exist when they are no longer present to their senses.

Traditional psychology has been very harsh towards any hint, in psychological writings, of anthropomorphism, the readings of feelings and thoughts into others without full justification. Piaget wanted to conjecture as to what the infants' experience of the world was, but he also wanted to do this in a scientifically acceptable and testable way. His solution to this difficult problem is another testament to his genius.

In *The Origins of Intelligence in Children*, Piaget describes the evolution of children's mental operations from the outside, as it were. In this book, he introduced some of the basic concepts of his theory of intelligence, including *accommodation* (changing the action to fit the environment) and *assimilation* (changing the environment to fit the action). Piaget could demonstrate these concepts by detailed accounts of infant behavior. When infants changed the conformation of their lips to fit a nipple, this provided one of many examples of accommodation. And when infants tried to suck upon every object that brushed their lips, this was but one of many examples of assimilation.

Other important theoretical concepts were also introduced. One of these was the *schema*.

A schema is essentially a structurized system of assimilations and accommodations—a behavior pattern. Sucking, for example, as it becomes elaborated, involves both assimilation and accommodation and the pattern gets extended and generalized as well as coordinated with other action patterns. When infants begin to look at what they suck and to suck at what they see, there is a coordination of the looking and the sucking schemata. Objects are constructed by the laborious coordination of many different schemata.

In *Origins*, Piaget thus emphasized description and concepts that, at every point, could be tied to behavioral observations. They are extremely careful and detailed and reflect Piaget's early biological training. I once had the opportunity to see his notebooks, and they were filled, page after page, with very neat notations written in a very small hand. Here is an example of one of Piaget's observations:

Laurent lifts a cushion in order to look for a cigar case. When the object is entirely hidden the child lifts the screen with hesitation, but when one end of the case appears Laurent removes the cushion with one hand and with the other tries to extricate the objective. The act of lifting the screen is, therefore, entirely separate from that of grasping the desired object and constitutes an autonomous "means" no doubt derived from earlier and analogous acts.

In *The Construction of Reality in the Child*, Piaget concerned himself more with the content of infants' thought than with the mental processes. He employed many of the same observations but from the perspective of the child's-eye view of the world. These inferences were, however, always tied to concrete observations and were checked in a variety of different ways. In this book, Piaget talked about infants' sense of space, of time and of causality, but at each point buttressed the discussion with many illustrative examples and little experiments such as the following:

At 0:3 (13) Laurent, already accustomed for several hours to shake a hanging rattle by pulling

the chain attached to it . . . is attracted by the sound of the rattle (which I have just shaken) and looks simultaneously at the rattle and at the hanging chain. Then while staring at the rattle (R) he drops from his right hand a sheet he was sucking, in order to reach with the same hand for the lower end of the hanging chain (C). As soon as he touches the chain, he grasps it and pulls it, thus reconstructing the series C-R.

Piaget used this example to demonstrate the infant's construction of a notion of practical time.

One of Piaget's important conclusions from the work presented in *Reality* is that for young infants (less than three months), objects are not regarded as permanent, as existing outside the infants' immediate experience. If, for example, one is playing with a young infant who is smiling and laughing up at the friendly adult, the child will not cease to laugh if the adult moves swiftly out of sight. To the young infant, out of sight is quite literally out of mind. By the end of the first year, however, infants cry under the same circumstances. One-year-olds have constructed, via the coordination of looking and touching, schemata, a world of objects that they regard as existing outside their immediate experience and that they can respond to in their absence.

Piaget's *Play, Dreams and Imitation in Childhood* is the third work in the infant trilogy and argues that the symbols with which we represent reality are as much constructions as the reality itself. Piaget found that symbols derive from both imitation (a child opens its mouth in imitation of a match box opening) and play (a child holds up a potato chip and says, "Look, a butterfly.") In Piaget's view, therefore, symbolic activities derive from the same developmental processes that underlie the rest of mental growth and are not separate from, but are part of, intellectual development. Piaget also found that the development of symbolic processes does not usually appear before the age of two. This coincides with the everyday observation that children do not usually report dreams of "night terrors" until after the second year. It is not until that age that most children have the mental ability necessary to create dream symbols.

Piaget's studies on infants were conducted during the 1930s, at which time he was also teaching, following new lines of research and writing theoretical articles on logic and epistemology. His fame attracted many gifted students to Geneva. One of these was Gertrude Szeminska, a Polish mathematician who did some fine work on mathematics and geometry. *The Child's Conception of Number* was one fruit of their collaboration. Another gifted graduate student was Bärbel Inhelder, whose thesis on the conservation and the intellectual assessment of retarded children was a landmark in the extension of Piagetian conceptions to practical problems of assessment and evaluation. Bärbel Inhelder became Piaget's permanent collaborator, and when Piaget retired, his university chair was given to Inhelder—a significant fact in a country where women still do not have the right to vote.

During the '30s, Piaget's lifelong academic affiliations and work patterns became fully established and solidified. Although he had a university appointment from the start of his career, the J. J. Rousseau Institute did not become an official part of the university until the 1940s. Piaget worked hard to insure that it was an interdisciplinary institute so that it would not be saddled with the stigma usually associated with schools of education at universities.

Largely because of Piaget's influence, teacher training is heavily weighted in the direction of child-development theory and research. In addition to the courses on child development offered by Piaget and his staff, students must participate in child-development research. With the aid of his student population, it was possible for Piaget and his graduate students to examine large numbers of children of all ages when they were conducting a particular research investigation. The assertion, which is sometimes made, that Piaget's studies were based on very few subjects, is true only for his infancy investigations. In all of his other explorations, Piaget employed hundreds of subjects.

Piaget's general mode of working is to set up a problem for a year or for several years and then to pursue it intensely and without distraction. Indeed, when Piaget is working, say on "causality," he does not want to talk about or deal with other research problems from the past. Once he has completed a body of work, he loses interest in it and all of his energies are devoted to the task at hand. Generally, Piaget meets with his colleagues and graduate students once a week, at which time the possible ways of exploring the problem are discussed and data from ongoing studies are presented. These are lively, exciting sessions in which new insights and ideas constantly emerge and serve as stimuli for still further innovation.

I have a rather vivid memory of one particular seminar meeting. It is usually the visiting scholars who are the most vocal while the Genevan graduate students tend to be rather quiet, although they are quite animated in their own meetings. In any case, Piaget had been talking about some of the research and I interjected, saying that I was playing devil's advocate, but why did he insist upon using the words *assimilation* and *accommodation*? After all, would not the American terms *stimulus* and *response* serve equally well? The question brought instant silence from the group, most of whom were aghast and waiting for lightning to strike me where I sat. Piaget, however, was most amused and a lively twinkle came into his eye. "Well, Elkeend," he said, "you can use *stimulus* and *response* if you choose, but if you want to understand anything, I suggest that you use *assimilation* and *accommodation*."

At the end of the year, Piaget gathers up all the data that has been collected and moves to a secret hideaway in the mountains. There he takes long walks, cooks loose omelets, thinks about the work that has been done and integrates it into one or several books that he writes in longhand on square pieces of paper. Piaget has a habit of writing at least four publishable pages every day, usually very early in the morning. The remainder of his mornings are spent teaching, meeting with students and staff or with a continuation of his early morning writing. In the afternoons, Piaget routinely takes a walk. It is then that he sorts out the ideas he is working on and thus prepares for the next day's writing. To this day, Piaget keeps to this routine as his health permits. It has been estimated that he has written the equivalent of more than 50, 500-page books.

Perhaps the major achievement of the 1930s and 1940s was the elaboration of Piaget's theory of intelligence into the four stages as we know them. This theory was articulated in close connection with his conservation experiments that provided the data base. The experiments, which resembled those on the permanence of objects in infants, enabled Piaget to compare children's performance on somewhat comparable tasks at many different age levels.

As a result of numerous investigations of children's conceptions of space, time, number, quantity, speed, causality, geometry and so on, Piaget arrived at a general conception of intellectual growth. He argues that intelligence, adaptive thinking and action develop in a series of stages that are related to age. Although there is considerable variability among individual children as to when these stages appear, Piaget does argue that the sequence in which the stages appear is a necessary one. This is true because each succeeding stage grows out of and builds upon the work of the preceding stage. At each level of development, children are again confronted with constructing or reconstructing reality out of their experiences with the world constructed during the previous stage. In addition, they must not only construct new notions of space, time, number and so on, they must also either discard or integrate their previous concepts with the new ones. From a Piagetian standpoint, constructing reality never starts entirely from scratch and always involves dealing with old ideas as well as with acquiring new ones.

In the last few decades, Piaget has extended his researches into new areas, such as memory, imagery, consciousness and causality. He has refined and consolidated his theoretical conceptions and has related them to different disciplines. While it is not really possible to review all of this work here, some aspects of it are significant for education.

One of the major research contributions during this period was the study of memory from the standpoint of Piaget's developmental stages. The research was published in a book under the joint authorship of Piaget and Inhelder.

As did Frederic C. Bartlett's book *Remembering*, this 1972 work by Piaget and Inhelder, *Memory and In-*

"What he has provided is much more valuable than tightly controlled experiments: challenging ideas that open whole new research areas."

telligence, has a good chance of becoming a classic in its field. As in the case of Bartlett's book, the Piaget and Inhelder work presents new data, new conceptualizations and fresh and innovative research approaches. While *Memory and Intelligence* provides no final answers to questions about memory, it offers a richness of hypotheses, and of experimental techniques, that will stimulate other researchers for years to come. Considering that this truly innovative book was written during Piaget's 70th year, one can only marvel at his unabated creativity and productivity.

The argument of the book is straightforward enough. What is the nature of memory? Is it passive storage and retrieval or does it involve intelligence at the outset and all along the way? Piaget's answer is that memory, in the broadest sense, is a way of knowing which is concerned with discovering the past. Although symbols and images are involved in memory, they do not constitute its essence. Rather, intelligence has to be brought to bear to retrieve the past. Hence, all "memories" bear the imprint of the intellectual schemata used to reconstruct them. Intelligence leaves its mark not only on the memory itself, but even upon the original registration that can only be coded within the limits of children's existing schemata.

All of this is not particularly new and could be derived from the work of Bartlett and other writers. What is new and what gives this book its special promise of becoming a classic is the repeated demonstration that children's memory of a given past experience changes with their level of intellectual development. A child, for example, who is shown a series of size-graded sticks before he or she can understand the relations involved, and who draws it poorly, may reproduce it correctly from memory six months later. The child's intel-

lectual understanding of the series modified the memory of it in ways that are predictable from cognitive developmental theory.

To be sure, there are many questions one can raise about the "experiments" themselves. Often the number of children involved is not very large and not all the children show the expected results. The procedures are not always clearly described and the results are presented in tables of percentage passing and without the imprimatur of significance tests. This is simply Piaget's style. There is no point in being annoyed by it or in demanding that he become more rigorous. What he has provided, in the end, is much more valuable than tightly controlled experiments: namely, ideas that challenge the mind and open up whole new areas for experimental research.

The work on memory is only one of a series of areas to which Piaget and his colleagues are applying this theory of intellectual development. In addition, work on imagery, learning, consciousness and causality have all been completed or are under way. Considering that much of this "creative" intellectual work has come during Piaget's eighth decade, one has to acknowledge that creative scientific work is not necessarily the province of the young.

Piaget has also published a number of books that serve to summarize and integrate much of the work that he has done over the past half-century. These books include a general text on child development that introduces the Piagetian work for a general audience. Then there is Piaget's book on biology and knowledge that relates the developmental findings regarding intelligence to more traditional biological conceptions and shows their underlying unity. A little gem of a book, *Structuralism,* outlines in a few brief chapters the central thrust of this movement, which unites many con-

temporary workers, including Piaget, Chomsky, Levi-Strauss' and Erving Goffman. Piaget makes clear that structuralism is a method of analysis and not a discipline or content area.

Of particular relevance to education is Piaget's *Science of Education and the Psychology of the Child,* which is essentially a critique of traditional education. The argument is that education is too concerned with the technology of teaching and too little concerned with understanding children. In Piaget's view, the overemphasis on the science of educating, rather than upon the science of the children being educated, leads to a sterile pedagogy wherein children learn by rote what adults have decided is valuable for them to learn. Basically, Piaget feels that teacher training and educational practice must have child development as their basic discipline. The psychology of the child should be the primary science of education.

These are but a few of the achievements of the last few decades of Piaget's work. And his energy and enthusiasm are unabated as he continues his work on physical and biological causality. Early this summer, he is participating in two conferences, an educational conference in New York and the annual Piaget conference in Philadelphia. At this writing, I am very much looking forward to seeing him again and hearing the latest ideas and research coming from Geneva. I have encouraged as many of my students who can attend to be present as well, since Piaget, to my mind, exemplifies more than genius. At least equally important is the example he presents of a man who, despite his early success, maintained an unwavering commitment to research, to intellectual independence and to the welfare of children all over the world.

Learning about Learning

Jane Stein

Preschool children are always counting, says Rochel Gelman, a University of Pennsylvania psychologist who studies cognitive development. It helps them to make sense of at least some of the puzzles—the arithmetic ones—they face in learning to cope with what for them must be an incredibly complex world. According to traditional developmental theory, however, they are not supposed to be able to count; they fail to comprehend the concept of conservation, or constancy, of numbers: that eight eggs in eight cups will still total eight when they are taken out of the cups and placed in a pile. Nevertheless, as Gelman observes, "They are always counting...practicing whatever they learn."

Those who, like Rochel Gelman, study cognitive development are dealing with the process by which children develop the ability to solve problems and to form complex concepts. That process is a basis for intelligent behavior and involves the child's interactions with the world. Until recently, preschoolers were viewed as cognitive incompetents. But it is now known that nonperformance—not being able to conserve numbers, for example—does not necessarily mean noncompetence.

Recent research into the development of cognition shows that children as young as six months of age are interested in numbers; by the time they are three, they have already developed a system of classifying sets of numbers. In addition, preschoolers can construct order, make inferences, explain cau-sality—all cognitive concepts once thought to be beyond them.

One reason for the growing awareness of preschoolers' cognitive competence is a change in research perspective. Instead of testing preschoolers' cognitive capacities against tasks they will be able to do when they are older, cognitive researchers are now focusing on what the youngsters are actually able to do at a given point in their development.

By observing how children solve problems—even when they solve them incorrectly—researchers are getting a better understanding of the wide range of strategies that are used, the complexities of the learning process and how cognitive abilities at one stage of development relate to abilities developed earlier and to those still to come. Though many studies address the development of cognition through mathematical concepts, many others do not; researchers in each area feel that they are all attacking something more fundamental, that the target is the nature of cognitive development itself, regardless of the route the investigator takes.

Early problem solving

"Children learn about problem solving even without direct instruction," says psychologist David Klahr of Carnegie-Mellon University. Although these early problem-solving capacities are typically called common sense, they are complicated processes involving considerable information and the use of multiple steps to reach a goal.

Consider the example of a four-year-old boy who is playing in the backyard. He wants to ride his bike, but he asks his father to unlock the basement door because "my socks are in the dryer." This seemingly bizarre statement is in fact, says Klahr, a fine example of logical problem solving. The problem is simple: The boy wants to ride his bike but is barefoot; he needs shoes so he won't hurt his feet on the pedals. He has to get his shoes—which is easy to accomplish since they are on the porch. He also has to get socks so his shoes won't rub his feet. This presents a secondary problem, that there are no more socks in his drawer. A search of memory produces a promising lead: A load of laundry was done recently, and the socks are likely to be in the dryer in the basement.

The boy seeks the most efficient way of getting the socks. He could walk around the house and go in by the front door. Or he could go in through the basement door, which is nearby, except that it is locked. He observes that his father is nearby and knows that Daddy usually has the keys with him. The final solution for the child is to ask his father to unlock the door so he can get his socks out of the dryer so he can put on his shoes in order to ride his bike.

This exercise in problem solving takes split seconds, and not one step is verbalized. It represents in capsule form a reasoning process we often call "intuitive."

To test the "intuitive" ability of preschoolers to solve problems, Klahr designed a game for young children in which they are asked to move one set of three inverted cans—each can of a different size and color—onto pegs so that they match the arrangement of another identical set of cans. Only one can is to be moved at a time, and a small can must not be placed on a larger one.

The youngsters come to Klahr with no prior knowledge of his puzzle. Yet, as his

Originally published in *Mosaic*, The Magazine of the National Science Foundation, November/December 1980.

ata show, all the youngsters tested—ranging n age from 3 to 6—came to the test already nowing the three basic strategies that they eeded to solve the problem:

- If they want object X to be in location B nd it is currently in location A, then try to nove it from A to B.
- If they want to move object X from A to and object Y is in the way, then remove bject Y.
- If the approach they are trying is too ard, then do some part of it that is easier.

The children used a combination of these trategies to solve the simplest of Klahr's uzzles, those that involved moving only ne object to achieve the correct configuraion. They also used the strategies in other ombinations for more complicated puzzles, ome involving at least seven moves. "There re many ways to skin a cat, even in this nini-task," Klahr observes.

Indeed, some children solve his puzzle nore effectively—with fewer steps—than thers. Although most of the youngsters id improve their skills during the course of he testing, this did not add to their prob- em-solving skills in general. "I don't think hey would be any better at playing chess ecause they know how to do this puzzle," Klahr says. But teaching specific problem- olving skills wasn't the point of his work. What Klahr did successfully was to show hat young children solve problems using vhat we call intuition and that the rich epertoire of problem-solving methods used y adults—means-ends analysis, search, valuation, planning—exist in rudimentary orms in youngsters as well.

Holistic to analytic

As children get older, their intuitive prob- em solving and other cognitive abilities ecome more sophisticated. Deborah Kem- er, a psychologist at Swarthmore College, is locumenting some of the cognitive changes hat take place in children. "Young children ave first the task of making sense of natu- al categories of objects in the world," she ays. "These tend to be categories of objects hat resemble one another, so overall sim- arity—mutual resemblance—is a useful cri- erion for establishing these categories." A oat, for example, is a natural category; all oats generally resemble each other. But oung children, unlike older children and dults, find it difficult, and at times impos- ible, to form categories of diverse objects y isolating a property they have in com- non. Scientific categories tend to have this tructure, says Kemler, but putting a boat nd a sponge together because they both oat may be beyond young children.

Kemler has followed the shifts in think- ing patterns among a group of elementary schoolchildren ranging from kindergarten to fifth grade. A test she uses consists of cards of varying sizes and shades of gray. She selects three cards from a pack; two of the cards are identical in size but are very different in shade (light gray and dark gray). The third card is close but not identical in size and close but not identical in shade to the dark gray card.

When asked which two cards go together, the kindergarten children uniformly pick the two that are similar overall but which are not identical in any way. Fifth graders naturally select the two cards that share one identical component. Data on the second graders were typically ambiguous; they were in transition between the holistic and an- alytical categorization strategies.

Why do youngsters make this cognitive shift? Perhaps, Kemler hypothesizes, there is a natural predisposition to deal with wholes first and switch to abstract properties later. She suggests that this way of grouping ob- jects may better prepare children to deal with complex concepts.

Daily experience and formal schooling in which children's learning becomes more and more governed by rules also play impor- tant roles in the switch. The rules that gov- ern reading, for example, are based in part on the sounds of common components of words. In an experimental test conducted by Rebecca Treiman and Jonathan Baron of the University of Pennsylvania, they report, kin- dergartners with minimal reading skills tended to group syllables like "poo" and "boo" because they sound alike overall. But at the same age, children with more advanced read- ing skills classify on the basis of a common segment. They put "bee" and "boo" to- gether instead because of the identical pho- netic component with which each begins.

Adapting to this propensity to see wholes first and then properties might point the way to strategies for the teaching of reading. Indeed, several existing strategies do take such a direction. But research into cognition is far from validating one approach or another. New educational tools are likely to develop, however, as more pieces of the cognitive- development puzzle are explained and as a better picture develops of what children know and how they learn.

More and less

There is a considerable range of things that children can do well, according to psy- chologist Rochel Gelman. She has focused much of her work on numbers because she believes they represent a uniquely human

cognitive ability, just as is early language. "There must be something innate in count- ing," she says. "Babies at six months detect a difference between two and three [objects] when different sets are flashed on a screen. Chimpanzees can't do it after hours and hours of training. Perhaps counting is re- stricted to humans...it is found universally, in all cultures, among young children and unschooled adults."

Complex rules about how to count and what to count develop without the benefit of specific formal instruction. Gelman has found that, by age three, preschoolers know addition and substraction, see the difference between more and less and can do simple mathematical reasoning.

A three-year-old participating in a sub- traction game with Gelman said, "There was three animals in the can." He looked around, pointed to her and said, "Took one cuz there's two now." In a game of "more and less," a four-year-old picked a five-item display as the winner "cause there's one, two, three, four, five." The loser was clearly the three- item display "cause only one, two, three." Not surprisingly, the counting rules and mathematical reasoning encompass more complex concepts as children get older. "You're not supposed to count something twice," said a five-year-old to Gelman. "You can't make it six if there are only five."

Gelman is focusing on what children can do well at given stages. In the process, she is also discovering what they are not very good at. A group of three-year-olds, for example, scored poorly—only 49 percent of their answers were correct—when asked to count a three-item display that was under clear plastic. The youngsters scored 87 percent correct, however, when they were asked to count the same items without the covering. According to Gelman, "stimulus variables affect counting....The younger the children, the more they need to point at or touch what they count."

Ingrained errors

But more to the point of cognitive devel- opment, Gelman's work shows that three- year-old children know how to count. They might count in an idiosyncratic way, such as 2, 6, 10 instead of 1, 2, 3, but they count in a stable order, thus proving they have mas- tered one of the basic how-to-count princi- ples. The one-one principle—each item in an array must be tagged with one and only one unique tag—is another index of emerging counting skills.

Gelman compares idiosyncratic errors in counting to the errors made by young lan- guage learners, such as "I runned." These errors show that the child's use of language—and numbers—is rule governed. The child adheres to these rules, regardless of their correctness, and may develop poor language and mathematic skills not only because the rule was learned incorrectly, but because of the consistency with which it is applied.

One of the problems facing cognitive researchers is to better understand the rules by which children learn and the ways those rules are applied. Most children will self-correct language and counting errors that develop spontaneously. But sometimes errors become ingrained when rules that inherently make sense but contain errors are learned and applied, or when valid rules are applied incorrectly. Lauren Resnick, co-director of the Learning Research and Development Center at the University of Pittsburgh, is studying the erroneous strategies that children use in solving arithmetic problems. She compares them to computer programs with bugs and calls them "buggy algorithms."

Several dozen buggy algorithms have been identified for subtraction, most of which develop because the children have never learned the complete standard strategy, or algorithm, or have forgotten parts of it. For example, when zero is taught, it may not be taught carefully enough or explained well enough. And when children do not fully understand it as a concept, they end up solving in unique ways problems in which it appears.

Resnick has found that children, when making a mathematical error, tend to follow a rule, but often one that incorporates their own modifications. In subtraction, for example, they might take a smaller digit from a larger one regardless of which one is on top. Or sometimes they follow only part of the rule. For example, when borrowing from a column whose top digit is a zero, the student writes "9" but does not continue to borrow from the column to the left of the zero. At other times, rules may be completely made up; whenever there is a zero on the bottom, zero is written as the answer. No matter how buggy the algorithm, it is systematically used, Resnick observes.

Resnick has had some success with three children who needed remedial instruction to help correct their buggy subtraction algorithms. She gave the youngsters a subtraction problem to solve:

$$\begin{array}{r} 300 \\ -139 \end{array}$$

To test the contention that understanding of an abstraction—in this case that a number can be built out of a variety of sets of

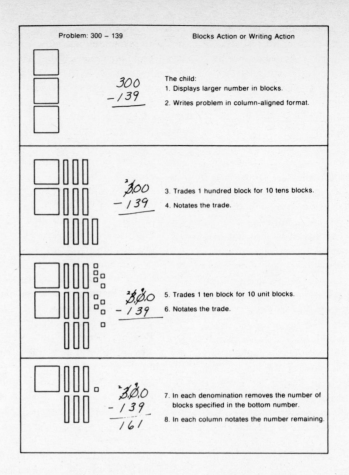

Problem: 300 − 139	Blocks Action or Writing Action
$\begin{array}{r} 300 \\ -139 \end{array}$	The child: 1. Displays larger number in blocks. 2. Writes problem in column-aligned format.
$\begin{array}{r} 300 \\ -139 \end{array}$	3. Trades 1 hundred block for 10 tens blocks. 4. Notates the trade.
$\begin{array}{r} 300 \\ -139 \end{array}$	5. Trades 1 ten block for 10 unit blocks. 6. Notates the trade.
$\begin{array}{r} 300 \\ -139 \\ \hline 161 \end{array}$	7. In each denomination removes the number of blocks specified in the bottom number. 8. In each column notates the number remaining.

quantities—is enhanced if there are at least two representations of the idea, Resnick supplemented the written subtraction problem with concrete forms such as number blocks, color coded chips, bundles of sticks or coins of different sizes. "They were not allowed to write anything down without making a corresponding move with a set of blocks," she explains. In order to do this, they had to understand that the top number in subtraction is made up of several components, each of a different value dictated by place, and that these quantities can be exchanged among each other as long as the total value remains the same. A hundred-block can be traded for 10 ten-blocks and a ten-block can be converted into 10 unit-blocks, for instance. "The possibility exists that we forced them to do it right because this system actually prohibits wrong moves," says Resnick.

The real test of whether or not there were cognitive changes will come during the next stage of her research, when the children are tested without the use of a second representation—the number blocks—several months after the remedial teaching. Resnick's research on buggy algorithms—knowing what the bugs are, what they represent in developmental cognition and how they get in—will ultimately lead to redesigned teaching manuals.

Persistent bugs

Not only young children, but college students majoring in science and mathematics too, and even graduate engineers, are plagued by buggy algorithms they carry into adulthood with them. Two University of Massachusetts investigators have found that many students have carried their bugs through four years of high school math without getting rid of them.

The scientists, John Clement and Jack Lochhead, discovered that though students could easily solve equations, they did not know how to formulate them from word problems. The most common error that Clement and Lochhead found was a simple reversal in translating word problems to equations. Students also made repeated errors in translating pictures or data tables to equations and words to graphs.

The errors, moreover, were not caused by carelessness but rather by self-generated and persistent misconceptions. After testing students with a variety of algebra problems, Clement and Lochhead concluded that the students have real cognitive difficulty with a semantic understanding of algebra.

Consider the following example: Freshmen engineering students were asked to write an equation that said the same thing as the English sentence: There are six times as many

...udents as professors at a university. Nearly [?]0 percent of the students wrote $6S=P$, which [sa]ys exactly the opposite of the sentence. [S]tudents often made a syntactical error by [m]echanically translating the words directly [in]to symbols in the same order as they [ap]peared in the sentence—six times the num[b]er of students, or $6 \times S$—or they made a [se]mantic error by assuming that six times [so]mething had to represent the largest group— [as]sociates with S. In either case, they acted [w]ithout fully understanding the most fun[d]amental concepts of algebra: variables and [e]quations.

Letters in algebra are called variables, and [th]ey stand for numbers that make an alge[br]aic equation true rather than for the names [of] objects. An equation is a statement in [w]hich one side is the numerical equal of the [ot]her side. "Because the students were con[f]used about the meaning of variable," says [C]lement, "many portrayed S as a symbol [fo]r 'student' rather than for 'number of stu[d]ents.' And since they were confused about [h]ow to write an equation, they had the equal [s]ign express an association rather than an [e]quivalence."

Students who understood the principles [in]volved wrote the correct equations, $S=6P$, [an]d translated it back into English as "The [n]umber of students is six times the number [of] professors." The key to understanding [th]e correct equation is to recognize that it [do]es not describe the situation in a literal [se]nse; instead it describes a hypothetical [re]lationship of equivalence.

The reason so many advanced students got [th]e problem wrong, according to Clement [an]d Lochhead, was that they had had little [ex]perience in testing their cognitive under[st]anding of algebra. High school textbooks, [th]ey say, usually give a formula or a choice [of] formulas and ask the student to solve an [al]gebra problem with it, rather than ask them [to] conceptualize the problem and then con[st]ruct the formula.

The University of Massachusetts profes[so]rs are concerned about enhancing the [in]tuitive reasoning processes in students. "I [a]m less interested in the procedure [of doing [th]e problem correctly] than in the discov[er]y of the procedure," says Lochhead. "If [st]udents are conscious of why they set up [a]n equation and how it works, then they can [u]se it for other kinds of problems. Then [th]eir knowledge will be more flexible." (See ["]Test yourself," accompanying this article, [fo]r additional algebra problems that stump [en]gineering students.)

[Pr]ofessionals, too

A curious finding for Clement and Loch[he]ad was that 17 professional engineers, each

---

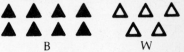
---

with between 10 and 30 years' experience, were given the translation problem in a slightly more difficult form, and eight missed it. When they were given the same problem to program on a computer, however, they all did it correctly. Chief among the reasons for the engineers' success with the problem on a computer is the unambiguous semantics of programming language. The interpretation of variables in computer language is clear; they stand for specific numbers. "College students working out algebra problems are not as precise [as computers]," says Lochhead. They tend to use a multi-purpose variable—for example, letting B in one part of a problem equal the number of books sold, then switching to let B equal the price of a book later on. "Poetically it's nice," Lochhead observes, "but scientifically it has to be precise."

In order to find out how to help students become more precise about their algebra reasoning, Clement and Lochhead conducted audio- and video-taped clinical interviews in which students were asked to think aloud as they worked through a series of algebra problems. One student, while grappling with a problem, said in frustration: "I think I'd like to kill my sixth-grade teacher because he didn't teach me this....All these little letters

that I've been working with for years in algebra and I can't [solve the problem]."

Clement and Lochhead use the tapes as a scientific tool—"just as a microscope is used in biology." As Clement explains: "We are actually recording evidence that can be used to construct maps of mental processes. We can study a ten-second reasoning segment in detail. We watch students explain an answer, hesitate, then change their minds." In more than one case, the two professors saw students write down the correct equation and then switch to an incorrect one. This illustrates how deep the cognitive misconception is; for these students, the incorrect solution is more compelling than the correct one. But by knowing the cognitive source of the problem, remedial teaching strategies could be designed to deal specifically—and perhaps finally—with the buggy equation.

Computers that learn

It is fascinating to speculate why one student intuitively solves a problem better than another. By identifying specific conceptual stumbling blocks, Clement and Lochhead have offered some clues. Additionally, computer models of mental processes are increasingly offering additional tools for knowing how people reason and ultimately learn.

4. CHILDHOOD

ABLE is a computer program at Carnegie-Mellon University that actually learns physics. When it starts off as a novice, the Barely ABLE model finds the right answer only after a considerable search—much the way a beginning student stumbles slowly and clumsily when solving textbook problems. But once a problem is solved, the computer "learns" and stores the new information or rules so that it can subsequently solve similar problems rapidly and directly, thus becoming a More ABLE model. The ABLE model is a kind of computer program incorporating a system of rules called a production system. Each production in the system encapsulates some small part of the knowledge and is recalled or evoked at just those times when it is relevant to the problem.

Real people learn essentially the same way, though more slowly, according to Jill Larkin, formerly a physicist but now a psychologist working in cognitive research at Carnegie-Mellon University. In fact, ABLE learns *too* quickly. Once ABLE has learned to measure the speed of a block on an inclined plane, then it can always do it. It is an automated piece of new knowledge. "People aren't as automated," says Larkin, "and they learn slowly."

Larkin compared the way first-year college physics students and physics profes-sors solve problems and found them to be very similar to the way that the novice Barely ABLE and the expert More ABLE solve them. "The novice student showed consid-erably more variability in seeking out prin-ciples to use," says Larkin, "while the experts knew what would be useful and didn't have to search [their minds] for everything to find it." (See "Programmed to Think," *Mosaic*, Volume 11, Number 5.)

Computers, however, are limited instru-ments. "They can take one small domain," Larkin says, "and solve problems in that area as well as very good students can. But the problems we are giving the computers are relatively simple—such as a lever prob-lem involving a ladder leaning against a bridge. Compare this with the complexity of what an engineer has to know to design a bridge. Computers fail to capture much of the rich-ness in human knowledge and learning."

Psychologist John Anderson, also at CMU, agrees that to get a computer to model human cognition, it must be provided with a large amount of prior knowledge. But Anderson does not consider this a barrier to develop-ing models. He is experimenting with a com-puter program that he calls ACT, which has been used to model geometry-problem solv-ing, the learning of concepts, and language acquisition. "ACT is true to life," Anderson says. "It is cautious and does not take leaps based on little evidence."

In short, ACT learns as if it were human. To demonstrate this, Anderson timed exper-imental subjects as they solved geometry problems. "It might take human subjects two to five minutes with the first applica-tion of a postulate in geometry," he recalls. "After several tries using it, they can solve the problem in less than five seconds."

The ACT model, according to Anderson, shows that a basic key to learning is the development of concepts of pattern recog-nition and the appropriateness of using a particular theorem. For example, "There are more than ten ways to prove that triangles are congruent, and the students have to learn which ones are relevant to a particular prob-lem," he notes. But one of the most common complaints from students about their prob-lem-solving ability is that they "don't know how to get started," or "don't know how to decide what to do." And if they don't learn which conditions are appropriate for apply-ing given principles, they too often learn inappropriate ones.

The National Science Foundation contrib-utes to the support of the research reported in this article through its Research in Sci-ence Education and Memory and Cognitive Processes Programs.

1,528 LITTLE GENIUSES AND HOW THEY GREW

For six decades, the famous Terman study has followed
the fortunes of a group of men and women with IQs above 135.
In some ways, "Terman's children," particularly the women,
seem to have been ahead of their time. Here's a preview
of what the granddaddy of all life-span research tells us
about how people of exceptional promise—as defined by IQ—
fare in their careers, marriages, and family lives.

DANIEL GOLEMAN

When Lewis B. Terman launched his famous study of high–IQ children in 1921, his aims were modest. Terman, a Stanford University psychologist who had developed the Stanford-Binet intelligence test, wanted to disprove the existing myth about bright kids. In those days, gifted children were little understood. There was a strong popular bias against them, as psychologist Pauline Sears explains: "Folklore had it that 'early ripe, early rot,' that precocious children were prone to insanity, physically weak, one-sided in their abilities, and socially inferior."

Within a few years, Terman had proved that such children were neither physically nor socially inferior, and that, as they grew older, they not only did not go to seed but surpassed their peers in accomplishment as well. From the start, Terman had also wondered: how would they fare in later life, as adults? Would their genius ultimately enable them to achieve greater success than their contemporaries, or was it destined to be thwarted? Even if their promise were fulfilled, moreover, would they be able to lead normal, happy lives?

The Terman study became the granddaddy of all life-span research.

For six decades, teams of investigators have been tracing the pains and pleasures of these gifted people, whose IQs are all above 135.* The most recent study was done in 1977, more than half a century after the original group of 1,528 young people, ages 3 to 19, was picked. On that occasion, three out of four surviving members dutifully filled out and returned their questionnaires. While the data will not be ready for full analysis until 1982, Robert and Pauline Sears, the Stanford psychologists who are the current stewards of the project, offered some of their impressions of it, along with reflections on the study as a whole, to *Psychology Today*.

Some of the results might have surprised Terman. Most of the men and women in the study have passed retirement age, and it now seems clear that exceptional intelligence does not preclude ordinary happiness or worldly success. But neither, apparently, does it guarantee extraordinary accomplishment. Although most of the

*IQ, or intelligence quotient, is derived by dividing a person's mental age—his or her score on the Stanford-Binet test compared with that of others in various age groups—by his chronological age. Scores that are above 135 represent the top 1 percent of the population—often considered genius IQ.

men and women in the study have done well in their careers, none appears to have achieved the summit of true genius. None has so far been awarded a Nobel prize or similar honor. There are few millionaires among them, and hardly any distinguished creative artists.(See box on page 143.)

If Terman's people have not turned out to be that exceptional, however, their lives do offer some fresh perspectives on a few contemporary issues. As some stress researchers have been suggesting recently, it appears that brilliant, hard-driving success-seekers do not inevitably succumb to early heart attacks or to other stress-induced ailments; indeed, the mortality rate of the least successful Terman subjects was twice that of the most successful. Nor is it necessarily true, as people often assert, that middle-aged men are obsessed with their work. To most of the men in the study, it was family that mattered most. Yet for the women, the absence of children made them more satisfied with their work—a particularly interesting finding, since many women today are feeling the pressure to have both children and a career.

Less unexpected was the finding that some Terman subjects did markedly better in life than others. It proved impossible to explain the difference with certainty. However, the most successful members of the group shared a special drive to succeed, a need to achieve, that had been with them from grammar school onward. "We have ratings on such things as desire to excel, persistence, from when the subjects were 10 years old, by both teachers and parents, and again at about age 18," Robert Sears says. "These were very high, compared with other children in their class."

A Classic Study: How It Was Done

Along with their review of the findings, the Searses offered a few caveats. Definitive conclusions, they warned, must await sophisticated analysis of the mountains of data that have been gathered so far, as well as the thousands of additional facts from studies of the Terman men and women in old age that are sure to pile up over the next couple of decades. Another difficulty is that the subjects may not be typical of gifted men and women in other generations. The Terman-Sears group lived through a very special period in history; in one way or another, the Great Depression and World War II must have marked them all.

In other ways, too, the sample will always represent a special case in human development. There is no way of knowing the effect of having a team of researchers looking over the subjects' shoulders to assess their growth and achievement. It may be that a subtle self-consciousness added special pressure to adjust and excel. There is a kind of reverse confirmation of observer influence: many of those who were least successful reported feeling guilty because they hadn't fulfilled their potential. Had they not been reminded every few years of their intellectual promise, they might not have borne that guilt.

To find his exceptional subjects, Terman had sifted through 250,000 schoolchildren to locate the thousand or so with the highest IQs. Because he taught at Stanford, he found it convenient to limit the selection to California public schools. His testing staff found 1,470 children with IQs of 135 or higher. Most of them were in grades three through eight, and their average age was 11 when they were selected in 1922. Another 58 children, younger siblings of the original group, were added in 1928. All in all, there were 857 boys and 671 girls.

The sample was by no means representative, even of California. For one thing, there were no Chinese, although the state had a high proportion of them. Latin American, Italian, and Portuguese groups were also underrepresented, and there were only two black children, two Armenians, and one American Indian child. Jewish children were overrepresented. While an estimated 5 percent of the California population was Jewish, the Terman sample was 10.5 percent Jewish.

There was also a social-class bias. Close to one out of three children were from professional families, although professionals made up only 3 percent of the general population. Only a smattering were children of unskilled laborers, compared with 15 percent in the general population.

Members of the sample were repeatedly surveyed. In 1922 and again in 1928, the youngsters were tested and interviewed, as were many of their parents and teachers. The group was assessed by questionnaires or interviews in 1940, 1950, 1955, 1960, 1972.

The major reports from the first 35 years of the study are contained in five volumes of *Genetic Studies of Genius* (Stanford University Press). Subsequent reports have appeared at intervals since then. Together, these reports add up to a classic of psychological research. The project is older than the so-called Grant study, a continuing examination of the lives of 95 Harvard graduates that began in 1938 and followed them into their 50s, under the supervision of psychiatrist George Vaillant. (See "The Climb to Maturity," in the September 1977 issue of *Psychology Today*.) It is far more comprehensive than Daniel Levinson's study at Yale of 40 men at midlife, whose biographies were reconstructed through extensive interviews.

The "A"s and "C"s: Job Success Compared

As adults, the Terman subjects varied widely in achievement. To find out what made the difference, Terman and his researcher, Melita Oden, compared the most and the least successful. The first comparison was made in 1940, followed by a second in 1960. (Since so few women in the sample were employed full time, this phase of the Terman study was restricted to men.) What the data from this phase reveal is that the most successful subjects were physically healthier and better adjusted as well.

Recognizing that no yardstick for achievement is universally accepted, the researchers chose work success as the best available measure. The investigators judged job success by asking whether a man "had made use of his superior intelligence in his life work, both in his choice of vocation and in the attainment of a position of importance and responsibility in an area calling for a high degree of intellectual ability." Two criteria for an "important" job were the income derived from it and the status generally accorded it by society (as measured by a standard scale).

In the 1960 comparison, the 100 most successful men were designated the *A* group, the 100 least successful, the *C* group. While the *C*s were the "failures" of the Terman kids, the judgment is relative: most of them equaled or exceeded the national average for job status and income. In 1959, the national median earned income was about $5,000; for the *C* group, it was $7,178. For the *A*s, though, median income was close to $24,000.

The *A* group included 24 university professors, 11 lawyers, 8 research scientists, and 5 physicians. Thirty were business executives; one was a farmer who operated large ranches. Only five men in the *C* group were professionals, and none was doing well. One, for example, who had done graduate work in mathematics and was employed as an engineer, actually worked at the technician level. The majority worked as clerks, salesmen, or in small businesses.

Over the years, the health records of the two groups were about the same, but by 1960, only 8 *A*s had died, compared with 16 *C*s. (The difference in mortality rates raises a question of whether lack of success may affect health.) Natural causes (heart disease and cancer) accounted for most deaths in both groups. Although more *A*s than *C*s served in World War II, only one *A* and three *C*s lost their lives in the war. One *A* committed suicide, compared with two Cs. And two Cs died in accidents, but no As did so.

Slight differences between the groups in childhood became bigger differences in adulthood. For instance, while both groups scored about the same on the intelligence tests in the original Terman study, the As skipped more grades in grammar school, graduated earlier from high school, and received more graduate training.

The As were a much livelier group than the Cs. As youngsters, they had more collections (of shells, rocks, stamps) and took part in more extra-curricular activities. As adults, As were members of more professional societies and civic groups than Cs

were. As were also more active phys-ically; at age 40, As favored sports in which they participated, while Cs pre-ferred to watch.

The As tended to come from more advantaged families. A parents—espe-cially the fathers—had had more edu-cation on the average than had the fa-thers of Cs. The educational edge even extended back to grandparents, the pa-ternal grandparents of As having com-pleted more years of college than those of Cs. Not surprisingly, then, more fathers of As than of Cs were professionals.

The home library of an A subject was more likely than that of a C to have 500 or more books. As came

from more stable homes. The parents of twice as many Cs were divorced by 1922, a trend that had become more marked by 1928 and 1940. The death rate among C parents was also higher, although the difference did not show up until 1940.

Finally, more than twice as many As as Cs came from Jewish families. Seventeen percent of the As but only 8 percent of the Cs were Jewish.

As children, As seemed better ad-justed than Cs. When rated by teach-ers and parents, As were given a slightly better evaluation for social adjustment, an advantage that be-came more pronounced in adulthood. As and Cs were about equally rebel-

WHERE ARE THE EINSTEINS AND PICASSOS?

The great majority of men and women in the Terman study have done well in their careers, which is not surprising for a group who as children were rated among the superintelligent. "But there's no-body in the group who's a real ge-nius—no Einsteins," reports psy-chologist Robert Sears.

By and large the Terman subjects were fast risers. At midlife, many were national figures within their own professions, widely known among bankers and scientists, for ex-ample. But their careers followed no single pattern. Their occupations ranged from postman to brigadier general, from sandwich-shop opera-tor to nuclear-laboratory director.

The group includes numerous sci-entists. There are a great many law-yers and some corporation heads. Since all members of the study were Californians, some ended up in the movie industry.

The life stories of a few Terman kids have been told in print, with identifying details expunged. In some of the cases, the subjects' iden-tities could probably be guessed with a little sleuthing. In 1959, when Ter-man sketched the unfolding careers of his group at the 45-year mark, one was described as "one of the coun-try's leading science-fiction and fan-tasy writers, who has produced some 60 short stories and novelettes, as well as 15 volumes of fiction and

nonfiction." Another was listed as "a motion-picture director who has made some of the most outstanding pictures of the last 10 years"; several of his pictures had won Oscars.

As of 1955, the average income for the group was $33,000, compared with a national average of $8,000. The highest income for the group was $400,000; there were no mil-lionaires at the time (it is a safe con-jecture that there are some now).

Although only one of the top six earners in 1955 had graduated from college, two-thirds of the whole sample were college graduates. A to-tal of 97 had doctorates, 92 had law degrees, 57 medical degrees, and 177 master's degrees. As a whole, the group was above the national aver-age in both occupational status and education. (Current information on income, education, and occupation-al status is unavailable, but the dif-ferences between the Terman group and the general population are pre-sumed to have held over the years.)

By midlife, 77 members of the co-hort (including 7 women)·were listed in American Men of Science, 33 in Who's Who. Together, members of the Terman group had—midway in their careers—produced nearly 2,200 scientific articles, 92 books and monographs, 235 patents, 38 novels, and 415 miscellaneous articles, ex-cluding the output of those em-ployed as journalists and editors.

One oddity: the group, Sears notes, "is low on artistic creation." It has produced no great musician and no exceptional painter, although it did include several musicians and artists who headed university de-partments, as well as some modestly successful weekend painters.

It may be that creative genius is too rare to have been part of the Ter-man sample. But perhaps it is unfair to expect that among this particular pool of intelligence there would hap-pen to be a Beethoven or an Einstein. Although the Terman group scored in the top 1 percent in intelligence in the country, so did two or three mil-lion other Americans. The apparent paucity of artistic genius in the sam-ple may mean that the tests them-selves were biased toward more pro-saic kinds of competence than cre-ative flair. While intelligence and creativity overlap, they are not the same, and they require different measures. The gifted test-taker may do well in school and work, but the creative genius may need something more than high IQ—something not ordinarily revealed in an IQ test.

But the final verdict is not in. There may well be some Terman ge-nius whose contribution has not yet been fully understood or appreciat-ed. "How do we know?" observes Pauline Sears. "They're not dead yet. It can take years to recognize an Einstein." —D.G.

lious in their youth, though. Nor was there any difference in either the degree of affection from, or rejection by, their parents.

The family background of the As apparently fostered ambition. When the Terman kids were rated by their teachers and parents in 1922, the one dimension that distinguished As from Cs was "prudence and forethought, will power, perseverance, and desire to excel." As reported a "strong liking for school." When, 30 years later, the Terman sample looked back at their childhood, the As recalled more frequently that their parents had encouraged initiative and independence. As also felt more parental pressure to excel: to get better grades, forge ahead in school, go to college.

The As were interested in everything. In 1940, a subgroup of the Terman sample took the Strong Vocational Interest Test. This group included 80 men who 20 years later, in 1960, were to be rated A and 77 who were to be rated C. The most notable difference between the groups was that As received a larger number of ratings showing them interested in a wider range of occupations, a pattern Sears interpreted as revealing "drive" or a high "level of aspiration."

The As seemed to get smarter as they got older. In both the 1940 and the 1950 follow-ups, As and Cs were given the Concept Mastery Test (CMT), an intelligence test for adults. In 1922, the average IQ for As was 157, for Cs, 150. (This difference, however, is too small to account for the discrepancy in later-life success between the groups. Small differences in score at the extreme upper end of the IQ curve stand for little actual difference in ability.) In adulthood, the spread grew, As scoring even better than Cs on the follow-up tests. The increased advantage of As may reflect in part their more advanced education. Even so, Cs were still a generally superior group. While the As' average CMT score was 147, and the Cs' was 130, the score for a group of Ph.D. candidates at a top university was only 119.

Precursors of Success

Why did Cs fail, compared with As? It seems to have been a matter of attitude and adjustment. The greater ambition of the As as children proved

a constant theme later in life. In 1940 and 1950, As were rated by parents, wives, and themselves as different from Cs on only three traits, all of which were related to ambition. As were more goal-oriented and had greater perseverance and more self-confidence. When Terman staff members visited them in 1950, the As seemed more "alert and attentive."

Cs attributed their lesser achievement to their own "lack of persistence," while As often named "persistence in working toward a goal" as the important factor in their success.

In 1960, at an average age of 50, the greater ambition of the As—which emerged in many different ways over the years—was clear. When asked to rate themselves, the As proved to have greater ambition to be recognized for accomplishment and to have more drive for vocational advancement. They also sought greater work excellence for its own sake, not just as a pathway to success.

The greater drive of As showed up in their social lives. In young adulthood, As more often than Cs expressed a strong interest in "being a leader, and having friends." This interest may have been compensatory; As more often recalled that as children they had "felt different from others" and had trouble making friends and entering into social activities. (There is no evidence for this from their childhood ratings, however.)

Overall, adult Cs admitted much discontent with their lot. For them, "making more money" was far more important than for As—but, of course, the Cs made less. Cs wished they had received more schooling. As were far more likely to have ended up in a vocation they preferred, while Cs more often came to their occupations as a result of chance or need. True to their goal-directed nature, As chose their careers far earlier than did Cs; many As had accurately predicted "what they would be when they grew up" by the time they reached 16. It is not very surprising, then, that many more As than Cs were satisfied with their careers, and that Cs were much more likely to feel their work gave them no satisfactory outlet for their mental capabilities. In fact, knowing they were part of the Terman study was at times a drawback for the Cs: they more often felt guilty for not living up to their potential.

The greater happiness of As over Cs extended to family life and personal adjustment. Fifteen percent of Cs reported having a problem with alcohol, while only 3 percent of As reported such a problem. An indicator that good marriage fosters achievement lies in the fact that while all the As have married, one in five of the Cs have not. The marriages of Cs were much more likely to fail; 16 percent of As were divorced by 1960; among Cs, the proportion was close to half.

As tended to marry better-educated women. They also had smarter kids, and more of them: an average of 2.5, compared with 1.6 for Cs. While the offspring of both groups were gifted, the average IQ of children of As was 140, for those of Cs, 132—about proportional to the original difference in IQ between the two groups.

The Terman study of who succeeds among the gifted sheds light on the recent report on determinants of success done by Christopher Jencks and a group of researchers at Harvard (see "Who Gets Ahead," *Psychology Today*, July 1979). Analyzing the findings of several national surveys, the Jencks group identified as key factors family background (including father's occupation and income, and parents' education), test scores in school, years of schooling completed, and teenage personality characteristics like studiousness. The As in the Terman sample—like some of the Cs—certainly had a winning combination of many or all of these factors. Yet the most successful among the gifted had something that the Jencks study did not have data to detect: a special drive to succeed that made a difference.

Life Satisfaction: Men

After more than half a century of tests, questionnaires, and interviews, the former Terman kids were asked in 1972 to reflect on the things that mattered most to them, their sources of satisfaction in life. Their average age at the time was 62, although some were as young as 52, others as old as 72. Results for men and women were reported separately, the men by Robert Sears, the women by Pauline Sears. "With a long life of accomplishment behind them," wrote Sears in his report on the men, "they would be in a position to evaluate its joys and sorrows, its successes and failures, its *might-have-been* as well as its *was*."

The men were asked to evaluate six areas of their lives: occupation, family life, friendship, richness of cultural life, service to society, and their overall joy in living. They were asked how important each area had seemed to them in early adulthood, as they were planning their lives; that is, what kinds of satisfaction they had hoped for from life. Then, they were asked how satisfied they had been in reality with each during the course of their lives. To judge total satisfaction, the actual degree of satisfaction for each was multiplied by the importance the person assigned to it. This meant that in measuring actual satisfaction, a particular pleasure had more weight if it had been sought-after all along.

What the men had most wanted in their younger days was a happy family life, and they were lucky enough to get it; in maturity, their greatest happiness came from their families.

In youth, work satisfaction was ranked second in importance, but in reality the men ranked it their third greatest source of actual pleasure. What came in second was something these men hadn't sought as a major goal: simple joy in living. At age 62, they felt they had gotten a fair share of pure pleasure out of life. However, joy moved down to third place in the final tally and work moved up to second, because work was more highly valued originally. (See chart below.)

Some men, of course, reported more overall life satisfaction than others. To see what factors in the past might have predicted satisfaction, Sears searched through the records back to 1922. Three sorts of data seemed promising: motivation, favorable life conditions, and expressed feelings.

The motivation that interested Sears was the need for achievement already discussed. The objective conditions that might have fostered satisfaction, Sears thought, were health, education, income, and occupational rung, all of which might produce good feelings about one's lifetime efforts. As it worked out, early feelings about oneself proved to be very important. Both an early liking for one's work and a feeling at age 30 of choosing one's career rather than drifting into it correlated with final occupational satisfaction, as did rating oneself as healthy from 30 onward. High-level training, status, and income had negligible importance for final career satisfaction. "Rather," concluded Sears, "it looks as if there were some continuing affective quality—an optimism about life, an enjoyment of occupational combat, and a feeling of self-worth—that characterized the more satisfied of these men at age 30 and persisted through the next three decades of their lives."

Pleasure in family life had different roots. One early predictor was the men's scores at about age 30 on a Marital Aptitude Test and a Marital Happiness Test devised by Terman. Other important factors included good mental health in 1940 and 1960, good social adjustment in grammar school, and sociability in high school. Having a favorable attitude toward fathers also predicted a satisfying family life. Having a wife who worked, though, correlated with a less satisfying family life.

A little more than 70 percent of the men had unbroken marriages by 1972 (approximately the national average for this age group). Sears compared these men with the 21 percent who had divorced, to see which early tests predicted marital stability as well as good mental health in general. Men in unbroken marriages also had better attitudes toward their parents, especially toward their mothers. This fact led Sears to speculate that "for the men with unbroken marriages, there had been final resolution of whatever conflicts may have existed in childhood, while for the divorced men, these conflicts were still remembered."

LIFE SATISFACTIONS OF THE GIFTED

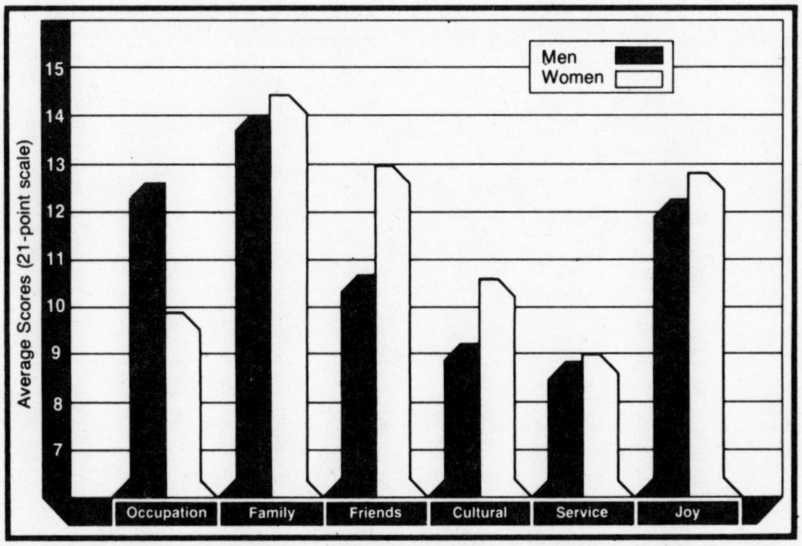

Chart shows the relative importance of six sources of satisfaction, as reported by the high IQ men and women in the Terman-Sears study in 1972. By that time, most of the subjects were near retirement age. "Joy" means overall pleasure in living.

Life Satisfaction: Women

The life patterns of the Terman women followed different paths from those of the men, so Pauline Sears and her colleague Ann Barbee did a separate analysis of the findings on women. However, the women were rated on their satisfaction in the same areas of life as were the men, and a search for predictors was made.

Feminists will be disappointed, but Sears and Barbee found their male and female subjects very different. Work was generally more important to men than it was to women. Women sought happiness in a great diversity of ways; they sought more satisfaction than did men from friends, culture, service to others, and joy in living. While both sexes saw family life as the most important area for achieving satisfaction, women valued it more highly.

When each source of satisfaction was weighted according to the importance placed on it, there were only three areas in which the sexes differed.(See chart, page 145). Men found their work far more satisfying than did women, while women found far more satisfaction than men did from both friends and cultural activities. There was no difference between men and women on the pleasures found in family, service to others, or joy in living.

The sexes were more alike when men were compared with working women, not with all women. The one sex difference then was that men gave more importance to work as a source of satisfaction. The working women (fewer than half of the women in the sample) derived just as much actual satisfaction as the men did. For women, as for men, money did not buy happiness. When the women with the highest incomes were compared with those with the lowest, there was no difference in happiness.

Not surprisingly, happy children often grew up to be happy adults, and satisfied female adults matured into satisfied older women. General satisfaction in 1972 was related to having positive relations with parents as far back as 1928. These women in midlife continued to look favorably on their own parents and had special admiration for their mothers. They also had favorable self-concepts from the early years on, rating themselves high in confidence and low in inferiority. "If a woman feels self-confident early in life," conjecture Sears and Barbee, "she is more likely to order her life in a way that promotes later satisfaction."

There were other antecedents of happiness in midlife. Good health and working in a profession were positive predictors, as were the education and occupation of a woman's husband. Being satisfied with her marriage, children, social contacts, and community service as far back as 1950 also predicted a woman's general satisfaction in 1972. Income was unimportant.

Career-oriented women had had more ambition for excellence in work during early and late adulthood—an attribute they shared with the most successful Terman men. Oddly, an early predictor for this group of women was having special math ability as children, in 1922.

In 1972, the women were asked to consider their lives as falling into one of four patterns: primarily that of a homemaker, pursuing a career, splitting her life between career and a family, or working part-time only. Then, they were asked whether that pattern fit the plans they had made in early adulthood and whether they wished they had chosen another pattern. Those who said their plan had been fulfilled and that they would not have chosen a different one were rated as highly satisfied in their lives. To the degree they wished it otherwise, satisfaction was rated lower. By these criteria, two thirds of the women were highly satisfied. The greatest satisfaction was reported by heads of households, whether or not they were single, divorced, or widowed.

Compared with a representative national sample of women who were part of a survey on the quality of American life by Angus Campbell, a survey researcher at the University of Michigan, the Terman women were better educated and had higher incomes and better jobs. On a measure of general happiness, the groups were comparable: married women, with or without children, were generally most happy, followed by the widowed, the single and, least happy of all, the divorced. When the psychologists looked at Terman women's satisfaction with their work patterns, however, they discovered a striking reversal in this order. Single women scored highest, then childless married women, divorced women, married women with children, and widows.

Groping for an explanation for the reversal, Sears and Barbee came up with a provocative idea. "For high-IQ women," they suggested, "independence from an unhappy marriage, the challenge of making one's own life as a widow or single person, activates, over time, feelings of competence rather than depression."

Indeed, perhaps the most important single conclusion to be drawn from the data on the Terman women is that a sense of being competent was tremendously important to them—making this retirement-age group seem surprisingly similar to women born a generation or two later.

They Learn The Same Way All Around The World

Dan I. Slobin

They harvested it down over there.
Way down over there?
Mmm. [yes]
Let's look for some too.
You look for some.
Fine.
Mmm.
[Child begins to hum]

The dialogue is between a mother and a two-and-a-half-year-old girl. Anthropologist Brian Stross of the University of Texas recorded it in a thatched hut in an isolated Mayan village in Chiapas, Mexico. Except for the fact that the topic was maize and the language was Tzeltal, the conversation could have taken place anywhere, as any parent will recognize. The child uses short, simple sentences, and her mother answers in kind. The girl expresses her needs and seeks information about such things as location, possession, past action, and so on. She does not ask about time, remote possibilities, contingencies, and the like—such things don't readily occur to the two-year-old in any culture, or in any language.

Our research team at the University of California at Berkeley has been studying the way children learn languages in several countries and cultures. We have been aided by similar research at Harvard and at several other American universities, and by the work of foreign colleagues. We have gathered reasonably firm data on the acquisition of 18 languages, and have suggestive findings on 12 others. Although the data are still scanty for many of these languages, a common picture of human-language development is beginning to emerge.

In all cultures the child's first word generally is a noun or proper name, identifying some object, animal, or person he sees every day. At about two years—give or take a few months—a child begins to put two words together to form rudimentary sentences. The two-word stage seems to be universal.

To get his meaning across, a child at the two-word stage relies heavily on gesture, tone and context. Lois Bloom, professor of speech, Teachers College, Columbia University, reported a little American girl who said *Mommy sock* on two distinct occasions: on finding her mother's sock and on being dressed by her mother. Thus the same phrase expressed possession in one context (*Mommy's sock*) and an agent-object

ACCORDING TO THE ACCOUNT of linguistic history set forth in the book of Genesis, all men spoke the same language until they dared to unite to build the Tower of Babel. So that men could not cooperate to build a tower that would reach into heaven, God acted to "confound the language of all the earth" to insure that groups of men "may not understand one another's speech."

What was the original universal language of mankind? This is the question that Psammetichus, ruler of Egypt in the seventh century B.C., asked in the first controlled psychological experiment in recorded history—an experiment in developmental psycholinguistics reported by Herodotus:

"Psammetichus . . . took at random, from an ordinary family, two newly born infants and gave them to a shepherd to be brought up amongst his flocks, under strict orders that no one should utter a word in their presence. They were to be kept by themselves in a lonely cottage. . . ."

Psammetichus wanted to know whether isolated children would speak Egyptian words spontaneously—thus proving, on the premise that ontogeny recapitulates phylogeny, that Egyptians were the original race of mankind.

In two years, the children spoke their first word: *becos*, which turned out to be the Phrygian word for bread. The Egyptians withdrew their claim that they were the world's most ancient people and admitted the greater antiquity of the Phrygians.

Same. We no longer believe, of course, that Phrygian was the original language of all the earth (nor that it was Hebrew, as King James VII of Scotland thought). No one knows which of the thousands of languages is the oldest—perhaps we will never know. But recent work in developmental psycholinguistics indicates that the languages of the earth are not as confounded as we once believed. Children in all nations seem to learn their native languages in much the same way. Despite the diversity of tongues, there are linguistic universals that seem to rest upon the developmental universals of the human mind. Every language is learnable by children of preschool-age, and it is becoming apparent that little children have some definite ideas about how a language is structured and what it can be used for:

Mmm, I want to eat maize.
What?
Where is the maize?
There is no more maize.
Mmm.
Mmm.
[Child seizes an ear of corn]:
What's this?
It's not our maize.
Whose is it?
It belongs to grandmother.
Who harvested it?
They harvested it.
Where did they harvest it?

"The basic operations of grammar all
are acquired by about age four, regardless of native
language or social setting."

relationship in another (*Mommy is putting on the sock*).

But even with a two-word horizon,
children can get a wealth of meanings
across:

IDENTIFICATION: *See doggie.*
LOCATION: *Book there.*
REPETITION: *More milk.*
NONEXISTENCE: *Allgone thing.*
NEGATION: *Not wolf.*
POSSESSION: *My candy.*
ATTRIBUTION: *Big car.*
AGENT-ACTION: *Mama walk.*
AGENT-OBJECT: *Mama book* (mean-
ing, "Mama read book").
ACTION-LOCATION: *Sit chair.*
ACTION-DIRECT OBJECT: *Hit you.*

ACTION-INDIRECT OBJECT: *Give papa.*
ACTION-INSTRUMENT: *Cut knife.*
QUESTION: *Where ball?*

The striking thing about this list is
its universality. The examples are
drawn from child talk in English, German, Russian, Finnish, Turkish, Samoan
and Luo, but the entire list could probably be made up of examples from
two-year-old speech in any language.

Word. A child easily figures out that
the speech he hears around him contains discrete, meaningful elements,
and that these elements can be combined. And children make the combinations themselves—many of their

meaningful phrases would never be
heard in adult speech. For example,
Martin Braine studied a child who said
things like *allgone outside* when he returned home and shut the door, *more
page* when he didn't want a story to
end, *other fix* when he wanted something repaired, and so on. These clearly
are expressions created by the child,
not mimicry of his parents. The matter
is especially clear in the Russian language, in which noun endings vary
with the role the noun plays in a sentence. As a rule, Russian children first
use only the nominative ending in all
combinations, even when it is gram-
matically incorrect. What is important
to children is the *word*, not the ending;
the *meaning*, not the grammar.

At first, the two-word limit is quite
severe. A child may be able to say
daddy throw, throw ball, and *daddy
ball*—indicating that he understands
the full proposition, *daddy throw
ball*—yet be unable to produce all
three words in one stretch. Again,
though the data are limited, this seems
to be a universal fact about children's
speech.

Tools. Later a child develops a rudimentary grammar within the two-word
format. These first grammatical devices
are the most basic formal tools of
human language: intonation, word
order, and inflection.

A child uses intonation to distinguish
meanings even at the one-word stage,
as when he indicates a request by a
rising tone, or a demand with a loud,
insistent tone. But at the two-word
stage another device, a contrastive
stress, becomes available. An English-
speaking child might say BABY *chair* to
indicate possession, and *baby* CHAIR to
indicate location or destination.

English sentences typically follow a
subject-verb-object sequence, and children learn the rules early. In the example presented earlier, *daddy throw ball*,
children use some two-word combinations (*daddy throw, throw ball, daddy
ball*) but not others (*ball daddy, ball
throw, throw daddy*). Samoan children
follow the standard order of pos-
sessed-possessor. A child may be sensitive to word order even if his native
language does not stress it. Russian
children will sometimes adhere strictly
to one word order, even when other
orders would be equally acceptable.

Some languages provide different
word-endings (inflections) to express

various meanings, and children who learn these languages are quick to acquire the word-endings that express direct objects, indirect objects and locations. The direct-object inflection is one of the first endings that children pick up in such languages as Russian, Serbo-Croatian, Latvian, Hungarian, Finnish and Turkish. Children learning English, an Indo-European language, usually take a long time to learn locative prepositions such as *on, in, under,* etc. But in Hungary, Finland, or Turkey, where the languages express location with case-endings on the nouns, children learn how to express locative distinctions quite early.

Place. Children seem to be attuned to the ends of words. German children learn the inflection system relatively late, probably because it is attached to articles (*der, die, das,* etc.) that appear before the nouns. The Slavic, Hungarian, Finnish and Turkish inflectional systems, based on noun suffixes, seem relatively easy to learn. And it is not just a matter of articles being difficult to learn, because Bulgarian articles which are noun suffixes are learned very early. The relevant factor seems to be the position of the grammatical marker relative to a main content word.

By the time he reaches the end of the two-word stage, the child has much of the basic grammatical machinery he needs to acquire any particular native language: words that can be combined in order and modified by intonation and inflection. These rules occur, in varying degrees, in all languages, so that all languages are about equally easy for children to learn.

Gap. When a child first uses three words in one phrase, the third word usually fills in the part that was implicit in his two-word statements. Again, this seems to be a universal pattern of development. It is dramatically explicit when the child expands his own communication as he repeats it: *Want that . . . Andrew want that.*

Just as the two-word structure resulted in idiosyncratic pairings, the three-word stage imposes its own limits. When an English-speaking child

wishes to add an adjective to the subject-verb-object form, something must go. He can say *Mama drink coffee* or *Drink hot coffee,* but not *Mama drink hot coffee.* This developmental limitation on sentence span seems to be universal: the child's mental ability to express ideas grows faster than his ability to formulate the ideas in complete sentences. As the child learns to construct longer sentences, he uses more complex grammatical operations. He attaches new elements to old sentences (*Where I can sleep?*) before he learns how to order the elements correctly (*Where can I sleep?*). When the child learns to combine two sentences he first compresses them end-to-end (*the boy fell down that was running*) then finally he embeds one within the other (*the boy that was running fell down*).

Across. These are the basic operations of grammar, and to the extent of our present knowledge, they all are acquired by about age four, regardless of native language or social setting. The underlying principles emerge so regularly and so uniformly across diverse languages that they seem to make up an essential part of the child's basic means of information processing. They seem to be comparable to the principles of object constancy and depth perception. Once the child develops these guidelines he spends most of his years of language acquisition learning the specific details and applications of these principles to his particular native language.

Lapse. Inflection systems are splendid examples of the sort of linguistic detail that children must master. English-speaking children must learn the great irregularities of some of our most frequently used words. Many common verbs have irregular past tenses: *came, fell, broke.* The young child may speak these irregular forms correctly the first time—apparently by memorizing a separate past tense form for each verb—only to lapse into immature talk (*comed, falled, breaked*) once he begins to recognize regularities in the way most verbs are conjugated. These over-regularized forms persist for years, often well into elementary

school. Apparently regularity heavily outranks previous practice, reinforcement, and imitation of adult forms in influence on children. The child seeks regularity and is deaf to exceptions. [See "Learning the Language," by Ursula Bellugi, PT, December 1970.]

The power of apparent regularities has been noted repeatedly in the children's speech of every language we have studied. When a Russian noun appears as the object of a sentence (*he liked the story*), the speaker must add an accusative suffix to the noun—one of several possible accusative suffixes, and the decision depends on the gender and the phonological form of the particular noun (and if the noun is masculine, he must make a further distinction on whether it refers to a human being). When the same noun appears in the possessive form (*the story's ending surprised him*) he must pick from a whole set of possessive suffixes, and so on, through six grammatical cases, for every Russian noun and adjective.

Grasp. The Russian child, of course, does not learn all of this at once, and his gradual, unfolding grasp of the language is instructive. He first learns at the two-word stage that different cases are expressed with different noun-endings. His strategy is to choose one of the accusative inflections and use it in all sentences with direct objects regardless of the peculiarities of individual nouns. He does the same for each of the six grammatical cases. His choice of inflection is always correct within the broad category—that is, the prepositional is always expressed by *some* prepositional inflection, and dative by *some* dative inflection, and so on, just as an English-speaking child always expresses the past tense by a past-tense inflection, and not by some other sort of inflection.

The Russian child does not go from a single suffix for each case to full mastery of the system. Rather, he continues to reorganize his system in successive sweeps of over-regularizations. He may at first use the feminine ending with all accusative nouns, then use the masculine form exclusively for a time, and only much later sort out the appropriate inflections for all genders. These details, after all, have nothing to do with meaning, and it is meaning that children pay most attention to.

Bit. Once a child can distinguish the

"Mothers in other cultures do not speak to children very much—children hear speech mainly from other children."

"Every normal child masters his particular native tongue, and learns basic principles in a universal order common to all children."

various semantic notions, he begins to unravel the arbitrary details, bit by bit. The process apparently goes on below the level of consciousness. A Soviet psychologist, D.N. Bogoyavlenskiy, showed five- and six-year-old Russian children a series of nonsense words equipped with Russian suffixes, each word attached to a picture of an object or animal that the word supposedly represented. The children had no difficulty realizing that words ending in augmentative suffixes were related to large objects, and that those ending in diminutives went with small objects. But they could not explain the formal differences aloud. Bogoyavlenskiy would say, "Yes, you were right about the difference between the animals—one is little and the other is big; now pay attention to the words themselves as I say them: *lar-laryonok*. What's the difference between them?" None of the children could give any sort of answer. Yet they easily understood the semantic implications of the suffixes.

Talk. When we began our cross-cultural studies at Berkeley, we wrote a manual for our field researchers so that they could record samples of mother-child interaction in other cultures with the same systematic measures we had used to study language development in middle-class American children. But most of our field workers returned to tell us that, by and large, mothers in other cultures do not speak to children very much—children hear speech mainly from other children. The isolated American middle-class home, in which a mother spends long periods alone with her children, may be a relatively rare social situation in the world. The only similar patterns we observed were in some European countries and in a Mayan village.

This raised an important question: Does it matter—for purposes of grammatical development—whether the main interlocutor for a small child is his mother?

The evidence suggests that it does not. First of all, the rate and course of grammatical development seem to be strikingly similar in all of the cultures we have studied. Further, nowhere does a mother devote great effort to correcting a child's grammar. Most of her corrections are directed at speech etiquette and communication, and, as Roger Brown has noted, reinforcement tends to focus on the truth of a child's utterance rather than on the correctness of his grammar.

Ghetto. In this country, Harvard anthropologist Claudia Mitchell-Kernan has studied language development in black children in an urban ghetto. There, as in foreign countries, children got most of their speech input from older children rather than from their mothers. These children learned English rules as quickly as did the middle-class white children that Roger Brown studied, and in the same order. Further, mother-to-child English is simple—very much like child-to-child English. I expect that our cross-cultural studies will find a similar picture in other countries.

How. A child is set to learn a language—any language—as long as it occurs in a direct and active context. In these conditions, every normal child masters his particular native tongue, and learns basic principles in a universal order common to all children, resulting in our adult Babel of linguistic diversity. And he does all this without being able to say how. The Soviet scholar Kornei Ivanovich Chukovsky emphasized this unconscious aspect of linguistic discovery in his famous book on child language, *From Two to Five:*

"It is frightening to think what an enormous number of grammatical forms are poured over the poor head of the young child. And he, as if it were nothing at all, adjusts to all this chaos, constantly sorting out into rubrics the disorderly elements of the words he hears, without noticing as he does this, his gigantic effort. If an adult had to master so many grammatical rules within so short a time, his head would surely burst. . . . In truth, the young child is the hardest mental toiler on our planet. Fortunately, he does not even suspect this."

NAVIGATING THE SLIPPERY STREAM OF SPEECH

Ream ember, us poke in cent tense all Moe stall ways con tains words knot in ten did. **If you read those words aloud, you'll discover one of the things psycholinguists have learned about spoken language: it "almost always contains words not intended." Not only that, the stream of speech also spawns some of our juiciest puns and riddles.**

Ronald A. Cole

Ronald A. Cole received his Ph.D. in experimental psychology from the University of California, Riverside, and is associate professor of psychology at Carnegie-Mellon University. He has published more than 30 articles on speech perception, and coauthored with Jola Jakimik, a chapter on understanding speech in *Strategies of Information Processing.* He is editor of *Perception and Production of Fluent Speech,* to be published by Lawrence Erlbaum Associates this fall.

All spoken language is a rapidly changing stream of sound, instantly decipherable only by those on intimate terms with its complex laws. Because speech is continuous, the possibility of confusion from slurred words and phrases is virtually unlimited, as in *half fast/half assed, more rice/more ice, some more/some ore.* B. F. Skinner created one of the most intriguing examples: *Anna Mary candy lights since imp pulp lay things.* Each word in the sentence can be understood, but the whole is nonsense. Read aloud quickly, however, the whole becomes a comprehensible sentence: *An American delights in simple playthings.*

Researchers in psycholinguistics have made some important strides in recent years in learning how people understand everyday speech. With the aid of new experimental techniques and sophisticated electronic equipment such as the spectrograph (shown on the following page), we can record the variations in energy expended in speech as a person talks. By picturing the stream of sound, the spectrograph

enables us to identify and measure the basic components of sound and analyze its patterns over time—a process that may yield knowledge helpful to deaf people trying to improve their speech, or to anyone learning a foreign language.

How do we hear and make sense of ordinary conversation? Most people have the mistaken impression that the words they "hear" are distinct, separate combinations of sound. The spectrograph demonstrates that words usually run together in more ambiguous sound patterns. Indeed, a word often cannot be heard correctly if it is taken out of context.

That fact was first shown experimentally more than 15 years ago at the University of Michigan by speech scientists Irving Pollack and James Pickett. From recorded conversations, Pollack and Pickett spliced out individual words, played them back to a group of people, and asked them to identify the words. Strangely, the listeners could understand only about half the words, on the average, although the same words were perfectly intelligible in the original conversation. The experiment has been replicated several times since with the same result: the particular sound pattern of a word, by itself, is often not enough to identify it.

Think of the difference between normal speech and the words on this page. Printed words are separated by space. They are the same wherever they occur. That isn't true of speech. There are no physical clues in the sound stream that consistently indi-

cate where one word ends and the next begins, and the acoustic structure (pronunciation) of a word can be quite different from one moment to the next. The result is that speech can often be interpreted in different ways, as illustrated by many jokes, riddles, and puns. Take the old favorite:

I scream, you scream.
We all scream for ice cream.

Or this riddle:

Q: If you have 26 sheep and one dies, how many are left?
A: 19 sick sheep.

The explanation is that *26 sheep* and *20 sick sheep* sound virtually the same. The riddle works because nearly everyone hears the former.

The same slurring and ambiguity make it possible to create perfectly rational sentences that are incomprehensible at first hearing. Try these on a friend: *In mud eels are, in clay none are* and *In pine tar is, in oak none is.* If you say either sentence without allowing for pauses, your friend will be unable to repeat either, although each word is real, and the sentence makes sense.

Separating out words from the stream of speech—segmentation is the technical term—is only part of the problem. Word recognition is also complicated by the fact that a word's acoustic structure can change with the context. Consider the pronunciations of *what* in: *What do you want?, Watcha want?, Whaddaya want?*

SPECTROGRAMS: VISUALIZING SPEECH

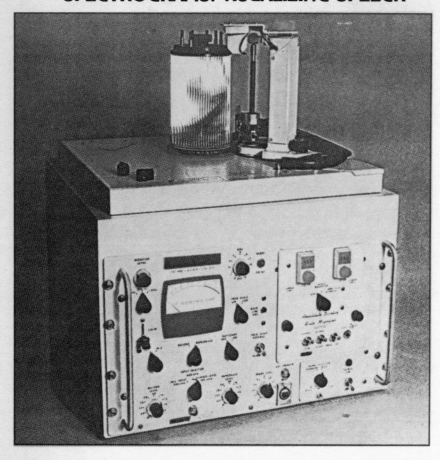

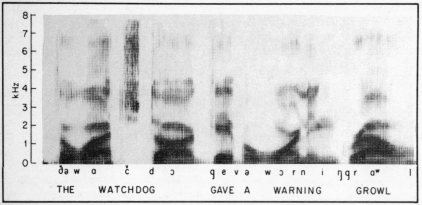

| THE | WATCHDOG | GAVE A | WARNING | GROWL |

A spectrograph—the most useful tool psycholinguists have for representing speech sounds—is shown above, along with a spectrogram. The drum at the top of the console is wrapped in heat-sensitive paper. Resting on the paper is a needle connected to a sound filter with a bandwidth of 300 cycles per second. The filter responds to sound by converting it to electric energy, which heats the needle. As the drum rotates, the needle moves up the heat-sensitive paper in response to the varying frequency of the sound. The marks on the lower part of the paper represent low-frequency energy produced by vowels and nasal sounds such as *m* and *n*; markings at the top of the paper are made by the high-frequency energy that accompanies sounds such as *ch* and *s*. The result is a spectrogram, a visual recording of speech energy that shows frequency along the vertical axis, time along the horizontal axis, and amplitude (loudness) by the darkness of the markings (loud sounds make the needle hotter).

The phonemes marked under the spectrogram, and the translation into words, are not part of the original spectrogram, but were added in order to show the sounds and words represented. It was once considered impossible to "read" spectrograms directly—that is, to pick out individual phonemes from the blurred continuum of a spectrogram. But, a few people now have this skill—the most fluent being Victor Zue, a phonetician at MIT.

People are as likely to say *Haya doin!* as *How are you doing,* or *Jeat jet!* as *Did you eat yet!*

Despite those problems, we usually understand one another. Indeed, we take our understanding of speech for granted. When people speak, we expect to know precisely what they are saying. Yet every time that happens, we have unknowingly made a series of complex decisions and, in the process, overcome numerous problems.

Sound and Syntax

Language is an intricately structured system. The grouping of the letters and words on this page, for example, depends on fixed rules of English syntax (the order of words), semantics (the meaning of words), and phonology (the structure of sound). Many sequences of phonemes—the consonant and vowel segments that are the basic sound components of speech—cannot occur in English words, and native English speakers are subconsciously aware of those restrictions. Playing Scrabble, we might bluff with *blit,* *shrats,* or *fren,* but we'd never try *dlit,* *srats,* or *vren.* We know, as our opponent knows, that English words cannot begin with *dl, sr,* or *vr.* That knowledge helps us group sounds into words.

Listening to someone talk, we can misconstrue *this lip* as *this slip* because *sl* at the beginning of a word is a permissible sound sequence. But we won't mishear *this rip* as *this srip,* because the *sr* sound cannot begin an English word.

Knowledge of these phonemic constraints—called phonotactics—is especially helpful in identifying words at the beginning of sentences, where they are most unpredictable. A sentence that begins *I'm a,* for example, can also be perceived as *I may,* or the start of *I'm making.* But a sentence beginning *I'm g. . .* can only be heard as *I'm* plus the start of a new word, since an *mg* sound can't begin a word in English.

The importance of syntactic and semantic knowledge in speech perception was demonstrated by psycholinguists George Miller and Stephen Isard at Harvard University in 1963. Miller and Isard began with normal sentences:

Gadgets simplify work around the house.

Accidents kill motorists on the highways.

Trains carry passengers across the country.

Bears steal honey from the hive.

Hunters shoot elephants between the eyes.

They constructed grammatical but meaningless sentences by combining the first word of one sentence with the second word of another sentence, the third word of still another, and so on: *Gadgets kill passengers from the eyes; Accidents carry honey between the house.* The researchers then produced ungrammatical strings of words by rearranging the same words haphazardly: *Around accidents country honey the shoot; Across bears eyes work the kill.*

After recording 50 sentences of each type, with noise in the background to simulate normal speech conditions, Miller and Isard asked their subjects to repeat each sentence. Under listening conditions in which subjects repeated normal sentences accurately 89 percent of the time, they repeated grammatical but meaningless sentences 79 percent of the time, and the random words 56 percent of the time. Obviously, the subjects used their knowledge of syntax and semantic structure to reconstruct what they had heard. When those clues were missing, they had trouble remembering the sentences.

One of the most dramatic illustrations of how context affects speech perception is the "phonemic restoration" experiment done by psychologist Richard Warren of the University of Wisconsin-Milwaukee. Warren replaced a phoneme—the first *s* in *legislatures*—with a coughing sound in the sentence *The state governors met with their respective legislatures convening in the capital city.* He then asked his subjects to tell him where in the sentence the cough had occurred. They couldn't tell, because they had unconsciously "restored" the *s* sound. People heard the *s* as if it were still intact; they perceived the cough as background noise. Under the right circumstances, then, listeners can use contextual information to generate sounds they have not actually heard because of noise or other distractions.

Psychologists John Long and John Morton, working in England, were the first to show that a word is recognized faster when it is predicted by previous words. Using a "phoneme-monitoring" technique invented by psychologist Donald Foss at the University of Texas, Long and Morton read sentences like the following to subjects, and asked them to push a signal button as soon as they heard a word beginning with *b*:

The sparrow sat on the branch singing a few shrill notes to welcome the dawn.

The sparrow sat on the bed singing a few shrill notes to welcome the dawn.

Morton and Long found that people responded more quickly to the *b* sound in predictable words, like *branch*, than they did in the unpredictable ones, like *bed*. They apparently combined what they heard (*The sparrow sat...*) with their knowledge of the world (*sparrows sit on branches*), to prime themselves for the rest of the sentence.

Extending that work, psychologist William Marslen-Wilson investigated how quickly and how accurately people can "shadow" speech (repeat it verbatim) while they listen to it. He found seven subjects who could speak intelligibly while shadowing a little more than a syllable behind what they were hearing. The subjects were able to say the beginnings of most words before they heard the finish. In addition, the errors they made were always syntactically and semantically appropriate. For example, in the sentence *It was beginning to be light enough so I could see*, two subjects inserted *that* following *so*. In the phrase *he heard at the brigade...*, five subjects replaced *heard at* with *heard that*. Marslen-Wilson concluded that our understanding depends strongly on preceding context, which includes the word immediately before the word we recognize.

Predictability and Patterns

In the past several years, my colleague Jola Jakimik and I have been investigating the role of knowledge in word perception by using a "listening-for-mispronunciations" task. Subjects listen to recorded sentences or stories and push a response key as soon as they detect a mispronounced word. We produce the mispronunciations by changing a single consonant to create an English nonsense word (for instance, *boy* to *poy*, *maybe* to *naybe*). The nonsense word is pronounced naturally, as if it were a real one. The listening-for-mispronunciations task has proved to be an extremely sensitive technique for examining the kinds of knowledge that listeners use automatically in understanding everyday speech.

We have found that the time it takes for a listener to detect a mispronunciation depends on what he has learned from the words preceding the mispronounced word. In one experiment, we constructed pairs of sentences, identical except for the word immediately preceding the mispronunciation. For example:

She wanted a mink poat (coat) to go with her Rolls Royce.

She wanted a pink poat (coat) to go with her Rolls Royce.

He wore a gold ling (ring) on his little finger.

He wore an old ling (ring) on his little finger.

People detected mispronunciations one syllable sooner when the word (*mink* or *gold*) preceding the word that was mispronounced primed them to expect a certain word (*coat* or *ring*). When the preceding word did not prime them, they needed to hear more of the words following to detect the mispronunciation.

The time it takes to recognize a word does not depend solely on its predictability; the sound pattern of the word itself is also important. If a word begins with a common first syllable, like *complain*, *contact*, *perform* or *provide*, there are many word possibilities we must consider (especially if the word is unpredictable). On the other hand, if the word begins with an unusual first syllable, like *shampoo*, *whisper*, *vampire*, or *cranberry*, we have fewer candidates to consider. We should therefore expect the reaction time to second-syllable mispronunciations to be faster in words with unusual first syllables like *shampoo (shamboo)* than in words with common first syllables like *complain (comblain)*, since in the first case, there are fewer possibilities to be considered. And, in fact, Jakimik confirmed that prediction as part of her doctoral dissertation.

To summarize, research shows that we make extremely rapid use of both sound and knowledge to recognize words, usually before a person has finished saying them. We rely both on the sounds that begin a word and on the information provided by the preceding words.

An Idiot-Savant Computer

Not even the most sophisticated of computers can match man's ability to use sound and knowledge to understand spoken language, despite numerous attempts to do exactly that. In 1971, the Advanced Research Projects Agency (ARPA) of the Department of Defense started a five-year, multimillion dollar project aimed at developing computer systems that can understand human speech. Only one system—HARPY, developed at Carnegie-Mellon University—met all of the project's performance goals. HARPY can recognize coherent speech in a quiet room if the speaker articulates carefully and uses a vocabulary limited to 1,011 words. Since HARPY is programmed to perform a specific task—document retrieval—it is especially good at recognizing questions like *Are there any articles on speech understanding by [Raj] Reddy?*

Though HARPY is the most successful speech-understanding system developed to date, it cannot even match the linguistic ability of a four-year-old child. HARPY is a kind of idiot savant, amazingly capable in a restricted area under ideal conditions. It works by storing all possible pronunciations of sentences that can be spoken in asking for documents, and then finds the best possible match to what it hears. For that reason, the computer system cannot recognize a novel sentence—something that four-year-olds do every day.

In its current form, it takes HARPY 10 times as long as it takes a human being to recognize speech. Thus, if a question takes two seconds to ask, HARPY will take 20 seconds to understand it. A college graduate has a vocabulary of about 20,000 words—20 times as large as HARPY's. In addition, a human being recognizes

speech under noisy conditions and understands many different accents. HARPY needs speakers who articulate carefully in a quiet room—and a Brooklyn accent can cause HARPY to draw a blank.

One of the computer's biggest problems occurs in recognizing phonemes. HARPY postulates several phonemes for each sound segment it identifies; it is correct on its first choice less than half the time. The computer's problem arises in comprehension because, in normal speech, the sound of a phoneme is sometimes changed by the neighboring phonemes: the *g* sounds different, for example, in the words *geese* and *goose*.

Reading a Spectrogram

Because of the complex relationship between sounds and phonemes, many psychologists and linguists believe that phonemes do not really exist as such in normal speech. Sounds are slurred and intermingled so completely, those scientists say, that they are indecipherable even when viewed on a spectrogram.

That argument presents a serious problem, however. If there are no independent sounds that correspond to phonemes (the building blocks of words), how do we understand one another? Psycholinguists who believe that phonemes do not exist in normal speech postulate a special "decoder," located in the left hemisphere of the brain, that takes speech as input and produces a sequence of intelligible phonemes as output. This decoder, the theory goes, is part of the brain's auditory system, inaccessible to the visual system. Therefore, according to those who take that position, no human being is capable of "reading" a spectrogram fluently: that is, capable of rapidly identifying individual phonemes on the tracings that represent speech energy.

The fact is, someone can. In 1971, Victor Zue of the Speech Communication Group at MIT decided to learn more about the relationship between sound and its representation on a spectrogram. Since then, working on his own, Zue has spent about an hour a day—more than 2,000 hours in all—studying spectrograms. He can now read them quickly and accurately. My colleague Alex Rudnicky of Carnegie-Mellon University and I

tested Zue's ability twice last year, and videotaped each session. We showed Zue spectrograms of normal English sentences, of semantically anomalous sentences (*Bears shoot work on the highway*), of random sequences of words, and of sequences of regular words mixed with nonsense words. His task was to identify all of the phonemes from the spectrograms. He first indicated the location of each phoneme, placing small vertical lines directly under it on the spectrogram, and then identified what the phoneme was.

We compared Zue's phoneme labels with those produced by experienced phoneticians who listened to tape recordings of the original sentences from which the spectrograms were made. The phoneticians identified 499 phonetic segments in the 23 sentences we used. Zue located 485 of them. He gave the phoneme the same label as the phoneticians on 85 percent of all segments—a level of agreement nearly as high as the average agreement (90 percent) among the phoneticians themselves.

Analyzing the videotapes, we discovered Zue's method. He identifies phonemes by recognizing characteristic visual patterns. He does not use his knowledge of the syntactic and semantic structure of English to interpret spectrograms—that is, he doesn't try to recognize words or otherwise make sense of the spectrograms he reads. After he labels a spectrogram, he often has no idea which words it contains. That fact was supported by our data, which showed he was as accurate in identifying phonemes in sequences of nonsense words as he was in identifying those in English sentences (about 85 percent in each). Although Zue doesn't usually "read" spectrograms in the normal sense—as words and sentences—I have seen him do so, and very quickly. When he reads words, he uses the context—the sense of the sentence—to group sounds into words just as we all do in normal conversation. I am optimistic that, given sufficient practice, Zue and others will one day be able to read spectrograms in "real time"—as the words are actually spoken.

More than thirty years ago, Bell Laboratories had some success with real-time speech reading, using equipment much less helpful than the spectrograms we now have. Participants in the Bell study—one of them a deaf engineer—attempted to recognize visual patterns for words on a "Direct Translator" (a device that produced a spectrographic display of speech as it was spoken). Although the display produced by the Direct Translator was not nearly so detailed as a modern spectrogram, the results were impressive. Most participants learned to identify about four words for every hour of study, and the deaf engineer learned to identify about 800 words in all. By the end of the project, after 90 hours of training, the participants were able to carry on conversations on the Direct Translator (without hearing one another), as long as participants spoke clearly and distinctly, with a common vocabulary, whenever they were using the device.

Victor Zue's accomplishment demonstrates a direct, learnable relationship between specific sounds and the visual patterns they produce on a speech spectrogram. That fact has important implications in teaching speech to the deaf or those with impaired hearing. The major problem the deaf have in learning to produce intelligible speech is that they can't hear what they say. Without that feedback, precise articulatory differences are almost impossible to learn or to retain. In fact, people who lose their hearing soon have trouble with their speech.

The problem may be solved by allowing deaf people to see the results of speech as soon as they have produced it. A real-time speech spectrogram now available commercially displays speech on a videoscreen as it is spoken. Using a split screen, the correct visual pattern of a sound can be shown on one half of the screen, and the results of each attempt to produce the sound can be reproduced on the other half of the split screen. Comparing the two patterns of sounds gives the speaker immediate feedback, helping him or her learn to say the words correctly—without ever having heard them.

The same procedure can be useful in learning the unfamiliar phonemic system of a foreign language. Think of the difficulty the Japanese have in pronouncing l and r in English. Since Japanese does not have an l phoneme, native Japanese speakers have never learned to hear the difference between the two sounds, and thus have great difficulty in articulating them in English. On a spectrogram l and r each look dramatically different. By trying to produce the correct *visual* pattern, with immediate feedback showing success or failure, a Japanese speaker can eventually learn to associate the correct sound with the articulatory movements that produce it.

Grasping the phonemic structure of a language is critical to understanding it. But that is only part of the knowledge we need to understand everyday speech. Even if we can identify all the phonemes in a sentence, additional knowledge is necessary if we are to recognize only the words the speaker intends us to hear. Remember, a spoken sentence almost always contains words not intended. Consider the sentence you have just read. Spoken, it contains the words *ream ember us poke in cent tense all Moe stall ways con tains words knot in ten did*. You might argue that only the intended words make sense. But that is just the point. To recognize the words we are meant to hear, we must make immediate sense of the sounds and hear only the intended combinations of those sounds.

To do that, we must use a knowledge of syntax and semantics—the same knowledge the speaker used to create the sound combinations in the first place. In a sense, we recognize words by recreating the other person's train of thought; the speaker and listener share in the process of putting the speaker's thoughts into words.

For further information, read:

Cole, R. A. *Perception and Production of Fluent Speech*, Lawrence Erlbaum Associates, in press.

Cole, R. A., J. Jakimik. "Understanding Speech: How Words Are Heard," in *Strategies of Information Processing*, G. Underwood, ed., Academic Press, 1978, $41.40.

Cole, R. A. "Listening for Mispronunciations: A Measure of What We Hear During Speech," *Perception and Psychophysics*, Vol. 13, No. 1A, 1973.

Marslen-Wilson, W. D., A. Welsh. "Processing Interactions and Lexical Access During Word Recognition in Continuous Speech," *Cognitive Psychology*, Vol. 10, No. 2, 1978.

Morton, J., J. Long. "Effect of Word Transitional Probability on Phoneme Identification," *Journal of Verbal Learning and Verbal Behavior*, Vol. 15., No. 1, 1976.

Warren, R. M., C. J. Obusek. "Speech Perception and Phonemic Restorations," *Perception and Psychophysics*, Vol. 9, No. 3B, 1971.

Education and Child Development

The family is the first major socialization force to which the child is exposed. From initial family interactions children acquire the social, emotional, and cognitive foundations upon which their encounters with the larger social environment will be based. During the period of childhood, children become involved with the second major socialization force, the school. New friendships develop as the child's peer network expands beyond his or her neighborhood. Children spend a considerable portion of each day under the supervision of authority figures other than their parents. Some children have been prepared for this separation event by their prior preschool or day care experience, but for many others school entrance represents the first extended separation from the home. Most children experience little difficulty adapting to school, but for others, school adjustment is more problematic.

For much of this century, educators believed that the development of cognitive, social, and emotional characteristics were determined exclusively by environmental factors. All infants were thought to enter the world on an equal footing with individual differences emerging as a function of differences in their rearing environment. Today we know that infants are not equal when they enter the world, and that individual differences emerge as a function of each individual's genotypic uniqueness in interaction with a set of individual environments. Even later born children in the same family cannot be said to have the same rearing environment as their earlier born siblings. The third born is genotypically different from his or her first and second born siblings. Parents of the third born are not the same as they were when they had their first born child. The family structure is different, the family economic status may have changed, and the interconnections of the family system with other social systems may have changed. Moreover, individual differences among babies will affect the nature of the relationships constructed with their caregivers. For example, increasing evidence points to important differences in temperament which appear to be present at birth. Thus, the optimal rearing environment for one child may not be optimal for another. Similarly, the optimal educational environment for one child may not be optimal for another.

Fortunately, some educators are coming to recognize that learning processes occurring during childhood are far more complex than previously imagined. Attempts to explain intellectual development independent of emotional development have proved too simplistic because they do not take into account the systemic nature of the organism. More and more, educators have argued that many children suffer from various learning "disorders," disturbances which prevent their functioning effectively in the traditional school environment. Seldom, however, do educators look within the educational system to assess the extent to which "learning disorders" are generated by the school environment itself rather than originating from within the child or within the home.

Recently, some child development specialists have advocated a radical change in the role of the school in order to facilitate optimal development of the child. This plan would establish child development centers in local schools with components for prenatal care, infant day care, nursery school, and child and family counseling. Another recent change in American schools involves attempts to teach moral behavior.

Looking Ahead: Challenge Questions

How might day care for infants affect the organization of affectional bonds with parents and other significant caregivers in the infant's world?

Can educational institutions change, or have they become such conservative forces that the only way to change them is to design an entirely new system for the education of children? If the latter, how would you begin?

What are our obligations to children who have difficulty adjusting to the traditional school environment? What are the benefits of early intervention programs? Of programs for the gifted? What should our educational goals be?

Do you believe that schools should attempt to teach children correct moral behavior? If so, whose definition of right and wrong conduct is to be used? Mine or yours?

Infant Day Care: Toward a More Human Environment

Arminta Lee Jacobson

In planning optimal environments for infant day care, the human element is often minimized. Desirable caregiving competencies which can be used as a basis for evaluating day care personnel are derived from research findings.

What implications for infant caregiving in day care can be drawn from mother-infant interaction research? Few studies have been done on interpersonal caregiving competencies or the effect of discrete caregiving variables on infant development in day care settings. Yet the spiraling increase of out-of-home care for infants demands research attention, both for professionals concerned with infant day care and for parents needing such services. The following discussion examines mother-infant interaction research in order to delineate current areas of exploration and to synthesize findings into a helpful format for administrators, educators, researchers, and others interested in upgrading the quality of infant day care.

Importance of the Primary Caregiver

The importance of the primary caregiver in an infant's development is often overlooked, although several research studies have given evidence of the relationship between early caregiving experiences and competencies in later childhood. In a longitudinal study of environmental determinants related to human competency upon entering school, White and Watts (1973) found that ratings of competency of children at age six varied very little from ratings of competency of those children at age three. Further investigation pinpointed the period of 10 to 18 months as the most cru-

cial in determining a child's later competency, especially in the areas of social skills and attitudes.

Yarrow et al. (1973) studied the relationship between mothering experiences during the first six months of life and selected intellectual and personal-social characteristics at 10 years of age. For boys, the Wechsler Intelligence Scale for Children IQ and several aspects of a child's relationship to others at 10 years of age were related to variables of maternal behavior at six months of age. In another study (Yarrow et al. 1972), the social environment was shown to be highly significant in influencing infant functioning, independent of dimensions of the inanimate environment.

In a report by Bayley and Schaefer (1964), results of an analysis of data collected in the Berkeley Growth Study between 1928 and 1954 showed maternal and child behaviors to be intercorrelated over an 18-year span of growth. The relationship to maternal behaviors was more significant for boys. Coping capacities of older children were found to be significantly related to early mother-infant interactions in a study by Murphy (1973) of 31 children and mothers. These studies and others contribute to the growing recognition of the impact of human relations in the earliest years of development.

Research studies of mother-infant interaction have varied in the type of interrelationships studied as well as in research design, data collection, and analysis. Categories of maternal and infant variables chosen to enter into relationship models have only gross similarities among investigations. Study of interpersonal behavior by its very nature requires direct observation of behavior and is susceptible to subjectivity in measurement. Behavior patterns are often quite complex, and interrelationships not included in the research model are often overlooked. Although infants have a limited repertoire of relationship responses, adult transactions are practically unlimited in variety and complexity. Nevertheless, important information discerned from such studies helps to identify those interrela-

tionship variables which effect optimal development in the early formative years.

Although the mother-infant dyad has been the focus of most research seeking to explain social determinants of infant behavior and development, findings from such research can easily be generalized in terms of appropriate behavior for any adult serving as a primary caregiver of an infant, even for a limited part of the day, as in a day care setting. What is conducive to optimal development in one setting could logically be conceived as appropriate in another setting.

An exception which must be made in generalizing research findings of mother-infant interaction to day care is the area of attachment of infant to caregiver. In a series of ongoing studies of mother-infant relations, Ainsworth and Bell (1972) have pinpointed quality of attachment between mother and infant as being related to other aspects of infant and maternal behavior. Mutual attachment of infant and caregiver cannot be assumed in a day care setting. Nature and degree of attachment of primary caregivers and infants in day care have not been investigated and would be difficult to study due to the common instability of such relationships over time and the confounding effects of other adults and infants.

Areas of Adult Influence

Given the limitations of research methodology and generalizations, what evidence from recent research can serve as guidelines for infant day care personnel in providing an optimal human environment for infants?

Infants' physical, emotional, social, and cognitive development is shaped to a great extent by the behaviors of the primary caregivers in relation to the children. Primary caregivers can be conceived of not only as determinants of infants' physical survival but also as social agents against which infants test their growing competencies and

Reprinted by permission from *YOUNG CHILDREN*, Vol. 33, No. 5 (July 1978), pp. 14-23. Copyright ©1978, National Association for the Education of Young Children, 1834 Connecticut Avenue, N.W., Washington, DC 20009.

conceptions of self and the world. The extent to which adults initiate interactions with or respond to infants, the affective nature of those behaviors, their content and context, all have an influence on infants' developing responses and behaviors and the context within which different facets of their development emerge.

Infant Competence. Personality characteristics, control, involvement, responsiveness, and attachment are some of the many types of maternal influences found to be related to infant development. In a study of the mother-infant dyad (Stern et al. 1969), a sequence of relationships between personality characteristics of mother, modes of maternal behavior, and responses and development of the infant was defined. The nine factors resulting from this composite appear to be distributed along a continuum ranging from child-centered to mother-centered maternal functioning. Effective mothers were defined as those whose infants were lovingly responsive to them and accelerated in development. The characteristics these mothers seemed to have in common were: (1) attentive, loving involvement with their infants; (2) high levels of visual and vocal contact; and (3) play involvement. The mothers producing the more accelerated infants were characterized as self-confident and skilled in their caregiving and individualistic in style.

White and Watts (1973) found that infants assessed in their study as highly competent had mothers who differed significantly from mothers of infants judged less competent. Mothers of the more competent infants involved themselves in more mother-infant interactions. Even when their infants were as young as 12 to 15 months, these mothers spent more time with "highly intellectual" activities and used interaction techniques which taught or were facilitative in nature. These mothers decreased their use of restrictive techniques as children grew older while mothers of the less competent infants increased their use. From the analysis of attitudes and values of mothers in the study, characteristics related to optimal development of children included a positive attitude toward life in general; enjoyment of infants in the one-to-three-year age range; an acceptance of the incompatibility of infant needs and preservation of posessions and household order; and the willingness to take risks for the sake of infants' curiosity and development.

Ainsworth and Bell (1972) studied infants' competence in direct dealing with the physical environment as measured by developmental competence on the Griffiths Scale. Positive relationships were shown between infant competence and maternal factors of sensitivity, acceptance, cooperation, and the amount of floor freedom allowed the infant. Amount of playing with the baby by the mother was also positively correlated with developmental scores of the infant. Frequency of punishment was negatively related to infant competence.

Level and variety of social stimulation (Yarrow et al. 1972) provided by a primary caregiver in the home have been found to be positively related to functioning of five-month-old infants. Infant functioning which related significantly to social stimulation included goal-directed behaviors, reaching and grasping, and secondary circular reactions. Adult responses, contigent upon infant distress, were found to be significantly related to goal-directed behavior in the infant.

Other studies (Murphy 1973; Stern et al. 1969) exemplify findings which support an optimal level of interaction, reporting a curvilinear relationship between development and degree of attention. In studying the development of coping ability in young children, Murphy (1973) found that optimal early mother-infant interactions were characterized by a balance of attention and autonomy, of interaction and letting the infant alone part of the time. Too much or too little attention, body contact, and talking to infants were found to be not good for infant development. These findings concur with findings (Stern et al. 1969) which characterize mothers of slow-developing infants as exhibitionist, vigilant, and including both high and low levels of physical contact. Murphy (1973) also found that patterns of mothering were related to individual infant temperaments in different ways, indicating the need for flexibility in interaction patterns.

Infant Vocalization. Mother responsiveness and infant vocalization have been examined in several studies. Clarke-Stewart (1973) reported a high relation of responsive maternal speech and children's competence in a longitudinal study of infants from 9 to 18 months of age.

Responsive mothers—those who ignore few episodes and respond with little delay—have infants with more variety, subtlety, and clarity of noncrying communication. During the second, third, and fourth quarters of the first year, infants of responsive mothers cried significantly less than infants of unresponsive mothers. Beckwith (1971b) also reported a positive relationship between mothers' ignoring of infants and frequency of infant crying. Infants who cried little had a wider range of differentiated modes of communication than did infants who cried often (Ainsworth and Bell 1972). Amount of maternal play behavior has also been found to be positively related to amount of infant vocalization (Clarke-Stewart 1973).

Perceptual-Cognitive Development. In the last decade, perceptual-cognitive development of very young children has interested researchers and parents. A study of perceptual-cognitive development in infants 12 weeks of age (Lewis and Goldberg 1969) also stressed the importance of maternal responses which are contingent upon the infant's behavior. Perceptual-cognitive development was found to be moderately related to the overall response of mother to infant's crying and vocalization and the amount of touching, holding, and smiling exhibited by the mother, and highly related to the amount of looking by the mother. These findings concur with other studies (Stern et al. 1969; White and Watts 1973) which characterize effective mothers as being very responsive to and involved with their infants.

An investigation of the relationships between maternal behaviors, infant behaviors, and individual differences in infant IQ (Beckwith 1971a) was made with the same infants at two interviews, during age ranges from 7.2 to 9.7 months and 8.5 to 11.3 months. This study revealed that low maternal verbal and physical contact within the home were significantly related to lower IQ on the Cattell Infant Intelligence Scale. Maternal restriction of infant exploration was found to be related to decreased interest in attaining speech during the last quarter of the first year and was significantly related to lowering of IQ scores.

Clarke-Stewart (1973) also reported maternal restrictiveness to be negatively related to scores on the Bayley Scale of Mental Development at 18 months. In this study the Bayley measure was highly correlated with the mother's nonphysical stimulation—looking and talking. Responsiveness of the mother was also related to the child's Bayley score and to the child's speed of processing information, schema development, language, and social and emotional competence. Stimulation by mother to promote achievement has also been found to be related to Cattell IQ scores at six months of age (Yarrow et al. 1973).

Ainsworth and Bell (1972) have studied cognitive development in White middle socioeconomic status (SES) infants and Black lower SES infants in terms of development of the concept of object permanence and scores on the Griffiths Development Scale. Infants who had harmonious interactions with mothers sensitive to their signals and who had developed attachment relationships of normal quality tended to develop the concept of person permanence in advance of object permanence. At 8 to 11

Table 1. Characteristics of Competent Infant Caregivers.

Desired Caregiver Characteristics	Cues to Desirable Caregiver Characteristics
I. Personality Factors	
A. Child-centered	1. Attentive and loving to infants. 2. Meets infants' needs before own.
B. Self-confident	1. Relaxed and anxiety free. 2. Skilled in physical care of infants. 3. Individualistic caregiving style.
C. Flexible	1. Uses different styles of caregiving to meet individual needs of infants. 2. Spontaneous and open behavior. 3. Permits increasing freedom of infant with development.
D. Sensitive	1. Understands infants' cues readily. 2. Shows empathy for infants. 3. Acts purposefully in interactions with infants.
II. Attitudes and Values	
A. Displays positive outlook on life	1. Expresses positive affect. 2. No evidence of anger, unhappiness, or depression.
B. Enjoys infants	1. Affectionate to infants. 2. Shows obvious pleasure in involvement with infants.
C. Values infants more than possessions or immaculate appearance	1. Dresses practically and appropriately. 2. Places items not for infants' use out of reach. 3. Reacts to infant destruction or messiness with equanimity. 4. Takes risks with property in order to enhance infant development.
III. Behavior	
A. Interacts appropriately with infants	1. Frequent interactions with infants. 2. Balances interaction with leaving infants alone. 3. Optimum amounts of touching, holding, smiling, and looking. 4. Responds consistently and without delay to infants; is always accessible. 5. Speaks in positive tone of voice. 6. Shows clearly that infants are loved and accepted.
B. Facilitates development	1. Does not punish infants. 2. Plays with infants. 3. Provides stimulation with toys and objects. 4. Permits freedom to explore, including floor freedom. 5. Cooperates with infant-initiated activities and explorations. 6. Provides activities which stimulate achievement or goal orientation. 7. Acts purposefully in an educational role to teach and facilitate learning and development.

months these infants were also advanced in the level of object permanence achieved. Harmonious attachment relationship, as well as floor freedom, were highly related to development scores.

Infant Play Behavior. Another area of consideration is the development of infant play behavior. According to findings by Clarke-Stewart (1973), the best single predictor of play behavior in infants was the amount of stimulation with toys and objects received from the mother at home.

Other researchers have studied quality of investigative behavior and exploratory play and its relation both to maternal behavior and to the quality of infant-mother attachment relationships (Ainsworth and Bell 1972). They found a significant relationship during the last quarter of the first year between frequent harmonious transactions with the mother, mother responsiveness to infant-initiated interaction, and the infant's greater exploration of toys and advanced behavioral schemata in play.

Social Development. Social development and play appear to be enhanced by some of the same maternal behaviors. In studying relations between the mother's behavior and the quality of the child's attachment, Clarke-Stewart (1973) found a number of nonlinear relationships. Optimally securely attached children—those able to use mother as a secure base from which to explore the environment and to which to return periodically at times of stress or for reassurance—where associated with homes where there was not constant exposure to a great number of people and where mothers were socially stimulating, responsive, and affectionate. In particular, the children's attachment was highly related to frequency of maternal social behavior.

In studying the use of mother as a secure base from which to explore, Ainsworth and Bell (1972) studied quality of infant attachment in relation to maternal ratings. Infants rated as highest in actively seeking proximity and interaction with mothers all had mothers above the median in sensitivity to infant signals, acceptance, cooperation, and accessibility.

The early manifestation of infant obedience indicates progress in social development. In a study by Stayton, Hogan, and Ainsworth (1971), maternal variables of sensitivity, acceptance, and cooperation were all highly intercorrelated with infants' compliance with commands during the last quarter of the first year. Frequency of verbal commands, frequency of physical intervention, and amount of floor freedom permitted the infant were not found to be related to compliance with commands.

Happiness. An obvious measure of effectiveness in interpersonal relations with infants is the degree to which positive affect or happiness is observed in the infant. Smiling and vocalizing and the absence of crying and fretting are seen as evidence of happiness. An infant's expression of happiness has been found to be most closely related to the mother's expression of positive emotions (Clarke-Stewart 1973). Mothers who vocalize and smile frequently have been found to have infants who vocalize and smile frequently. The more positive the maternal behaviors, the less frequently the infants fret and cry (Lewis and Wilson 1972). Infant fretfulness has been observed to be related to maternal rejection and self-control. Lower levels of infant fretfulness are associated with maternal effectiveness in physical, social, and instrumental behaviors (Clarke-Stewart 1973).

Implications for Infant Day Care Workers

Despite the inconsistency of focus and the nebulous nature of desirable maternal behaviors, mother-child interaction studies provide a sound research base for determining desirable caregiving attributes. In view of the empirical evidence on the importance of human interactions to infant development, it is clear that infant day care workers must be highly competent in interpersonal skills for quality caregiving. It is imperative that day care administrators hire and train infant caregivers on the basis of their attitudes and behaviors in interpersonal relations with infants.

Table 1 represents a synthesis of characteristics which provide an optimal human environment for infant caregiving, as generalized from the research findings. The categorization of caregiving behaviors can be used for further development of competency profiles for infant caregivers. Administrators of infant care centers will be most interested in those items helpful to selecting and evaluating infant caregivers; characteristics reflected in Table 1 provide possibilities for structuring interview or evaluation schedules. More specific attention to developmental levels of competence within behavior indexes could lead toward individualized training experiences for infant day care staff.

Further research, directed toward specification, assessment, and integration of infant caregiving behaviors, is needed, since only through delineation of these important human behaviors can child care personnel plan knowledgeably for the optimal care of infants.

Other implications for the placement of caregivers in day care settings come from research findings which highlight cultural, racial, and SES differences in maternal expectations for infants and in mother-infant interaction behaviors (Goldberg 1972; Lewis and Ban 1973; Lewis and Wilson 1972; Tulkin and Cohler 1973). Developmental differences in infants have been associated with caregiver differences. Consideration should be given to placing caregivers in day care settings where their cultural and SES values and expectations are similar to those of the families served.

Recognition of wide variations in caregivers' attitudes, sensitivities, and behaviors should also prompt day care professionals to work cooperatively with parents in setting caregiving goals for infants. Parents can help caregivers define the infant's nature and needs and the kind of·environmental variables most effective in maximizing the infant's potential.

The crucial importance of the earliest experiences of life need continual emphasis and investigation. It is hoped that persons responsible for planning day care experiences for infants will be creatively sensitive to ways in which the quality of life for infants can be improved.

References

Ainsworth, M. D. S., and Bell, S. M. "Mother-Infant Interaction and the Development of Competence." ERIC Document Reproduction Service No. ED 065 180, 1972.

Ainsworth, M. D. S.; Bell, S. M.; and Stayton, D. J. "Individual Differences in Strange-Situation Behavior of One-Year-Olds." In *The Competent Infant: Research and Commentary*, edited by L. J. Stone, H. T. Smith, and L. B. Murphy. New York: Basic Books, 1973.

Bayley, N., and Schaefer, E. S. "Correlations of Maternal and Child Behaviors with the Development of Mental Abilities: Data from the Berkeley Growth Study." *Monographs of the Society for Research in Child Development* 29, no. 6 (1964), serial no. 97.

Beckwith, L. "Relationships Between Attributes of Mothers and Their Infants' IQ Scores." *Child Development* 42, no. 4 (1971a): 1083-1097.

Beckwith, L. "Relationships Between Infant's Vocalizations and Their Mother's Behaviors." *Merrill-Palmer Quarterly* 17 (1971b): 211-226.

Caudill, W., and Frost, L. "A Comparison of Maternal Care and Infant Behavior in Japanese-American, American, and Japanese Families." ERIC Document Reproduction Service No. ED 057 153, 1971, Honolulu, Hawaii.

Clarke-Stewart, K. A. "Interactions Between Mothers and Their Young Children: Characteristics and Consequences." *Monographs of the Society for Research in Child Development* 38, nos. 6-7 (1973), serial no. 153.

Goldberg, S. "Infant Care and Growth in Urban Zambia." *Human Development* 15 (1972): 77-89.

Lewis, M., and Ban, P. "Variance and Invariance in the Mother-Infant Interaction: A Cross-Cultural Study." ERIC Document Reproduction Service No. ED 084 006, 1973, Princeton, New Jersey.

Lewis, M., and Goldberg, S. "Perceptual-Cognitive Development in Infancy: A Generalized Expectancy Model as a Function of the Mother-Infant Interaction." *Merrill-Palmer Quarterly* 15 (1969): 81-100.

Lewis, M., and Wilson, C. D. "Infant Development in Lower-Class American Families." *Human Development* 15 (1972): 112-127.

Murphy, L. B. "Later Outcomes of Early Infant and Mother Relationships." In *The Competent Infant: Research and Commentary*, edited by L. J. Stone, H. T. Smith, and L. B. Murphy. New York: Basic Books, 1973.

Stayton, D. J.; Hogan, R.; and Ainsworth, M. D. S. "Infant Obedience and Maternal Behavior: The Origins of Socialization Reconsidered." *Child Development* 42 (1971): 1057-1069.

Stern, G. G.; Caldwell, B. M.; Hersher, L.; Lipton, E. L.; and Richmond, J. G. "A Factor Analytic Study of the Mother-Infant Dyad." *Child Development* 40 (1969): 163-181.

Tulkin, S. R., and Cohler, B. J. "Childbearing Attitudes and Mother-Child Interactions in the First Year of Life." *Merrill-Palmer Quarterly* 19 (1973): 95-106.

Tulkin, S. R., and Kagan, J. "Mother-Child Interaction in the First Year of Life." *Child Development* 43 (1972): 31-41.

White, B. L., and Watts, J. C. *Experience and Environment: Major Influences on the Development of the Young Child*. Englewood Cliffs, N.J.: Prentice-Hall, 1973.

Yarrow, L. G.; Goodwin, M. S.; Manheimer, H.; and Milowe, I. D. "Infancy Experiences and Cognitive and Personality Development at Ten Years." In *The Competent Infant: Research and Commentary*, edited by L. J. Stone, H. T. Smith, and L. B. Murphy. New York: Basic Books, 1973.

Yarrow, L. G.; Rubenstein, J. L.; Pedersen, F. A.; and Jankowski, J. J. "Dimensions of Early Stimulation and Their Differential Effects on Infant Development." *Merrill-Palmer Quarterly* 18 (1972): 205-218.

Diet and Schoolchildren

Like medical doctors, educators have been slow to realize how often there is a direct relationship between the kinds of food kids consume today and their behavior and academic achievement.

Fred L. Phlegar and Barbara Phlegar

"Let thy food be thy medicine and thy medicine be thy food."
— Hippocrates

Recently a Radford University professor said to us, "We have taken our son off sugar-coated cereals. He is much calmer now and is getting along much better in school."

Did this professor know what he was talking about? If so, how many of our young people are adversely affected by eating popular, highly processed foods? How many students are creating problems for themselves and their teachers, school administrators, and parents?

Although definitive studies are still scarce, suspicion is aroused by cases like the following:

In San Jose, California, a mother reported that one of her foster children became aggressive and hostile after she stopped her home cooking and started using bakery products. Then other foster children in her home started having the same behavior problems. One had to be put on tranquilizers. After she learned about the possible effects of some foods on behavior, this foster mother adjusted the diets of all of her children. She reports that the changes were "unbelievable." All the children became much more serene and made better grades in school.

Hugh W. S. Powers, Jr., a physician in Dallas, Texas, has adjusted the diets of many students with a significant success

FRED L. PHLEGAR (Radford Virginia Chapter) is chairman, Secondary Education Department, Radford University, Va. BARBARA PHLEGAR (Radford Virginia Chapter) teaches learning-disabled children in the Radford City Schools.

rate. One 16-year-old boy had emotional problems and little energy; he was failing some of his classes. Seven months after he was given a special diet he was doing very well in school.[1]

One of our learning-disabled 15-year-olds was so hyperactive he was constantly moving, pecking, drumming his fingers, and annoying other students. After we discussed the problem with him and his parents, his diet was adjusted. He has now calmed down, his attention span has increased, and he is achieving more academically.

What foods and substances influence behavior? The list is long — and loaded with commonly used items. Unfortunately, individual reactions to these substances vary greatly, so that it is impossible to establish a forbidden list. For most young people the effects are generally negligible and are tolerated without much trouble. However, these same foods and substances may be devastating for others.

One couple reported a daughter with constant physical problems, including infected ears. A son was hysterical and destructive. The daughter's problem was milk; the son's was wheat. After dietary adjustments, the mother reported a happy home and regarded the change as miraculous.

Sugar, milk, eggs, corn, wheat, citrus products, beef, pork, caffeine, and additives (in the form of flavorings, coloring, preservatives, and stabilizers) are on the list of troublemakers. William G. Crook, a physician in Jackson, Tennessee, states that he has treated over 160 new hyperactive patients by adjusting diet. When certain foods are eliminated, most patients show significant improvement — usually within one week. He says, "I know, beyond any shadow of a doubt, based on what parents of my patients tell me, that many, and perhaps most, hyperactive children can be helped by changing their diets."

Dr. Crook also believes that as many as 50% to 75% of the population suffer from food allergies and that most of them don't know it. Common symptoms in children are headaches, abdominal pains, runny noses, fatigue, bed-wetting, and hyperactivity. Food-related allergies also affect the nervous system, causing aggressiveness, temper tantrums, depression, and poor coordination, Dr. Crook reports.

One of Dr. Crook's patients was thought to be mentally retarded at the age of 3. Now he is a normal teenager and is getting good grades in school. His problem was chocolate. One candy bar caused serious effects for two days. He became jittery and nervous and could not concentrate.[2]

Benjamin Feingold, a physician at Kaiser Permanente in San Francisco, believes that dramatic changes can be made in student behavior and achievement through diet. When synthetic food coloring and flavoring are eliminated from patients' diets, he says, remarkable personality and behavioral changes occur. Hyperactive children become calmer and more responsive, have a longer attention span, and are better able to cope with their environment. These changes are followed by an improvement in scholastic achievement. Drugs are not needed to cure these students.[3]

In March 1976 the 2,000 delegates to the New York State United Teachers Association (NYSUT) thought so much of Dr. Feingold's theory that they passed a resolution ending as follows:

Resolved, that NYSUT go on record in opposition to the use of any artificial food coloring or food flavoring that contributes to hyperkinesis; and further be it

Resolved, that NYSUT bring all possible and necessary pressure to bear on food processing and food distributing companies (including the sponsorship of legislation) to cease and desist from the use of any artificial food colorings and food flavorings that have been shown to contribute to hyperkinesis; and further be it

Resolved, that all NYSUT members be encouraged to refrain from purchasing and/or use of food products that contain artificial food colorings or artificial food flavorings that have been shown to contribute to hyperkinesis.[4]

Barbara Reed, chief probation officer of the Municipal Court of Cuyahoga Falls, Ohio, discovered that young offenders consumed great quantities of sugar, soft drinks, and starch. She recommended sugar-free, low-starch, no-junk-food diets. Those who followed her recommendations were never back in court. One judge was so amazed at the changes in these offenders that he began to order defendants to eat nutritional diets.[5]

Leonard J. Hippchen, a professor at Virginia Commonwealth University, reports biochemical research showing that the brain is affected by molecular substances that are normally there. The optimum amounts differ from person to person. Abnormal levels can lead to a variety of pathological thought and behavior patterns. For example, a student with a vitamin deficiency, a dependency disease, can become violent.

Hippchen also found a significant relationship between hypoglycemia (low blood sugar) and criminal behavior. Studying juvenile delinquents, he found that most of them suffered from hypoglycemia. They were arrested for disorderly conduct, assault and battery, attempted suicide and homicide, cruelty, embezzlement, larceny, and arson. These delinquents were former hyperactive students who were too restless to learn. They were behavior problems in school and at home. They grew up without a salable skill. Many became failures, truants, dropouts, and, finally, criminals.[6]

The above discussion should, we think, persuade educators to do something about the diet of hyperactive students. Many of these students end up in the principal's office, in alternative schools, in special education classes, or in jail. However, if more evidence is needed, then our discussion should include what is most important in many schools (whether educators admit it or not) — athletics.

At the University of Montreal, the Department of Nutrition and Dietetics conducted a year-long experiment with an amateur hockey team. The team was divided into two groups. One group was free to eat candy and chocolate bars. The second group was to eat sugar-free foods. The sugar eaters' play deteriorated as more and more sugar and sugar-related products were consumed. This group had weakened metabolism. Its members were physically inferior to those in the sugar-free group. Concentration, resistance to illness, and overall ability decreased, even with small amounts of sugar. The sugar-free group's performance improved as the year progressed.[7]

Some classroom teachers give candy as a reward for good behavior or correct responses to questions. The device may be counterproductive. Fortunately, some teachers are aware of what happens when students eat lots of candy. In Roanoke, Virginia, an attempt was made in 1976 to move Halloween from Saturday night back to Friday night so that students would have two days to get the sugar out of their systems. One council member who is also a teacher observed that the candy in the children's bloodstream made them uncontrollable in the classroom.[8]

Alan C. Levin, director of the New York Institute for Child Development, says that his office was chaotic the day after Halloween. His patients, who were hyperactive and learning disabled, could not participate in the therapy sessions after they had eaten so much candy. Their attention spans were limited; they were almost unmanageable.[9]

Nearly all surveys of problems in education, including those made by Phi Delta Kappa, list student discipline as number one. Motivation is usually number two. Both of these problems have a direct bearing upon scholastic achievement. But poor discipline and bad motivation are only symptoms of a deeper malaise. It will do little good to treat the symptoms alone. We use conferences, detention, corporal punishment, and suspension to help the student conform. Sometimes counselors and school psychologists help find academic, psychological, and/or sociological causes. Referrals are then made to physicians — and students end up on tranquilizers.

For some of these students, the cause of disruptive behavior is the food they are eating. If we changed their diets, their behavior and scholastic achievement should improve. Does this mean that most of the students who are referred to the principal should have their diets checked?

We believe it does. All students and their parents should be made aware of the importance of foods for physical and mental health. Special attention should be given to students in special education, to behavior problems, and to those students with physical problems, especially allergies. Furthermore, it should be emphasized that it is the students' and parents' responsibilities to do something about changing the diet. Psychologically, it is much better for students to take responsibility for their own actions and their own health.

Interestingly enough, typical school lunch menus contribute to poor nutrition. The usual fare in school cafeterias is as follows: devitalized white bread with additives; polished white rice; canned sweetened fruits; chocolate milk; hot dogs and luncheon meats loaded with nitrates, nitrites, and artificial colors; artificially flavored fruit punches; chemically made ice cream; instant foods, such as potatoes; and prepared frozen and canned foods preserved and doctored with additives.

Some school systems and state education authorities have made changes in the foods served in lunch programs and in vending machines. On 13 November 1975, for example, the West Virginia Board of Education passed a resolution prohibiting the sale of candy, soft drinks, chewing gum, and flavored ice bars in the public schools of the state.[10]

In Denver, Colorado, school lunches include freshly baked breads, fresh produce, and meat entrees low in fat. Junk foods are not available. A la carte alternatives include vegetarian plates. The program includes inservice training for food service employees, nutrition education, and a student advisory council.[11]

In Fulton County, Georgia, under the direction of Sara Sloan, the food program includes inservice training for employees.

5. EDUCATION AND CHILD DEVELOPMENT

Elementary schools conduct weekly mini-classes on nutrition. The lunch programs include freshly prepared foods free of additives and sugar. They are also low in cholesterol. Thirty of the 79 county schools now offer lunches prepared from natural foods.[12] Reportedly, students who eat these lunches regularly are absent less often than other students. The coaches report increased stamina and endurance among their athletes.

At Helix High School, in La Mesa, California, students on a similar luncheon regimen excelled in athletics and won more trophies and scholarships than before. Insurance rates were lowered because the students had fewer accidents.

In Montreal, Canada, the school council has banned the sale of junk foods in vending machines in all the schools. The school lunches are prepared "from scratch" and include freshly prepared baked goods.[13]

The Milwaukee Public Schools have improved the foods they serve. Thomas J. Farley, food technologist and director of food services there, makes several worthwhile points about the program. First, in January 1977 the Board of Directors banned all foods and beverages of any kind except school lunches between the hours of 8 a.m. and 3:30 p.m. The Type A lunch was effectively upgraded by tailoring menus to the students' tastes. Students were involved in workshops for a year, helping to work out procedures and menus. Good Type A lunches are essential if junk foods are to be eliminated.

Students in Milwaukee serve on taste test panels, participate in menu planning, and serve on committees that write menus. Only one meal a day is served in all 160 Milwaukee schools. There are no à la carte menus, no vending machines, and no alternative menus. No bread, rolls, muffins, cakes, or cookies have been purchased for over 16 years. Everything comes fresh from the system's own ovens at half the cost of purchased products.[14]

There are many reasons why it will be difficult for all school systems to do much about the foods they serve, and about student eating habits. First, there are the multi-billion-dollar food-packing companies, soft drink corporations, and vending machine operators. They don't want the eating habits of students changed. As a rule, profits are greatest where nutritional values are least. These corporations have the resources to advertise, publish propaganda, and lobby at both state and national levels. Furthermore, they contribute money to certain departments in large universities, so that research will be done to support their position.

Second, there are huge profits in these products for the schools themselves. Many districts profit from their lunch programs, and they profit especially from the vending machines. So it is the administrators who clamor for carbonated drinks, peanuts, potato chips, chewing gum, candies, and new junk foods. In West Virginia school administrators opposed the ruling that eliminated junk food in their schools. No doubt the profits are used for a variety of good causes. In 1974-75 over $1.4 million worth of candy was sold in the junior and senior high schools of Los Angeles. When dentists expressed a concern about cavities, the sale of candy was defended by school officials because the profits were needed to buy band uniforms.[15]

A third reason why it will be difficult to change eating habits is that parents often work outside the home and are busy with outside activities. Packaged and processed foods save time. Fast-food restaurants are also popular with busy people. Most of these restaurant foods contain preservatives, flavorings, food colors, and stabilizers. Many frozen foods contain large quantities of chemical additives. Ice cream, in most cases, is a chemical mixture. Too much candy, sugar, and salt and too many colas and snacks are consumed. If the family eats out, the chances are that they will patronize a fast-food restaurant. But these foods are as bad as or worse than the convenience foods purchased in grocery stores.

Fourth, most people do not believe that products that contain sugar, caffeine, additives, salt, corn, and wheat can do much damage. Moreover, they say, "What if they do? We want to enjoy what we eat. We eat foods because they taste good, not because they are good for us. We are used to nice packages of processed foods that have been sugared and colored." Even baby foods have added sugar. It is no wonder that our tastes and addictions begin early.

Finally, the medical profession itself does not understand or support the primary thesis that behavior can be changed by changing the diet. Orthomolecular physicians agree wholeheartedly, but the hypotheses we have offered here are not well accepted by the medical establishment in general.

Orthomolecular medicine is basically the treatment and the prevention of disease by the adjustment of the natural constituents of our bodies. It emphasizes these natural substances in the body and not the chemicals and drugs that are foreign to us.

The term "orthomolecular" was first used by Linus Pauling in 1968. It means "right molecule." Orthomolecular physicians attempt to create the optimum molecular environment for the cells of the body, which helps prevent disease. The ideal goal of medicine should be disease prevention. Curative treatment is an admission of failure. Yet the medical establishment seems to be more interested in treating symptoms than in dealing with the causes of disease.[16]

Clinical nutrition is not even taught in most medical schools. In fact, one survey found that the typical physician knew no more about nutrition than his receptionist. However, when she had a weight problem she knew more about nutrition than he did.

Orthomolecular physicians and psychiatrists have successfully treated patients with most of the degenerative diseases by using changes in nutrition. Last fall the CBS *60 Minutes* program featured the Longevity Center in Santa Monica, California, directed by Nathan Pritikin. It successfully treats very serious heart patients with diet, vitamin and mineral supplements, and exercise.[17]

Orthomolecular doctors have also helped many students who are hyperactive, hostile, learning disabled, and schizophrenic. Students with dyslexia, hyperkinesis, phobias, obsessions, hallucinations, delusions, time distortions, and gross perceptual distortions also have been helped.

What are the implications of this for educators? First, students, teachers, administrators, and parents need to be educated about nutrition and about the relationships that exist between nutrition and physical and mental health. The very popular maxim, "You are what you eat," is often quoted in articles about natural foods. Yet few people select their foods rationally. Teaching people what to eat while they are children should make a powerful impact upon their health. Cooperation from the media and food industry would be beneficial, of course. Without it, successful nutrition education will be very difficult. But we must do what we can.

Nutrition minicourses should be conducted in the schools. Students need to have regular input into the menu planning for the cafeteria. The formation of nutrition clubs in the schools can foster further growth in wholesome eating habits. These clubs can sponsor communitywide nutrition workshops to educate parents and the general public.

Second, state legislatures and state boards of education need to pass laws and adopt policies that will restrict the distribution of soft drinks and junk foods in schools. Education lobbies that have been regarded as hypocritical when they say they seek more money "because children need good schools" might regain some credibility by sponsoring such laws and policies.

Third, since educators are concerned about the health and behavior of all students, they should change the foods that are available in school. More juices, whole grain products, raw fruits, and raw vegetables should be served. Sugar, soft drinks, candy, chocolate milk, processed convenience foods, prepared frozen and canned foods, flavored ice bars, and vending machines for dispensing junk foods should *not* be used in the schools. Freshly prepared foods and baked goods should be served in the cafeterias. Preparing food "from scratch" tends to keep them additive free and low in cholesterol. If vending machines are used, they should dispense snacks such as nuts, seeds, and fruit juices.

Fourth, all students being considered for special education should have a physical examination from an ortho-molecular physician, or at least from a physician who is aware of the values expressed and the diagnostic procedures used by orthomolecular-trained physicians.

Finally, the 2% to 5% of the students who are chronic discipline problems and students who continuously fail school subjects should be checked by an ortho-molecular physician or a physician using the orthomolecular approach. Many students with discipline problems will also have scholastic problems, and they may be the same ones who are being considered for special education.

---

1. Hugh W. S. Powers, "Dietary Measures to Improve Behavior and Achievement," *Academic Therapy*, Winter 1973-74.

2. William G. Crook, "Adverse Reactions to Food Can Cause Hyperkinesis," 12 December 1977 letter to the editor of the *American Journal of Diseases of Children*.

3. Benjamin F. Feingold, *Why Your Child Is Hyperactive* (New York: Random House, 1975).

4. New York State United Teachers Association, "Hyperkinesis," Resolution 89, March 1976.

5. Barbara Reed (as cited by Timothy Schellhardt), "Can Chocolate Turn You into a Criminal? Some Experts Say So," *Wall Street Journal*, 2 June 1977.

6. Leonard J. Hippchen, "Contributions of Biochemical Research to Criminological Theory," Department of Administration of Justice and Public Safety, Virginia Commonwealth University in 31 January 1978 letter.

7. Robert Rodale, "Winning by Eating Right," *Prevention*, October 1977, pp. 25-30.

8. Fran Coumbs, "County's Halloween Scheduled Saturday," *Roanoke Times and World-News*, 28 October 1976.

9. Alan C. Levin, "Kid Food — Key to Problem Behavior," *Woman's Day*, 23 August 1977.

10. Faith Gravenmier, Hearings before the Subcommittee on Elementary and Secondary Education, U.S. House of Representatives, Washington, D.C., August 1976.

11. Fran Smith, "School Lunch: Local Activists Make Headway," *Nutrition Action*, October 1977, pp. 6-8.

12. Sara Sloan, *A Guide for Nutra-Lunches and Natural Foods* (Atlanta: Fulton County Schools Food Service Program, 1977).

13. Jane Kinderlehrer, "A Tale of Two Cities," *Prevention*, February 1977, pp. 60-63.

14. Personal communication.

15. "Candy Profits Buy Band Uniforms for L.A. Kids," *National Health Federation Bulletin*, April 1976, pp. 16, 17.

16. Linus Pauling, "On the Orthomolecular Environment of the Mind: Orthomolecular Medicine," in *A Physician's Handbook on Orthomolecular Medicine* (New York: Pergamon Press, 1977).

17. Nathan Pritikin, *Pritikin Diet Book* (New York: Grosset and Dunlap, 1979).

THE MISMATCH BETWEEN SCHOOL AND CHILDREN'S MINDS

Because neither theorists nor teachers understand children's minds, many students learn to hate school.

MARGARET DONALDSON

Margaret Donaldson, *professor of psychology at the University of Edinburgh, has worked at Piaget's Research Institute in Geneva and with Jerome Bruner at Oxford. Her book,* Children's Minds, *was published by W.W. Norton.*

Visitors to any elementary school would notice that most children in the kindergarten and first-grade classrooms are excited, happy, and eager to learn. But if they were to continue their visit to classrooms of the higher grades, they would find many who are unhappy, unresponsive and bored. Yet from infancy all normal human beings show signs of a keen desire to learn—a desire that does not appear to depend on any reward apart from the satisfaction of achieving competence and control. This desire is still strong in most children when they enter school. How is it that something that starts off so well regularly ends up so badly? Why do many children learn to hate school?

The answer cannot be that most children are stupid after the age of six, nor can it be that teachers enjoy making children miserable. Recent research into the nature of children's language and thinking can help us to see what goes wrong.

It is now clear that we have tended both to underestimate children's competence as thinkers and to overestimate their understanding of language.

The underestimations are in large measure a result of the theories of the most influential of all students of child development, Jean Piaget. From his experiments with young children he concluded that until the age of about seven, though competent in practical skills, children are extremely limited thinkers.

The overestimation of children's understanding of language is, in part, a result of the theories of linguist Noam Chomsky. In the 1960s Chomsky caused a wave of excitement among psychologists by drawing attention to the significance of one simple fact: Children who are only two or three years old can utter complicated sentences *that they have never heard before.*

Since these sentences generally conform to the rules of syntax, the children must, in some sense, know them. And even sentences that are not fully correct by adult standards still show that rules of some kind are at work in their construction. In fact the errors frequently reveal rules, as when "bringed" is used instead of "brought." We can be fairly sure that children who say "bringed" have not heard adults say it. They must have generated the form for themselves by applying the rule for forming past tenses in weak verbs in English.

The implications of these facts presented psychologists with a highly challenging question: How is it possible for a young child to master such a complex system of rules? There seemed to be only two possibilities: The child either has remarkable skills as a thinker, or some very special skills as a language learner. Chomsky argued for the second of these explanations. He proposed that human beings are endowed with a highly specialized faculty for learning language, which he called a language-acquisition device. This idea enjoyed widespread popularity for a while, and at least part of the reason was that the other explanation seemed implausible. There was evidence around—weighty, respectable evidence obtained by careful systematic study and experiment—that appeared to imply that the young child is not much of a thinker.

Yet there was other evidence, freely available but largely neglected, that pointed to a different conclusion. Anyone who talks with young children seriously and attentively knows that they say a great many thoughtful and seemingly intelligent things. The curious thing is that while so much attention was being paid to the grammatical sophistication of children's speech, very little attention was being paid to its meaning.

The following examples are remarks made by very young children (all younger than five, and some barely three years old) as they were listening to stories:

"The nails will tore his trousers." (A prediction about what will happen to a character who is putting nails in his pocket. Uttered in tones of concern.)

"You can't sew a turnip." (A confusion of "sew" and "sow." Uttered with scorn.)

"He's got sharp teeth and sharp claws. He must be a wild cat."

These examples are typical. Yet they are remarkable for the awareness of possibility and impossibility, of contingency and necessity revealed. They establish beyond doubt a well-developed ability to make sense of a highly complex world. And the impression of good sense gleaned from these isolated re-

Reprinted from *Human Nature*, March 1979. Reprinted by permission of A.P. Watt, Ltd., England.

marks becomes even greater when we are aware of the context in which they were made.

The problem is to reconcile these observations with experimental evidence that seems to show that children of this age are quite limited in their ability to deal with possibility, to reason inferentially, and to think intelligently in general. Although the evidence comes from diverse sources, the source that has been the most influential over the past few decades is the work of Piaget and his colleagues in Geneva. The reason for Piaget's great influence is not only his ingenious studies of children, but also the fact that he has woven his findings into a theory of great internal consistency and beauty, so that the mind is dazzled.

Once adopted, a theory tends to make us disregard evidence that conflicts with it. This is true even of a theory that is not very impressive, let alone one like Piaget's. But the most entrenched theory can be dislodged in the face of overwhelming evidence that it is wrong.

In the case of our theories about children's thinking, such evidence has been mounting for some time. Recent research, much of it concerned with the comprehension of language, throws new light on the reasons children can seem so limited as thinkers when they are tackling Piaget's tasks, yet so skilled when we watch them and listen to them in their spontaneous behavior.

In the early days after Chomsky's revolution in the field of linguistics, almost all those who were doing research on child language concentrated on studies of what children said. The reason was simply that the evidence needed in order to work out the rules of grammar that the children in some sense knew and used was evidence about language production. As long as this was the main concern, the question of what children could understand was largely ignored. But over the past few years there has been a marked shift of interest from syntax to semantics, and studies of comprehension have come into their own.

My own interest in language comprehension began in 1968 with a study of the ways in which children interpret the words "less" and "more." For this research we built two large cardboard apple trees, each equipped with six hooks on which beautiful red apples could be hung. Putting different numbers of apples on each tree, we asked which tree had more apples, which less. Then the children were asked to put more (or less) on one tree than on the other. The results surprised us. We had thought it likely that they would understand "more" better than they understood "less." What we found was that the two words appeared to be treated as synonyms. No matter which word we used, we tended to get the same responses—and they were the responses that were correct for "more."

This result was provocative in many ways, and it led to a series of further studies. One of these, by David Palermo of Pennsylvania State University, replicated the original results; but others, including research by Susan Carey of M.I.T., used different methods and cast doubt on the interpretation of these first findings.

My own view now is that we were asking the wrong question. Instead of asking how the children understood the words "less" and "more," we should have been looking at their interpretation of the utterances in which these words were used. I was led to see this by the results of a series of further studies, the first of which was still planned as an investigation of the development of word meanings. I wanted to study the children's understanding of words like "all" and "some." Naïvely, as it now seems to me, I hoped to discover the meanings of these words for the children by inserting them in statements like "All the doors are shut" or "All the cars are in the garages," and presenting the statements along with objects that rendered them true or false. What I discovered was that the children seemed to have no single meaning for "all." They judged all the doors to be shut if there was no door open, but often they judged all the cars to be in the garages when one car was sitting in full view outside. In the latter case the question they seemed to be considering was not where the cars were but whether all the garages were occupied. So we tried another configuration. We used two rows of garages, one row having four and the other six; and two rows of cars, one with four and one with five. In this study we again looked at the interpretation of sentences with the word "more" in them.

We were going to ask the children to make a comparison, and we wanted to make it as easy as possible for them, so we arranged the cars one above another on two shelves, with the extra car in one row always projecting on the right-hand side.

When the two rows of cars were presented alone, without garages, most of the children said there were more cars in the row of five than in the row of four. However, we also presented the cars enclosed in garages: four cars in the row of four garages, and five cars in the row of six garages (so that one garage was empty). About one third of the children now changed their judgments, saying there were more cars in the row of four. Again the question the children appeared to be answering was whether all the garages were filled: This was the thing that seemed to stand out, and no matter what we asked, this was the question they answered. Clearly "fullness" was what they expected to be asked about.

An important feature of this last experiment was that the task the children were given was, in its logical structure, the same as a Piagetian conservation test. Of all the tests that Piaget devised to reveal the nature of children's thinking in the preschool and early school years, the conservation tests are probably the best known. There are many such tests—tests of conservation of number, weight, volume, and length—but they all make use of the same key elements. Conservation of number will serve as an example.

First, the experimenter shows the child two rows of objects, equal in number and laid out opposite one another in one-to-one correspondence. The child is asked whether there is the same number in the two rows. Unless the child agrees that this is so, the test cannot proceed.

In the next step the experimenter destroys the perceptual equivalence of the two rows, by moving the objects in one row closer together, for instance. Usually this action is accompanied by a remark like "Now watch what I do," which ensures that the child is paying attention.

Once the new configuration is in place, the original question is repeated, usually in the same words. Children who continue to say that the two rows are equal are said to conserve and are called conservers. If they claim that one row now has more objects than the other, they are not conserving and are called nonconservers. According to Piaget, a child's ability or inability to conserve is an indication of his or her stage of mental development.

Clearly the task involving two rows of cars, with and without enclosing garages, has the structure of a conservation task, although it is an unorthodox one. First the child is asked a question that calls for a comparison, then something irrelevant to the meaning of the words in the question is changed, then the original question is asked again.

Some may say that many of the children in our study were not conserving. But *what* were they not conserving? The most plausible and generally applicable notion seems to be that what the children were failing to conserve was their interpretation of the words of the experimenter. The same question was put to them twice, but *in a different context*. Adults would have discounted the shift in context. They would have known that in this kind of formal task they were meant to discount it and to base their replies on the meaning of the words. But perhaps the children did not know they were meant to do this, or perhaps they could not do it because the context was too powerful for them. In either event, it looked as though some nonlinguistic feature was strong enough to cause a shift in interpretation of what, for an adult, was a repetition of the same question.

At this stage my attention was drawn to the possibility that part of the reason for the shift in interpretation had to do not with the physical context of the experiment, like fullness of garages or length of rows, but with what the children thought was the intention underlying the experimenter's behavior. I owe this insight to James McGarrigle, who devised an ingenious experiment to test it.

The experiment made use of a number-conservation task and a small teddy bear called Naughty Teddy. The task proceeded in the usual way up to the point where the child agreed that the number in the two rows was the same. But before the experimenter went on to the next stage, Naughty Teddy would emerge from hiding, swoop over one row, and disarrange it. Once the one-to-one correspondence had been destroyed the child was invited to help put Naughty Teddy back in his box (an invitation that was usually accepted with glee) and the questioning was resumed: "Now, where were we? Ah yes, is the number in this row the same as the number in this row?" and so on. What happened was that many more children between the ages of four and six conserved number in the Naughty Teddy version of the task than in the traditional version (50 out of 80, compared with 13 out of 80).

Piaget's account of the reasons for failure cannot deal with this finding. His explanation makes use of several related arguments, putting the emphasis now here, now there. But at the heart of them all is the notion that children fail to conserve because they cannot sufficiently "decenter," that is, they are not flexible about shifting their point of view. Typically nonconservers are held to "center," or concentrate attention, on a particular state or feature, failing to take account either of transformations between states or of other features of the object. They center on the fact that one row of objects is longer than the other and fail to notice that the latter is more dense. In general, they are believed to center on the present moment and to make judgments on the basis of how things *now* appear, with no relation to how they were a moment ago.

It is quite possible to fit these arguments to the finding that children will sometimes change their judgments about the numbers of cars in two rows after the addition of enclosing garages. We have only to say that children who do so "center on fullness." However, there seems to be no way to fit them to the findings of the Naughty Teddy study (and these findings have already been replicated twice, by Julie Dockrell of Stirling University and by Irene Neilson of the Glasgow College of Technology). Nothing in Piaget's theoretical account of conservation suggests that it should matter who changes the arrangement of the objects.

But the Naughty Teddy results do fit very well with the idea that nonconserving children fail to answer the experimenter's question in the same way on the second occasion because, for them, it is not the same question. It seems different because it is not sufficiently detached, or disembedded, from the context of what the child believes the experimenter wants.

So disembedding will explain more of the findings than decentering will. But this does not establish that the decentering argument is false. It is possible that the child has difficulty with both decentering *and* disembedding. We must look at other evidence to see whether children are as limited in their ability to decenter as Piaget would have us believe.

In Piaget's view, inability to decenter is a feature of young minds that shows itself in a wide variety of ways. Some of these have already been considered. Another, perhaps the most fundamental, is the inability to appreciate the relativity of one's own point of view in space and time. A simple example of this is the inability to understand that one's own view of an object is not the same as that of someone looking at it from another side.

In a famous experiment, Piaget established that children presented with a three-dimensional model of a group of mountains have great difficulty choosing a picture of how the model would look to a doll viewing it from another position. For the most part, young children given this task choose the picture that shows exactly what they themselves see. It seems, then, that they are notably lacking in mental flexibility, bound by the egocentric illusion that what they see is the world as it really is. If this were true, it would certainly have far-reaching implications for the ability to think and reason.

Recent research has called this conclusion into question. My own thinking on the subject has been influenced by the work of Martin Hughes. Hughes placed before a group of children a configuration of two walls intersecting to form a cross. At the end of one of the walls he placed a wooden doll, representing a policeman. The children were then given another wooden doll, representing a boy, and were asked to "hide the boy so that the policeman can't see him." (The policeman was not tall enough to look over the walls.)

The arrangement made it easy to tell whether the children were able to escape from the domination of their own point of view, and the results were clear. Even three year olds were highly competent at the task. They showed no sign of a tendency to hide the doll from themselves, as would have been predicted from Piaget's theory, and they showed every sign of understanding what the policeman would be able to see from where he stood. Even when there were two policemen, placed so that the only effective hiding place was one where the boy doll was clearly visible to the child, about 90 percent of the responses from three- and four-year-old children were correct.

The policeman task differs in many ways from Piaget's mountain task, but one is particularly significant in light of what we now know. In the policeman task there is an interplay of motives and intentions that is entirely comprehensible, even to a child of three. For this reason the task makes *human sense* to the children: They understand instantly what it is all about. The verbal instructions are so well supported by the context that no difficulties of disembedding arise. As soon as the doll is handed over, the children's faces light up, they smile, they latch on.

The mountain experiment on the other hand does not make immediate human sense in this way. There is no interplay of motives and intentions, no intelligible context. The task is as disembedded as the one given the American Indian who was asked to translate into his native tongue the sentence, "The white man shot six bears today." The Indian was baffled. "How can I do that?" he asked. "No white man could shoot six bears in one day."

Now we can reconcile the disparity between children's skills as thinkers in everyday situations and their limitations when confronted with formal tasks. Most formal tasks are geared to minds that are capable of a high degree of disembedding of thought and language—minds that are able to dispense with the support of human sense—and these tasks make demands of a quite special kind.

When we first learn to think and to use language, it is within situations where we have purposes and intentions and where we can recognize and understand similar purposes and intentions in others. These humanly meaningful contexts sustain our thinking.

Precisely how they do this is of the greatest theoretical interest, but it is still mysterious. One thing is clear: When thought and language are functioning smoothly in real-life contexts, we are normally aware of the ends to which our activity is directed but not of the mental means that are needed to get there. We do not stop to think about our thinking or about the words we are using.

A formal task interrupts the flow of life. It demands deliberation, mental awareness, and control. It is by definition a thing to be considered out of context. We must set our minds to it. We must accept the premises, respect the constraints, direct our thought. This activity is difficult and, in a sense, unnatural. But that does not mean it should be avoided or abandoned—only that it will not happen spontaneously. We must recognize this fact so that we do not label our children "stupid" or "backward" if at first they find it hard.

The ability to take a problem out of context and consider it in its own right is the product of long ages of a particular kind of culture. It is closely linked to the development of literacy because written language, unlike speech, is by its very nature disembedded. Speech is transient, elusive, entangled in happenings. A written page, or a clay tablet, is physically separate and permanent: You can take it with you and go back to it. It is scarcely possible to learn to handle written language without becoming aware of it as a system and as a tool of the mind.

Disembedded intellectual skills underlie all our mathematics, all our science, all our philosophy. It may be that we value them too highly in comparison with other human skills and qualities, but we are not at all likely to renounce them. We have come to depend on them. And as schooling progresses, the emphasis on them becomes harder to evade or postpone. The student who can solve problems, as problems, divorced from human sense, is the student who will succeed in the educational system. The better a student is at it, the more awards he or she will receive, and the better that student's self-image will be. But large numbers of students never achieve even a moderate level of competence in these skills and leave school with a sense of failure.

Seen in the context of human history, universal compulsory schooling is a new social enterprise, and it is a difficult one. We should not be surprised or ashamed if we do not yet know how to manage it well. At the same time, if we are going to persist in it, there is urgent need for us to learn to manage it better. We must not forget how grave a responsibility we assume when we conscript children for these long, demanding years of service. And when the outcome is not all that we would wish, we must not resort to blaming this on the shortcomings of the children. Since we impose the demands, it is up to us to find effective ways of helping children to meet them.

Many children hate school because it is a hateful thing to be forced to do something at which you fail over and over again. The older children get, the more they are aware that they are failing and that they are being written off as stupid. No wonder many of our children become disheartened and bored.

What are we to do about it? There is no simple formula, but there are a number of guiding principles.

5. EDUCATION AND CHILD DEVELOPMENT

The first takes us back to the topic of decentering. Although research has shown that children are better at this than Piaget claims, it is true that human beings of any age can find it hard. As adults we often fail to understand the child's point of view. We fail to understand what perplexes a child and why. In *Cider with Rosie*, Laurie Lee gives an account of his own first day at school: "What's the matter, Love? Didn't he like it at school, then?"

"They never gave me the present."

"Present? What present?"

"They said they'd give me a present."

"Well, now, I'm sure they didn't."

"They did! They said: 'You're Laurie Lee, aren't you? Well, you just sit there for the present.' I sat there all day but I never got it. I ain't going back there again."

The obvious way to look at the episode is to say that the child didn't understand the adult. But if we are to get better at helping children, it is more profitable to say that the adult failed to make the imaginative leap needed to understand the child. The story carries a profoundly important moral for all teachers and parents: The better you know something yourself, the greater the risk of not noticing that children find it bewildering.

When Jess Reid of the University of Edinburgh studied children who were learning to read, she found that some did not have the least idea of what reading was. They could not say how the postman knew where to deliver a letter. They did not understand the relationship between the sounds of speech and the marks that we make on paper, or that these marks are a means of communication.

It would help greatly if children told teachers when they felt perplexed. Many do not. But if they are explicitly encouraged to ask questions, they can often do so effectively, and the act of asking helps children become conscious of their own uncertainty.

It is also important to recognize how greatly the *process* of learning to read may influence the growth of the mind. Because print is permanent, it offers special opportunities for reflective thought, but they may not be taken if the reading child is not given time to pause. Once children gain some fluency as readers, we can help them notice what they are doing as they extend their skills and begin to grapple with possibilities of meaning; for it is the thoughtful consideration of possibility—the choice of one interpretation

among others—that brings awareness and control.

One final principle is implicit in all that has been said: If we want to help children to succeed at school and to enjoy it, it is not enough to avoid openly calling them failures. We must respect them as thinkers and learners—even when they find school difficult.

If we respect them and let them know it, then the experience of learning within a structured environment may become for many more of our children an opening of new worlds, not a closing of prison bars.

For further information:

Bruner, J. S. "The Ontogenesis of Speech Acts." *Journal of Child Language*, Vol. 2, 1975, pp. 1-19.

Donaldson, Margaret. *Children's Minds*. W. W. Norton, 1979.

Grieve, R., R. Hoogenraad, and D. Murray. "On the Child's Use of Lexis and Syntax in Understanding Locative Instructions." *Cognition*, Vol. 5, 1977, pp. 235-250.

Lempers, J. D., E. R. Flavell, and J. H. Flavell. "The Development in Very Young Children of Tacit Knowledge Concerning Visual Perception." *Genetic Psychology Monographs*, Vol. 95, 1977, pp. 3-53.

Macnamara, J. "Cognitive Basis of Language Learning in Infants." *Psychological Review*, Vol. 79, 1972, pp. 1-13.

Olson, D. R. "Culture, Technology and Intellect." *The Nature of Intelligence*, edited by L. B. Resnick. Halstead Press, 1976.

Learning Right From Wrong

Ellen Sherberg

Ellen Sherberg is a free-lance writer and a reporter for KMOX (CBS radio in St. Louis).

Ten-year-old Eddie was feeling pretty lucky. It was the last day of school, and a special lunch—hamburgers and ice cream—was on the cafeteria menu. Since he had gym class before lunch period, and the gym was right next to the cafeteria, he could rush to be among the first in the lunch line. His friend Tom, however, had to walk much farther to the cafeteria, and found more than 60 students waiting by the time he got there.

"Can I cut in front of you?" Tom asked Eddie. "I'll give you my ice cream."

Should Eddie let Tom cut in line? Would Tom still be Eddie's friend if he said no? What about the other kids who had been waiting? What about that extra ice cream? What if Tom told Eddie he'd miss an important ride to a doctor's appointment if he didn't cut in line? Should that matter?

Ethical dilemmas.

"Cutting in line" is one of many minor discipline problems that crop up during the average school day. But it is not only a discipline problem; for the child, it is also an ethical problem.

"Commonplace situations are often moral dilemmas," explains Peter Scharf, co-author of *Growing Up Moral* (Winston Press, Inc.). "And it's critical that children learn to confront these dilemmas if they're going to grow into morally responsible adults."

Discussions about morality used to be reserved for parents and the pulpit, but today's schools are playing a larger and larger role in children's moral education.

Two approaches to teaching about values.

Of course, questions of moral behavior are not new to the classroom setting. Sensitive teachers have always been aware of the moral issues raised by their curricula, current events, and children's personal experiences.

But today's educators rely on two specific approaches to moral education. The more widely used approach is based on the ideas of educator Louis Raths, who built upon the thinking of John Dewey to formulate his system of *values clarification*. The other approach, based on the theories of Harvard educator/psychologist Lawrence Kohlberg, is known as *cognitive moral development*.

Values clarification: the process is the key.

The values-clarification approach developed by Raths focuses on the *process* people use to develop their beliefs and behavior. Through a variety of analytical techniques applied to classroom discussion, teachers point out to their students the three major aspects of decision-making, all of which are part of what Raths calls "the process of valuing."

First, students are made aware of their own values (for example, "It is wrong to cooperate with the enemy because it is disloyal to your own country"); then they're asked to identify their priorities ("Saving people's lives, especially your parents', is more important than being loyal to your country"); finally, they are brought to recognize the relation of their individual beliefs to the values held by the culture in which they live ("After the war, members of your family may consider you a hero, but others may call you 'traitor' ").

Looking at old lessons a new way.

With the values-clarification techniques, even material that appears to be universally accepted can be used to stir classroom controversy. For example, a history lesson about the Pilgrims at Plymouth Colony could lead a teacher to ask, "Is there anything you value so strongly that you would leave this town or country if it were taken away?"

If a student says yes, it doesn't matter if he prizes his dog, his family, or his freedom. What is important is how he arrived at that decision and how consistent it is with his other beliefs and actions. It's also important for him to understand how other people perceive his decision.

Other students might snicker if a ten-year-old says his dog is as important to him as his friends are. But if the student goes on to explain how the dog's presence keeps him from being scared at night, his decision takes on new merit in the class. Others begin to see that his preference for his dog is consistent with his values and with qualities that most people consider important, such as loyalty, companionship, and protectiveness.

According to the values-clarification system, it doesn't really matter if Eddie—the little boy in the cafeteria line—decides to let his friend "cut in" or not. If, during his decision-making process, he said loyalty is more important to him than anything and for that reason he is willing to let his friend in line despite the glares of other students, his decision would be acceptable. If, on the other hand, Eddie said that ice cream is more important to him than anything and his friend has promised him extra ice cream—and he doesn't care if his actions make other kids angry—his decision would be just as acceptable because, again, it would be consistent with his values.

While the process of decision-making is the key to the values-clarification approach, the *decision itself* is what counts in Kohlberg's philosophy. In a classroom where cognitive moral development is taught, Eddie's decision to let his buddy in line might not be acceptable.

Cognitive moral development: an offshoot of Piaget's work.

Kohlberg's theory is based directly on the philosophy of Jean Piaget, who was the first modern psychologist to observe that a child's conception of social rules evolves in stages.

Piaget's research consisted of watching children of different ages playing, and then questioning them about the rules of their games. He found that toddlers, who tend to play by themselves, have no sense of rules as social obligations. As children begin to play together, around the time they enter school, they play games according to rules that must be strictly obeyed. They consider these rules to have emanated from adults and, as such, to be unchangeable. Piaget noticed that preadolescents also play according to a defined set of rules. But these rules are viewed as being based on the mutual consent of the group of players, and must be followed if one wants to be loyal; one can change the rules, however, if the majority of the group agrees to the change.

Kohlberg takes Piaget's observation that a child's view of rules evolves as he grows older and extends it to the concept that a child's moral consciousness develops through stages. According to this theory, if Eddie were six or seven years old, he might not let his friend cut in line because he would be afraid his teacher would yell at him. At this point, Eddie would be at the first stage of moral development.

At stage two, when Eddie is slightly older, he begins to understand sharing in a very pragmatic ("you scratch my back, I'll scratch yours") sense; at this stage he might let his friend in line in exchange for the ice cream.

Stage-three development has a connotation of peer approval. "Right" is defined by what others expect, and behavior is often judged by intentions. At this stage, Eddie might reason that it's okay to let his friend cut in because all the other children in line would expect him to let their friends cut in.

Stage four is known as "the law-and-order orientation"; now judgments of right and wrong are based on rules and on respect for formal authority. Because cutting in line is against the rules, Eddie at this stage might turn his buddy down.

Following Kohlberg's Piaget-based system, as Eddie grows older and becomes a teenager he should begin making his own moral decisions. He might, for example, decide to let his friend in line, not because of the ice cream, but because he knows his pal has to eat in a hurry or miss his ride.

At this point Eddie is measuring the needs of the individual against the rules of the school. He considers issues of fairness and justice and makes a decision he thinks he can live with. Consequently, he considers how fair it would be to those who have been waiting in line if his buddy cut in. If his friend is actually going to miss a ride that will take him to the doctor or to a job where he makes money to support his family, Eddie might decide to allow him to cut in line; if the ride is to a movie, Eddie might decide to favor the feelings of the other students waiting in line.

That most teenagers would consider all these factors may sound a bit farfetched. Most adolescents—and adults as well—would simply say to a pal, "It's okay, stand in front of me."

That's why Kohlberg's theory is far more than a description of moral development. It's based on a strong belief that discussions of moral dilemmas led by a well-trained teacher can bring students to a higher level of moral reasoning, where they will consider—and act on—the issues of fairness applied to society as a whole.

While recognizing that students must progress naturally through the various stages of moral development, Kohlberg also believes that constructive discussions can sometimes hasten progress by increasing moral consciousness.

This facet of the Kohlberg philosophy was tested by one of his graduate students, who tried to raise the level of reasoning in his Sunday-school class through debate of dilemmas such as Eddie's. At the end of the year, the graduate student reported that his pupils had advanced by an average of one-third of a stage.

Moral education is making a difference.

In a much less structured environment, Sarah Wallace, former chairman of the values committee at John Burroughs School, a private coeducational school in St. Louis, Missouri, says she's seen definite changes in her teenage children and their friends since the moral-education program was instituted three years ago.

"The kids are learning a bit more maturity," Mrs. Wallace says. "They're more understanding of why somebody acts a certain way."

Moral growth can also be encouraged in younger children. Bill McCoy, who co-authored *Growing Up Moral*, uses film, written material, and discussions to explore moral dilemmas with his fourth, fifth, and sixth graders at Top of the World Elementary School in Laguna Beach, California.

"Students who at one time could only talk about stage-two, or person-to-person, relationships on an 'I'll do this for you if you'll do this for me' basis start to consider other people. They suddenly begin to ask, 'What's expected of us?' " he reports.

In McCoy's homeroom, fifth graders arrange their chairs in a circle and spend 20 to 40 minutes discussing problems such as cutting in the lunch line.

In his social-studies classes, McCoy uses California history to illustrate moral dilemmas. For example, the Chinese migration in the 1860s, when the expanding railroads inspired a wave of settlement, presents an opportunity to discuss racial issues.

Even more dramatic discussions take place about the Donner Party, a legendary group of California settlers who resorted to·cannibalism to survive a harsh winter. Among the questions posed: "Was it right or wrong for the survivors to eat the bodies of the people who died?" "Why?" "Would it ever be right to kill a person for food in order for a larger group to survive?"

Teachers following Kohlberg's moral-development approach will lead the discussion toward questions affecting the good of the entire group; their goal is to create a consciousness that will lead to a more caring, just society.

What the critics say.

But that's a tall order, and although a 1975 Gallup poll showed that more than 75 percent of the parents queried favored school instruction in moral behavior, there have also been critics of moral education. They include some religious organizations that question what morals are being taught in today's secular classrooms.

But even supporters of moral education acknowledge that teachers must be wary of some of its pitfalls. Because the theory deals with the cognitive domain—or how a child thinks—sometimes teachers forget that how a child feels is equally important. For example, little Eddie in the cafeteria line might decide it's not right to let his friend cut in line, but burst into tears because he's afraid of hurting his buddy's feelings. Or maybe he will cry because he isn't sure if he made the right decision. His fears and self-doubts must be reckoned with as well as his decision-making process.

According to Dr. Ann DiStefano, assistant professor of education at Washington University in St. Louis, another drawback to both the values-clarification approach and Kohlberg's theories (indeed, to much of today's education) is the possibility of creating a classroom situation that "rewards kids who are especially adept at debating and can think clearly on their feet." A teacher's ability to direct a discussion is crucial to avoid the creation of a verbal elite among students, she feels. When the emphasis is on how well a decision is articulated, it may not be on how well it is carried out. The teacher must also remind his or her students that morality is more than just saying what's "right" and espousing democratic values; it's acting in a concerned and consistent manner.

One of the strongest criticisms of moral education is that parents aren't part of the process. Some schools are trying to correct that. When moral education was introduced in the Brookline, Massachusetts, public schools, for instance, great care was taken to include presentations to parents' groups and to explain the philosophy during parent-teacher meetings. The adult evening school offered a course on moral education, but parents were slow to develop interest.

At Crossroads School, a private junior high school in St. Louis, Missouri, parents and children go to school together for an evening. They're divided into groups of about ten persons (with no parent in the same group as his or her child), and discussions center on curfews, reporting to parents, drugs, and relationships. As with discussions among students in the classroom, the purpose of this is not to change minds or to decide who is right. Instead, the evening gives parents and children an opportunity to examine one another's thinking. It's been so popular that what started out as an interesting experiment has become a tradition at Crossroads.

But Crossroads is the exception, not the rule. For the most part, schools—especially public schools—only have time to deal with their students. That means problems can arise when parents want only one moral position presented, and the teacher is presenting alternatives.

Parental disapproval.

If a parent believes a child should always agree with the President, for example, conflicts can erupt if classroom conversation dwells on how to disagree in a democracy and students explore the possibility of differing with government policy. Supporters of moral education in the schools say the same parent who objects to this would also object to a traditional history lesson if it concerned the Bay of Pigs fiasco and focused on President John Kennedy's misjudgments.

According to those who support moral education, it is intended to help young people think clearly, know what they believe, and understand how their beliefs fit into or affect the society as a whole. Supporters of moral education emphasize that it does not aim to promote values that are in opposition to those a child learns from his family. However, it does aim to provoke discussion; and, at the very least, the kinds of stimulating conversation it can inspire in a home where open communication is encouraged will provide a thought-provoking and beneficial alternative to watching television.

Toward a Nonelitist Conception of Giftedness

David Feldman

*A concern for the gifted child is reemerging. Mr. Feldman proposes
that we broaden our concept of giftedness, that we embrace a developmental view
of it, and that we avoid elitism. All children are gifted, he says.*

*DAVID FELDMAN is associate professor,
Eliot-Pearson Department of Child Study,
Tufts University, and the author of "The Child
as Craftsman," which appeared in the
September, 1976, Kappan.*

I shall call her Jenny. She sat in the
middle of the fourth row. I noticed
that she seemed to be reading instead of
listening to me. In fact, she was so preoccupied that her pale blue eyes did not stray
from her social studies notebook as I approached her desk. When I got closer to
her, I realized that Jenny was hiding a
book inside the notebook I thought she
was studying. She looked up as I stood
beside her desk, as if to ask, "And what
can I do for you now that you have disturbed my reading?" Disconcerted, I
mumbled something about Albert Camus
being an interesting writer (for she was
reading Camus' *The Rebel*), to which she
replied: "What makes you think that this
is the only book by Camus I've read?"
Startled by her reply, I returned to the
security of the front of my classroom and
tried to carry on with my lesson on
political and economic differences between the U.S. and the Soviet Union.

Sound familiar? What teacher has not
had a similar experience? The incident I
just described happened nearly 15 years
ago when I was a new social studies
teacher in a California high school. Over
the years I have had occasion to reflect on
my experience with Jenny and to wish I
had been better equipped then to educate
this academically gifted young woman.
Indeed, I have spent much time and
energy between then and now trying to
better understand giftedness in the hope
that Jenny's younger sisters and brothers
would have a more challenging and satisfying experience with school than I was
able to provide for her in 1964.

I am sorry to say that my efforts and
those of others who have undertaken
similar tasks have had precious little impact on the lives of gifted children in

school. And yet there seems to be a stir-
ring of interest, a flurry of activity, a hint
of renewed commitment to the education
of more able youngsters. In my own state
of Massachusetts (which has historically
had little interest in gifted children, but
assumes it has more of them than any
other) grassroots organizing is under way,
bureaucrats in the State Department of
Education are making promises, and a
blue-ribbon panel is going to produce — yes — a position paper on the
education of gifted children. In this article
I'll examine the renewed interest in gifted
children and suggest what I think must be
done if significant progress toward better
education of the gifted is to be achieved.

I might as well make my biases clear
right up front. I believe that the ways in
which giftedness has been defined, although broader than in the past, are still
much too narrowly confined to academic
talent. I also believe that commitment to
gifted children is hampered by elitist
assumptions that seem to go with notions
of giftedness. If efforts on behalf of gifted
children are to be increased, nonelitist
arguments will have to be mounted. Finally, I believe that the gifted should be better educated not because the country (or
the world) needs them, but because education adapted to their needs will mean that
these children will lead happier and more
satisfying lives. These beliefs or biases will
play a key role in the suggestions I shall
make later in this article about how to
capitalize on the momentum toward better education for gifted children. But let
me return to the question I asked earlier:
Why the sudden interest in giftedness?

To see why gifted children have begun to be noticed again, it is necessary to take a look back to the early
years of this century when the whole idea
of a "class" of gifted children was accepted. Lewis Terman, the father of the
gifted-child movement and the person

who devised the first practical IQ test for
American children, believed that youngsters who scored well on his IQ test were
to be the future leaders in all fields of
endeavor. Terman defined giftedness as
an IQ of 130 or above and genius as an IQ
of 150 or above. This definition burdens
us to the present day. I still hear parents at
cocktail parties casually dropping the remark that their child has a "genius IQ."

The idea was simple enough. If one
could measure innate intelligence (as Terman thought his IQ test did), and if intelligence was a general capacity for excellence removed from any specific field
(as Terman believed it was), then an
estimate of a child's IQ was all one needed
to be pretty sure which children were
gifted. This is precisely what Terman and
his co-workers thought they were doing in
the early twenties when they launched the
study that was to become the monumental
work, *Genetic Studies of Genius*. It
became a five-volume report on more
than 1,000 high-IQ children followed
from school age into their sixties and
beyond.

It is impossible to calculate the full impact of Terman's view of giftedness on
educational policy and practice, but it
seems that giftedness and genius came to
be defined in IQ terms not just among
educational researchers but in the public's
mind as well. As early as the 1920s, critics
such as the late Paul Witty were arguing
that much more than IQ was involved in
giftedness, but as Witty told me a few
years ago, he was little listened to at the
time and even less thereafter. Thus, before we had a chance to find out what
would actually become of Terman's
1,000-plus "geniuses," a movement was
launched to create special classrooms for
children who scored high on IQ tests.

I must also mention that the gifted-
child movement was part of a larger
cultural phenomenon, the eugenics movement, which supported the use of IQ tests
as a way to select "superior" from "in-

ferior" humanity for purposes of immigration, sterilization, and classification into occupational and educational categories. I bring up this fact not to embarrass or reprimand those who have been identified with the gifted-child movement; the country was in a different state of mind in those days, and the gifted-child movement was probably no more elitist than many other segments of our society. I mention eugenics because the decline of interest in giftedness that began in the early 1950s cannot be fully explained without acknowledging a general turning away from prewar values and assumptions concerning birth origin, status, opportunity, and the role of American leadership in the world. It also happened that Lewis Terman died in 1956 and the "termites" (as his subjects called themselves) had grown up.

In short, at the same time as the U.S. was recovering from a war to defeat the most totalitarian regime in modern history, it was forced to reflect on how close it too had come to instituting a class system based not on ethnicity but on "innate intelligence." It had been necessary to use IQ tests for classifying military personnel during the war, but when the threat subsided, so did the need to rationalize using the tests. At a more practical military level, it became clear during the war that IQ was a poor predictor of leadership, innovation, creativity, or initiative among military personnel. Therefore, for both the loftiest and most mundane of reasons, the IQ came under serious and repeated attacks as the basis for judging innate potential for high-level performance.

And what of the Terman group? It became clear by the 1950s (and it is much clearer today) that, far from becoming "geniuses" in all fields, Terman's subjects grew up to represent "the American Dream." While the "termites" cannot count among them a single name familiar to all of us, they are nonetheless a blessed group. They are happier, healthier, richer, and more productive than their peers. But they did not come close to fulfilling the hopes Terman had attached to them. If the purpose of Terman's study had been to select successful, happy people, then it would have achieved its aim — but it wasn't and it didn't. At the risk of being trite, I would say that what Terman achieved was to define giftedness and genius as the likelihood of achieving "the good life." While few would question the desirability of achieving affluence, a happy and healthy family, richly rewarding relationships, and a challenging, productive work life, many have begun to question whether the capacity to achieve these things is limited to a small proportion of the (white) population. The gifted-child

movement began to die as the country turned away from preparing an elite that would lead the world to providing access to the fruits of capitalism and democracy for all sectors of society. It was not really a coincidence that the *Brown* v. *Board of Education* decision (1954) coincided with Terman's last major appeal for gifted children.

Two events — one external and the other internal — kept the movement alive through the 1960s. The first event came in 1957, when Russia launched Sputnik I into orbit. This touched off a frantic campaign to upgrade science education that lasted until we beat Russia to the moon a decade later. The second event was not so much an event as an attempt to keep the gifted-child movement alive by capitalizing on the commitment the country had made to its poor and disenfranchised. A new field sprang up: the "gifted-disadvantaged."

Armed with their trusty IQ tests and a few newfangled instruments (mislabeled "creativity tests"), many of the same people who were leaders of the gifted-child movement and their students began searching for the disadvantaged-gifted. It took a critic with a penetrating mind, Edgar Friedenberg, to point out the essential narrowness of the conception of giftedness that guided the efforts of this new/old group of educators. Friedenberg wrote in *Change* magazine in 1970:

> Educational measurement is an inherently conservative function. It depends on the application of established norms to the selection of candidates for positions in the existing social structure, and on the terms and for the purposes set by that structure. The testing process usually cannot muster either the imagination or the sponsorship needed to search out and legitimate new conceptions of excellence which might threaten the hegemony of existing elites. On the contrary, educational measurement is at present wholly committed to the assumption that legitimate forms of learning are rational and cognitive, and that such learning is the proper goal of the academic process. This is an ideological, not a technical, difficulty.

Within the past year or so, signs of renewed interest in gifted children have become ever more visible: legislation at the federal level authorizing an "Office for the Gifted" in Washington has been on the books since 1969, although only funded since 1974; the *Phi Delta Kappan* recently (November, 1976) devoted a section to "The Many Faces of Giftedness"; one of the two yearbooks published by the National Society for the Study of Education for 1979 deals with "The Gifted Child" (the first yearbook since 1958 to be organized around this theme); and the powerful Social Science Research Council in New York has constituted a national committee of leading researchers to try to stimulate greater interest in the gifted among other investigators. The number of articles, news stories, media segments, and parent groups seems to be increasing rapidly. What is it all about?

Frankly, I am not sure what it is all about. One guess is that people are beginning to realize how high the price has been to pursue equality of educational opportunity. Without wanting to give up the gains achieved in making our society more open and responsive to all individuals, increasing numbers of parents, educators, and researchers are arguing for more resources for the promotion of excellence. It is not so much a backlash as it is a forward look. The time has come, they seem to be saying, for a policy that will provide a more even distribution of resources between the groups often called "disadvantaged" and the extraordinarily able children who have had to wait patiently while their more needy brothers and sisters were served.

Thus all the arguments in favor of more funds for bright children used over the last five decades have been dusted off and brought back into public debate. The gifted, we are told, are bored and languish in regular classes. They are more likely to be disturbed and distressed (i.e., to become "school problems") because the curriculum does not challenge them sufficiently. Gifted children are our future leaders, and they should be well prepared to take their places in important positions. And most compelling, our security depends upon gifted individuals using their

---

> *"All the arguments in favor of more funds for bright children used over the last five decades have been dusted off and brought back into public debate."*

talents to the fullest for the good of their country.

It was the last of these arguments that spurred the latest surge in resources for education of gifted children between 1950 and 1965. The Cold War loosened congressional purse strings, and the dollars thus liberated were used to upgrade science and engineering education. The problem is, of course, that all of these arguments — even the security argument — have an anachronistic ring. Though the arguments are perhaps as valid as they ever were, it seems to me unlikely that a renewed commitment to gifted children is likely to be launched on the buoyancy of traditional proposals made in conventional ways. If not these, then, what pleas for the gifted will be heard? And heeded?

Again, I am not sure, but it seems clear that a few issues must be addressed if the gifted-child movement is to build a broader constituency. There are three changes in our views about giftedness that seem to me essential if more children are to be served. These changes may be introduced briefly as follows: 1) The concept of giftedness must be *broadened* to encompass many more kinds of gifts; 2) a shift from a trait to a *developmental* view of giftedness is needed; and 3) whatever concept of giftedness we embrace, it must be fundamentally *nonelitist*. Let me explain a bit further what I mean by these three points.

1. Traditionally, as I tried to show earlier, giftedness has meant a high IQ score. In the past 20 years this narrow definition has been broadened to include high creativity (as assessed, usually, by so-called creativity tests) and specific talents. The federal legislation now in force includes the criteria I have just mentioned. While it represents progress toward a more varied notion of giftedness, I think the concept should be pushed much further. The fact is that most selection criteria for gifted-children programs still use IQ tests plus a smattering of other measures. Until a genuine reorientation in thinking about giftedness takes place, it will be difficult to transcend the narrow view that still leads most administrators to choose IQ test as the primary basis for selection. I hope that my second and third propositions will show how such a reorientation in attitudes and beliefs about giftedness can be brought about.

2. Even the more flexible definition incorporated into federal legislation sees giftedness as a *trait*, a quality that is part of the biological makeup of the child. It is of course true that children differ in their natural abilities and inclinations, but these have been conceived in such general terms that they are of little use in guiding pro-

gram selection or program development. Once you know that a child scores high on an IQ test, for example, what kind of program for that child do you create? Or consider the reverse situation. Suppose you wanted to launch a program in earth science for elementary school children. How would you use your knowledge of children and their capabilities to build a program with appropriate challenges for the children?

I believe that the most promising source of new knowledge for the education of gifted children to have emerged in the past few decades is what I shall call a *developmental* framework. Differing from trait views, the developmental view sees each child as proceeding through several sets of stages, each stage succeeding the one before it. Stages may be either very general (such as logical reasoning) or more specific within fields (such as ballet or mathematics). Because the developmental framework emphasizes progress within specific fields or domains of knowledge, it leads to selection criteria and program features directly related to particular kinds of excellence. Rather than search for a general trait or several general traits of giftedness, a developmental framework suggests that giftedness is as varied as the fields in which human beings pursue excellence.

In a recent speech in Massachusetts, Dorothy Sisk, head of the federal Office for the Gifted and Talented, said that sometimes children are identified as gifted in one school district and not in another. This is of course because school districts use different screening and selection criteria. If a developmental framework were adopted, the choice of selection criteria would *follow* from decisions about what kinds of giftedness the community values. If the arts were chosen for emphasis, then selection and program decisions would be guided by this choice. Sisk's dilemma would be resolved, because no one would expect the criteria for a program in mathematics to be the same as the criteria for a program in geology, let alone for a program in music. The

developmental framework recognizes that resources are limited, but places the responsibility for how to allocate them where it belongs: on those who make educational policy decisions in the community. Parents still may find that their child is selected for a program in one school district and not selected in another, but the decision is one that makes sense. School District A may value science, School District B music. Parents and voters accept decisions that make sense to them, and these decisions make sense.

3. Finally, I turn to the most difficult but most important change in a new view of giftedness: To generate support, giftedness must be conceived in nonelitist terms. Nearly 20 years ago, John Gardner faced the issue squarely in his book, *Excellence: Can We Be Equal and Excellent Too?* Gardner argued that recognition and support of giftedness had to be done in such a way as to avoid hostility in those who are not gifted. The only way to do this, he concluded, was to promote an educational system that provides "opportunities and rewards for individuals of every degree of ability so that individuals at every level will realize their full potentialities . . ." (p. 115).

In the years since *Excellence* appeared, the search for a way to genuinely value excellence in an egalitarian society has continued. I offer the following two propositions in the hope that a truly humane view of excellence can be incorporated into our way of thinking, and that this, in turn, will lead to genuine enthusiasm for the qualities that make a person uniquely suited to be excellent in one field even as we are enthusiastic about other human qualities that lead to excellence in other fields.

Proposition I: *All children are gifted.* This proposition can never be proven, but it can also never be disproven. I simply choose to assume the more positive of the alternatives, because it places the burden on the education system to find out what a child's gifts are and

---

Twenty-seven states now make some statutory provision for the education of children who are exceptional "by virtue of giftedness."* Of these, only Pennsylvania requires the same formal IEP (individualized educational program) process for the gifted that is mandated for the handicapped in the federal Education for All Handicapped Children Act (P.L. 94-142). However, Arizona, California, Florida, and North Carolina are moving in that direction, according to Phil Juska, IEP coordinator in the Montgomery County Intermediate Unit, Norristown, Pennsylvania.

*See F. A. Karnes and E. C. Collins, "State Definitions of Gifted and Talented: A Report and Analysis," *Journal of the Education of the Gifted*, February, 1978, pp. 44-62.

---

then to provide an environment in which those gifts may be expressed. Proposition I is admittedly a radical departure from traditional conceptions of giftedness, but it is not without empirical support.

A few years ago a student of mine (Joseph Bratton) and I did a study in which we used 18 of the most common criteria for selection of gifted children to choose fifth-grade students for a hypothetical special program. Our results showed that 92% of the students were selected on the basis of one or more of these criteria; only four of the 49 students were selected by half or more of the criteria; and no student was selected by all criteria.* And remember that only criteria currently used for selection into special programs were included in the study. Therefore, farfetched though it may sound, I think it not at all implausible that

*See *Exceptional Children*, vol. 38, 1972, pp. 491, 492.

a broadened view of giftedness would reveal that every child is gifted in some socially valued way.

Proposition II: The purpose of education is to promote excellence in as many forms as possible. Another idealistic-sounding phrase, perhaps, but I think it is reasonable in the context of this article. Granted that there is a long way to go, it seems to me an inevitable conclusion from what I have written that an educational system that is humane and that values excellence could try to do nothing less than what is encompassed in Proposition II. The phrase "as many forms as possible" recognizes that there are limitations both in our ability to know giftedness when we see it and in the resources available to nurture it. But the burden of these limitations is placed where it belongs: on our lack of knowledge about the nature and nurture of giftedness and on the inadequacy of

resources to support even those forms of giftedness we do understand. This leads me back into that classroom where I faced a gifted student so many years ago.

I do not know what my student Jenny's experience in school would have been if the two propositions I have suggested had been guiding educational policy and practice in 1964. Jenny is in her early thirties now, and I can only hope that she has affirmed her gifts and managed to find expression for them. Some people say that the anguish caused by adversity, neglect, boredom, callousness, and arbitrariness is a springboard for creativity; to be sure, Jenny found many of these things in school. My view is that there is enough that is frustrating in the world at large to spur even the most complacent individual to action; the mission of education should reflect the other side of the coin.

Child Rearing and Child Development

Although child-rearing advice has changed over the ages, continuities can be identified. John Watson's early 1900 view that children should be reared strictly in order to correctly shape their behavior can be traced to John Locke's advice of the late 1600s. Similarly, post World War II advice to rear the child in a democratic and permissive atmosphere can be linked historically to Rousseau's 1700s concept of the child as a being of nature, free of original sin, and full of natural curiosity. Today, child-rearing advice seems to have struck a middle road between Locke's "authoritarian" approach and Rousseau's "permissive" approach. Thus, parents are encouraged to provide their children with ample love, to cuddle their infants, to use reason as the major disciplinary technique, and to encourage verbal interaction. However, they are to do these things in an environment where rules are spelled out clearly and children are expected to obey them. Suggestion, persuasion, and explanation have become the techniques of rule enforcement, rather than spanking or withdrawal of love.

Perhaps our modern day advice on child rearing merely reflects the ebb and flow of advice that has occurred over the ages. On the other hand, modern day child-rearing advice may reflect a growing awareness of effective child rearing based upon the accumulated knowledge gained from the scientific study of human development. Moreover, modern day child-rearing advice may reflect a growing aversion to the excessive violence, aggression, and alienation of contemporary American society. Finally, the decline in the size of the American family may have expanded our awareness of the needs of individual children. Extended families have become rare, with the typical nuclear family consisting of two adults and three or fewer children.

However, even the nuclear family is in a state of flux. Increasing numbers of children are being reared in single-parent families, due to a higher divorce rate and an increasing number of unmarried mothers. Changes in the traditional economic roles of women mean that more and more young children are receiving supplementary rearing in day care centers or preschools. And, an increasing number of children are receiving their primary caregiving from single-parent fathers.

Although the demands made of parents are more confusing and pressing than ever before, fewer parents seem to be prepared to assume responsibility for child rearing. Parenthood often is thrust on an individual who has had minimal opportunity to acquire child-rearing expertise. Moreover, when inexperienced caregivers seek outside help, often they are confronted with conflicting advice with no standard against which to judge its validity.

To be sure, parenting is not an easy task in contemporary American society. At minimum, parents must be flexible and willing to try different approaches to child rearing, while constantly evaluating them against "expert" opinion and against their own common sense. Perhaps parents should take the same consumerism attitude toward child-rearing experts that they take toward manufacturers of other goods and services.

Looking Ahead: Challenge Questions

How do you view the notion that fathers can be as competent in child rearing as mothers? Do your opinions apply to infancy as well as childhood?

Increasing numbers of America's next generation of child rearers are reared in abusive and neglecting environments. What are the implications for American society if this social problem is not eliminated?

Behavior modification techniques reflect a direct approach to the management of child behavior. Do these techniques reflect a novel approach to child rearing, or are all techniques of child rearing forms of behavior modification?

America has been described as one of the most violent societies in the world. How does the media contribute to this image, or is it an image at all? If you were responsible for television programming for one of the major networks, what changes would you make in the types of programs shown in order to counter violence and sexism?

A NEW LOOK AT LIFE WITH FATHER

Glenn Collins

Researchers have lately probed the father-infant relationship and found few significant differences in the way children relate to fathers and to mothers. Their studies may sharply alter our concepts of what parenting is all about.

For in the baby lies the future of the world:
Mother must hold the baby close so that the baby knows
that it is his world;
Father must take him to the highest hill so that he can
see what his world is like.
—MAYAN INDIAN PROVERB

The tiny white room on the ground floor of the John Enders Research Building is only a block away from the quiet green quadrangle of the Harvard Medical School. The plate on the door reads, "RESEARCH LABORA-TORY I," and the room is a jumble of television cameras, tape decks and video cables. There are, however, some unexpected items: boxes of Kimbies are stacked under a table; atop the table, a container of Wet Ones Moist Towelettes, and next to the Wet Ones, propped in a sturdy alumimum seat, a 96-day-old baby named Eddie. He is intently studying James, his father, standing before him. Two television cameras are capturing these moments on half-inch magnetic tape.

"Bet you're glad to see me!" says James, smiling. "Were you good with Mommy? You know, I missed you all day. . . ." As he talks to his son, he taps him, tickles him, and smiles, his eyebrows moving in a language of their own. Eddie arches forward in his jumpsuit, kicks his feet in their little red socks, coos and giggles. After exactly two minutes, James leaves the room. The cameras observe Eddie for another 30 seconds, and then they are turned off.

Although a pediatrician on the staff of the Child Development Unit of Children's Hospital Medical Center in Boston will later play back the tape, James and young Eddie aren't patients: They are participants in one aspect of current research on fathering. Investigators will play back the tape at one-seventh speed, and will conduct a microanalysis of the facial expressions of James and Eddie, recording their vocalizations and their body movements on matching graphs. This information will be used to create a saw-tooth chart that plots a father's typical interaction with a child; its signature is distinctive, different from the characteristic pattern of a mother's interaction.

☐

Near Princeton, N.J., a father holds open the front door and waits with his wife and two children as a team of researchers hefts a videotape camera and assorted television equipment into his house. It is just about dinnertime, and the investigators, from the Infant Laboratory at the Educational Testing Service, ask the parents where they normally eat their evening meal.

"Tonight, in the dining room," says the father. The E.T.S. technicians start the camera running, and leave the house.

Tentatively, the parents call their children to dinner. The 3-year-old waves at the television eye. The 6-year-old sticks out his tongue. The father seems a bit unnerved. They start eating, and, before any of them might have expected, they forget about the technological presence in the dining room. After dinner, the E.T.S. researchers return for their equipment; later they analyze the behavior they view on the tape.

After studying 50 families, the E.T.S. investigators can generalize about what they have seen. Fathers talk more to their sons than they do to their daughters. Children talk less to fathers than to mothers, or to each other. And fathers, in their dinner behavior, tend to ask questions.

□

In a laboratory at the University of Wisconsin, a father sits in an easy chair four feet away from a television monitor. He is about to see something unpleasant, but he doesn't know it. Electrodes from an eight-channel polygraph recorder—a lie detector—have been attached to his index and middle fingers, and a rubber cuff has been inflated on his left bicep; his heart rate, skin conductance and blood pressure are being monitored. Soon the television screen glows with a six-minute videotape of a 5-month-old baby boy. The infant looks around gravely and makes a sound; then he squirms, and soon he begins to cry. Loudly. Insistently. Interminably. Even though the baby on the screen isn't his, the father feels ever more uncomfortable under the assault; he moves, tenses, his heart rate rises and his blood pressure soars. Later, after testing 148 subjects, the investigators are able to report that there is no physiological difference between the reaction of a father or a mother to the sight of a squalling baby. To both, it is equally distressing.

Research examining what fathers do and how they do it has been booming in recent years. Not all of it employs computer analyses and electronic bric-a-brac. Much of it involves nothing more complicated than placing a trained observer in a room with a father who is playing with, or talking about, his child.

The impact of all this father-watching is beginning to be felt in courts of law, in hospitals and in universities, which face the task of redirecting the training of a new generation of doctors, pediatricians, psychotherapists, health-care professionals and teachers. "Our whole society has had the notion that a biological bond between mother and child made fathers less able, less interested and less important than mothers in caring for children," says James A. Levine, a Wellesley College researcher. "Courts have based decisions on that notion, therapists have treated patients on the basis of it, and men and women have made life choices because of it."

In fact, 44 percent of the mothers of children under the age of 6 in this country are working, only 24 percent of existing families are traditional nuclear families, and the "two-paycheck" marriage is the norm for nearly half of all two-parent families in America. In the changing society reflected by these statistics, the new knowledge about fathering has important implications for how children will be raised and educated, and will help to shape the kind of nation we inhabit in the 1980's.

□

Fathers haven't always been a fashionable research subject for social scientists. "When I started out 17 years ago, there just wasn't much data," says Henry Biller, professor of psychology at the University of Rhode Island and a pioneering researcher in the field. "The recent increase in data collection on fathers is amazing. We have something of a revolution in thinking among those involved in early childhood development."

In past decades, researchers focused on the father as a role model, or studied him inferentially: by examining the impact of his *absence,* in families where the father had died, divorced or gone to war. Fathers have also long had a place in psychoanalytic theory, becoming important in the Oedipal stage, when the son competes with him for the mother.

Social-development theorists viewed the mother-infant relationship as unique, vastly more important than subsequent relationships; it was even termed the prototype for all close relationships. In 1958 and then again in 1969, John Bowlby, the British psychiatrist, published his elegant and influential theories of attachment, a word that is usually defined by behaviorists as the preference for, or desire to be close to, a specific person. "But the real synonym for 'attachment' is love," says Dr. Michael Lewis, a developmental psychologist at the Educational Testing Service. Bowlby, drawing on the animal-study work of ethologists and the parental-deprivation observations of cognitive psychologists, suggested that there was an evolutionary advantage to a unique bond between mother and infant; he reasoned that this bond was an imperative of the very growth and development of the species.

Subsequently, many researchers investigated the mother-child interaction, revealing the nature of the infant's early relationship with its caretaker. However, fathers weren't even present in most of the studies. "A major reason that fathers were ignored was that fathers were inaccessible," says E. Mavis Hetherington, a University of Virginia psychology professor who has studied family-related questions for 25 years. "To observe fathers you have to work at night and on the weekends, and not many researchers like to do that."

Studies of humans and of nonhuman primates began to suggest that infants had strong attachments to persons who had little to do with their caretaking and physical gratification; nor were these relationships necessarily derived from the child's bond with its mother. In a classic study, primatologist Harry Harlow demonstrated that the attachment process was not limited to a feeding context. Investigators also showed that the actual amount of time an infant and his mother spent together was a poor predictor of the success of their relationship. Consequently, a child's tie to its mother continues to be viewed as crucially important; however, its exclusivity and uniqueness have been challenged. The new research emphasizes the complexity of an infant's social world.

Researchers have now identified some of the ways in which fathers are important to children. Henry Biller sums up the findings: "The presence and availability of fathers to kids is critical to their knowledge of social reality, their ability to relate to male figures, to their self-concepts, their acceptance of their own sexuality, their feeling of security. Fathers are important in the first years of life, and important throughout a child's development." Frank A. Pedersen of the National Institute of Child Health and Human Development has also demonstrated that mothers can perform better in their parenting roles when fathers provide emotional support.

Researchers from a number of disciplines using ingenious new methods now suggest that the father-infant relationship is not what we thought it was: that, for example, there are few significant differences in the way children attach to fathers and to mothers; that fathers can be as protective, giving and stimulating as mothers; that men have at least the potential to be as good at taking care of children as women are; and that the characteristic interplay of father and infant, when scrutinized minutely, is distinctive in many fascinating ways. The new fathering research offers fresh insights about the "distant" father and about fathers' roles across disparate cultures; it reveals that fathers have been ignored in research and in medical practice in curious and interesting ways; and it offers a synthesis of the relationship between fathers, children, families and society.

James Herzog, M.D., a psychiatrist who teaches at Harvard, says this of the new findings: "We're in what I call the post-competency phase now. We don't need to prove that fathers 'can do it, too.' The question now is, what is the specific role of the male parent, and what is the difference between being a father and being a mother?"

□

In 1970 a Harvard Ph.D candidate named Milton Kotelchuck began a study of fathers that created a stir when it was presented in Philadelphia at the 1973 meeting of the Society for Research in Child Development. Kotelchuck, now director of health statistics and research for the Commonwealth of Massachusetts, had set up a classic "separation-protest" situation—a test of attachment—in studying the reactions of 144 infants when their fathers and mothers walked out of a play-

room and left them with a stranger. Previous studies had observed the effects of a mother's departure on her child; Kotelchuck was able to determine that infants were just as upset when a father left them.

In four other studies, Kotelchuck and his associates found few significant differences in the way the infants attached to fathers and to mothers. They demonstrated that, in fact, children have extended social worlds and can attach equally well to siblings, peers and other figures.

Michael E. Lamb, research scientist at the University of Michigan's Center for Human Growth and Development, has carried on his investigations—including the crying-baby experiment described earlier—at Michigan, the University of Wisconsin at Madison, and at Yale, and was the editor of an influential 1976 anthology, "The Role of the Father in Child Development." His first key study of attachment appeared in 1975; in it, 7- and 8-month-old boys and girls and their parents were viewed in the home setting. An observer dictated a detailed account of the behavior he saw into a tape recorder. That narrative was then analyzed by applying 10 measures of attachment and affiliation; whether the baby "Smiles," "Vocalizes," "Looks," "Laughs," "Approaches," "Is in proximity," "Reaches to," "Touches," "Seeks to be held" or "Fusses to." Lamb and his co-workers found that no preferences were evident for one parent over the other among these infants, at the age when they should, according to Bowlby's theory, be forming their first attachments.

Lamb and his colleagues reported that when mothers held their infants, it was primarily for things like changing, feeding or bathing; fathers mostly held their children to play with them, and initiated a greater number of physical and idiosyncratic games than mothers did. This paternal play tended to be boisterous and physically stimulating. Furthermore, boys were held longer than girls by their fathers; fathers start showing a preference for boys at one year of age and this preference increases thereafter.

Currently, Lamb and his associates are studying 100 families from the time of pregnancy until their children attain the age of 18 months. The sample includes families where there are working wives, also represented are a few fathers who are primarily responsible for infant care. In Sweden, for the past six months, they have been observing role-sharing fathers and both mothers and fathers who have primary responsibility for child care.

If Michael Lamb has tended to focus on the child in his work, Ross D. Parke, professor of psychology at the University of Illinois at Champaign-Urbana, has centered his research on fathers themselves. In a 1972 study that is a classic in the literature of fathering, Parke and his colleagues haunted a hospital maternity ward in Madison, Wis., and observed the behavior of both middle-class and lower-class parents of newborn babies. They found, most strikingly, that fathers and mothers differed little in how much they interacted with their children. Fathers touched, looked at, talked to, rocked and kissed their children as much as their mothers did. The study suggested that they were as protective, giving and stimulating as the mothers were —even when the fathers were alone with their babies.

In later work, Parke and his collaborators measured the amount of milk that was left over in a baby's bottle after feeding time; infants consumed virtually the same amount of milk whether fathers or mothers did the feeding. They found that fathers were equally competent in correctly reading subtle changes in infants' behavior and acting on them; fathers reacted to such infant distress signals as spitting up, sneezing and coughing just as quickly and appropriately as mothers did. Parke asserted that men had at least the potential to be as good at caretaking as women. However, fathers tended to leave child care to their wives when both parents were present.

In the last three years, Parke, Douglas Sawin of the University of Texas and their collaborators have conducted two major studies involving 120 families. They observed family interactions, and used high-speed electronic "event recorders" with 10-button keyboards and solid-state memories to tap out four-digit codes that recorded behaviors as they saw them. Although there are very few differences in *quality* between mothers' and fathers' interactions with their children, one observed disparity is that fathers are more likely to touch and vocalize to first-born sons than to daughters, or to later-born children.

□

Eddie and James, whose close encounter began this article, were participants in a continuing investigation of children's early learning abilities and communication patterns at the Child Development Unit of Children's Hospital Medical Center in Boston. Originally, father-infant and mother-infant pairs were videotaped periodically during the first six months of babies lives. The unit's newer work involves the father-mother-infant triad.

In Laboratory I, where young Eddie became something of an intramural television celebrity among social scientists, the research continues. Infants are placed in an alcove created by a blue-flowered curtain, and are taped with father or mother. These laboratory situations, though artificial, place the maximum communicative demand on the parent and child, the researchers say; they bring out the kinds of intense play situations that normally occur only during brief periods during an ordinary day.

Two trained observers play back the videotapes of the sessions and perform a "microbehavioral analysis" of the interaction of both the parent and the baby. The researchers assign numerical scores to rate such facial expressions as frowns, pouts and smiles; sounds like gurgles or coos; motions of hands and feet, and even eye

movements. Ultimately, the observers note clusters of these behaviors and chart them during each second of elapsed time over the entire interaction.

Graphs of fathers' and mothers' behavior show distinctive patterns. In all of the families studied by the Child Development Unit, the chart of the mother's interaction is more modulated, enveloping, secure and controlled. The dialogue with the father is more playful, exciting and physical. Father displays more rapid shifts from the peaks of involvement to the valleys of minimal attention.

There are other characteristic differences: Mothers play more verbal games with infants, so-called "turn-taking" dialogues that are composed of bursts of talking or cooing that last four to eight seconds, and are interrupted by three- to four-second pauses. Fathers tend to play more physical games with infants; they touch their babies in rhythmic tapping patterns or circular motions.

To provide conceptual models for the way babies interact with adults, the Boston researchers have employed the theories of cybernetics, the discipline that studies the control and regulation of communication processes in animals and machines. Researchers have broken with the traditional lexicon of rat psychology, and talk about the "interlocking feedback of mutually regulated systems" and "homeostatic balances between attention and nonattention." The baby, in its reciprocal interaction with an adult, modifies its behavior in response to the feedback it is receiving. Infants, they say, seem to display periods of rapt attention followed by recovery intervals, in an internally regulated cycle that maintains the balance of the infant's heart, lung and other physiological systems.

"It's important to say that father doesn't offer some qualitatively better kind of stimulation; it's just different," says T. Berry Brazelton, M.D., director of the Child Development Unit, a pioneer in the study of family interactions. "Mother has more of a tendency to teach the baby about inner control, and about how to keep the homeostatic system going; she then builds her stimulation on top of that system in a very smooth, regulated sort of way. The father adds a different dimension, a sort of play dimension, an excitement dimension, teaching the baby about some of the ups and downs—and also teaching the baby another very important thing: how to get *back* in control."

There are also interesting similarities in infants' relationships with both parents, says Michael Yogman, M.D., the pediatrician who videotaped James and Eddie and who has specialized in the study of fathers at the Child Development Unit since 1974. "With both parents," he says, "we see that behavior is mutually regulated and reciprocal, that there is a meeting of behaviors."

Dr. Brazelton says that "there's no question that a father is essential to children's development. Our work shows that babies have this very rich characteristic model of reaction to at least three different people—to father, to mother and to strangers. It shows me that the baby is looking for richness, that he's looking for at least two different interactants to learn about the world." For Dr. Brazelton, to whom Mayan Indians told the saying that preceded this article, its poetry is exceedingly descriptive.

"It seems to me," says Dr. Brazelton, "that the baby very carefully sets separate tracks for each of the two parents—which, to me, means that the baby wants different kinds of people as parents for his own needs. Perhaps the baby is bringing out differences that are critical to him as well as to them."

The Boston researchers plan to explore the later development of the paternal and maternal dialogues with children. They also hope to refine their procedures to the point where they may be useful as a diagnostic tool for practitioners.

□

Fathers are being studied from other perspectives. Although psychiatric clinicians, those who see patients, had always noted that the father played an important role in the psychological development of children, as late as 1973 the psychoanalytic literature bemoaned the lack of theorizing about the father's role during the first two years of life.

Building on Margaret Mahler's ideas on the successive stages of an infant's "psychological birth," psychoanalyst Ernst Abelin and others focused on the father's role in helping infants separate from mothers.

Some behavioral psychologists can't take the efforts of the psychoanalytic theorists very seriously, since the data for such work are often derived from the study of a single patient who may be going through the process of becoming a father, or coping with difficulties of parenthood. "It's better to observe what's going on," says Alison Clarke-Stewart, a University of Chicago psychologist. "It's not distorted by retrospective recollection or the perceptions of the person who's being studied."

Psychoanalysts reply that the observational method is limited. "How people behave is highly determined by their fantasies, conflicts and unconscious processes," says Dr. Herzog. "These are the causes of the behavior that others observe. I have nothing against documenting this behavior, but we need to look at the inner life, too."

Part of that inner life is a well-documented clinical phenomenon, the so-called "womb-envy" —the envy of women's capacity to give birth—among some expectant fathers and even among male children. Perhaps a societal counterpart of this is the "couvade" phenomenon that anthropologists have noted in many cultures, where men undergo elaborate rites of passage paralleling their wives' pregnancies and birth-giving.

"We know that the time of pregnancy and becoming a father is extremely important to men, a crucial and stressful time," says Alan R. Gurwitt, M.D., a psychiatrist, analyst and associate clinical professor at the Yale Child Study Center. "Yet the astounding thing in this society is that the father has come to be a subject of ridicule—there is no end to the cartoon and the movie stereotypes portraying the expectant father, and fathers in general, as bumbling fools."

He says there is still a tendency to ignore fathers on the part of obstetricians, pediatricians, nurses and even child psychiatrists. "This failure to involve fathers even in the treatment of their children runs very deeply," says John Munder Ross, a Manhattan psychotherapist and clinical assistant professor at Downstate Medical Center, who is coediting an anthology of the new psychoanalytic views of fathering. "It may have to do with the relations of clinical workers to their own fathers. There seems to be an awful lot of stereotyping of fathers as 'absent and ineffectual,' or 'tyrannical and sadistic.' "

To the psychoanalysts, the process that is fathering continues. "The middle-aged father frequently finds himself in a painful situation," says Stanley Cath, M.D., a psychoanalyst and associate clincial professor at Tufts Medical School. "His adolescent children may be rebellious and challenging to him; he himself may be trying to separate from his own father, who may be aged or dying; and the grandfather himself may be looking for support" as he faces the debilitation of old age. "Of course, a man can be the father to his children, and also the father to his parents," says Dr. Cath. "We rediscover the father, and the definition of fathering, throughout our lifespan."

□

As a social phenomenon, the evolution of fathering in man and various primate precursors is a matter of sheer conjecture. Paleontology provides little data on social interaction. Some cultural historians have tried to make inferences from the study of recent "primitive" societies, by which they mean complex societies that have not received the blessings of technology.

Margaret Mead's famous 1930 study of the Manus people of New Guinea reported that, at the age of a year, children were given from the mother's care into the father's. He would play with the baby, feed it, bathe it, and take it to bed with him at night.

Fathers in the Thonga tribe in South Africa, observed during the last century, were ritually prevented from having almost anything to do with infants until the babies were 3 months old. However, fathers in the Lesu culture of Melanesia commonly took care of babies while their wives were busy cooking or gardening. And among the !Kung bushmen in northwestern Botswana today, fathers have a great deal of contact with children, holding and fondling even young infants.

In analyses of all known cultures, anthropologists have suggested that, in about two-thirds of societies, wives, and children accord the paterfamilias deference, that husbands exert authority over their wives and that most cultures trace descent through the father's line. In nonindustrial cultures, these analyses suggest, fathers generally play a small role in relating to young children. In other words, the similarities of men's roles outweigh the fascinating differences that may exist.

Male figures—though not necessarily fathers—are involved in child care in most cultures. Just how involved is another question. Applying a measurement called the Barry and Paxson Father-Infant Proximity Scale, researchers Mary Maxwell West and Melvin J. Konner at Harvard University found that social and cultural conditions are related to the level of involvement of fathers with their children, and suggest that there is the potential among males for caring for their young if other conditions encourage it. West and Konner found that fathers observed in cultures with monogamous nuclear families were generally involved parents. So were fathers in "gathering" societies—the form of society that existed during 98 percent of human history. They suggest that distant fathering is associated with warrior cultures ("hunting" societies) and with societies where men's agricultural or military activities take precedence.

Of course, the political and economic equivalents of warfare exist in modern industrial cultures, and it can be debated how much they affect males' involvement in fathering. There is conjecture that the tradition of the Roman pater-familias had some influence on current patterns, as well as the Christian concept of the Old Testament God. The few attempts at compiling histories of fathering show the Industrial Revolution to be a major disrupter of family life as it existed when many fathers were tradesmen or farmers working in the presence of their children.

The cross-cultural evidence shows clearly that the father has been many things in many societies; it suggests that, if the culture allows, fathering can be whatever fathers want to make of it.

□

Is there any answer to the question posed earlier by Dr. James Herzog: Is fathering the same as mothering? And, if not, is one parenting style superior to the other?

"You don't want to imply from these studies that people are interchangeable," says Alan Sroufe, a University of Minnesota child-development professor who is doing studies of attachment there. "Sure," he says, "an infant can attach to a woman or a man. But women have natural advantages in parenting. It's not just nursing—for example, mothers lactate as soon as a baby cries. But mothers also have the experience of carrying the baby for nine months, and if the business of attachment comes from sensitivity to being tuned into a baby, mothers have the advantage."

"But there is a crucial distinction to be made here," says Milton Kotelchuck. "Yes, pregnancy and lactation can make it easier for a mother to attach to a child. But the essential thing is that infants don't know that they are supposed to relate more to the mother than to the father."

"It is my speculation—and I want to emphasize that word," says Michael Lamb, "that we will find that biological differences are very small, and that they are exaggerated and magnified by the rituals and the roles that societies build around those distinctions. But are these differences genetic? My answer is 'Yes, but'— where the *but* is more important that the *yes*.

"Aside from the question of genetics," he says, "there is good evidence to believe that mothers and fathers can be equally effective as parents. They just have different styles. Perhaps it's really not fathering or mothering—it's parenting."

□

One researcher attempting to synthesize the relationship between child, family and society is Michael Lewis, director and senior scientist at the Institute for the Study of Exceptional Children at the Infant Laboratory of the Educational Testing Service. (It was his investigators who conducted the videotaped observations of Princeton fathers at dinner.) Lewis holds that different people—mothers, fathers, peers, siblings, grandparents, uncles, aunts and other relatives—serve the child's needs in different cultures in different ways: "I am saying that a father's role is cultural and historic rather than biological and evolutionary."

"There's no good data on any of this," Lewis says, "but my impression is that, to an extent in the general culture, fathers are defining their functions in new ways—the 'new fathering' we hear about." He adds that "we haven't assessed the basic question of values here yet, and that's what we need to do. If the cultural matrix is changing, is it assisting the values of our culture?"

To an extent, society has legitimized the needs of parenting men. "In a sense, fathers have come out of the closet," says Mavis Hetherington. "They feel more comfortable about being parents, and are more actively fighting for their rights." Recent revolutionary changes in the way society views men are now treated by the media as commonplace: men's improved position in child-custody cases or men's right to single-parent adoptions in most states.

Nevertheless, it is James Levine's hunch that women are more aware of the issue of fatherhood than men are. "I think it is becoming more of a question for women as more of them are working outside the home. Women make demands on men to parent in a way that fits in with their new concepts of how they will live their lives," says Levine, a research associate at the Wellesley College Center for Research on Women who wrote an influential 1976 book on male parenting options, "Who Will Raise the Children?"

However, Levine says, "the biggest push for change is coming from the economic pressures—the necessity for both parents to work. I think the bottom line in all of this is the economic situation of women."

6. CHILD REARING AND CHILD DEVELOPMENT

Michael Lamb believes that "it's a depressingly small number" of fathers who take on a large share of all that is involved in bringing up a child. He does not view the recent research about fathers' abilities as a new panacea. "But," he says, "I think we must realize that, in general, the average male won't be better than the average female as a caretaker. Yes, babies can attach to father. But that isn't to say that they won't be closest to the primary caretaker, which is usually mother."

Says Levine, "Where we really miss the boat is when we say the male role is changing, and cite as evidence the fact that men are changing diapers, bottle feeding, etcetera." The truly important part, Levine feels, involves a man's sense of emotional responsibility: "It's not just the taking *care* of kids, but it's who carries around that inner *sense* of caring, that extra dimension of emotional connection."

It is possible that there will be competition in parenting. "At this point," says Dr. Brazelton, "everyone is goading men on to do more, but the second that men get good at it, and really enjoy nurturing, it may cause problems that'll have to be faced. Fathers who are taking an equally nurturant role may threaten some mothers."

For Levine, looking ahead, the most interesting question is, "research for what?" It seems to him that the next step, theoretically and practically, is to give some guidance to medical and mental-health practitioners: "The most interesting area for research has to do with total family interaction, the family systems perspective."

Virtually all of the father-watchers are wary, however, of being prescriptive—of saying that fathers should parent in a specific, more "nurturant" way.

"The crucial impact of the new research," says Douglas Sawin, "should be that a father's role ought to be an optional choice—and that, with a little support and training and education, they can be primary parents —but only if they want to be. For they have the basic competence and warmth and nurturance abilities. Whether they implement them or not is their decision."

When Mommy Goes to Work...

What happens to her kids' emotional development... her husband's ego... her own self-esteem?

SALLY WENDKOS OLDS

Sally Olds has three daughters, aged 15, 18 and 20, and has worked part time or free lance in public relations and journalism ever since her youngest was a year old.

It used to be easy to diagnose the problems of children whose mothers worked outside the home—a group of youngsters that today totals more than 27 million in this country alone. Is Mary overly dependent and whiny? That's because she doesn't see enough of her Mommy. Does Billy do badly at school? Poor thing, he doesn't have the loving attention of a mother who could help him with his homework. Is Freddy stealing candy bars from the corner store? He wouldn't if he had Mom's guidance at home!

Such assumptions may seem logical, but they just don't hold up when scrutinized under the research microscope. As social scientists delve more deeply into the effects on children of their mothers' working, their findings are turning out to be quite different from long-accepted beliefs.

Let's take a moment for a brief history lesson. Twenty-five years ago, only 1.5 million mothers were in the labor force. Today, 14 million are. As late as 1940, only one female parent in ten worked outside the home. Today, four in ten do. Before 1969, most women with children between the ages of 6 and 17 spent their days at home. Today, the United States Department of Labor reports that a record nine million women with children 6 to 17 years old are working. In fact, nearly three million have little ones aged three to five, and over two and half million have babies under three!

With employment patterns shifting so dramatically, it's only logical that we reevaluate our long-held beliefs about child care and babies' needs in general—beliefs that for years have kept mothers tied to their babies' cribs for fear of sparking emotional and psychological traumas later on. Most of our baby-care gospel (example: "children need a loving mother at home") is based on studies of hospitalized youngsters conducted during the 1940's and 50's. Not surprisingly, researchers found that infants in understaffed institutions, who were cut off from familiar people and places and who were cared for by a bewildering succession of hospital nurses, eventually suffered severe emotional problems. Valid as these studies may be, they tell us nothing about babies who, though looked after by competent baby-sitters or day-care workers during the day, are reunited with their own loving parents come evening. Fortunately, studies of the last decade have sharpened and reinforced this distinction.

In 1973, for example, Harvard University pediatrician Dr. Mary C. Howell surveyed the voluminous literature on children of working mothers. After poring over nearly 300 studies involving thousands of youngsters, she concluded: "Almost every childhood behavior characteristic, and its opposite, can be found among the children of employed mothers. Put another way, there are almost no constant differences found between the children of employed and nonemployed mothers." To wit: Researchers found both groups equally likely to make friends easily or to have trouble getting along with their peers, to excel at their studies or to fail, to get into trouble or to exhibit model behavior, to be well adjusted and independent or to be emotionally tied to the apron strings, to love and feel loved by their parents or to reject them.

Just recently, Harvard psychologist Jerome Kagan and two researchers from the Tufts New England Medical Center, Phillip Zelazo and Richard Kearsley, zeroed in on the possible effects of day care on the emotional and developmental progress of infants whose mothers worked, as compared to children raised by their mothers at home. As the yardstick for his evaluation, Kagan used three characteristics considered "most desirable" by parents: intellectual growth, social development and ability to achieve a close relationship with the mother. His results? Provided the center was well staffed and well equipped, Kagan and his colleagues were unable to find *any* significant differences between the two groups of children.

Since a mother's working per se is no longer considered a crucial factor in a child's development, what factors *are* important? To find out, let's examine the problem from a different perspective. Instead of thinking in terms of working and stay-at-home mothers, we'll divide women according to whether or not they *enjoy* whatever it is they are doing, and here we can see the differences emerge.

Back in 1956, psychologist Jack Rouman traced the progress of 400 California school children and found that the emotional problems they suffered were related not to their mothers' employment status but, rather, to the state of their mothers' emotions. He concluded: "As long as the child is made to feel secure and happy, the mother's full-time employment away from the home does not become a serious problem."

Take Linda Farber, a Philadelphia city clerk who hates her job, is bitter at her ex-husband for leaving her, making it necessary for her to work, and who feels tied down by her six-year-old son, Greg. He, in turn, is wetting his bed again, gets stomachaches every morning before school and is withdrawing from other children. On the other hand, Marjorie Gorman would love to return to the personnel office where she worked before her kids were born, but her husband insists, "It's your duty to stay home with the children." Marjorie is bored and restless. Annie, her oldest daughter, has run away from home three times, has thrown a kitchen knife at her parents and is habitually truant.

Of course, these children's problems are not triggered simply by their mothers' attitudes about work. But maternal unhappiness and resentment is easily communicated to other members of the family, and can, indeed, influence the quality of home life.

Studies undertaken by University of Michigan psychologist Lois Wladis Hoffman bear this out. She found that employed women who enjoy their jobs are more affectionate with their children and less likely to lose their tempers than mothers who are disenchanted with their daily work. Furthermore, those who are content with their situations are more likely to have sons and daughters who think well of themselves, as measured on tests of self-esteem, than are resentful workers or unhappy homemakers. Following a 1974 review of 122 research papers on working mothers and their children, Dr. Hoffman concluded, "The dissatisfied mother, whether employed or not and whether lower class or middle class, is less likely to be an adequate mother." Norwegian psychologist Aase Gruda Skard agrees: "Children develop best and most harmoniously when the mother herself is happy and gay. For some women the best thing is to go out to work, for others it is best to stay in the home."

For Ellen Anthony, staying at home to care for her small baby was stifling. "I need to work," she insists. "Without some outside stimulation and a way to discharge pent-up energy, I become bored and aggressive. Now that I'm back at my public relations post, I don't overpower my daughter and husband so much and we're all happier." Carol Brunetti, on the other hand, left a good job as a department store buyer to devote full attention to her infant son. "I haven't missed my job for a minute," she says. "I love the flexibility of making my own hours. And whenever I want to go somewhere, I just take Jason along with me."

But Mom's attitude is not the only one that must be taken into consideration. No one will argue the fact that the happiness of both mother and children also depends on the father: How a husband feels about his wife's working is crucial to the emotional climate within the home. And his attitude is a distillation of many things—whether he considers himself a success or a failure at his own profession, what the basic marital relationship is like and how willing he is to assume a fair share of the management of the household and the children if his wife takes a job.

Obviously, the woman whose husband approves of her working is lucky: Balancing job and family is never easy, but when a wife has to do the juggling herself, as well as contend with a husband's opposition, it's twice as difficult.

Happily, many a man who was originally opposed to his wife's working has discovered that he likes spending more time getting to know his children, that money problems have lessened and that he and his wife have more to talk about now that she's also exposed to new people and situations.

Although many psychoanalysts continue to stress the need for an exclusive relationship between mother and baby, recent research has shown that such re-lationships are probably the exception rather than the rule, even in families where the female parent does not go out to work. For one thing, most fathers today are vital figures in their children's lives. A 1974 study by Milton Kotelchuck of Harvard University found that one- and two-year-olds are just as attached to their fathers as to their mothers. And for another, the typical baby in our society is cared for by several other people in addition to its parents.

According to anthropologist Margaret Mead, who has examined child-rearing patterns in societies around the world, the notion that a baby must not be separated from its mother is absurd. Babies are most likely to develop into well-adjusted human beings, she says, when they are cared for "by many warm, friendly people"—as long as most of the loved ones maintain a stable place in the infants' lives.

There's the rub. For many working mothers, finding these "warm, friendly people" to care for their children on a long-term basis is often a frustrating and expensive proposition. Experts agree that the following scenarios are probably the most stable (and, in turn, most successful), especially for babies and toddlers:

• A father who is able to dovetail his work schedule with his wife's so that their child can be looked after by one parent or the other.
• A grandmother, other relative, friend or neighbor who cares for a child in his or her own home.
• Family day care—an arrangement similar to the one above but between people who have not previously met, often arranged by a public agency.
• A full-time babysitter who comes to the house five days a week and may perform housekeeping chores, too.
• A well-run, well-staffed day-care center.

But once the parents have made the decision that Mommy should work, what about the kids? How will *they* take to their mother's new role—and if they don't, what can you do to make them understand?

Most likely, children will have mixed feelings about Mommy's new job. David, nine, whose mother is the only working mother on the block, sometimes asks her, "Why can't you be home when I get home from school like Mark's mother? She always gives us milk and cookies." But the day David's class visited the dress factory where his mother works, he proudly explained her role in designing the clothes they saw being produced.

One woman met her child's resentment head-on. After ten-year-old Lisa had asked for the umpteenth time, "Oh, why do you have to work, anyway?" her mother stopped what she was doing, sat down with her daughter and explained just how important her job was to her. She let Lisa know that she understood the child's annoyance but she made it clear—without getting angry—how unhappy, bored and restless she would be staying home.

A group of 11-year-olds told an investigator that they loved the responsibility of using their own keys to let themselves in and out and they relished the privilege of having the house to themselves for a few hours after school.

What can both parents do to help children more readily accept their mother's employment? Child-care experts suggest that you:

• Plan your schedules so that at least one parent is with the baby for half his or her waking hours during the first three years of life.
• Institute new child-care arrangements a week or so before you start a job, so that your child has a chance to get used to the new set-up.
• Don't take a full-time job for the first time or make a big change in child-care arrangements when your baby is between six months and a year old, or between one-and-a-half and two-and-a-half. Try to wait a couple of months after any major upheaval—such as a move to a new home, a long illness or the break-up of a marriage.
• Keep in close touch with whoever is caring for your child and consider her or him a partner in nurturing.
• Plan "child time" into your schedule when your youngsters can depend on having some uninterrupted time with you. It need not be long, but it should be regular.
• Let your children know how much they mean to you, and that they mean more to you than your job.

"The mother who obtains satisfaction from her work, who has adequate arrangements so that her dual role does not involve undue strain, and who does not feel so guilty that she overcompensates, is likely to do quite well and, under certain conditions, better than the nonworking mother," insists Dr. Hoffman.

In other words, it's not a matter of "whether" or "where"—but of "how" the woman who works balances the seemingly conflicting elements in her life. As one magazine editor explains, "I feel I have the best of both worlds— I love my family and I love my work, and every day in every way I feel a little better about being me."

Suffer the children

Child abuse: one of the world's most horrifying social problems.

Gary Turbak

RON MADDUX killed Melisha Gibson. Because she had wet her bed, Maddux forced the four-year-old to walk for hours between the kitchen and her bedroom. Occasionally, he made her drink a spoonful of hot sauce. Each time she vomited. Once, he promised her a glass of water if she'd down another gulp of the sauce, but when she complied, he drank the water instead. In the evening, he began hitting her with a stick as she walked. After she was finally allowed to sleep, he awakened her and put her in a cold shower because she had again wet the bed. Before morning, Melisha died in her sleep.

Was Ron Maddux some crazed stranger who had spirited Melisha away from her loving parents? No, he was her stepfather, and her mother was not unaware of what was happening.

The abuse of children—most often by parents—has become a social ill of major proportions. "If you had a disease that affected so many children annually," says Dr. Douglas Besharov, director of the U.S. National Center on Child Abuse and Neglect, "you'd have an epidemic." Dr. Vincent Fontana of the New York Foundling Hospital Center for Parent and Child Development believes child abuse "may be one of the most common causes of death in children." In the U.S.A., 2,000 youngsters are abused to death each year. Only cancer, accidents, congenital abnormalities, and pneumonia take more young lives there. West German parents batter 1,000 of their progeny to death each year; England is not far behind, with 700. In Canada, about 100 children die every year as a result of ill treatment. Statistics are lacking for most other nations, but there's no reason to believe their children are any safer. Only recently has any country taken a hard look at the cruelty being inflicted upon its children.

In 1961, a U.S. doctor, C. Henry Kempe, coined the term "battered child syndrome" to explain the repeated fractures, burns, cuts, and bruises suffered by thousands of U.S. children. Only through doctors willing to testify in such cases can a path be blazed through the taboos against interfering with the rights of parents.

Taking the wraps off child abuse has unveiled a nightmare. Babies are dunked in boiling water. Youngsters are locked in boxes or closets for years. Bed-wetters are chained to their beds. Infants with soggy diapers are set on hot steam radiators. Little fingers are held in open flames and forced onto hot burners. Lighted cigarettes are pressed against tender young skin. Children are beaten with fists, electric cords, auto fan belts, pool cues, baseball bats, and chair legs. And they are stabbed, shot, shocked, drowned, and have plastic bags tied over their heads. Usually, parents are responsible.

Surprisingly, these abusive mothers and fathers are not psychotics on the rampage; 90 percent of them fall into the broad classification of "normal." They are the people next door, the couple sitting next to you in church. They are—potentially—any parents.

Perhaps the factor most often leading a parent to abuse his child is the abuse he himself suffered as a youngster. Parents nearly always rear their children the way they were raised, and if their history includes abuse, they pass it on to the next generation. One study in England discovered a family with five generations of abuse, including six deaths, three batterings, ten cases of cruelty or starvation, and 11 children left unsupervised. Only seven offspring were not obviously mistreated.

West German sociologists have uncovered other reasons for parental abuse of children: the child thwarts the plans parents may have for their own lives; the youngster doesn't live up to their expectations; the parent feels threatened by the child and comes to regard the child as a rival; the child resembles an unliked relative; he or she is of the wrong sex; the child is the result of one parent's former marriage and is resented by the new spouse; or the youngster is deformed or retarded and thus thought to bring disgrace to the family.

Although child abuse ignores political boundaries, race, income, and social, educational, ethnic, and class distinctions, a profile of the potentially abusive parent has emerged. He is immature. He feels isolated and distrustful of others. He often lacks self-control and has a poor self-image. He exhibits a low tolerance for day-to-day frustrations. Most abuse, says U.S. physician Dr. Claude Frazier, "is a result of the inability of rather ordinary people to cope with life's stresses."

And so, for reasons perhaps deeply buried in their psyches, these parents lash out at their children. The ill treatment may come in several forms. Physical abuse is the most common and is limited only by the cruel ingenuity of the adult. Emotional abuse, such as the ignoring, rejecting, berating, or abandoning of a child, is more subtle; this kind of treatment can permanently scar the child's developing personality. Sexual abuse is greatly underdiagnosed but is known to be perpetrated frequently, particularly against young girls. Finally, parents may neglect to properly feed and clothe a youngster, or they may withhold needed medical help.

What is being done about this glaring social evil? Programs vary from well-developed to non-existent. West Germany now has a government-sponsored 24-hour abuse hotline and information center located in Hamburg. Poland uses its system of "protective courts" to separate abused or neglected children from their parents; about 1,000 sentences a year are passed there curtailing or suspending parental rights.

Some Australian states provide a 24-hour reporting center, and offer extensive protection to those who register complaints. France's Association for Defense of Child Abuse Victims and England's National Children's Bureau are two of the many organizations working on the problem in those nations. The Israeli justice system employs a "youth examiner" to ferret out the truth in suspected incidents of abuse and neglect; the examiner supplants the police in many cases and is allowed to testify in court on the child's behalf.

The U.S. National Center for the Prevention and Treatment of Child Abuse and Neglect in Denver, Colorado, provides residences where troubled families may live as a unit while they work together on the problem. Other Denver programs include therapy for parents, a round-the-clock crisis nursery, and a therapeutic play school for abused children. Many similar organizations and programs are functioning in other U.S. cities.

One of the biggest roadblocks to effective prevention or treatment is in finding the families who need it. A few governments legally require doctors, nurses, and teachers to report suspected cases; but in most countries, reporting is purely voluntary. France has recently begun pushing for more citizen participation in locating abused children. The magazine Paris-Match urges its readers to report abuse, assuring them they will remain anonymous. It suggests this procedure: If you suspect that harm has been done to a child, talk to him and try to determine the source of his injuries; report suspected abuse or neglect to a social worker or teacher; or summon a policeman when you spot an abused child in a public place.

Recognizing the signs of abuse is not always easy. The U.S. Department of Health, Education, and Welfare says that abused children—in addition to their obvious injuries—may seem unduly afraid of parents, fear contact with all adults, have normal injuries that are improperly treated, display poor general care, exhibit extremes of behavior, have unexplained learning problems, go to school early and stay late, and be improperly clothed. Such symptoms should be reported to authorities—by anyone who observes them.

Like their children, abusive parents often follow a behavioral pattern. They may remain isolated or mix

Healing the damaged child

"THE RIGHT to affection, love, and understanding. . . .
The right to be brought up in a spirit of peace. . . ." These are two of the 10 rights cited in the United Nations Declaration of the Rights of the Child, the humanitarian document which this year, on its 10th anniversary, inspired the International Year of the Child. But many children will never attain these rights unless concerned adults, like ourselves, help to change their circumstances.

How do we do this? How do we invade the sanctum of the family, the basic, and the hardest to penetrate of all social units, surrounded as it is by a wall of respectability, regardless of its internal conditions? Public exposure of abusive parents is disgracing, embarrassing to accuser and accused. Yet, when a child's safety—perhaps his life—is in question, how do we NOT respond? What CAN we do?

For one thing, we can become advocates of the rights of children. As advocates, we can be on the lookout for abused children, and we can report their condition to local authorities or child welfare agencies. We can make sure that counseling is available for abusive parents. And we can offer our services as teachers, surrogate parents and grandparents—and friends—to children who have known the terrors of brutality.

"The hearts of small children are delicate organs," wrote U.S. novelist Carson McCullers in "The Member of the Wedding." "A cruel beginning in this world can twist them into curious shapes. The heart of such a child may fester and swell until it is a misery to carry within the body. . . ."

The International Year of the Child*—the first year of Rotary's 3-H program—is the time to reach out to our damaged children—through awareness, through advocacy, through active evidence of love.   —THE EDITORS

*A/E Editor's note: The United Nations proclaimed 1979 the International Year of the Child to recognize the human rights of children.

poorly with neighbors. They may offer inconsistent explanations for what could reasonably be normal childhood injuries in their youngsters. They may refuse professional help for their children. Usually, they are overly critical of and rarely exhibit pride in their children. They keep their children confined at home for long periods. They make unrealistic demands.

It should be the responsibility of every individual to report suspicious circumstances to proper officials. To report a case of abuse when, in fact, none exists may cause hard feelings and resentment; but failure to report an actual case will result in a child's suffering needlessly, sometimes even in his death.

Child abuse is certainly not new. Roman fathers were allowed to kill their unwanted children. Various other cultures have also practiced infanticide, and corporal punishment is a disciplinary tactic in many societies. It is still shocking, though, to realize that—in the words of U.S. psychologist David Bakan—"children are being tortured daily in their homes by the very people who gave them life."

"It may well be," says Bakan, "that a long time ago in the cave when it was cold and there was not enough food, the father—irritated by the sound of his crying child—picked up the baby and hurled it against the wall. But we are no longer in the cave."

For some children, unfortunately, a modern habitat makes little difference.

Rotarians and other concerned citizens in your community can help, however, by discovering these mistreated youngsters and offering them—and their parents—the kind of social assistance that has proven effective elsewhere.

The battering of helpless children is a sick—and sickening—social practice. Saving the life, the health, the sanity, of just one child is reason enough for any adult to take a stand whenever he can give assistance. Fear, indifference, the desire to "mind our own business," should not be allowed to interfere with the natural, humane desire to help whenever we hear the screams of children or see their battle scars. Where the children suffer—and no one defends them—the society suffers even more.

The Curse of Hyperactivity

Johnny wouldn't go to sleep. Instead, he spent most of the night tearing around the house. When he was tall enough to unhook the screen door, he began to explore the neighborhood, and his frantic parents once found him wandering down the middle of the street in his diapers. On another occasion, he turned on the clothes dryer and climbed inside. At the age of 2, he was expelled from nursery school.

One-year-old Hugh had to be strapped into his highchair, but he still managed to fall over the side. When he began to talk, the words came out so fast no one could understand him. He was a mass of bruises from bumping into anything that stood in his frenetic path. By the time he was 7, he had dislocated his thumb, broken his wrist and fractured his collarbone twice. "He looked like an abused child," his mother recalls.

Steven was the terror of the neighborhood. Once, he went after the boy next door with a golf club. Another time, he tried to strangle a little girl with a jump rope. By the age of 9, he had been expelled from three schools. "We couldn't take him anywhere, and we couldn't leave him alone," says his mother. "I thought I was going crazy."

All three of these children suffer from a disorder so baffling and complex that doctors have had trouble naming it. It's been called "child hyperactivity syndrome," "hyperkinesis," "minimal brain dysfunction" and "attention deficit disorder." By whatever name, it afflicts an estimated 2.5 million American youngsters, or about 5 per cent of the school-age population. Its victims make life a shambles for their parents, turn classrooms into chaos, get into trouble with the police and, as adults, often lead disastrously unhappy lives.

Diets: Hyperactivity has been blamed on everything from brain damage to low levels of sugar in the blood. Treatments range from stimulant drugs to psychotherapy and special diets. Neither the cause nor the cure has yet been found. But because the disorder is so common and serious, researchers are mounting a broad new effort to identify cases and bring them under control. "These kids typically bounce around to all types of professionals," says psychologist Keith Conners of Washington's Children's Hospital National Medical Center. "They need coordinated, centralized long-term management, like epileptics or diabetics."

Much of the current research has been devoted to dispelling misconceptions. The problem was first described by German physician Heinrich Hoffman in a collection of moral tales for children published in 1845. Hoffman epitomized hyperactivity in "Fidgety Phil," the "naughty, restless child" who "grows still more rude and wild." But doctors came to realize that hyperactives aren't simply naughty. A worldwide epidemic of encephalitis during World War I left some children with residual brain damage that caused a distinctive behavior pattern—including impulsivity and agitation. Subsequently, doctors assumed that any child with those symptoms was brain-damaged. When studies failed to show obvious neurologic brain lesions in most cases, the experts covered their embarrassment by using the term "minimal brain dysfunction." "They meant, 'We know the brain damage must be there, but we can't detect it'," says psychiatrist Dennis Cantwell of the University of California, Los Angeles. "It was a wastebasket term to cover everything under the sun."

Researchers recently revised their terminology when they realized that agitated behavior isn't really the hyperactive child's main problem. The hallmark of the disorder actually is a severe inability to focus attention and concentrate. Normally, children learn to disregard most extraneous sights and sounds. But the so-called hyperactive remains aware of the slightest distraction, whether it is a ticking clock or the fluttering of a curtain. To describe such children, the American Psychiatric Association adopted the term "attention deficit disorder" (ADD).

ADD children usually suffer from several specific deficiencies. They may have poor coordination and have difficulty throwing a ball, writing or drawing. They usually have learning disabilities. They also tend to be impulsive and unable to weigh the consequences of their behavior. Finally, they have trouble developing social skills and many become aggressive. Craig, a Salt Lake City boy, disrupted his reading class by whispering loudly to children nearby. On the playground, he got into fights nearly every day. He wasn't invited to play games because he insisted on imposing his own rules. Poor performance in school, inability to make friends and criticism from parents and teachers eventually give the hyperac-

tive child a grotesquely poor self-image. "Underneath the hyperactive behavior you may see a very depressed child," says Dr. Donald J. Cohen of Yale University's Child Study Center.

Squirming: The tell-tale signs of hyperactivity often appear shortly after birth. A normal one-month-old infant will stop moving when he sees someone's face or is presented with a toy. The hyperactive baby moves about and fusses constantly. "Recently, I held a colleague's two-month-old child and he didn't stop squirming for ten minutes," says psychologist Rachel Gittelman-Klein of New York's Columbia-Presbyterian Medical Center. "For the first time, I had the subjective experience of what a hyperactive infant is like."

The early diagnosis of hyperactive behavior is critical. "The family physician has the best chance of diagnosing cases before the secondary problems of low self-esteem and depression become deep-seated," says Cantwell. Unfortunately, pediatricians and family doctors often are too busy to detect the signs of ADD in an infant, or may not be fully aware of them. The pediatrician, Cantwell points out, should be alert for the mother who says her child is running her ragged.

If it hasn't been spotted earlier, hyperactive behavior becomes obvious when the child starts school. His impulsivity and inability to concentrate conflict with the order of the classroom. The teacher's demand that the child pay attention is fruitless, because the inability to concentrate is the main problem. "Before they enter school, most kids learn that when someone talks to you, you stop and listen," says Cohen. "But hyperactive kids have difficulty living up to rules because they are constantly distracted."

Pediatricians once comforted parents by assuring them that their hyperactive youngster would "outgrow" the problem. Experts now reject the theory. A child with the disorder does tend to be less restless as he gets older, but many of his other troubles remain. Drs. Gabrielle Weiss and Lily Hechtman of McGill University followed 64 ADD children over a five-year period. At the ages of 12 to 16, 70 per cent had repeated at least one grade in school, and a third had repeated two or more grades.

The disorder continues through adulthood. Dr. Paul Wender of the University of Utah found that hyperactive men and women tended to be impulsive and unable to make friends or hold jobs. Typically, they were hot tempered, many had drinking problems and some became child abusers as parents. "Their lives were no hits, no runs and plenty of errors," says the Salt Lake City psychiatrist.

Hyperactivity often runs in families. "It isn't only Johnny who's hyperactive," a mother once confided to UCLA's Cantwell. "My husband can't sit still, he won't listen to anything you say, he blows his top and he can't hold a job for more than a year." Such anecdotes prompted Cantwell to look into the family backgrounds of 50 hyperactive youngsters. He found that 16 per cent of the fathers, 10 per cent of the uncles and 12 per cent of all male relatives had, themselves, suffered from the disorder. To rule out the chance that the household environment had brought on the child's behavior problems, Cantwell studied the families of hyperactives who had been adopted. He discovered no such clustering of hyperactive case histories among the adoptive parents, suggesting that nature rather than nurture plays the larger role.

By a ratio of five to one, hyperactivity is most common in boys. "I think it has something to do with the general vulnerability of males to brain damage," says Conners. He theorizes that girls are better protected during the critical stages of development in the womb. Other intra-uterine factors may also cause hyperactivity. Women who smoke more than fourteen cigarettes a day, says Conners, increase their risk of having a hyperactive child.

Neurotransmitters: Such evidence implies that hyperactivity is a physical rather than purely emotional problem. The disorder may involve neurotransmitters, the chemicals that conduct nerve impulses in the brain. Yale's Cohen and Dr. Bennett A. Shaywitz have examined the spinal fluid of hyperactive children and found evidence of a deficiency of the neurotransmitter dopamine. Similarly, Dr. Walid Shekim of the University of Missouri finds signs that the neurotransmitter norepinephrine may also be lacking.

Since there is no cure, the treatment of hyperactivity is both difficult and controversial. Since the 1930s, stimulant drugs have been one of the most important means of controlling the disorder. The simulants—amphetamines, methylphenidate (Ritalin) and pemoline—have the paradoxical effect of reducing restlessness and improving the child's ability to concentrate. What the drugs seem to do is to increase levels of dopamine and norepinephrine in the brain. According to Gittelman-Klein, 90 per cent of the children she treats show some improve-

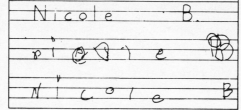

STIMULATING SKILLS

These handwriting samples show how stimulants can help a hyperactive child. The teacher wrote the top line, the child the middle one before taking a stimulant. After drug treatment, the child wrote the bottom line—an obvious improvement.

ment, and 60 per cent respond dramatically.

But the drugs have drawbacks. They curb a child's appetite and may also interfere with his sleep. Many parents are understandably wary about their use. "A lot of them say no 'speed' for my kids," says Conners. Indeed, experts charge that the drugs are sometimes misused. If a doctor prescribes enough drugs to totally quiet the child, he may dull his senses. Even when used properly, the drugs do not improve the child's ability to learn. "There is no magic that will make the child catch up on his reading," says psychologist Robert Sprague of the University of Illinois. "He still has to be taught."

Psychologist James Swanson of the Hospital for Sick Children in Toronto believes that 40 per cent of hyperactive kids shouldn't get drugs at all. Working with neurologist Dr. Marcel Kinsbourne, he tested rote memorization among 600 children taking Ritalin. They found that 60 per cent of the group performed better on the test after taking the drug, but the rest actually fared worse. Swanson thinks that only those hyperactives who have a dopamine deficiency will respond to stimulants. The rest become lethargic when given the drugs because of an excess of dopamine.

Food Additives: One of the most widely publicized treatments is the special diet originated by allergist Dr. Benjamin F. Feingold. According to his theory, espoused in a best-selling book, "Why Your Child Is Hyperactive," hyperactivity is often caused by food dyes, flavoring and preservatives. He claims that a diet that eliminates such additives relieves the symptoms in half the children who follow it. "It's an approach that removes components that were never proven safe in the first place," says the 80-year-old doctor.

The diet is hard to follow for the simple reason that it forbids practically everything that children like. Among the proscribed items are most packaged cereals, commercial baked goods, bologna, hot dogs, manufactured ice creams, candy, mustard and catsup. The child is supposed to restrict himself to such "natural" foods as homemade baked goods, fresh meats, poultry,

fish and plain yogurt. An estimated 200,000 children are on the diet and many parents assert that it works. "Life was hell," recalls Mrs. Gudrun Petersen, whose 17-year-old son was severely hyperactive. "In three days, he was a different boy. Had it not been for the diet, Richard would probably be in jail by now."

About 100 "Feingold Associations" have sprung up around the country to promote the diet. Like Alcoholics Anonymous, they also act as a support group where parents can share their experiences and comfort one another. Meetings help relieve the pervasive guilt parents often feel about having a hyperactive child. "It's great for parents," says one mother, Ann Borchers of San Rafael, Calif. "You can blame it on the additives and not on yourself."

There is little scientific evidence to support the Feingold diet. In one study, Washington's Conners gave children cookies laced with artificial colors for two weeks, followed by two weeks of pure cookies. Neither their parents nor teachers noticed any difference in the youngsters' behavior. A number of other studies have produced negative, or inconclusive, results. Swanson, however, found that children taking capsules that contained food dyes did less well on memory tests than did those getting placebos. At best, says Conners, only about 5 per cent of hyperactive children are sensitive to artificial color.

What hyperactive children need most, according to the experts, is a broad training program that involves the parents as well as teachers. At the simplest level, the parents should supervise the child and try to provide some order in his otherwise chaotic environment. "They can't just tell a hyperactive child to clean up his room and then go shopping," says Dr. Esther Wender, who works with her husband Paul at the University of Utah. "They have to break the task down into smaller units—like telling the child to make the bed, then clean off the dresser and, finally, to pick up his toys."

Shapes: Classroom learning tasks also must be broken down into small segments for the hyperactive child. To help a youngster look at a picture without being confused by it, the teacher might first show him a triangle and then ask him to pick out all the triangular shapes in the composition. By moving slowly, step by step, the teacher enables the child to recognize various shapes and gradually distinguish significant differences. Classes should always be kept small to prevent distraction, the experts agree.

Some of the special schools for such youngsters employ elaborate electronic training aids. The Foundation School in Orange, Conn., uses computers that allow pupils to move at their own pace. Each child sits in his own cubicle before a video-display screen. If he has a problem with visual perception, for example, he may be shown two pictures of boxes that are the same size but of different color. If he recognizes the differences and pushes the appropriate button, the computer screen flashes a "Yes" or "Yeah, You Did It" in bold colors. If he makes a mistake, the machine flashes "Boo."

Behavior therapy, in which the child is rewarded when he performs well, is one of the most effective ways of dealing with hyperactives. Andrew Ackley, 13, of Lake Ronkonkoma, N.Y., is sent to his room for five minutes the instant he misbehaves. But when he does well in school and around the house, his father may buy him a toy. "The number-one priority for parents is to be consistent," says psychologist Susan G. O'Leary of the State University of New York at Stony Brook. "You can't be sloppy with a hyperactive child."

Researchers clearly have a long way to go before hyperactivity can be fully controlled. In the meantime, the burden of handling the millions of children with the disorder will require the diagnostic skill of family doctors, the dedication of teachers and, most particularly, the patience of loving parents.

MATT CLARK with DAN SHAPIRO, MARY HAGER
in Washington, JANET HUCK in Los Angeles,
PAMELA ABRAMSON in San Francisco
and bureau reports

WE WANT YOUR ADVICE

Any anthology can be improved. This one will be—annually. But we need your help.

Annual Editions revisions depend on two major opinion sources: one is the academic advisers who work with us in scanning the thousands of articles published in the public press each year; the other is you—the person actually using the book.

Please help us and the users of the next edition by completing the prepaid article rating form on the last page of this book and returning it to us. Thank you.

The Children of Divorce

I remember it was near my birthday when I was going to be 6 that Dad said at lunch he was leaving. I tried to say, "No, Dad, don't do it," but I couldn't get my voice out. I was too much shocked. All the fun things we had done flashed right out of my head and all the bad things came in, like when he had to go to the hospital with his bad back and when he got mad at me. The bad thoughts just stuck there. My life sort of changed at that moment. Like I used to be always happy and suddenly I was sad.

—An 8-year-old girl

On the surface, the children of divorce don't seem any different from kids whose families are still intact. They wear the same tattered jeans, smile with the same metallic braces, spend mindless hours listening to The Who. But they are different—for divorce, though no longer a stigma, is nonetheless a wrenching series of crises that sets these children apart. While much attention has been devoted to the plight of the single mother, and more recently to the travails of the single father, professionals finally are addressing themselves in a systematic and intelligent way to the dilemma of children of divorce. And the results are startling. Says Albert Solnit, director of the Yale Child Study Center in New Haven, Conn.: "Divorce is one of the most serious and complex mental-health crises facing children of the '80s."

The problem is formidable in numbers alone. There are currently 12 million children under the age of 18 whose parents are divorced—and all in all, around 1 million children a year suffer through the dissolution of their families. The sheer magnitude of this marital falling-out has impelled sociologists, psychologists, family-court judges and educators to explore just how gravely the children are affected. In spite of the comforting old saw that children are flexible enough to learn to "cope" and "bounce back," the evidence suggests that the impact of divorce and the resulting period of adjustment can be both painful and damaging. "The trauma of divorce is second only to death," says child psychologist Lee Salk. "Children sense a deep loss and feel they are suddenly vulnerable to forces beyond their control."

Counseling groups have sprung up in schools, in the courts and in private practice to help children through the transition from two parents to one. New attention is also being focused on the unanticipated problems of step-families (page 198), now referred to in sociological jargon as "reconstituted" or "blended" families.

Scary Stereotypes: The innocent victims of divorce have always been a popular subject. In 1927, Clara Bow and Gary Cooper starred in the movie "Children of Divorce," a melodrama in which the grown-up offspring descend into debauchery and, finally, suicide. The contemporary treatment of the subject deals less with the scary stereotypes and more with the realistic heartbreak of both parents and child. In the climactic custody battle for a 6-year-old boy in the film "Kramer Vs. Kramer," there are no "good guys" or "bad guys," no winners—only losers. The children of divorce have also become a staple of prime-time television serials and a market for specialized books. The ranks of current fiction and nonfiction for children now boast no fewer than 46 titles dealing with family breakups, including a read-aloud picture book written for pre-schoolers called "Mommy and Daddy Are Divorced."

The country's institutions are finding themselves forced to address the problem in new ways. Schools, traditionally reluctant to get involved in a child's life outside the classroom, are now beginning at least to take notice of the special needs of children from divorced families. Many boys take out their frustration and anger in school, becoming bullies or disrupting the classroom. Girls, on the other hand, are apt to become withdrawn and silent. The burden has fallen on teachers to help these students because divorcing parents are often not up to dealing with their children's stress. "When a mother can't get out of bed in the morning or remember to put make-up on, she can't go to a PTA meeting or listen to kids' problems," explains Mary Hirschfeld, a Beverly Hills divorce lawyer-turned-family counselor.

Courts are also struggling with the new complexities of child custody. Whereas 100 years ago children invariably remained with the father after divorce, traditional custody later shifted to the mother, unless she was proved seriously unfit. Now, with more and more parents negotiating for joint custody, with fathers fighting for sole custody and with distressing numbers of parents resisting any custody, family courts are backlogged with long and painful battles. The children are caught brutally in the middle.

Such strain and pressure is spawning a whole new specialization in divorce among family counselors. Research has shown that the brunt of the shock can be lessened, for example, if the children are told about the divorce in a realistic way, if they are reassured about keeping in contact with both parents and if their daily routines are disturbed as little as possible. Professionals also encourage children to air their problems instead of bottling them up. In Lexington, Mass., 30 teen-agers in a self-help group called Divorced Kids Group meet weekly under the supervision of a high-school guidance counselor to sort out their complex feelings and give one another support. "Sometimes I feel I've got the worst deal. My family is so screwed up," says one 17-year-old girl. "Then I come here and someone has far more problems."

Though divorce is not easy on any child, some come through it with few scars. Called "the survivors" by family professionals, such children can become more self-reliant and relaxed with the removal of marital stress. "They get along well with both parents," says Nancy Weston, director of the Divorcing Family Clinic in Santa Monica, Calif. "They have a great time with each of them."

Ultimate Impact: For most of divorce's children, however, the prognosis is uncertain. It is estimated that 45 per cent of all children born in any given year will live with only one of their parents at some time before they are 18. No one is sure what the ultimate impact will be. "Traditionally, the family has been the transmitter of social values," observes Arthur Norton, chief of the Marriage and Family branch of the Census Bureau. "Will instability in the family make social institutions unstable? The jury is still out."

In a way, I thought I'd made it happen. I thought maybe I'd acted mean to my mother and my sister and I was being punished by God. So I tried to be really good by not waking Mom before schooltime and getting my own breakfast and maybe God would change His mind.

6. CHILD REARING AND CHILD DEVELOPMENT

But it's been three years now, and I'm used to it all. Sometimes, when I make a wish with an eyelash, though, I still wish for Dad to come home.

—A 9-year-old girl

Pervasive as divorce may be, few children are prepared for it to happen in their families. A child's world suddenly collapses, and one expert estimates that for 80 per cent of the children, there is no warning at all. Even when the tidings come gently, the reaction is almost universal: shock, followed by depression, denial, anger, low self-esteem and, among preteens, the feeling that they are somehow responsible for the divorce. "The child learns the rules of human relationships in the immediate household," says Donald A. Bloch, director of the Ackerman Institute for Family Therapy in New York. "When the child sees that world splitting up, he feels his world is shattered. His learned rules no longer make sense or are true."

A child's response to divorce varies according to age. Toddlers between the ages of 2 and 4, for example, often regress in their development to a more dependent level, demanding to be fed instead of feeding themselves, reverting to diapers. In this age group, when sexual interest runs high, the removal of the parent of the opposite sex is thought by some psychiatrists and psychologists to be particularly detrimental to the child's sexual development. At the same time, children this age often wish parents of the same sex out of the competition, and when the parent actually leaves, the child is convinced he or she has caused it.

Children between the ages of 6 and 8 also take on the responsibility for the split-up, but they have the additional fears of abandonment and often of starvation. "They are old enough to realize what is going on, but they don't have adequate skills to deal with it," explains John Tedesco, chief psychologist at the Des Moines Child Guidance Center in Iowa. Many experts agree that this is the most critical age for children of divorce—and it is the one with the largest number of children affected.

Some children react by trying to gain control over what they feel to be chaos. One little boy, who used to delight in derailing his toy trains, began carefully running the cars over the tracks day after day. Others, especially boys whose fathers have left, try to become the missing parent, assuming responsibilities that are not appropriate for their young age. One 7-year-old in East Hampton, N.Y., worked himself into exhaustion by staying up all night to protect his mother and sister from burglars. In a study of 26 children of divorce between the ages of 5½ and 7 in Marin County, Calif., social worker Judith Wallerstein was struck by their pervasive sense of sadness. Unlike toddlers who could fantasize that their families were still intact, these slightly older children could not, and they tended to dream up fanciful reasons for the divorce. "They only knew each other two days before getting married and they should have known each other at least nine," one 7-year-old explained to Wallerstein.

Father Figure: Between the ages of 8 to 12, the children's most distinguishing emotion is anger directed at whatever parent is thought to be the initiator of the divorce. A 9-year-old girl accused her mother of kicking her father out of the house. "She's acting like a college kid," the girl told Wallerstein in disgust. "At age 31—dancing and dating and having to be with her friends." The anger can erupt in classrooms and alienate friends just when they are most needed. At the same time, children at this age often form a very close relationship with one friend, or with a teacher or another adult, transferring emotions from the noncustodial parent. But other effects may come to the surface later, cautions Hilary Anderson who founded CHILD (Children Helped in Litigated Divorce) in Chicago. "Often a girl will look for a father figure in people she dates. She looks for someone to take care of her. She gives up."

Teen-agers have a different set of problems. Unlike younger children, they feel little sense of blame for the separation of their parents, but they are saddled with what Tedesco calls the "loyalty dilemma." "Mom doesn't want me to like Dad, and vice versa," says 14-year-old Hilary Brodley of Chicago, eighteen months after her parents' separation. "She tells me bad stories. He tells me others. I'm always in the middle."

The sex of the child also plays a part in the impact of divorce. According to University of Virginia psychologist E. Mavis Hetherington, who recently completed a study on 72 divorced middle-class families, boys are the harder hit. More is expected of them, she explains, and they receive far less support from their mothers, teachers and peers as a result. The boy may then begin a destructive circle of bullying other children, then crying when they hit back. As a consequence of alienating boys his own age, the child turns to younger boys or to little girls, learning feminine rather than masculine play patterns. A little girl, on the other hand, vents her sadness by crying—literally—for attention. "When she whines, she is usually helped," says Hetherington.

Cutting through all age and sex distinctions is an obsessive desire to reunite the parents. One 2½-year-old in Chicago spent fretful hours trying to place his father's hand in his mother's hand. A 9-year-old New York girl spent all winter without a jacket on, trying to get sick enough so her parents would have to care for her—together. "All I did was get a lot of colds," she says ruefully. Even though her parents have been separated for four years, a 13-year-old girl in Virginia is adamant that they not get a divorce. "Once all the papers are signed, I wouldn't have a chance," she says. "Now I have hope."

You never feel permanent anymore. I feel like an animal with a mind. You have to spend so much time with each person. You go from place to place. And I don't feel at home at Dad's. I feel very strange when his girlfriend is around. I think of it as being her fault.

—A 15-year-old girl

Adjusting to postdivorce life between two homes and two parents, often in reduced circumstances, is complicated and confusing. The costs of litigation and maintaining two households can reduce the family's income enormously, and a recent Census Bureau report found that only 25 per cent of divorced, separated or single mothers receive child support. One financially pressed divorcee in Los Angeles had to move her three children across town into her parents' one-bedroom house. In that single move, her oldest child lost not only his father and his home, but also his school and his friends. After constantly getting into fights at his new school and failing at his schoolwork, he entered therapy just in time. "The child was suicidal because he had swallowed all the guilt and couldn't handle all the change," explained Los Angeles counselor Jim Larson.

Children involved in custody battles are the most torn. Easily swayed by appeals from either parent, children are often duped into switching allegiances. One 13-year-old California girl was told to choose which parent she wanted to live with. She chose her mother because she had been promised her own phone. But the phone never materialized. "Christmas came and there was no phone," says the girl who is now in a counseling group. "She bribed me. Dumb. So I went back to Dad. Then we all started fighting. I was switching back and forth. The court finally said I was totally confused."

To avoid the no-win pain of custody fights, more and more parents are turning to joint custody, sharing equal responsibility in caring for the children. Eight states now have joint custody provisions. Ordinarily, this means that the children shuttle between their parents' homes. In Houston, Texas, Heath Ruggles, 6, and his older brother Tracy, 9, spend one night with their mother, the next with their father and stepmother, and alternate weekends. The four-year-old experiment seems to be working—almost. Heath's nursery school report card last year found him exemplary in all ways except for one confusion: he wasn't sure where he lived. Chuck and Joan Mathison of Long Beach, Calif., came up with an imaginative solution to this problem: they move and the children stay. They spend alternate years living with their two children Todd, 10, and Lisa, 11, in their old house, relinquishing one night a week and weekends to the off-year parent. Psychologists and the courts are divided about the wisdom of joint custody. "One group of psychologists tells us the best course is to shuttle the child back and forth between the parents on an equal basis," says Marvin Freeman, recently retired supervising judge at the Los Angeles Family Law Court. "The others think they should stay put in one home. At least they both agree that the kids should have their own bed wherever they go." The arrangement also leads to problems when the two parents disagree over a matter such as schooling or medical treatment: at times the courts have to break these deadlocks.

Embittered parents who don't think the courts will give them a fair shake sometimes resort to kidnapping their own children. About 25,000 children a year are snatched or hidden from one parent by the other. Children's Rights, Inc. (CRI), a Washington-based clearinghouse for information on parental kidnapping, estimates that if snatched children aren't found within six months, they probably won't be found for years. Parents on the run move frequently and often don't enroll the children in school for fear of being traced. "Child-snatching still amazes me," says Rae Gummel, CRI co-director. "It's like a bad soap opera."

Each man my mother went out with I considered my next stepfather. And with every one I'd try to be that much more caring so that he'd like me and we'd get off to a good start. I finally realized she was just having fun and didn't want to get married again right now. So I bagged the whole thing. I was exhausted trying to be a son to each one. Her boyfriends became just people.
—A 15-year-old boy

The return to the dating game is invariably a chaotic time for both parent and child. Both men and women feel compelled to join a social whirl to re-establish their attractiveness, leading to a temporary neglect of the children. The children can feel abandoned as a result—but may store up their resentment. "It can burst out later," says Chicago divorce attorney Burton Zoub. "Years later they'll say, 'You never came to see me during that time'."

Parental dating often poses a threat to the children who see the new partner as replacing the departed parent. The children tend to feel as if they're being disloyal—and may be jealous over the newcomer's place in their parent's affections. But over a period of time, children can enjoy—and adjust to—their parent's friends if they feel included in the relationship. "I'd be furious with you if you had a secret lover we didn't know about," one 9-year-old girl told her mother. "But Larry is a friend to all of us."

Juggling sexual activity and parenting also poses a challenge, both to divorced parents and their curious offspring. One divorced mother in Washington worked it out by trading her sons off every other weekend with another divorced woman, leaving one house child-free. But on the whole, children seem far less disturbed by their parents' sexual activity than the parents do. "I caught one guy sneaking out of the house at 5 a.m.," said one 12-year-old boy in disgust. "Why didn't he want to have breakfast with me? It was tacky." Indeed, many children hope for even the appearance of a normal family. Eleven-year-old Eddie Coleman in New York City is happy his father's girlfriend has moved in. "Before, all we had was a grownup, a boy and a little black cat in this big house," says Eddie. "It feels nice to have a family."

No matter how hard I try to erase the idea, I really want to get married. Even with my parents and all my grandparents divorced, I believe in commitment. In fact, I want a huge wedding with bridesmaids, a partner for my whole life and a family. It's a

challenge and the optimism I have is funny to me. But if I'm lucky, I'll have that sense of continuity.
—A 23-year-old woman

The verdict is still out on how this generation of divorced children will fare in their own marital lives. Hetherington, an acknowledged expert in the field, is pessimistic after her work with divorced families. She predicts three out of four children of divorce will repeat the pattern. Wallerstein is more optimistic because the children she sees still believe a mother, father and children living under the same roof is the norm—and a worthwhile goal.

Muddling Through: How long is the period of adjustment for children of divorce? That also varies with the age of the child and the stability of the child's life following separation. Hetherington concludes that the impact dwindles away a year or two after the separation. "All of the family members tend to get worse before they get better," she says. "An important message to understand is that, even though the situation seems to be deteriorating, it improves very suddenly." The adjustment takes longer, according to Wallerstein. In the study she co-authored with psychologist Joan Kelly, she found that five years after the separation, a third of the children seemed to be resilient; an equal number seemed to be muddling through, coping as they could, and the rest were bruised, looking back to life before the divorce with intense longing.

Some institutions and professionals are now trying to ease the painful passage through divorce for the children. A training program for teachers on how best to cope with divorced parents and their children is being offered through the McLean Institute for School Consultation in Belmont, Mass. Many schools, such as the public schools in East Hampton, N.Y., have instituted programs to spot troubled children early. Utilizing dolls, clay and drawing to reach the children, the East Hampton program has proved very beneficial. "Jake," a second-grader, refused to believe his father had left home even though the father had moved to California. "He withdrew by becoming a machine in 'Star Wars' rather than Jake at home," says counselor Polly Haessler. Finally Jake admitted to himself that his father was indeed gone when he built a house out of blocks, put in mother and father dolls, then took the father doll out.

Best Interest? Courts are turning more and more to outside professionals for help in deciding custody and visitation cases. Instead of the old bitter adversary proceeding where lawyers fought for their clients regardless of the best interest of the child, some states are either referring such cases to conciliation courts, where the families can work out a compromise with an impartial third party, or appointing a guardian to represent the child. In contested custody cases in Wisconsin, the court is required by law to appoint an attorney for the child. The lawyer interviews teachers, clergy, neighbors and friends for guidance as to which living arrangement is best for the child. The attorney's fee is paid by the parents if they are able, or by the community.

In Los Angeles, where the conciliation court successfully mediates 55 per cent of the cases, the court employs five therapists for custody evaluations as well as a staff of custody investigators—social workers with master's degrees—who do family studies to determine the child's best interest. "They solve more problems than all the lawyers and judges put together," says Judge Freeman.

Kidnapping one's own children may well be on its way to being a Federal offense. States now vary widely on child-snatching laws. While 39 states have enacted the Uniform Child Custody Jurisdiction Act, the remaining eleven are potential havens for parental kidnappers. The Parental Kidnapping Prevention Act, now in Senate hearings, would honor and enforce custody and visitation decrees of other states, and make it a Federal misdemeanor to restrain or conceal a child in violation of a custody or visitation decree. Some think the cure may be worse than the disease. Criminalization "may increase the potential for violent confrontation and emotional trauma, if not physical danger to the child," FBI executive assistant director Lee Colwell warned in Senate hearings last week.

Mock Trial: Counseling programs are also helping to mend familial fences and build support systems for children who withdraw into emotional isolation. In Hennepin County, Minn., the Department of Court Services offers a free program called The Divorce Experience for families about to be dissolved. The first meeting explains the court system and divorce laws to parents, the next two explain the process to the children, having them take part in a mock trial. In Rolling Meadows, Ill., a peer-group counseling program for divorced kids uses such exercises as falling backward off a 6-foot-high "Trust Log" into the hands of the group and scaling a 14-foot-wall using nothing but each other for help. The program has been so successful that social worker Toby Landesman experimented with a mother-daughter group. "A lot of kids think they had something to do with the divorce, so we build bridges back home," says Landesman.

The brunt of divorce can be eased from the beginning if the children are properly told and reassured about their futures. Child psychologist Lee Salk advises parents to tell the child about their plans together if possible and not to hide their own distress. "The most important thing parents can do is to tell the child that even though they are divorcing each other, they are not divorcing the child," says Salk, himself the divorced father of two. Richard A. Gardner, a New Jersey psychiatrist and the author of five books on divorce, suggests that the children be told the real reasons for the split-up. "Just telling the child that 'We don't love each other anymore,' is a cop-out," says Gardner. "Tell them the real reasons, like, 'Your father drinks too much,' or 'I've met someone else I care about more.' But no matter what deficiencies are presented to the child, the parent must present the other parent's assets as well."

Though divorce is upsetting to everyone involved, it is equally disturbing, perhaps worse, for children to live in an embattled household. And there are some who profess that divorce can be positive for children. "All crises provoke tension and behavioral difficulty, but they can also be learning and growing experiences," says sociologist Lenore Weitzman, director of the California Divorce Law Research Project. "No one knows how much disruption it causes for unhappy families to remain intact. No one's done that study."

No Role Models: For better or worse, divorce continues to split families at an alarming rate. The number of children involved in divorce has tripled in the last twenty years. And though parents, children and professionals are struggling to deal with such new domestic realities as single-parent families, there are no longstanding precedents, no established role models to draw from. Divorce and its aftermath can be a labyrinth of confusion and conflict, some of which may never be resolved. To divorce lecturer and author Rabbi Earl Grollman of Temple Beth El in Belmont, Mass., divorce can be even more traumatic then death. "The big difference is, death has closure, it's over," says Grollman, who performs divorce ceremonies for families. "With divorce, it's never over."

LINDA BIRD FRANCKE with DIANE SHERMAN in Washington,
PAMELA ELLIS SIMONS in Chicago, PAMELA ABRAMSON
in San Francisco, MARSHA ZABARSKY in Boston, JANET HUCK
in Los Angeles and LISA WHITMAN in New York

After Remarriage

There are four of hers and three of his and one of theirs. His ex-wife has remarried and gained two of her new husband's. Her ex-husband has also remarried into a ready-made family, adding two more stepsons and stepdaughters. And there are four sets of grandparents, not to mention assorted aunts, uncles and cousins. On the "geneogram" at the Stepfamily Foundation, Inc., in New York, the family structure looks like the organization chart of a multinational corporation. But to counselor Jeannette Lofas, president of the foundation, this is the norm. "There are only 22 characters in this cast," laughs Lofas. "This is going to be an easy one to work out."

The nation's swelling step-population now includes some 12 million stepparents and 6.5 million stepchildren under 18. And though people like to think that such blended families live in rewedded bliss, the bliss can be short-lived. The impact of remarriage on a family, regardless of how high the expectations, is second only to the crisis of divorce. Because there are no guidelines for acceptable step-family behavior, at least one expert attributes the higher rate of divorce in second marriages—40 per cent as against 33 per cent in first marriages—to the strain of trying to work it all out. Says Andrew Cherlin, assistant professor of Social Relations at Johns Hopkins University: "It takes a real emotional toll to try to cope with the problems that people in first marriages take for granted."

Brunt: The problems that arise in remarriage are all the more devastating because, for the most part, they are unexpected. The children's secret and often subconscious longing to reunite their real parents is shattered and they are faced with a new parent they neither want nor have room for emotionally. The unsuspecting stepparent can bear the brunt of anger the children have stored up toward the parent they feel deserted them. "We had a terrific time until Sam's mother and I were married," says a New York attorney. "Then slam. Sam all but pretended I was invisible."

Children entering a step-family can feel twice defeated, once for having been unable to prevent the divorce, and again for not being able to prevent the remarriage. Children often form especially strong bonds with a single parent. The arrival of a new partner, particularly for the father, weakens these bonds and the children feel abandoned yet again. "There is a tendency for the new wife to take priority over a man's time, his children, his money," says anthropologist Paul Bohannan, who conducted a survey on stepfathers for the National Institute of Mental Health. "She says the kids can't come for the weekend and he goes along with it."

Juggling different rules and values while still trying to please both sets of parents also places an exhausting burden on the children. "What always happens is that if I get into trouble at school or if my grades are bad," says 14-year-old Janey Harris, who lives in Marin County, Calif., "I'll go through the whole hassle with Mom and stepfather. Then I'll think it's all over and I have to go through it again with Dad. I get tired of it." Children can also feel torn by a divided sense of loyalty. "If we go to Disneyland or something like that with our real dad," says Alex Grishaver, 12, "we don't tell our stepfather because we don't want him to feel bad for not taking us to places like that."

To add to the confusion, children of remarriage often inherit an instant set of new stepsisters and stepbrothers, relationships they are not prepared for. "Not only do the kids not get the pay-

(continued on following page)

(continued from preceding page)

offs in the remarriage that the parents get, but they pay a price," says Dena Whitebook, director of counseling at the American Institute of Family Relations in Los Angeles. "The long-hoped-for brother turns out to be a pig and they find themselves sharing a bathroom with a stranger." At the other extreme, some older step-siblings hit it off too well and start relationships of their own, a tricky dilemma that even the experts find confusing. At a seminar attended by 70 marriage counselors last year the participants were asked to vote on whether step-siblings should be allowed to date. The audience refused to declare either way.

'Super Mom': To help step-families through the maze of both expected and unexpected problems, workshops, discussion groups and candid how-to books are becoming part of the family therapy network. In her book, "Instant Parent," stepmother Suzy Kalter confesses she ended up with an ulcer and a failing marriage until she stopped trying to be "Super Mom." The nearly disastrous experience of New York stepmother Jeannette Lofas, whose four stepdaughters stopped talking to her after she married their father eight years ago, led her to co-author the book "Living in Step" and to found the Stepfamily Foundation, Inc., in 1975. Lofas has worked with 120 step-families and lost only two to divorce. "The usual problem is that the adults have different parenting styles, resulting in the total nonstructuring of the family," says Lofas, whose relationship with her stepdaughters is now strong. "There are not clear-cut chores. No one knows where they fit."

Despite the problems that stepfamilies face, five out of six divorced men and three out of four women remarry within three years. At the Stepfamily Foundation of California, Inc., psychologist Emily Visher and her husband, John, a psychiatrist, are cautiously optimistic about the prospects of these remarriages. "Children have seen the disruption of adult relationships either through death or divorce and now they have the opportunity of seeing a couple working together in a positive way," says Visher, co-author with her husband of the book "Stepfamilies." "That can give them faith in their own future." The way to the future may be rough. "In remarriage there is no honeymoon," says John Visher, himself a stepfather of twenty years' standing. "There is instant pandemonium rather than the gradual progression of first-family marriages. Remarriage does not mean that you'll live happily ever after."

LINDA BIRD FRANCKE with
MICHAEL REESE in San Francisco

Go Get Some Milk and Cookies and Watch the Murders on Television

Daniel Schorr

Daniel Schorr, formerly CBS national and foreign correspondent, is now senior correspondent of Ted Turner's Cable News Network. Donna Rockwell contributed to this article.

I believe television is going to be the test of the modern world, and that in this new opportunity to see beyond the range of our vision we shall discover a new and unbearable disturbance of the modern peace or a saving radiance in the sky. We shall stand or fall by television—of that I am quite sure.
　　　　　　　　　　—E.B. White (1938)

John W. Hinckley Jr. causes me to reflect, having recently turned 65, on what the media age has wrought. Hinckley's unhappy lifetime of some 26 years coincides roughly with my life in television. Whatever else made him want to shoot a President, Hinckley epitomizes the perverse effects of our violence-prone culture of entertainment.

Hinckley weaves together strands of media-stimulated fantasy, fan frenzy, and the urge to proclaim identity by starring in a televised event. His success is attested to by everything that has happened since March 30, when he managed to disrupt the regular programs listed in his copy of *TV Guide* to bring on command performances by Dan Rather, Frank Reynolds, Roger Mudd, and the other news superstars. Since November 22, 1963, these electronic special reports—the modern equivalent of the old newspaper extra—have been America's way of certifying a "historic event."

Much has been shown to Hinckley's generation to lower the threshold of resistance to violent acts. When the time came for Hinckley to act—to plug himself into this continuum of television and movie violence—the screenplay was easily written, the roles nearly preassigned. The media-conscious "public" President, Ronald Reagan, attracted the cameras, which attracted the crowds, which provided both the arena and the cover for the assailant. The network cameras routinely assigned, since the Kennedy assassination, to "the presidential watch" recorded the "actuality" and showed it in hypnotic, incessant replays. The audience tingled to the all-too-familiar "special report" emblazoned across the screen.

To nobody's surprise, the celebration of violence stirred would-be imitators. The Secret Service recorded an astonishing number of subsequent threats on the President's life. One of them came from Edward Michael Robinson, 22, who had watched the TV coverage and later told police that Hinckley had appeared to him in a dream, telling him to "bring completion to Hinckley's reality."

Psychiatrist Walter Menninger examined Sara Jane Moore, who tried to kill President Ford in 1975, and found it no coincidence that two weeks earlier a well-publicized attempt on Ford's life had been made by Squeaky Fromme.

"There is no doubt," Dr. Menninger told me, "of the effect of the broad, rapid, and intense dissemination of such an event. The scene in front of the Washington Hilton must have been indelibly coded in everybody's mind with an immediacy that does not happen with the print media. We have learned from the studies of television that people do get influenced by what they experience on television."

The broadcasting industry says it can't help it if occasionally a disturbed person tries to act out depicted violence—fictional or actual. In 1975, a Vietnam veteran in Hyattsville, Maryland, who had told his wife, "I watch television too much," began sniping at passersby in a way he had noted during an episode of *S.W.A.T.*—and, like the fictional sniper, was killed by a police sharpshooter.

The American Medical Association reported in 1977 that physicians were telling of cases of injury from TV imitation showing up in their offices and hospitals. One doctor treated two children who, playing Batman, had jumped off a roof. Another said a child who had set fire to a house was copying an arson incident viewed on television.

No court has yet held television legally culpable for the violence it is accused of stimulating. In Florida in 1978, fifteen-year-old Ronny Zamora was convicted—after a televised trial—of killing his elderly neighbor despite the novel plea of "involuntary subliminal television intoxication." The parents of a California girl who had been sexually assaulted in 1974 in a manner depicted three days earlier in an NBC television drama lost their suit against the network.

That's as it should be. I support the constitutional right of the broadcasting industry to depict violence, just as I support *Hustler* magazine's right to depict pornography—with distaste. As Jules

Feiffer, the cartoonist and civil libertarian, has noted, one sometimes finds oneself in the position of defending people one wouldn't dine with. What troubles me, as I reflect on the case of John Hinckley, is the reluctance of television to acknowledge its contribution to fostering an American culture of violence, not only by the way it presents fantasy but by the way it conveys reality—and by the way it blurs the line between the two.

Violence is one of the manifestations of the quest for identity. When you've lost your identity, you become a violent person looking for identity.
—Marshall McLuhan (1977)

In 1974 Reg Murphy, then editor of the *Atlanta Constitution* (he is now publisher of the *Baltimore Sun)*, was kidnapped. He says his abductors immediately sped to an apartment and turned on a TV set to see whether their act had made the evening news.

In 1971 prison rioters in Attica, New York, listed as a primary demand that their grievances be aired on TV.

In 1977 in Indianapolis, Anthony George Kiritsis wired a sawed-off shotgun to the neck of a mortgage company officer, led him out in front of the police and TV cameras, and yelled: "Get those goddamn cameras on! I'm a goddamn national hero!"

In 1974 in Sarasota, Florida, an anchorwoman on television station WXLT said on the air, "In keeping with Channel 40's policy of bringing you the latest in blood and guts in living color, you're going to see another first—an attempt at suicide." Whereupon she pulled a gun out of a shopping bag and shot herself fatally in the head.

These incidents—the list could go on and on—were all aspects of the phenomenon of the mass media as grand arbiter of identity, validator of existence. Descartes might say today, "I appear on television, therefore I am."

One becomes accustomed, after working a long time in the medium, to hearing strangers remark, without elaboration, "I saw you on television!" One even gets inured to being hauled over to meet somebody's relatives. It is as though the TV personality has an existence of its own. I experienced the other side of this phenomenon in 1976 when I stopped broadcasting for CBS. People asked, solicitously, if everything was all right—as though, being off the air, I had ceased to be in some existential sense.

"Getting on television" has become a preoccupation of people in govern-

ment, politics, and industry, not to mention all manner of single-issue advocates. Candidates will fashion their campaigns around "photo opportunities." Senators will be drawn by the presence of cameras to legislative hearings they otherwise would skip.

Many people will do almost anything to get on TV. Some will even kill.

Anthony Quainton, former head of the State Department's Office for Combating Terrorism, associates the increase in casualties during hijackings and hostage-takings with the desire of terrorists to insure news-media attention. Deliberate acts of horror—like the tossing out of slain victims—are planned as media events. On the other hand, the failure of the hijacking of a Turkish plane to Bulgaria in May was at least partly due to the fact that two of the terrorists had left the plane to give a press conference.

Sometimes the aim is to hijack television itself. When the radical Baader-Meinhof gang in West Germany kidnapped a politician in 1975 as hostage for the release of five imprisoned comrades, it forced German television to show each prisoner boarding a plane and to broadcast dictated propaganda statements. "For 72 hours we lost control of our medium," a German television executive later said.

When Arab terrorists seized the Vienna headquarters of OPEC in 1975, killing three persons and taking oil ministers hostage, the terrorists' plan called for them to occupy the building until TV cameras arrived.

A central feature of the plan of the San Francisco "Symbionese Liberation Army," which kidnapped Patricia Hearst, was the exploitation of the media—forcing radio and television to play its tapes and carry its messages.

The Hanafi Muslims' hostage-taking occupation of three locations in Washington in 1976 was a classic case of media-age terrorism. The leader, Hamaas Abdul Khaalis, spent much of his time giving interviews by telephone, while his wife checked on what was being broadcast.

"These crimes are highly contagious," warns Dr. Harold Visotsky, head of the department of psychiatry at Northwestern University. "Deranged persons have a passion for keeping up with the news and imitating it."

It does not seem to matter much if they are keeping up with "the news" or with "entertainment," for more and more the distinction is thinly drawn. A real attempt on the President's life produces a rash of threats. A prime-time drama

about a bomb on an airplane produces a rash of reports of bombs on airplanes.

In all of this, television claims to be innocent—a helpless eyewitness, sometimes even a hostage. It's not that simple.

To begin with, television has helped blur the lines between reality and fantasy in the general consciousness.

Television news itself—obliged to co-exist with its entertainment environment, seeking to present facts with the tools of fantasy—ends up with a dramatized version of life. Everything that goes into making a well-paced, smoothly edited "package" subtly changes reality into a more exciting allegory of events. The confusion is compounded by the use of "cinéma réalité" techniques in fictional dramas, and the modern forms of fact-and-fiction "docudramas" and "reenactments" of events.

It began to come home to me that audiences were blurring the distinction between reality and entertainment when I received telephone calls from several persons, during the 1973 Senate Watergate hearings that preempted soap operas, asking that the networks "cancel" a boring witness and "put back John Dean and his nice wife." Moreover, some friends of mine praised a "documentary" shown by NBC, *The Raid at Entebbe*, and had to be reminded that it was a reenactment.

The gradual erosion of the line between fact and fantasy, between news and theater, can have serious consequences. People slow to react to accidents and muggings may be experiencing the existential question of whether these things are really happening. A woman wrote columnist Abigail van Buren of being bound and gagged by a robber who told the victim's four-year-old boy to watch television for a while before calling for help. The child looked at TV for the next three hours, ignoring his mother's desperate efforts to get his attention. Perhaps, to the child, the show was more real than his mother's muffled screams.

Having obscured the difference between fantasy and reality, television offers incentives to people who are seeking emphatic ways of getting recognition. Innocent hand-waving, as an attention-getting device, yields to demonstrations, which in turn yield to riots.

In my own experience, covering urban unrest for CBS in the 1960s, threatening rhetoric tended to overpower moderate rhetoric and be selected for the network's *Evening News* because it made "better television." I have no doubt that television helped to build up militant blacks

6. CHILD REARING AND CHILD DEVELOPMENT

like Stokely Carmichael and H. Rap Brown within the black community by giving them preferred exposure. Nonviolent leaders found themselves obliged to escalate the militancy of their own rhetoric. When Martin Luther King Jr. came to Washington in 1968 to discuss plans for the "poor people's march" that he did not live to lead, he told me he had to allude to possibilities for disruption as a way of getting media attention.

At a community meeting after the first night of rioting in the Watts area of Los Angeles in 1965, most of those who spoke appealed for calm. But a teenager who seized the microphone and called for "going after the whiteys" was featured on evening TV news programs. A moderate commented, "Look to me like he [the white man] want us to riot." Another said, "If that's the way they read it, that's the way we'll write the book."

In recent years, television news, compelled to come to terms with its own potency, has sought to enforce guidelines for coverage of group violence.

Television tries to guard against being an immediate instigator of violence, but its reaction is too little and too late to overcome the cumulative consequences of a generation of depicted violence. It is like trying to control proliferation of nuclear weapons after distributing nuclear reactors over a prolonged period.

The most important thing is that a causal relationship has been shown between violence viewing and aggression.
—Dr. Jesse Steinfeld, Surgeon General of the United States (1972)

At St. E's, the Patients Thought Hinckley "Was Nuts."

The average American watches television for four hours and 30 seconds every day, according to A.C. Nielsen figures. Women watch the most: four hours and 47 minutes a day. Men watch four hours and six minutes. Children age two to eleven watch three hours and 52 minutes a day, and children age twelve to seventeen watch the least: three hours and seventeen minutes.

For many Washingtonians, television is kept in its proper place and perspective: Research shows that Washingtonians read more and watch less television than residents of any other major city in the country. But television is used increasingly as a babysitter or an opiate in institutions. To find out how much television is watched by those who might have trouble discriminating between television and real life, we surveyed the TV habits at five area institutions:

At St. Elizabeths Hospital, mental patients are permitted to watch unlimited television. Social worker Helen Bergman, who deals with men and women aged 25 to 35, says the television is on in the patient lounge all day long. Patients watch soap operas during the day, and in the evening they vote when there's a conflict over which show to watch. Bergman says that many patients are upset by excessive violence, and that some of the more disturbed patients talk to the television and laugh inappropriately at it. She personally dislikes television because it discourages patient interaction. One staff member says the employees watch as much TV as the patients and would be unhappy if its use were restricted.

Patients are encouraged to watch news events, and they were particularly interested in the coverage of the Reagan shooting. Bergman recalls that one patient remarked, "Boy, was he nuts,"

in reference to John Hinckley.

The Cole Residence in Northeast DC is a group home for boys 16 to 18 who are awaiting trial for minor offenses. Rick Bricher, assistant administrator, says no restrictions are placed on television viewing. Bricher says the staff encourages residents to watch special programs, particularly those that focus on black issues. Sports programs are popular, as well as network programs featuring black actors, such as *The Jeffersons*. What will be watched is determined by majority rule.

Inmates at DC's Lorton Reformatory are permitted to watch unrestricted television. The set is on every day from around noon until 11 PM, except when inmates are being counted. Salanda Whitfield, a Lorton administrator, says each dormitory has a 25-inch color set and the inmates vote on what to watch. Because inmates work on different schedules, someone is watching television all the time. Soap operas, sports, police, and adventure shows are the most popular. Some of the inmates watch the local news to find out who got caught doing what, because they often know the people involved in area crime. Occasionally they speculate on who might be the perpatrator of an unsolved crime. When the Supreme Court is in session, many inmates watch the Monday-night news to see if any decisions affecting their cases have been handed down.

Whitfield says inmates admire the "flashy types" in action shows. He doesn't think Lorton inmates are sophisticated enough to pick up any new ideas from television criminals, though they might get a new "wrinkle."

Dr. Martin Stein, an administrator at the Dominion Psychiatric Treatment Center in Falls Church, says the use of television is an area of great concern to the facility's staff. The patients, primarily adolescents, are not restricted in what they watch. However, a busy schedule, which includes a full day of school, leaves little time for television. Stein adds that the center does not want to shelter patients from normal activities and that the time and effort of monitoring television could be put to better use by the staff. Like Bergman at St. Elizabeths, Stein expresses concern that television hinders patient interaction.

Stein says the patients prefer comedies such as *M*A*S*H* and *Fantasy Island* to drama and action shows. They tend to avoid programs that contain excessive violence, and become anxious when such programs are on. According to Stein, schizophrenic patients often think the television is talking to or about them or sending them special messages.

For children aged four to ten at the Fairfax Brewster School, a private school for normal students at Bailey's Crossroads, the *Dukes of Hazzard* is the overwhelmingly favorite show. Nearly all named a character on that show when asked who they would be if they could be a television character. Sports were also popular, along with *Bugs Bunny*, *Woody Woodpecker*, and *The Greatest American Hero*. The children disliked the news (boring), soap operas, and *The Incredible Hulk* (dumb). Out of seven children, only one had a parent who specified the programs she could and could not watch. Most watched some programs with their families and more than half frequently ate dinner in front of the television.

When asked the type of program he enjoyed most, one nine-year-old said he liked shows in which stuntmen were shot or pushed over cliffs because "it's neat how they don't bleed or get hurt."
—HEATHER PERRAM

For three decades, since the time when there were 10 million TV sets in America, I have watched efforts to determine objectively the effects of televised violence while the TV industry strove to sweep the issue under the carpet.

What television hated most of all to acknowledge was that violence on TV was not incidental or accidental but a consciously fostered element in the ratings race. In 1976 David Rintels, president of the Writers Guild in Los Angeles, where most of the blood-and-guts scripts are spawned, told a congressional committee: "The networks not only approve violence on TV, they have been known to request and inspire it.

"There is so much violence on television," he said, "because the networks want it. They want it because they think they can attract viewers by it. It attracts sponsors. Affiliate stations welcome it."

A personal experience brought home to me the industry's sensitivity to the subject. In January 1969 my report for an *Evening News* telecast, summarizing the interim findings of the National Commission on the Causes and Prevention of Violence, was altered shortly before air time at the direction of Richard N. Salant, president of CBS News, to eliminate a comment about television. The passage cited the commission's view that while "most persons will not kill after seeing a single violent television program, . . . it is possible that many learn some of their attitudes about violence from years of TV exposure and may be likely to engage in violence." For management to override the news judgment of the "Cronkite show" was extremely rare.

Riots and assassinations would bring the issue periodically to the fore, but the research had been going on for a long time. For more than a quarter of a century social scientists have studied the effects of violence-viewing—especially on children.

■ At Stanford University, Professor Albert Bandura reported that children three to six years of age whose toys were taken away after they had seen films showing aggression would be more likely to pound an inflated doll in their frustration than children who had not seen such films.

■ A Canadian study by R.S. Walters and E. Llewellyn Thomas found that high school students who had viewed aggressive films were more likely than others to administer strong electric shocks to students making errors on an exam.

■ An experiment conducted in Maryland for the National Institute of Mental Health found serious fights in school more common among high school students who watched violent TV programs.

■ Bradley Greenberg and Joseph Dominick, studying Michigan public-school pupils, found that "higher exposure to television violence in entertainment was associated with greater approval of violence and greater willingness to use it in real life."

■ Drs. Dorothy and Jerome Singer of Yale University concluded from an exhaustive series of interviews that the children who watched the most television were likely to act most aggressively in family situations. Although they could not produce a "smoking gun" that would influence the TV industry, they argued that they had eliminated every other factor that could account for the high correlation between aggressive behavior and viewing of "action-oriented" shows.

■ Dr. Leonard Berkowitz of the University of Wisconsin, in two experiments ten years apart, found that third-graders watching a great many violent programs were likely to be rated by other pupils as high in aggressive behavior and that, at nineteen, most of them were still described as "aggressive" by their peers. In fact, reported Dr. Berkowitz, the amount of television viewed at the age of nine is "one of the best predictors of whether a person will be found to be aggressive in later life."

Congress took an early interest in the question of violence in TV programs. In 1952 the House Commerce Committee held hearings on excessive sex and violence on television. Senate hearings on TV violence and juvenile deliquency, conducted by Senators Estes Kefauver of Tennessee and Thomas Dodd of Connecticut, stirred episodic public interest. The hearing transcripts make a tall stack, adding up to fifteen years of congressional alarm over television, and industry reassurance that it was addressing the problem.

The controversy over television assumed a new dimension of national concern in the wake of the urban riots and assassinations of the 1960s. In 1968, after the assassination of Robert Kennedy, President Johnson named a commission, headed by Dr. Milton Eisenhower, to inquire into the causes of violence and how it might be prevented.

Between October and December 1968, the Eisenhower Commission held hearings on television, questioning social scientists and industry executives about the extent to which the medium might be the instigator or abettor of violent acts.

One commission member, Leon Jaworski, later to be the Watergate prosecutor, expressed the belief that television might have "a tremendous responsibility" for violence in America.

The television networks acknowledged no such responsibility. When Commissioner Albert E. Jenner asked whether "the depiction of violence has an effect upon the viewer," Dr. Frank Stanton, president of CBS, replied: "It may or may not have. That is the question we don't have the answer to."

Nevertheless, the commission decided to formulate an answer. After a long debate—from which Lloyd N. Cutler, the executive director, disqualified himself because of his law firm's TV-industry clients—the panel declared in its final report that it was "deeply troubled by television's constant portrayal of violence . . . pandering to a public preoccupation with violence that television itself has helped to generate."

The panel's report concluded: "A constant diet of violence on TV has an adverse effect on human character and attitudes. Violence on television encourages violent forms of behavior and fosters moral and social values in daily life which are unacceptable in a civilized society. We do not suggest that television is a principal cause of violence in our society. We do suggest that it is a contributing factor."

A two-volume report of the commission's "Task Force on Mass Media and Violence" concluded that, as a short-range effect, those who see violent acts portrayed learn to perform them and may imitate them in a similar situation, and that, as a long-term effect, exposure to media violence "socializes audiences into the norms, attitudes, and values for violence."

The Eisenhower Commission's report on television had little impact—it was overshadowed in the news media by its more headline-making findings about riots, civil disobedience, and police brutality. The networks acted to reduce the violence in animated cartoons for children and killings in adult programs, and the motion-picture industry quickly compensated by increasing the incidence and vividness of its bloodletting.

However, Congress, on the initiative of Rhode Island Senator John O. Pastore, a long-standing critic of television, moved to mandate a completely new investigation, calling on the US Surgeon General for a report on TV and violence that would, in effect, parallel the report associating cigarette smoking with cancer.

Worried about what might emerge from such a study, the television industry lobbied with President Nixon's Secretary of Health, Education, and Welfare, Robert Finch, to influence the organization and conduct of the investigation. It successfully opposed seven candidates for appointment to the committee, including the best-known researchers in the field. The Surgeon General's Committee on Television and Social Behavior, as consituted, comprised five experts affiliated with the broadcasting industry, and four behavioral scientists innocent of mass-media background.

Three years and $1.8 million later, the committee produced its report, "Television and Growing Up: The Impact of Televised Violence," supported by five volumes of technical studies. The full report, read by few, provided telling data on the role of TV violence as instigator of aggression in young people, but the nineteen-page summary that would determine the public perception emerged opaque and ambiguous, after an intense struggle within the committee.

"Under the circumstances," it said, watching violent fare on television could cause a young person to act aggressively, but "children imitate and learn from everything they see." The research studies, it said, indicated "a modest association between viewing of television and violence among at least some children," but "television is only one of the many factors which in time may precede aggressive behavior."

The summary danced around the crucial issue of causation: "Several findings of the survey studied can be cited to sustain the hypothesis that viewing of violent television has a causal relation to aggressive behavior, though neither individually nor collectively are the findings conclusive."

The ambiguity was mirrored in the pages of the *New York Times*. A front-page story on January 12, 1972, based on a leak, was headlined TV VIOLENCE HELD UNHARMFUL TO YOUTH. But when the report was officially released a week later, the *Times* story said, "The study shows for the first time a causal connection between violence shown on television and subsequent behavior by children."

"It is clear to me," said Surgeon General Jesse Steinfeld, presenting his report at a hearing conducted by Senator Pastore, "that the causal relationship between televised violence and antisocial behavior is sufficient to warrant appropriate and remedial action."

There was no significant remedial action. As the decade of urban violence and assassination ebbed, the issue of television violence faded, to come back another day. And another day would bring another report.

Even before the latest incidents of violence, a new inquiry had started. Dr. Eli A. Rubinstein had first come to the Surgeon General's committee as a vice chairman fresh from the National Institute of Mental Health. His experience with the investigation led him to make the study of the mass media his career.

In 1980, Dr. Rubinstein, now professor of psychology at the University of North Carolina, persuaded President Carter's Surgeon General, Dr. Julius Richmond, to assemble an ad hoc committee to prepare an updated version of the 1972 Surgeon General's report on its tenth anniversary. Two volumes of new technical studies have already been compiled. The conclusions are yet to be written, but there is no doubt that they will reinforce and expand the original timidly stated findings.

One thing the new report will do, Dr. Rubinstein said, is to lay to rest the theory that depicted violence can actually decrease aggression by serving as a "cathartic"—the cleansing and purging of an audience's emotions that Aristotle held to be the highest test of tragedy. Advanced by some behavioral scientists studying television, the theory was examined during the 1972 study for the Surgeon General, which concluded that there was "no evidence to support a catharsis interpretation." The updated report, citing new empirical studies, will make that point more strongly.

"A tremendous amount of work has been done over the past ten years, and the volume of literature has probably tripled," Dr. Rubinstein says. "If any mistake was made ten years ago, it was to be too qualified about the relationship between TV violence and aggressiveness. We have a lot of new evidence about causality, and about what constitutes causality. We know much more about how television produces aggressive behavior. We know more about how fantasy can crowd out reality, and the specific influences of television on disturbed minds.

"The fundamental scientific evidence indicates that television affects the viewer in more ways than we realized initially. You will recall that the original smoking-and-health study was limited to the lungs, and later it was learned how smoking affects the heart and other parts of the body. In the same way, we now know that the original emphasis on TV violence was too narrow. Television affects not only a predisposition towards violence, but the whole range of social and psychological development of the younger generation."

How Many Murders Can Your Kids Watch?

The National Coalition on Television Violence says these are the most violent programs on national television. The data was compiled between February and May of 1981, and the scores for each program are in violent acts per hour.

Prime-time Shows	Network	Acts of Violence
Walking Tall	NBC	25
Vegas	ABC	18
Lobo	NBC	18
Greatest American Hero	ABC	18
Incredible Hulk	CBS	14
Magnum P.I.	CBS	14
Hart to Hart	ABC	14
Dukes of Hazzard	CBS	14
B.J. & the Bear	NBC	14
Fantasy Island	ABC	11
Enos	CBS	11

Saturday Morning Cartoons	Network	Acts of Violence
Thundarr the Barbarian	ABC	64
Daffy Duck	NBC	52
Bugs Bunny/ Roadrunner	CBS	51
Superfriends	ABC	38
Richie Rich/ Scooby Doo	ABC	30
Plasticman, Baby Plas	ABC	28
Heathcliff & Dingbat	ABC	28
Fonz	ABC	28
Tom & Jerry	CBS	27
Popeye	CBS	26
Johnny Quest	NBC	25
Drak Pak	CBS	23
Batman	NBC	19
Godzilla/Hong Kong Phooey	NBC	18
Flintstones	NBC	13
Tarzan/Lone Ranger	CBS	13

The new Surgeon General's report, scheduled for release by the Reagan administration in 1982, is likely to be challenged by the TV industry with all the vigor displayed by the tobacco lobby when opposing the report on smoking and cancer. Inevitably, it will be read for clues to violent behavior of people like John Hinckley.

In the absence of family, peer, and school relationships, television becomes the most compatible substitute for real-life experience.
—National Commission on the Causes and Prevention of Violence (1969)

What made Hinckley different, what made him shoot the President are ultimately matters for psychiatry and the law to determine. But the "media factor" played a part.

As Hinckley withdrew from school and family life, he retreated progressively into a waiting world of violent fantasy, spending more and more time alone with television—an exciting companion that made no demands on him.

But television was not the only part of the media working to merge fact and fantasy for Hinckley. He was strongly influenced by *Taxi Driver*, a motion picture about a psychopath who found the answer to his anxieties through his obsession with violence. Like the taxi driver, Hinckley oscillated between wanting to kill a public figure to impress the object of his affections, and wanting to "rescue" her from "evil" surroundings. Paul Schrader, author of the screenplay, tells me that the moment he heard that President Reagan had been shot, his reaction was, "There goes another taxi driver!"

Hinckley was also affected by fan frenzy, a special manifestation of the media culture. It focused not only on Jodie Foster, the female lead in *Taxi Driver*, but also on former Beatle John Lennon, whose music he played on the guitar. Last New Year's Eve, after Lennon's murder, Hinckley taped a monologue, in his motel room near Denver, in which he mourned: "John and Jodie, and now one of 'em's dead.

"Sometimes," he said, "I think I'd rather just see her not . . . not on earth than being with other guys. I wouldn't wanna stay on earth without her on earth. It'd have to be some kind of pact between Jodie and me."

And the influences working on Hinckley extended beyond the visual media. The idea of a suicide pact was apparently drawn from *The Fan*, a novel by Bob Randall that Hinckley had borrowed—along with books about the Kennedy family and Gordon Liddy's *Will*—from a public library in Evergreen, Colorado. In the book, the paranoid fan of a Broadway star, feeling rejected in his advances by mail, kills the actress and himself as she opens in a theater production. Early last March, as Foster was preparing to open in a New Haven stock-company play, Hinckley slipped a letter under her door saying, "After tonight John Lennon and I will have a lot in common."

The plan that finally congealed this welter of media-drawn inspirations and impelled the young misfit to action was a presidential assassination. Before setting out, he—like the fictional fan—left behind a letter to be read posthumously. It was to tell Foster that he intended, through "this historical deed, to gain your respect and love."

As though to document his place in the media hall of fame, he dated and timed the letter and left behind, in his room in the Park Central Hotel, tapes of his guitar playing, his New Year's Eve soliloquy, and a telephone conversation with Foster.

A failure at most things, Hinckley was a spectacular media success who had survived to enjoy his celebrityhood—a lesson that won't be lost on other driven persons.

No one could doubt his importance or challenge his identity as the news cameras clustered around the federal courthouse when he arrived for his arraignment in a presidential-size limousine heralded by police sirens.

In the great made-for-TV drama, participants more "normal" than Hinckley seemed also to play assigned roles, as if caught up in some ineluctable screenplay. The TV anchors were reviewed for smoothness, composure, and factual accuracy under stress. Secretary of State Haig, making a gripping appearance in the White House press room, was panned for gasping and for misreading his lines. President Reagan, with considerable support from White House aides and from the smoothly reassuring Dr. Dennis O'Leary, himself an instant hit, won plaudits for a flawless performance as the wisecracking, death-defying leader of the Free World.

The effect was to reinforce the pervasive sense of unreality engendered by a generation of television shoot-outs—the impression that being shot doesn't really hurt, that everything will turn out all right in time for the final commercial.

One can understand the desire to assure the world that the government is functioning. But Dr. David Hamburg, the psychiatrist and former president of the Institute of Medicine of the National Academy of Sciences, believes it harmful to imply that a shooting can be without apparent physical consequence.

"Getting shot is not like falling off a horse," Dr. Hamburg says. "To sanitize an act of violence is a disservice. It is unwise to minimize the fact that a President can get hurt and that he can bleed."

One more contribution had been made to obscuring the pain and reality of violence, to blurring the critical distinction between fiction and fact. The media President was, in his way, as much a product of the age of unreality as was John Hinckley, the media freak. In the media age, reality had been the first casualty.

HOW I STOPPED NAGGING AND STARTED TEACHING MY CHILDREN TO BEHAVE

Paul S. Graubard

An embattled psychologist turned househusband tells how he learned to use his professional skills to run his home, train his children—and save his sanity.

I had always felt that taking care of a house would be a breeze. Becoming a homemaker on a full-time basis would give me more time with my children, and still leave room for all the personal projects I had been putting off. My job was getting routine, and I was sick of commuting. I wanted a change.

My wife, Joy, was delighted when I suggested switching roles. During the 18 years I had worked, she said, she had probably cleaned the toilet 3,000 times, mopped the kitchen floor at least twice that often, and fixed over 15,000 meals. Our children were beginning to grow up—Amy was 16, Risa was 17, Michael was six and in school now also, and even little David was two. Joy had long wanted to go back to teaching elementary school and to have more time for playing her flute. She was more than happy to turn the house over to me.

It seemed an ideal arrangement. I felt I was a good father, and this would really give me a chance to show what I could do. I am a child psychologist and a college professor; it would be easy enough to give up my practice for a while, and I could take off a year or more from teaching. This new arrangement would give me time to write a children's play I had in mind, to perfect my French cooking, to practice the guitar.

Disillusionment set in early—the first morning, to be exact. I wasn't used to changing diapers, writing notes to the teacher, looking for lost sneakers, cooking breakfast and matching up socks—all at once. No matter what I asked the children to do, their sleepy voices would answer, "In a minute." A minute had never seemed like much, but try to get a school-bus driver to wait that long! Michael was always the last one out. He usually made it, but sometimes he didn't and I would have to drive him to school, muttering words he never heard in "Dick and Jane."

The first evening I cooked my secret chicken recipe and brought it to the table. My first surprise came when I found that the children assigned to set the table hadn't done their jobs. My wife smiled understandingly and told me it had happened before and would happen again. Then the children told me frankly that they preferred Kentucky Fried Chicken to mine.

The evenings that followed were not much better. Instead of sitting down with a martini and the newspaper at five o'clock, I was in the kitchen. I peeled and chopped and basted—and snapped at the kids. Joy sat in the den with a mountain of schoolwork. She hardly noticed what we ate.

During the next few weeks there were mornings when I found myself thinking about having a drink instead of a cup of coffee. The only bright spot in my daily routine was that brief period every afternoon when David took a nap. Once, when he refused to sleep, I was tempted to put some—just a little—gin in his orange juice. Fortunately, a neighbor dropped by just in time.

The house, as well as my mental stability, seemed to be coming apart at the seams. Dust, clutter and dog hairs filled the family room, toys and clothing inched their way up the stairs, dishes overflowed in the sink. How, I asked myself, could one person hope to accomplish all that had to be done in the course of a normal day? And if that wasn't bad enough, the children's behavior was. Not only did I have to nag them into doing what little they did around the house, but most of the time they didn't even seem to be listening to me. I never knew whether they would follow orders or not. Would they take the bicycle out of the driveway before it was run over or after? Would they walk the dog in time to prevent an accident? Would they stop fighting before or after I lost my temper?

When I wasn't settling fights, I was tying shoes. I scrubbed hands and faces, did the laundry, set the table, washed dishes and everything else I thought four children required. As slow a learner as I was, I soon realized there was a relationship between my lack of free time and my children's lack of responsibility around the house. By the time I came to this, however, they'd not only strengthened their existing bad habits but had picked up some new ones.

I pleaded with them. It didn't help. I hollered at them. They hollered back. I called family meetings, explained my position over and over again, and ended up talking to myself. Their passive-resistance campaign, wherein they did what they had to do, but ever so slowly, was giving me indigestion. My self-image as a father was in bad shape.

The household had functioned well when my wife was in charge of things. I began to realize the price of that pleasant state. Joy had often said that she'd never had a minute to herself—I now knew what she was talking about. Frankly, I didn't want to put in that much time or energy; taking care of the family was my first priority, but I wanted a life of my own, too. When I figured out that for the rest of the year I would be doing three loads of wash every Tuesday and Thursday morning, I knew I had to do something—fast.

At that point, I started looking for all the help I could get. The most help came from my wife, but she found, as I had found out before her, that with the pressures of a job and commuting there was precious little time left over. Prepared foods, disposable diapers and Dr. Spock were helpful, too, but I was still overwhelmed.

I finally ended up using what I knew best—psychology. In my practice as a child psychologist, I'd worked with hundreds of families and thousands of children. My specialty had been behavior modification. I'd used it to help children overcome social and academic difficulties and to help families learn to work together.

It is a well-known adage that the shoemaker's children often go without shoes. I had not been applying skills to my own children's development. Now, faced with my own survival needs, I began to apply behavior-modification techniques at home.

Dr. B.F. Skinner of Harvard University is usually credited with inventing behavior modification. The real credit should go to Grandma. She started the method when she formulated her famous law: "First finish your spinach. Then you may have your ice cream." I began using it whenever I could.

To apply Grandma's Law, I had to be specific about what I wanted. I started out by making a list of the most important problems. The first list read: toilet train David, get Michael off to school on time, get Michael and Amy to clean up after themselves, have Amy come home on time in the evening, help Risa and Amy stop squabbling, and get everyone to do a share of the chores.

Once I was satisfied that achieving those goals would be to the children's benefit as well as my own, I had to find a way of motivating them, of making them want to reach the goals. I thought of it—as Grandma might have—as a *when-then* proposition. *When* I saw the desired behavior (or even anything that came close to it), *then* I would reinforce it. "Reward improvement" became my motto.

Reinforcers, in psychologist's terms, are simply rewards that work. Reinforcers increase the number of times a given behavior occurs. You have to observe the child closely in order to know when and what to reinforce, and which reinforcer is appropriate to use. For example, some of my children would have given almost anything to go on a fishing trip with me. The others couldn't have cared less. Only the child, through his or her behavior, can tell you whether the reward works. If it doesn't, it obviously isn't a reinforcer.

Behavior modification sounds simple, and in many respects it is. But to repeat: Behavior that is reinforced will be strengthened; behavior that is not reinforced will be weakened. That is the essence of the method. Reward a child for the good things he does, and those good things are likely to be repeated. Conversely, if you do *not* reinforce good behavior, it is not likely to be repeated.

I discovered (actually my wife pointed it out to me) that I had fallen into a trap. I had been rewarding my children's misbehavior by paying so much attention to it, and I was taking the good things for granted. Everyone likes to be praised or rewarded when he or she does a good job, and children are no exception. In the beginning, I used some material rewards to help us get back on the track; after a while I found my own interest and encouragement to be the greatest reinforcers of all.

I used the following three-step method:

1. *Specify* what you want to change; for example, dawdling over homework or coming home late.

2. *Make a change.* See if you can solve the problem before it gets out of hand. For example, I found serving dinner earlier cut down on the squabbling. Make up a *when-then* sentence; *When* Amy comes home on time, *then* she can use the car. *When* David uses the potty, *then* he can go swimming.

3. *Evaluate.* Check off on a calendar every day in which the desired behavior occurs. Both you and the child will be encouraged by the progress. If there is no progress, you had better think up another approach.

Here are some examples of how I used this method:

Toilet training. When David's movements were becoming regular, I used the when-then proposition. When you use the potty, I told him, then we can go swimming. I read to him while he was on the potty. When he got off, I didn't say anything, but I stopped reading the book. When he got back on, I resumed reading. Maybe it was the book, maybe it was the promise of the beach or maybe he was just ready, but it took him only two days to master toilet training.

Getting up in the morning. It was harder getting Michael to the bus stop on time. I knew he loved to watch TV, so I told him he could watch a program after supper if he earned a ticket. He could earn his ticket by being *at* the bus stop by 8:10, which would give him, and me, five minutes to spare. I also told him I would wake him up once and once only.

The first day Michael didn't make it to the bus stop until 8:14. That night he tried to stay up late anyway. No ticket, no late TV, I told him. Tears welled up in his eyes, and he begged for another chance. I held firm and repeated the deal to him.

On the second day the same thing happened. The third day he asked if he could watch a late program if he earned his ticket. I assured him he could. It was the easiest morning I had with him. I kept the system going for another two weeks, and I let him know how proud I was of him whenever I could. When I felt he could manage on his own, I dropped the ticket system.

Clean up, curfew (and finding someone to clean the house). After Amy's cheerleading stage, which consisted of her leaping through the house and shouting oaths to spur her team on, she entered her sloppy and defiant stage.

The condition of the kitchen was hard on the eye after her snacks of garlic powder and assorted concoctions on Ritz crackers. She also had trouble meeting her curfew, and I was getting tired of hearing how "everybody else's father" let their daughters stay out later than I did.

Fortunately, about that time Amy received her driver's permit. So I made a deal with her. For starters she had to be in by 9:30 on school nights. She also had to clean up any messes she made in the kitchen. I told her that when she met those conditions, she could borrow the car on Saturday or Sunday afternoons. She agreed.

I also told her she had to pay for her own gas. When she said she had no money, I offered her a job in the house doing the heavy cleaning at standard wages. She accepted.

Amy followed the rules and got her rewards.

I also had trouble teaching Michael to clean up. Somehow, he always "forgot." He was young, he said, and God hadn't finished with him yet, so how could I expect him to be perfect?

Michael was a baseball fanatic. I made a deal with him. When he kept the kitchen neat for three days in a row, then I would take him to the Batting Cage, which featured a machine that pitched real hardballs right over the plate. It was a place he very much wanted to go.

I checked off whether he kept the kitchen neat. Every time he did what he was supposed to, I thanked him. I also set a good example, because by this time I had learned that children will do what you do and not what you say should be done.

Michael's memory improved 100 percent. The third day I took him to the Batting Cage. I then required him to keep the kitchen clean for a week in exchange for another trip there. By this time cleaning up after himself was becoming a habit. I continued taking him to the

Batting Cage every week or so because he liked it so much, and I told him I appreciated how much he'd changed.

Squabbling. My daughters seemed to bicker constantly, and it didn't help when I added to the din by yelling at them to stop.

We had recently eaten in a Chinese restaurant and the children had thoroughly enjoyed it. I told them we would return to the restaurant on the condition that they reduce their bickering. I said I was going to keep a record of every fight, which we finally defined as any discussion that could be heard two rooms away. They could argue as much as they wanted to in the cellar, but only there. When they decreased their arguments in the house to fewer than five a week, we could have Chinese food again.

Every time they argued in the house, no matter who started it I checked it off on the calendar. The first day I counted 12 arguments, the majority of which concerned who was wearing whose clothes.

They exchanged clothes so frequently they could no longer remember which item of apparel was whose. They also lent each other's clothes to friends. Together we sorted out the clothes and worked out the rule: no borrowing, no lending. Risa and Amy agreed to try it out for a month. The arguments dramatically decreased, and within three weeks they met their goal.

Pitch in. After a while, when I was beginning to get the children's attention fairly regularly, I decided it was time to become a little more organized. I listed all the household chores. The children took turns picking their favorite jobs, and the rest were placed on an alternating system so that everyone ended up taking turns cleaning pots or mopping the floor.

To the greatest extent possible I used Grandma's Law. Spurred on by a desire to watch "The Wonderful World of Disney" and "Animal Kingdom," Michael really became efficient at dishwashing and sweeping. Risa became a whiz at finishing her homework so she could use the phone, and even David picked up his toys more readily so he wouldn't miss out on our nightly wrestling match. The allowances were given out *after* Saturday chores, and I did the little ones' laundry only if it had been put in the hamper. The big ones had to do their own. I also discovered that routines are to children what a foundation is to a house. Once they were on a solid basis—knew what was expected of them—things held up very well.

Modifying my own behavior. At the beginning I needed a lot of help. Not only was I tired and overworked, I was resentful. I'd wanted to write plays, not grocery lists. I'd wanted to cook gourmet dinners, not hamburgers and rice with ketchup. I had to begin ridding myself of my own bad habits.

I knew that behavior modification would work with the children, but I also had to motivate myself. I used rewards here, too. *When* David kept dry for seven days, *then* I would exchange baby-sitting duties with a neighbor and spend an afternoon out. When Amy learned to clean up after herself in the kitchen, then I would spend a few hours working on my writing. I also tried to use reinforcers—skiing, swimming, telling stories—that I as well as the children would enjoy.

One day, when I lapsed back into a nagging and threatening stage, I kept track of the number of times I nagged them; there were 18 in a ten-hour period. The deal I made with myself was that *when* I had fewer than five a day I would go fishing by myself on a Sunday.

I told everyone in the family what I was doing and asked them for their help. I told them that when I reached my goal, then I would make them their favorite meal—spaghetti with French fries on the side. I asked everyone to praise me when I settled disagreements reasonably.

Habits are hard to break, our own as well as our children's, and most of us need all the help and support we can get. More eyes, or in this case ears, helped me become more aware of myself. If I did a good job, the children told me so, and that made me feel good.

Questions from friends. Some friends voiced concern over my use of behavior modification. They were concerned that I was "bribing" my children. I did not feel that way at all. Rewarding is not bribing. The children earned every reward they got, and they worked hard. Complaining and becoming indignant usually didn't help me correct a situation. Reinforcement did.

The year comes to an end. During the course of the year we all had to learn to give and take a little, and that took time and work. I certainly ended up appreciating the job my wife had done in the house, and she appreciated what it was like to work full time and commute. And we all began feeling good about life in the household. Behavior modification allowed me to get my work done as well as to be with and to take care of my family. It also gave me an alternative to nagging and punishing. More important, the approach helped the children develop self-confidence, independence and skills.

As the year ended, Joy and I admitted to each other that we both found our career shifts disquieting. Missing work and adult company was only part of the picture for me. I felt economically dependent. Joy felt displaced. At the beginning especially her feelings were hurt when the children came to me to heal a wound or comfort them. She also found that the pressures of a full-time job kept her away from the music she liked so much.

I didn't want to go back to work full time. I enjoyed being with the children. Besides, I flattered myself that they needed me at home more than a full-time job would allow. Joy didn't want to spend all her days away from home although she liked some time out of the house.

So now we both work part time and share household responsibilities. I have a small private practice, consult and write. Joy works as a free-lance musician and teacher in a day-care center. There are certain problems that go along with this arrangement. Sometimes we have too much work and sometimes not enough. Sometimes we have enough money and sometimes we don't. But, overall, we feel it is a good arrangement, for the children as well as us. Each of us can now share more fully in their day-to-day lives, which pleases the children as much as it does us. And isn't that what family life is all about?

Development During Adolescence and Young Adulthood

In 1904 G. Stanley Hall published his extensive work, *Adolescence: Its Psychology and Its Relation to Physiology, Anthropology, Sociology, Sex, Crime, Religion and Education.* In this work Hall advanced his "ontogeny recapitulates phylogeny" theory of development. Hall believed that the development of the individual organism (ontogeny) repeated the various phases of the evolution of the species (phylogeny). The adolescent repeated the 18th century's idealism marked by revolt against authority, passion and emotionality, and commitment to goals. In short, Hall envisioned adolescence as a period of "storm and stress." Whereas Hall's recapitulation theory no longer is regarded as having scientific merit, his characterization of adolescence as a time of turmoil and rebelliousness has persisted.

To be sure, adolescence is a time of marked physical, social, emotional, and cognitive change. Secondary sex characteristics appear accompanied by rapid changes in height, weight, and body proportion. Social development involves reduced dependence on the family and greater dependence on peers. Cognitive achievements include abstract reasoning, hypothesis testing, and inductive and deductive reasoning. The adolescent's thinking becomes liberated from the concreteness of the child's reasoning. The pressures of peer group, school, and family may produce conformity, or may lead to rebellion or withdrawal directed against friends, parents, or society at large.

Other researchers argue that much of the storm and stress of adolescence is a myth, created from an overemphasis on adolescent fads and rebelliousness and an underemphasis on obedience, conformity, and cooperation. Overemphasis on the "negative" aspects of adolescent behavior may create a set of expectations which the adolescent strives to achieve. In any event, adolescence is a major transitional period in human development. It is a time for casting off the dependence of childhood and for assuming the independence and responsibilities of adulthood. Adolescence also leads to separation from the family, a topic that has received surprisingly little attention from developmentalists.

Whereas the onset of adolescence is demarcated by the emergence of secondary sex characteristics and the achievement of reproductive maturity, the onset of adulthood is more difficult to distinguish, particularly in modern technological societies. In some cultures a ritualistic ceremony marks the transition to adulthood—a transition that occurs quickly, smoothly, and relatively problem free. In our culture the transition is vague indeed. Does one choose the age at which the adolescent achieves the right to vote, the privilege of driving a car, or perhaps the right to sign one's own informed consent form in order to participate in behavioral research? In any event, the status of adulthood usually is granted shortly after the teen years. Developmentalists are finally addressing themselves to the study of problems of "middle" and "old" age, but they have shown far less interest in the early years of adulthood. Since this is the time when adults are giving birth to the next generation, perhaps it is time for developmentalists to focus attention on those who are parenting as well as those who are receiving caregiving.

Looking Ahead: Challenge Questions

How might our society help to make the transition from childhood to adulthood less problematic for teenagers?

What changes in our society would help to reduce adolescent pregnancy, delinquency, teenage suicide, and excessive obedience to authority?

How would you define adolescence? Adulthood? To what extent would your definition apply universally?

Do you agree that adolescence is a time of "storm and stress?" If you were participating in a survey of adolescents' attitudes about their current family relationships, would your answers support the "storm and stress" view?

Adolescents and Sex

James E. Elias

James E. Elias is assistant professor of sociology at California State University at Northridge. He has directed several research studies and is the author of numerous publications in the field of human sexuality.

Probably more misinformation exists about adolescent sexuality than about any other sexually related topic. Questions about adolescent sexuality can be grouped into three categories: (1) Is a "sexual revolution" occurring among adolescents? (2) What influences do erotic stimuli and pornography exert in the life of the adolescent? (3) What is the relationship between sex education and adolescent attitudes and behavior patterns?

I would like to relate some of the findings of four different studies that explored the above questions. All were conducted during the past four years at the Institute for Sex Research at Indiana University.

The Sexual Revolution

Perhaps the most important finding was that increased premarital coitus among our adolescent population indicative of a sexual revolution is *not* supported by the data. An analysis of the etiology of this belief may provide some answers and bring the situation into proper perspective. In determining whether a sexual revolution is occurring among adolescents, three "levels" must be considered: the *media level,* where products, programs, and performances project the image of a sensate and sexually oriented society; the *attitudinal level,* which includes the feelings and opinions expressed by adolescents; and the *behavioral level,* which includes activities adolescents actually engage in.

The revolution obviously exists in the media, which bombard the public with sexual stimuli. If young people responded to these stimuli, and if the promises of sexual success were pursued, then a real sexual revolution would be taking place. The trend toward sexual explicitness that has occurred in the media has been rapid, exposing the population to a candidness that in the past was only implied. Parents are shocked by many media presentations, yet they are the prime consumers of this material, not their children. Thus, a "schizophrenia" between the *normative* system and actual practices is evident.

Adolescents have become more permissive in their attitudes toward sexual behavior and openly discuss sexuality and sexual relationships as a natural part of everyday life. Parents are often startled by such openness because, in their childhood, sex was taboo as a discussion topic. Therefore, frank curiosity, questions about nonmarital intercourse, or even the mention of a word like *homosexuality* at the dinner table give parents the idea that their children have completely escaped their grasp and gone astray. Parents fail to realize that the frankness of dialogue is not automatically accompanied by a corresponding frankness in behavior.

It is in this sphere that possible ambiguity between parental and adolescent attitudes is most evident. The past generation engaged in nonmarital and extramarital intercourse more than they apparently admitted. In other words, their professed attitudes were more conservative than their behaviors. The present generation illustrates the reverse: attitudes are more liberal than behavior. Unfortunately, parents tend to equate attitudinal positions with behavior.

In terms of actual sexual behavior, adolescents have not changed radically from their parents. The frequency of nonmarital coitus has been increasing for females. Kinsey reported a 20 to 25 percent nonvirginity rate for college females. A more recent study at the Institute for Sex Research shows that this rate has risen to 33 percent. This increase has occurred over a twenty-five-year period and certainly cannot be called "revolutionary." The fear that the availability of oral contraceptives might encourage adolescent promiscuity is unfounded. It appears that the use of oral contraceptives has not had a significant effect in this age group. In addition, among adolescents the availability or nonavailability of oral contraceptives is not the chief factor in their decision to have intercourse.

For males there has been little change in the frequency of nonmarital coitus—remaining around the 60 percent level for the college population. There is a definite increase in petting, including petting to orgasm, which may be a substitute for coital behavior. However, a high nonmarital coital frequency, indicative of a promiscuous society, is not in evidence. Those engaging in nonmarital intercourse are doing so more often and enjoying it more than their elders did, without attendant guilt feelings. Therefore, the idea of a sexual revolution is encouraged and perpetuated by the mass media, *not* by research data.

Erotic Stimuli

It seems most strange that our society chooses to restrict, repress, and condemn any material that tends to arouse sexual feelings. Those members of the middle class most critical of erotic materials are among its chief consumers. These conflicting behaviors, creating disparate public and private norms, are reflected in the inconsistent enforcement of "community standards." No agreement on the evaluation of erotica exists.

My studies of the place erotica occupies in adolescent sexual development show some interesting results. One study consisted of depth interviews with adolescents from varying social classes. Respondents were questioned about what material they would consider "unacceptable." The range of

This article first appeared in THE HUMANIST March/April 1978 and is reprinted by permission.

responses was considerable. The most frequent response from males reveals that they do not view erotic materials as objectionable. (Erotic materials include nudist books, sex action graphics, and stag or "blue films" showing explicit sexual activity.) In their definitions of unacceptability, the social class and sex differences of the respondents are evident. Higher social classes equate acceptability with "tastefulness" or aesthetic values. A more rigid standard exists among females, who seem to find even the marginal materials unacceptable.

In determining what is "erotic," material often has been assessed in terms of "significant others"—individuals or groups that include peers, parents, teachers, and the community. Values differ according to generation, sex, and socioeconomic levels. Males, breaking away from home ties earlier than females, seek their peers for support, and this group becomes the norm-interpreting agency for many of them. Their definitions of erotica are seldom similar to those of their parents with the possible exception of upper and upper-middle class males. It is significant that *none* of the males saw their definitions as similar to those of their teachers or the community.

Females, on the other hand, strongly identify with the family value system, and over two-thirds of the respondents indicated definitions similar to those of their families. The other one-third of the females see their definitions as similar to those of their peers.

Any discussion of exposure to erotica must deal with the types available: general material of a sexual nature, such as *Playboy*; marginal erotica; erotica; and hard-core erotica.

"Those engaging in nonmarital intercourse are doing so more often and enjoying it more than their elders did, without attendant guilt feelings. Therefore, the idea of a sexual revolution is encouraged and perpetuated by the mass media, not by research data."

General sexual materials contain no exposed sex organs or explicit sexual activity. *Marginal erotica* includes pictures, magazines, and books that show sex organs but not sexual activity. Also belonging in this category are live performances of the striptease variety that do not involve sexual activity. *Erotica* includes those types of material that depict sexual activity—coitus, fellatio, or cunnilingus. Live performances that involve masturbation or "fondling" of the sex organs belong here. *Hard-core erotica* is a category reserved for "blue films" and live performances where sexual acts, particularly coitus, are explicitly depicted. (None of the adolescent sample had ever seen a live show of this type.)

Nearly all adolescents had been exposed to *Playboy*, the most popular magazine of its kind on the current market. *Playboy* allows "fantasy to replace fact" in the depiction of female nudes. (Males in higher socioeconomic groups tend to show a greater interest in the articles than in the pictures.)

Interest in marginal erotica occurs primarily in the male sector of the adolescent subculture. Nearly all males in this group have been exposed to this type of material because products are widely distributed and openly displayed. Since most "nudist" magazines are male oriented, females have a low exposure. Erotica, on the other hand, is not as easily available and must be sought out or found in private rather than in public displays. Widest exposure to this type of material occurs among middle-, lower-middle-, and lower-class males. Decks of cards depicting sexual behavior, cartoon books, and photographs are not uncommon in this sector of the young male's world, where sexual arousal and peer acceptance are all-important.

Most frequently, a male's first exposure occurs when a member of his peer group shares the erotica in his possession. The progression seems to be from pictures showing nude females to those depicting heterosexual activity. Adolescents profess great interest in these materials, and curiosity about sexuality seems to be strong among members of this subculture. They report that their first encounter with erotica occurs early—around age seven or eight. The majority voice a positive reaction to the materials.

Females show a curiosity toward erotic material, but they label the material "bad" or "nasty." Their exposure seems limited more to marginal erotic materials, probably because most material is directed toward the male market. Their reactions range from embarrassment to "It makes me feel dirty inside," and these responses probably reflect their tendency to adhere to family standards.

Erotica appears to serve several purposes for the adolescent male: arousal, and in some cases, subsequent masturbation; sex education, especially when there has been no other source; and peer-group status through the ownership of the medium of information. Many of the males interviewed indicated that the value of the materials was enhanced when they discovered that possession brought not only peer group acceptance but increased status.

Sex Education

Sex education (education about human sexuality) begins as a matter of course very early and continues throughout the life-span of the individual. Much of this education is indirect. In addition, parents frequently turn aside questions that a child asks or offer no explanation for the negative sanctioning of interest in the sexual. Formal sex education should start as early as the child begins school because he has already asked questions and, for the most part, they remain unanswered or answered by poor analogies or with reprimands that are psychologically harmful.

As the child reaches puberty, he looks to his peers for interpretations, and peer-group norms persist. Direct questions (often pretending prior knowledge) about human sexuality are common, especially among males. In contrast to the direct sexual approach taken by males, young adolescent females seem heavily oriented toward the romantic aspect of love. The sexual information shared by adolescents seems to consist of "pooled ignorance." The individual with "the answers," whether accurate or not, gains status in the eyes of his peers, and the quality of his information frequently goes unchallenged.

My study of high school students and their sources of sex

education reinforces the findings of past research: *the peer group is the major source of sex information.* The source that provided a "great deal of information" was, for males, other males of the same age, and for females, other females of the same age. The second major source of sexual information is peers of the opposite sex. Teachers, who in the past have provided little, if any, instruction in this area, have emerged as another source of information—for adolescents who have taken a course in sex education and for those who have not.

The parental role in sex education appears to be lacking except in the mother-daughter relationship. Sixty-five percent of the females reported that their mothers had discussed sex with them. Discussions often centered around menstruation and the "negative" aspects of premarital sexual intercourse. The father served as a source of information in only 2 percent of the cases, and 33 percent of the females reported that neither parent had talked with them.

Parents evidently do not provide an equivalent kind of instruction for males. Sixty-four percent of the males reported that neither parent had discussed sex with them, and 26 percent indicated that they had discussed sex with their fathers. Only 10 percent answered positively to a mother-son discussion. Interestingly enough, males indicated that neither parent was the first source of information on any of the nine topics included in the study. For females the mother is the first source of information regarding menstruation, contraception, pregnancy, and menopause. However, in areas such as sexual intercourse, male erection, masturbation, prostitution, and homosexuality, peers provide the first source of information.

In summary, the sources of learning about human sexuality are various, the most frequent being the peer group. Much of the information is learned at ages earlier than previously expected, and the "innocence of youth" is largely a myth.

Attitudes Toward Sex Education

From a short series of questions asked in the post-test situation concerning sex education in the schools, it is clear that adolescents have a positive attitude toward it. Over 90 percent of the students who had taken such a course indicated that they felt sex education should be taught in the schools. Religious teachings are often cited as being in conflict with the teachings of sex education. However, 95 percent of the students in a predominantly Catholic community, after having taken part in the program, did not find sex education in disagreement with their religious beliefs.

Another prevailing view is that sex education in school conflicts with that given at home. Over 90 percent of the students who had taken the sex education course disagreed. The course offered in this particular school system did not assume an overt "moral" stance with regard to sexual matters, and this may explain why students reported little conflict among school, home, and church. Since the home and the church are not major sources of sex information, the conflict issue appears to be a myth.

The sex education course seemed to effect significant changes in adolescent attitudes. Students declared that they would handle sex education with their own children differently than they themselves had experienced it. What to look for in a mate and "how far to go" on a date are great concerns of the adolescent, and the benefit of a sex education course helped provide a frame of reference where otherwise only vague peer-group norms would apply. Mate selection seems especially important for those who will not seek education beyond high school and will probably marry soon after graduation.

As currently presented in the educational system, sex education offers "too little too late," and in many instances the best it can do is present accurate information to correct the misinformation students have acquired. Sexual learning begins long before the advent of formal education and informal sources remain the primary vehicle of information.

Pregnant Children:
A Socio-Educational Challenge

U.S. teen-age childbearing rates are among the world's highest.
Pregnancy is the most common reason why teen girls leave school, at high cost to taxpayers.
Ms. Hendrixson suggest changes in our treatment of our pregnant children.

Linda L. Hendrixson

LINDA L. HENDRIXSON is a senior health education major at Montclair State College, N.J. She is certified as a sex educator by the American Association of Sex Educators, Counselors, and Therapists, and has taught human sexuality courses for teens and adults. She has served as a guest lecturer in human sexuality classes at Montclair State College.

Close to 13 million of the 60 million women who became mothers in 1975 became parents before they became adults. . . . Early childbearing is increasing everywhere, is emerging as a serious problem in many countries, and has reached alarming levels in others.[1]

In 1976, 160 people from 39 nations gathered at Arlie Center in Virginia for the First Interhemispheric Conference on Adolescent Fertility. They cited problems of poverty, fear, loneliness, boredom, uncertainty, vocational discontent, and lack of both job opportunities and education as significant factors leading to the increased number of adolescent pregnancies around the world. Their report calls for changes in certain social policies, laws, customs, and services in order to reduce the high rate of unintended pregnancy in the world's teen population.

How does the U.S. rank in relation to the rest of the world? Our teen-age childbearing rates are among the highest. Of 22 selected countries, including both industrialized and underdeveloped nations, the U.S. ranks fourth, with a rate of 58 births per 1,000 females aged 15-19. Romania ranks third and New Zealand second. East Germany heads the list. Japan ranks lowest, with five births per 1,000 in this age group. The U.S. ranks higher than many less-developed nations, such as the Philippines, Tunisia, and East Malaysia.[2]

More U.S. teen-agers are sexually active at earlier ages than has been realized.

At present there are 21 million adolescents in the U.S. between the ages of 15 and 19. Of these, 11 million are estimated to have had sexual intercourse at least once. Of the eight million 13- to 14-year-olds in this country, over 1.5 million are believed to have had sexual intercourse at least once. Despite the traditional portrayal of increased early sexual activity among minorities and the poor, recent evidence points toward increasing sexual activity at earlier ages among middle- and upper-class white teen-agers. More is known about patterns of sexual activity among teen girls than boys.[3] Female adolescent sexual activity is of a more visible nature; we can count those who present themselves for physician-prescribed contraception, abortion, prenatal care, and for social services as a result of pregnancy.

Only a fraction of the 4.3 million sexually active 15- to 19-year-old females in the U.S. desire pregnancy at this age. Studies show that two-thirds do *not* want pregnancy. Some use contraception consistently; others use it sporadically or not at all. One in six who are "at risk" become pregnant each year. This totals over one million pregnancies in the 15-19 age group, of which 667,000 are unintended. Moreover, 30,000 girls aged 13 or 14 conceive each year. This increases total teen pregnancies to 1,030,000. There are few reliable statistics on the number of sexually active girls under 12 years old, although pregnancies in this age group are becoming more common.[4]

How do teen-agers resolve their pregnancies? Of the one million 15- to 19-year-old females who became pregnant in 1974, 59% carried to term (38% legitimate, 21% illegitimate), 27% had abortions, and 14% had miscarriages. Despite the availability of abortion as an option, almost 600,000 teen-agers gave birth that year.[5]

In response to the growing number of teen pregnancies brought to light by the Guttmacher Institute Study, President Jimmy Carter proposed a "teen pregnancy initiative" as part of the 1978-79

budget. The resulting Adolescent Health, Services, and Pregnancy Prevention and Care Act of 1978 (AHSPPC) was passed last September as an amendment to an omnibus health services bill. It authorizes an initial $60 million in grants for fiscal year 1979 for community-based adolescent pregnancy programs. Core services must include pregnancy testing; maternity and adoption counseling; prenatal and postpartum care; nutrition information; and sexuality, family life, and family planning education. Referral services must include family planning; venereal disease and pediatric care; educational and vocational services; and additional health services. None of the authorized funds may be used to pay for abortions — indeed, even counseling for abortion as an option is not mandated by the bill. Core services also exclude infant day care that would permit adolescent mothers to return to school or seek employment.[6]

Critics feel that AHSPPC places undue emphasis on carrying to term, with its inherently increased risks for both the teen mother and child. A facility opposing abortion need only mention that the adolescent might go elsewhere for other counseling, not even indicating what those services are or where they can be found.

New Jersey is the most densely populated state in the U.S. Its rate of teen-age pregnancy continues to increase.

There were approximately 332,000 girls in grades 7-12 in New Jersey schools during 1977. Of these, 130,000 were estimated to be sexually active. Of this group, 60,000 were expected to become pregnant during that year. Dr. Robert L. Johnson, director of adolescent medicine at the New Jersey College of Medicine and Dentistry in Newark, estimates that two out of every 10 girls in New Jersey junior and senior high schools become pregnant each year. Dr. Johnson says that the pregnancy rate in the inner cities may be as high as 30%, with an alarming increase in the under-15 age group.[7]

The New Jersey Department of Health reported 12,000 births to 15- to 19-year-olds during 1977. Of these, 59% were illegitimate, an increase from 39% in 1970. Girls under 14 delivered 238 babies, 94% of which were illegitimate.[8]

Dr. Johnson observes that very young girls are not physically able to give normal birth. "Their bone structure makes delivery very difficult, and 70% of these girls have to be delivered by Caesarean section."[9]

Babies born to teen mothers are more likely to die before their first birthday. They have statistically lower birth weights and are more often born prematurely than babies born to mothers in their twenties.[10] Stillbirths are more than twice as frequent, and the rate of birth defects is significantly higher in the under-20 age group. These defects include mental retardation, spinal deformities, respiratory problems, clubfoot, and epilepsy. Both biological immaturity and poor nutrition among U.S. teens are seen as factors leading to the high incidence of birth defects.[11]

A national study conducted in 1974 indicates that younger mothers tend to have children with lower IQs than older mothers.[12] Contributing factors include inadequate prenatal and early childhood nutrition and living in poverty-level families with parents or others who have low levels of education. In addition, the rate of child abuse and neglect is high for children born to younger parents.[13]

The risks to teen mothers are also great. The maternal death rate from pregnancy and its complications is 60% higher for females under 15 and 13% higher for females 15-19 than for mothers in their early twenties. Other hazards include toxemia, nonfatal anemia, and nutritional deficiencies that may lead to other diseases.[14]

Pregnancy is the most common reason why female students drop out of school. Studies throughout the U.S. show that one-half to two-thirds of all female school dropouts do so because of pregnancy.[15] Those who leave school usually lack the skills necessary to find reliable employment. They are unlikely to return to school after their babies' births.

A pregnant teen-ager who leaves school and remains single is often alienated from her friends. She faces negative social pressures and perhaps parental abuse. She may be uncomfortable in dating situations after her baby's birth and may have difficulty adjusting sexually in subsequent relationships. Her guilt feelings may remain for many years. If she chooses abortion as the solution to her problem-pregnancy, she is often faced with emotional and financial trauma as well as social disapproval.

Little research has been done on the effects of a nonmarital pregnancy on the father of the child. One 1973 study showed that 40% of the black fathers interviewed married their pregnant girlfriends. Another 40% provided some financial support at least for a time, but the remaining 20% deserted and left their pregnant partners to fend for themselves.[16] The proportion of young white males who marry their pregnant girlfriends or who support their illegitimate children may be higher because of better employment opportunities.

How much do state social services cost taxpayers? In New Jersey the average Medicaid payment for prenatal care and delivery is $1,000. Day care, per child per year, costs $2,500, while foster care, per child per year, costs approximately $1,670. A mother and child on welfare, per year, exclusive of food stamps and Medicaid, cost approximately $3,000. However, the average family planning cost in New Jersey clinics is only $54 per patient per year. Clearly, society bears a heavy financial burden in supporting a child and its young, unmarried mother who, lacking marketable skills, must live on welfare in order to survive.[17]

Title IX of the Education Amendments of 1972 and New Jersey's Title VI mandate that sex discrimination be eliminated in all federally assisted education programs. In 1974 Section 86:40 was added to Title IX to protect the rights of pregnant students in U.S. schools. Accordingly, a pregnant student 1) may not be expelled from school, 2) may not be required to attend a special school for pregnant students, 3) may not be barred from any program, course (including physical education), or extracurricular activity, 4) may not be required to take special courses in child care or related topics unless those courses are required of every other student in the school, 5) may not be required to leave school at a certain time before the birth of the child or be required to remain out of school for a certain length of time afterwards, and 6) may not be required to furnish notes from her physician that she be allowed to continue or reenter a course of study unless such notes are required of all students in relationship to illness or intended surgery. In addition, a special instructional program for "homebound" students must be made available to the pregnant student, based on the same criteria as programs for other students.[18]

Many schools throughout the U.S. have not acquainted female students with their rights in regard to pregnancy and have neglected to inform the staff of their responsibilities. Despite both federal and state laws that encourage pregnant students to continue their education, most of them drop out and never return.

As of 1975, only one-fifth of all states required mandatory sexuality education as part of their health education or other school programs. New Jersey is one of the majority of states where sexuality education is optional, even though alcohol, tobacco, and drug education are required.

In January, 1967, the New Jersey State Board of Education adopted a policy on sex education that "recommends that each local Board of Education make provisions in its curriculum for sex education programs."[19] By leaving the decision to local boards, the way was paved for districts to avoid developing educational programs in this vital sphere. Most have done just that. Now, however, the number of teen pregnancies in New Jersey has reached "epidemic" proportions (a curious term to describe a natural process). The State Department of Health is urging mandatory sex education and the teaching of "parenthood" in all New Jersey schools, beginning no later than grade 5.[20] P. Paul Ricci, president of the State Board of Education, has appointed a five-member committee to review the situation and make recommendations.

Under the New Jersey Administrative Code, Chapter 212, each school district may voluntarily plan a program for its pregnant students. The response has been less than enthusiastic. Programs of varying status exist in Passaic, Jersey City, Elizabeth, Monroe Township, Paterson, New Brunswick, and Ocean City. Funding is complex and not always adequate. Some programs are financed by local boards; others are supported, in part, by the Division of Vocational Education of the State Department of Education. The division's Home Economic and Consumer Education Project provides important assistance to some programs in the areas of prenatal nutrition, child development, hygiene, consumer education, and homemaking skills. Social services are provided by other agencies as well as by local school guidance departments.

Of the existing programs for pregnant teens in New Jersey schools, the Family Learning Center of the New Brunswick schools is the most outstanding. The program was established in 1969 with ESEA Title III funds. It is now financially supported by the New Brunswick Board of Education. This project has been "validated as innovative, successful, cost-effective, and exportable by the standards and guidelines of the U.S. Office of Education."[21]

The staff consists of one full-time head teacher, four part-time academic teachers, a home economist, a part-time counselor, and a nurse. Instruction includes a full range of academic courses, plus education in foods, clothing, nutrition, consumer economics, maternal and child health, reproductive biology, growth and development, physical and emotional changes during pregnancy, heredity and environment, hygiene, and labor and delivery. The center also offers typing, bookkeeping, stenography, and group counseling. Classes are small and individualized. Prenatal care is arranged through the outpatient departments of local hospitals or through private physicians. The teen mother continues in the program for six weeks after her baby's birth. During this time she is referred to Planned Parenthood for contraceptive counseling.

Between 40 and 50 young women pass through this program each year. They develop a more positive self-image and become more self-assured and independent. Former disciplinary problems disappear when they return to their home schools to finish their high school education. Most of them leave the program with marketable skills.*[22]

The problem of teen-age pregnancy is complex. Solutions must take into consideration two separate groups: 1) those teen-agers who, though sexually active, do not want pregnancy to result and 2) those who, for various reasons, desire pregnancy. Considering the myriad factors involved, I propose the following as suggestions for change:

1. A positive, nonjudgmental attitude toward the psychosexual development of adolescents, their need to love and be loved by persons within and without the family, and the realization that many adolescents express these basic needs in sexual ways, as adults do, despite society's restrictive attitudes.

2. Mandated family life, parenting, and sexuality education by all state departments of education. These programs must encompass grades K-12 and must emphasize responsibility, ethical relationships, and decision-making skills. At the appropriate levels, they should include discussions of contraception and abortion. "Non-parenthood" should be presented as a viable, socially acceptable option. Student involvement in curriculum development and evaluation is essential.

3. Enforcement of the provisions of Title IX (and in New Jersey, Title VI) as they pertain to pregnant students.

4. Coordination of all existing school- and community-based prenatal education programs.

5. Development of voluntary prenatal education in all school systems for prospective teen mothers *and* fathers.

6. Improved high school guidance and counseling services that will encourage female students to pursue educational and/or vocational opportunities beyond high school, including those fields that have been traditionally male-oriented.

7. School- and community-sponsored programs for parents as the primary sex educators of their children.

8. Church-sponsored programs in family life and sexuality education.

9. Extension of present public family planning services to *all* adolescents, regardless of age and ability to pay.

10. Availability of abortion as a voluntary option, regardless of age and ability to pay.

11. Continued biomedical research to find new, safe, and effective birth control methods, both natural and artificial, to meet adolescent and adult needs.

12. National health insurance coverage for all health and social services related to contraception, abortion, pregnancy, and childbearing for adolescents.

I agree with Daniel Callahan, director of the Institute of Society, Ethics, and the Life Sciences, who says:

At the very least, teen-agers should have as much knowledge of sex, as many and as good services available, and as many choices open to them as do adults. Adults hardly have all the knowledge they should have or all the services they need. But whatever they have at least should be shared equally with teen-agers. Like adults, teen-age girls can get pregnant, and like adults, teen-age boys can impregnate girls. Why then should they remain ignorant, unserved, and uncared for?[23]

*Information about the Family Learning Center may be obtained by writing to E. Cassandra Jordan, Director of Pupil Personnel Services, New Brunswick Public Schools, 24 Bayard St., New Brunswick, NJ 08901.

1. *Findings and Recommendations of the First Inter-hemispheric Conference on Adolescent Fertility* (Arlie Center, Arlie, Va., 1976), p. 1.

2. Alan Guttmacher Institute, *11 Million Teenagers: What Can Be Done About the Epidemic of Adolescent Pregnancies in the United States* (New York: Alan Guttmacher Institute, 1976), p. 7.

3. Ibid., p. 9.

4. Ibid., pp. 16, 17.

5. Ibid., pp. 10, 11.

6. U.S. Senate, *A Bill To Establish a Program for Developing Networks of Community-Based Services To Assist Pregnant Adolescents and Adolescent Parents. . . ,* S. 2910, Ninety-fifth Congress, Second Session, 1978, pp. 17, 18.

7. Alfonso A. Narvaez, "Pregnancies Rising in 10-15 Age Group," *New York Times,* May 15, 1977, 11-8, 9.

8. Family Planning Public Affairs Office, "Facts of Life in New Jersey," *New Jersey Family Planning News,* November/December, 1978.

9. Narvaez, op. cit.

10. J. Walters, "Birth Defects and Adolescent Pregnancies," *Journal of Home Economics,* November, 1975, p. 24.

11. F. Ivan Nye, *School-Age Parenthood: Consequences for Babies, Mother, Fathers, Grandparents, and Others* (Pullman, Wash.: Washington State University, 1977), p. 2.

12. Walters, op. cit.

13. Nye, op. cit., p. 4

14. Guttmacher Institute, op. cit., p. 23.

15. Ibid., p. 25.

16. Nye, op. cit., p. 9.

17. Family Planning Public Affairs Office, op. cit.

18. Project on Equal Education Rights, *Cracking the Glass Slipper: PEER's Guide to Ending Sex Bias in Your Schools,* 1977, p. 2. Available from NOW Legal Defense and Education Fund, 1029 Vermont Ave., N.W., Washington, DC 20005.

19. New Jersey State Department of Education, *Guidelines for Developing School Programs in Sex Education* (Trenton: 1967), p. 1.

20. New Jersey State Department of Education, "Mandatory Sex Education Classes Urged by Health Department Official," *Interact,* March, 1978, p. 6.

21. New Brunswick Public Schools, *Educational Services for School-Age Parents* (New Brunswick: 1974), p. 3.

22. E. Cassandra Jordan, "Educational Options in New Brunswick," *Today's Education,* February/March, 1978, pp. 66-68.

23. Guttmacher Institute, op. cit., p. 59.

THE MANY ME'S OF THE SELF-MONITOR

Is there a "true self" apart from the social roles we play?
Perhaps not for people identified in studies as high
self-monitors, who are keenly aware of the impression they are
making and constantly fine-tuning their performance.

MARK SNYDER

Mark Snyder is professor of psychology at the University of Minnesota in Minneapolis, where he teaches a graduate-level course called "The Self." In addition to his research on self-monitoring, he is studying stereotypes and the effect of stereotypes on social relationships.

"The image of myself which I try to create in my own mind in order that I may love myself is very different from the image which I try to create in the minds of others in order that they may love me."

—W. H. Auden

The concept of the self is one of the oldest and most enduring in psychological considerations of human nature. We generally assume that people are fairly consistent and stable beings: that a person who is generous in one situation is also likely to be generous in other situations, that one who is honest is honest most of the time, that a person who takes a liberal stance today will favor the liberal viewpoint tomorrow.

It's not always so: each of us, it appears, may have not one but many selves. Moreover, much as we might like to believe that the self is an integral feature of personal identity, it appears that, to a greater extent, the self is a product of the individual's relationships with other people. Conventional wisdom to the contrary, there may be striking gaps and contradictions—as Auden suggests—between the public appearances and private realities of the self.

Psychologists refer to the strategies and techniques that people use to control the impressions they convey to others as "impression management." One of my own research interests has been to understand why some individuals are better at impression management than others. For it is clear that some people are particularly sensitive to the ways they express and present themselves in social situations—at parties, job interviews, professional meetings, in confrontations of all kinds where one might choose to create and maintain an appearance, with or without a specific purpose in mind. Indeed, I have found that such people have developed the ability to carefully monitor their own performances and to skillfully adjust their performances when signals from others tell them that they are not having the desired effect. I call such persons "high self-monitoring individuals," and I have developed a 25-item measure—the Self-Monitoring Scale—that has proved its ability to distinguish high self-monitoring individuals from low self-monitoring individuals. (See box on page 219.) Unlike the high self-monitoring individuals, low self-monitoring individuals are not so concerned about taking in such information; instead, they tend to express what they feel, rather than mold and tailor their behavior to fit the situation.

My work on self-monitoring and impression management grew out of a long-standing fascination with explorations of reality and illusion in literature and in the theater. I was struck by the contrast between the way things often appear to be and the reality that lurks beneath the surface—on the stage, in novels, and in people's actual lives. I wanted to know how this world of appearances in social relationships was built and maintained, as well as what its effects were on the individual personality. But I was also interested in exploring the older, more philosophical question of whether, beneath the various images of self that people project to others, there is a "real me." If we are all actors in many social situations, do we then retain in any sense an essential self, or are we really a variety of selves?

Skilled Impression Managers

There are striking and important differences in the extent to which people can and do control their self-presentation in social situations: some people engage in impression management more often—and with greater skill—than others. Professional actors, as well as many trial lawyers, are among the best at it. So are successful salespeople, confidence artists, and politicians. The onetime mayor of New York, Fiorello LaGuardia, was particularly skilled at adopting the expressive mannerisms of a variety of ethnic groups. In fact, he was so good at it that in watching silent films of his campaign speeches, it is easy to guess whose vote he was soliciting.

Of course, such highly skilled performances are the exception rather than the rule. And people differ in the extent to which they can and do exercise control over their self-presentations. It is the high self-monitoring individuals among us who are particularly talented in this regard. When asked to describe high self-monitoring individuals, their friends say that they are good at learning which behavior is appropriate in social situations, have good self-control of their emotional expression, and can effectively use this ability to create the impression they want. They are particularly skilled at intentionally expressing and accurately communicating a wide variety of emotions both vocally and facially. As studies by Richard Lippa of California State University at Fullerton have shown, they are usually such polished actors that they can effectively adopt the mannerisms of a reserved, withdrawn, and introverted individual and then do an abrupt about-face and portray, just as convincingly, a friendly, outgoing, and extraverted personality.

High self-monitoring individuals are also quite likely to seek out information about appropriate patterns of self-presentation. They invest considerable effort in attempting to "read" and understand others. In an experiment I conducted with Tom Monson (then one of my graduate students), various cues were given to students involved in group discussions as to what was socially appropriate behavior in the situation. For example, some of them thought that their taped discussions would be played back to fellow students; in those circumstances, I assumed they would want their opinions to appear as autonomous as possible. Others believed that their discussions were completely private; there, I assumed they would be most concerned with maintaining harmony and agreement in the group. High self-monitoring individuals were keenly attentive to these differences; they conformed with the group when conformity was the most appropriate behavior and did not conform when they knew that the norms of the larger student audience would favor autonomy in the face of social pressure. Low self-monitoring individuals were virtually unaffected by the differences in social setting: presumably, their self-presentations were more accurate reflections of their personal attitudes and dispositions. Thus, as we might have guessed, people who are most

MONITOR YOUR SELF

On the scale I have developed to measure self-monitoring, actors are usually high scorers, as are many obese people, who tend to be very sensitive about the way they appear to others. For much the same reason, politicians and trial lawyers would almost certainly be high scorers. Recent immigrants eager to assimilate, black freshmen in a predominantly white college, and military personnel stationed abroad are also likely to score high on the scale.

The Self-Monitoring Scale measures how concerned people are with the impression they are making on others, as well as their ability to control and modify their behavior to fit the situation. I believe that it defines a distinct domain of personality that is quite different from the traits probed by other standard scales.

Several studies show that skill at self-monitoring is not associated with exceptional intelligence or with a particular social class. Nor is it related, among other things, to being highly anxious or extremely self-conscious, to being an extravert, or to having a strong need for approval. They may be somewhat power-oriented or Machiavellian, but high self-monitoring individuals do not necessarily have high scores on the "Mach" scale, a measure of Machiavellianism developed by Richard Christie of Columbia University. (Two items from the scale: "The best way to handle people is to tell them what they want" and "Anyone who completely trusts anyone else is asking for trouble.") The steely-eyed Machiavellians are more manipulative, detached, and amoral than high self-monitoring individuals.

The Self-Monitoring Scale describes a unique trait and has proved to be both statistically valid and reliable, in tests on various samples.

At left is a 10-item abbreviated version of the Self-Monitoring Scale that will give readers some idea of whether they are low or high self-monitoring individuals. If you would like to test your self-monitoring tendencies, follow the instructions and then consult the scoring key. —M.S.

These statements concern personal reactions to a number of different situations. No two statements are exactly alike, so consider each statement carefully before answering. If a statement is true, or mostly true, as applied to you, circle the T. If a statement is false, or not usually true, as applied to you, circle the F.

1. I find it hard to imitate the behavior of other people.   T F
2. I guess I put on a show to impress or entertain people.   T F
3. I would probably make a good actor.   T F
4. I sometimes appear to others to be experiencing deeper emotions than I actually am.   T F
5. In a group of people I am rarely the center of attention.   T F
6. In different situations and with different people, I often act like very different persons.   T F
7. I can only argue for ideas I already believe.   T F
8. In order to get along and be liked, I tend to be what people expect me to be rather than anything else.   T F
9. I may deceive people by being friendly when I really dislike them.   T F
10. I'm not always the person I appear to be.   T F

SCORING: Give yourself one point for each of questions 1, 5 and 7 that you answered F. Give yourself one point for each of the remaining questions that you answered T. Add up your points. If you are a good judge of yourself and scored 7 or above, you are probably a high self-monitoring individual; 3 or below, you are probably a low self-monitoring individual.

skilled in the arts of impression management are also most likely to practice it.

Although high self-monitoring individuals are well skilled in the arts of impression management, we should not automatically assume that they necessarily use these skills for deceptive or manipulative purposes. Indeed, in their relationships with friends and acquaintances, high self-monitoring individuals are eager to use their self-monitoring abilities to promote smooth social interactions.

We can find some clues to this motive in the way high self-monitoring individuals tend to react to, and cope with, unfamiliar and unstructured social settings. In a study done at the University of Wisconsin, psychologists William Ickes and Richard Barnes arranged for pairs of strangers to spend time together in a waiting room, ostensibly to wait for an experiment to begin. The researchers then recorded the verbal and nonverbal behavior of each pair over a five-minute period, using video and audio tapes. All possible pairings of same-sex undergraduates at high, moderate, and low levels of self-monitoring were represented. Researchers scrutinized the tapes for evidence of the impact of self-monitoring on spontaneous encounters between strangers.

In these meetings, as in so many other aspects of their lives, high self-monitoring individuals suffered little or no shyness. Soon after meeting the other person, they took an active and controlling role in the conversation. They were inclined to talk first and to initiate subsequent conversational sequences. They also felt, and were seen by their partners to have, a greater need to talk. Their partners also viewed them as having been the more directive member of the pair. It was as if high self-monitoring individuals were particularly concerned about managing their behavior in order to create, encourage, and maintain a smooth flow of conversation. Perhaps this quality may help self-monitoring people to emerge as leaders in groups, organizations, and institutions.

Detecting Impression Management In Others

High self-monitoring individuals are

WILLIAM JAMES ON THE ROLES WE PLAY

A man has as many social selves as there are individuals who recognize him and carry an image of him in their mind . . . . But as the individuals who carry the images form naturally into classes, we may practically say that he has as many different social selves as there are distinct *groups* of persons about whose opinions he cares. He generally shows a different side of himself to each of these different groups. Many a youth who is demure enough before his parents and teachers swears and swaggers like a pirate among his 'tough' young friends. We do not show ourselves to our children as to our club companions, to our masters and employers as to our intimate friends. From this there results what practically is a division of the man into several selves; and this may be a discordant splitting, as where one is afraid to let one set of his acquaintances know him as he is elsewhere; or it may be a perfectly harmonious division of labor, as where one tender to his children is stern to the soldiers or prisoners under his command."

—**William James**
The Principles of Psychology, 1890

---

also adept at detecting impression management in others. To demonstrate this finely tuned ability, three communications researchers at the University of Minnesota made use of videotaped excerpts from the television program "To Tell the Truth." On this program, one of the three guest contestants (all male in the excerpts chosen for the study) is the "real Mr. X." The other two who claim to be the real Mr. X are, of course, lying. Participants in the study watched each excerpt and then tried to identify the real Mr. X. High self-monitoring individuals were much more accurate than their low self-monitoring counterparts in correctly identifying the real Mr. X. and in seeing through the deception of the other two contestants.

Not only are high self-monitoring individuals able to see beyond the masks of deception successfully but they are also keenly attentive to the actions of other people as clues to their underlying intentions. E. E. Jones and Roy Baumeister of Princeton University had college students watch a videotaped discussion between two men who either agreed or disagreed with each other. The observers were aware that one man (the target person) had been instructed either to gain the affection or to win the respect of the other. Low self-moni-

toring observers tended to accept behavior at face value. They found themselves attracted to the agreeable person, whether or not he was attempting to ingratiate himself with his discussion partner. In contrast, high self-monitoring observers were acutely sensitive to the motivational context within which the target person operated. They liked the target better if he was disagreeable when trying to ingratiate himself. But when he sought respect, they were more attracted to him if he chose to be agreeable. Jones and Baumeister suggest that high self-monitoring observers regarded agreeableness as too blatant a ploy in gaining affection and autonomy as an equally obvious route to respect. Perhaps the high self-monitoring individuals felt that they themselves would have acted with greater subtlety and finesse.

Even more intriguing is Jones's and Baumeister's speculation—and I share their view—that high self-monitoring individuals prefer to live in a stable, predictable social environment populated by people whose actions consistently and accurately reflect their true attitudes and feelings. In such a world, the consistency and predictability of the actions of others would be of great benefit to those who tailor and manage their own self-presentation in social situations. From this perspective, it becomes quite understandable that high self-monitoring individuals may be especially fond of those who avoid strategic posturing. Furthermore, they actually may prefer as friends those comparatively low in self-monitoring.

How can we know when strangers and casual acquaintances are engaged in self-monitoring? Are there some channels of expression and communication that are more revealing than others about a person's true, inner "self," even when he or she is practicing impression management?

Both scientific and everyday observers of human behavior have suggested that nonverbal behavior—facial expressions, tone of voice, and body movements—reveals meaningful information about a person's attitudes, feelings, and motives. Often, people who engage in self-monitoring for deceptive purposes are less skilled at controlling their body's expressive movements. Accordingly, the body may be a more revealing source of information than the face for detecting those who engage in self-monitoring and impression management.

More than one experiment shows how nonverbal behavior can betray the true attitude of those attempting impression management. Shirley Weitz of the New School for Social Research reasoned that on college campuses where there are strong normative pressures supporting a tolerant and liberal value system, all students would avoid saying anything that would indicate racial prejudice—whether or not their private attitudes supported such behavior. In fact, she found that among "liberal" white males at Harvard University, the most prejudiced students (as determined by behavioral measures of actual attempts to avoid interaction with blacks) bent over backwards to *verbally* express liking and friendship for a black in a simulated interracial encounter. However, their *nonverbal* behaviors gave them away. Although the prejudiced students made every effort to say kind and favorable things, they continued to do so in a cool and distant tone of voice. It was as if they knew the words but not the music: they knew *what* to say, but not *how* to say it.

Another way that prejudice can be revealed is in the physical distance

people maintain between themselves and the target of their prejudice. To demonstrate this phenomenon, psychologist Stephen Morin arranged for college students to be interviewed about their attitudes toward homosexuality. Half the interviewers wore "Gay and Proud" buttons and mentioned their association with the Association of Gay Psychologists. The rest wore no buttons and simply mentioned that they were graduate students working on theses. Without the students' knowledge, the distance they placed their chairs from the interviewer was measured while the interviews were going on. The measure of social distance proved to be highly revealing. When the student and the interviewer were of the same sex, students tended to establish almost a foot more distance between themselves and the apparently gay interviewers. They placed their chairs an average of 32 inches away from apparently gay interviewers, but only 22 inches away from apparently nongay interviewers. Interestingly, most of the students expressed tolerant, and at times favorable, attitudes toward gay people in general. However, the distances they chose to put between themselves and the interviewers they thought gay betrayed underlying negative attitudes.

Impression Managers' Dilemmas

The well-developed skills of high self-monitoring individuals ought to give them the flexibility to cope quickly and effectively with a diversity of social roles. They can choose with skill and grace the self-presentation appropriate to each of a wide variety of social situations. But what happens when the impression manager must effectively present a true and honest image to other people?

Consider the case of a woman on trial for a crime that she did not commit. Her task on the witness stand is to carefully present herself so that everything she does and says communicates to the jurors clearly and unambiguously her true innocence, so that they will vote for her acquittal. Chances are good, however, that members of the jury are somewhat skeptical of the defendant's claims of innocence. After all, they might reason to themselves, the district attor-

ney would not have brought this case to trial were the state's case against her not a convincing one.

The defendant must carefully manage her verbal and nonverbal behaviors so as to ensure that even a skeptical jury forms a true impression of her innocence. In particular, she must avoid the pitfalls of an image that suggests that "she doth protest her innocence too much and therefore must be guilty." To the extent that our defendant skillfully practices the art of impression management, she will succeed in presenting herself to the jurors as the honest person that she truly is.

It often can take as much work to present a truthful image as to present a deceptive one. In fact, in this case, just being honest may not be enough when facing skeptical jurors who may bend over backwards to interpret any and all of the defendant's behavior—nervousness, for example—as a sign of guilt.

The message from research on impression management is a clear one. Some people are quite flexible in their self-presentation. What effects do these shifts in public appearance have on the more private realities of self-concept? In some circumstances, we are persuaded by our own appearances: we become the persons we appear to be. This phenomenon is particularly likely to occur when the image we present wins the approval and favor of those around us.

In an experiment conducted at Duke University by psychologists E. E. Jones, Kenneth Gergen, and Keith Davis, participants who had been instructed to win the approval of an interviewer presented very flattering images of themselves. Half the participants (chosen at random) then received favorable reactions from their interviewers; the rest did not. All the participants later were asked to estimate how accurately and honestly their self-descriptions had mirrored their true personalities.

Those who had won the favor of their interviewers considered their self-presentations to have been the most honest of all. One interpretation of this finding is that those people were operating with rather pragmatic definitions of self-concept: that which produced the most positive results was considered to be an accurate reflection of the inner self.

The reactions of other people can make it all the more likely that we become what we claim to be. Other people may accept our self-presentations at face value; they may then treat us as if we really were the way we pretend to be. For example, if I act as if I like Chris, chances are Chris will like me. Chris will probably treat me in a variety of friendly ways. As a result of Chris's friendliness, I may come to like Chris, even though I did not in the first place. The result, in this case, may be beneficial to both parties. In other circumstances, however, the skilled impression manager may pay an emotional price.

High self-monitoring orientation may be purchased at the cost of having one's actions reflect and communicate very little about one's private attitudes, feelings, and dispositions. In fact, as I have seen time and again in my research with my former graduate students Beth Tanke and Bill Swann, correspondence between private attitudes and public behavior is often minimal for high self-monitoring individuals. Evidently, the words and deeds of high self-monitoring individuals may reveal precious little information about their true inner feelings and attitudes.

Yet, it is almost a canon of modern psychology that a person's ability to reveal a "true self" to intimates is essential to emotional health. Sidney Jourard, one of the first psychologists to hold that view, believed that only through self-disclosure could we achieve self-discovery and self-knowledge: "Through my self-disclosure, I let others know my soul. They can know it, really know it, only as I make it known. In fact, I am beginning to suspect that I can't even know *my own soul* except as I disclose it. I suspect that I will know myself "for real" at the exact moment that I have succeeded in making it known through my disclosure to another person."

Only low self-monitoring individuals may be willing or able to live their lives according to Jourard's prescriptions. By contrast, high self-monitoring individuals seem to embody Erving Goffman's view of human nature. For him, the world of appearances appears to be all, and the "soul" is illusory. Goffman defines social interactions as a theatrical performance in which each individual acts out a "line." A line is a set of carefully chosen verbal and nonverbal acts that express one's self. Each of us, in Goffman's view, seems to be merely the sum of our various performances.

What does this imply for the sense of self and identity associated with low and high self-monitoring individuals?

I believe that high self-monitoring individuals and low self-monitoring individuals have very different ideas about what constitutes a self and that their notions are quite well-suited to how they live. High self-monitoring individuals regard themselves as rather flexible and adaptive people who tailor their social behavior shrewdly and pragmatically to fit appropriate conditions. They believe that a person is whoever he appears to be in any particular situation: "I am me, the me I am right now." This self-image fits well with the way high self-monitoring individuals present themselves to the world. It allows them to act in ways that are consistent with how they believe they should act.

By contrast, low self-monitoring individuals have a firmer, more single-minded idea of what a self should be. They value and strive for congruence between "who they are" and "what they do" and regard their actions as faithful reflections of how they feel and think. For them, a self is a single identity that must not be compromised for other people or in certain situations. Indeed, this view of the self parallels the low self-monitoring individual's consistent and stable self-presentation.

What is important in understanding oneself and others, then, is not the elusive question of whether there is a quintessential self, but rather, understanding how different people define those attributes of their behavior and experience that they regard as "me." Theory and research on self-monitoring have attempted to chart the processes by which beliefs about the self are actively translated into patterns of social behavior that reflect self-conceptions. From this perspective, the processes of self-monitoring are the processes of self—a system of operating rules that translate self-knowledge into social behavior.

For further information, read:

Gergen, Kenneth. *The Concept of Self*, Holt, Rinehart & Winston, 1971, paper, $4.50.

Goffman, Erving. *The Presentation of Self in Everyday Life*, Doubleday (reprint of 1959 edition), paper, $2.50.

Snyder, Mark. "Self-Monitoring Processes," in *Advances in Experimental Social Psychology, Vol. 12*, Leonard Berkowitz, ed., Academic Press, 1979, $24.

Snyder, Mark. "Cognitive, Behavioral, and Interpersonal Consequences of Self-Monitoring," in *Advances in the Study of Communication and Affect, Vol. 5: Perception of Emotion in Self and Others*, Plenum, 1979, $24.50.

Snyder, Mark. "Self-Monitoring of Expressive Behavior," *Journal of Personality and Social Psychology*, 30(1974): 526-537.

Why Johnny Can't Disobey

Sarah J. McCarthy

Sarah J. McCarthy has taught in the Pittsburgh Public Schools and is now studying psychology and creative writing at the University of Pittsburgh, where she has taught the course Psychology of Women.

Few people are too concerned about whether Johnny can disobey. There is no furor or frantic calls to the PTA, as when it is discovered that he can't read or does poorly on his S.A.T. scores. Even to consider the question is at first laughable. Parents and teachers, after all, are systematically working at developing the virtue of obedience. To my knowledge, no one as yet has opened a remedial disobedience school for overly compliant children, and probably no one ever will. And that in itself is a major problem.

Patricia Hearst recently said that the mindless state of obedience which enveloped her at the hands of the Symbionese Liberation Army could happen to anyone. Jumping to a tentative conclusion from a tip-of-the-iceberg perspective, it looks as though it already has happened to many, and that it has required nothing so dramatic as a kidnapping to bring it about.

Given our experience with various malevolent authority figures such as Adolph Hitler, Charles Manson, Lieutant Calley, and Jim Jones, it is unfortunately no longer surprising that there are leaders who are capable of wholesale cruelty to the point of directing mass killings. What remains shocking, however, is that they are so often successful in recruiting followers. There seems to be no shortage of individuals who will offer their hearts and minds on a silver platter to feed the egos of the power-hungry. This becomes even more disturbing when one ponders the truism that society's neurotics are often its cultural caricatures, displaying exaggerated manifestations of its collective neuroses. There are enough examples of obedience to horrendous commands for us to ask if and how a particular culture sows the seeds of dangerous conformity.

Political platitudes and lip service to the contrary, obedience is highly encouraged in matters petty as well as profound. Linda Eton, an Iowa firefighter, was suspended from her job and catapulted to national fame for the radical act of breast-feeding at work. A dehumanized, compartmentalized society finds little room for spontaneity, and a blatantly natural act like breast-feeding is viewed as a preposterous interruption of the status quo.

Pettiness abounds in our social relationships, ensuring compliance through peer pressure and disapproval, and enforced by economic sanctions at the workplace. A friend of mine, a construction worker, reported to his job one rainy day carrying an umbrella. The foreman was outraged by this break from the norm, and demanded that the guy never again carry an umbrella to the construction site, even if the umbrella *was* black, since it "caused his whole crew to look like a bunch of faggots."

Another friend, though less scandalizingly visible in his job as a security guard during the wee hours for a multinational corporation, was caught redhanded playing a harmonica. Mercifully, he was given another chance, only to be later fired for not wearing regulation shoes.

Ostensibly, such firings and threats are deemed necessary to prevent inefficiency and rampant chaos at the workplace. But if employers were merely concerned about productivity and efficiency, it certainly is disputable that "yes-people" are more productive and beneficial than "no-people." Harmonicas may even increase efficiency by keeping security guards sane, alert, and awake by staving off sensory deprivation. A dripping-wet construction worker could conceivably be less productive than a dry one. And the Adidas being worn by the errant security guard could certainly have contributed to his fleetness and agility as opposed to the cumbersome regulation shoes. The *real* issues here have nothing to do with productivity. What is really involved is an irrational fear of the mildly unusual, a pervasive attitude held by authorities that their subordinates are about to run amok and need constant control.

These little assaults on our freedom prepare us for the big ones. Having long suspected that a huge iceberg of mindless obedience existed beneath our cultural surface, I was not particularly surprised when I heard that nine hundred people followed their leader to mass suicide. For some time we have lived with the realization that people are capable of killing six million of their fel-

This article first appeared in *THE HUMANIST*, September/October 1979 and is reprinted by permission.

low citizens on command. Jonestown took us one step further. People will kill themselves on command.

In matters ridiculous and sublime, this culture and the world at large clearly exhibit symptoms of pathological obedience. Each time one of the more sensational incidents occurs—Jonestown, the Mai Lai massacre, Nazi Germany, the Manson murders—we attribute its occurrence to factors unique to it, trying to deny any similarities to anything close to us, tossing it about like a philosophical hot potato. We prefer to view such events as anomalies, isolated in time and space, associated with faraway jungles, exotic cults, drugged hippies, and outside agitators. However, as the frequency of such happenings increases, there is the realization that it is relatively easy to seduce some people into brainwashed states of obedience.

Too much energy and time have been spent on trying to understand the alleged compelling traits and mystical powers of charismatic leaders, and not enough in an attempt to understand their fellow travelers—the obedient ones. We need to look deeper into those who *elected* Hitler, and all those followers of Jim Jones who went to Guyana *voluntarily*. We must ask how many of us are also inclined toward hyperobedience. Are we significantly different, capable of resisting malevolent authority, or have we simply had the good fortune never to have met a Jim Jones of our own?

Social psychologist Stanley Milgram, in his book *Obedience to Authority*, is convinced that:

> In growing up, the normal individual has learned to check the expression of aggressive impulses. But the culture has failed, almost entirely, in inculcating internal controls on actions that have their origin in authority. For this reason, the latter constitutes a far greater danger to human survival.

Vince Bugliosi, prosecutor of Charles Manson and author of *Helter Skelter*, commented on the Jonestown suicides:

> Education of the public is the only answer. If young people could be taught what can happen to them—that they may be zombies a year after talking to that smiling person who stops them on a city street—they may be prepared.

Presumably, most young cult converts have spent most of their days in our educational system, yet are vulnerable to the beguiling smile or evil eye of a Charles Manson. If there is any lesson to be learned from the obedience-related holocausts, it must be that we can never underestimate the power of education and the socialization process.

Contrary to our belief that the survival instinct is predominant over all other drives, the Jonestown suicides offer testimony to the power of cultural indoctrination. Significantly, the greatest life force at the People's Temple came from the children. Acting on their survival instincts, they went kicking and screaming to their deaths in an "immature" display of disobedience. The adults, civilized and educated people that they were, lined up with "stiff upper lips" and took their medicine like the followers they were trained to be—a training that didn't begin at Jonestown.

When something so horrible as Jonestown happens, people draw metaphors about the nearness of the jungle and the beast that lurks within us. It seems that a more appropriate metaphor would be our proximity to an Orwellian civilization with its antiseptic removal of our human rough edges and "animal" instincts. On close scrutiny, the beast within us looks suspiciously like a sheep.

Despite our rich literature of freedom, a pervasive value instilled in our society is obedience to authority. Unquestioning obedience is perceived to be in the best interests of the schools, churches, families, and political institutions. Nationalism, patriotism, and religious ardor are its psychological vehicles.

Disobedience is the original sin, as all of the religions have stated in one way or another. Given the obedience training in organized religions that claim to possess mystical powers and extrarational knowledge and extoll the glories of self-sacrifice, what is so bizarre about the teachings of Jim Jones? If we arm our children with the rationality and independent thought necessary to resist the cultist, can we be sure that our own creeds and proclamations will meet the criteria of reason? The spotlight of reason which exposes the charlatan may next shine on some glaring inconsistencies in the "legitimate" religions. Religions, which are often nothing more than cults that grew, set the stage for the credulity and gullibility required for membership in cults.

A witch hunt is now brewing to exorcise the exotic cults, but what is the dividing line between a cult and a legitimate religion? Is there a qualitative difference between the actions of some venerated Biblical saints and martyrs and the martyrs of Jonestown? If the Bible contained a Parable of Guyana, the churches would regularly extoll it as a courageous act of self-sacrifice. Evidently saints and martyrs are only palatable when separated by the chasm of a few centuries. To enforce their beliefs, the major religions use nothing so crass as automatic weapons, of course, but instead fall back on automatic sentences to eternal damnation.

Certainly there must be an optimal level of obedience and cooperation in a reasonable society, but obedience, as any other virtue that is carried to an extreme, may become a vice. It is obvious that Nazi Germany and Jonestown went too far on the obedience continuum. In more mundane times and places the appropriate level of obedience is more difficult to discover.

We must ask if our society is part of the problem, part of the solution, or wholly irrelevant to the incidents of over-obedience exhibited at Jonestown and Mai Lai. Reviewing social psychologists' attempts to take our psychic temperatures through empirical measurements of our conformity and obedience behavior in experimental situations, our vital signs do not look good.

In 1951 Solomon Asch conducted an experiment on conformity, which is similar to obedience behavior in that it subverts one's will to that of peers or an authority. This study, as reported in the textbook *Social Psychology* by Freedman, Sears, and Carlsmith, involved college students who were asked to estimate lines of equal and differing lengths. Some of the lines were obviously equal, but if subjects heard others before them unanimously give the wrong answer, they would also answer incorrectly. Asch had reasoned that people would be rational enough to choose the evidence of their own eyes over the disagreeing "perceptions" of others. He found that he was wrong.

When subjects were asked to estimate the length of a line after confederates of the experimenter had given obviously wrong answers, the subjects gave wrong answers about 35 percent of the time. Authors Freedman, Sears, and Carlsmith stress:

It is important to keep the unambiguousness of the situations in mind if we are to understand this phenomenon. There is a tendency to think that the conforming subjects are uncertain of the correct choice and therefore are swayed by the majority. This is not always the case. In many instances subjects are quite certain of the correct choice and, in the absence of group pressure, would choose correctly 100 percent of the time. When they conform, they are conforming despite the fact that they know the correct answer.

If 35 percent of those students conformed to group opinion in unambiguous matters and in direct contradiction of the evidence of their own eyes, how much more must we fear blind following in *ambiguous* circumstances or in circumstances where there exists a legitimate authority?

In the early sixties, Yale social psychologist Stanley Milgram devised an experiment to put acts of obedience and disobedience under close scrutiny. Milgram attempted to understand why thousands of "civilized" people had engaged in an extreme and immoral act—that of the wholesale extermination of Jews—in the name of obedience. He devised a learning task in which subjects of the experiment were instructed to act as teachers. They were told to "shock" learners for their mistakes. The learners were actually confederates of the experimenter and were feigning their reactions. When a mistake was made, the experimenter would instruct the teacher to administer an ever-increasing voltage from a shock machine which read "Extreme Danger," "Severe Shock," and "XXX." Although the machine was unconnected, the subject-teachers believed that they were actually giving shocks. They were themselves given a real sample shock before the experiment began.

Milgram asked his Yale colleagues to make a guess as to what proportion of subjects would proceed to shock all the way to the presumed lethal end of the shockboard. Their estimates hovered around 1 or 2 percent. No one was prepared for what happened. All were amazed that twenty-six out of forty subjects obeyed the experimenter's instruction to press levers that supposedly administered severely dangerous levels of shock. After this, Milgram regularly obtained results showing that 62 to 65 percent of people would shock to the end of the board. He tried several variations on the experiment, one of which was to set it up outside of Yale University so that the prestige of the University would not be an overriding factor in causing subjects to obey. He found that people were just as likely to administer severe shock, whether the experiments occurred within the hallowed halls of Yale or in a three-room walk-up storefront in which the experimenters spoke of themselves as "scientific researchers."

In another variation of the experiment, Milgram found that aggression—latent or otherwise—was not a significant factor in causing the teacher-subjects to shock the learners. When the experimenter left the room, thus permitting the subjects to choose the level of shock themselves, almost none administered more than the lowest voltage. Milgram concluded that obedience, not aggression, was the problem. He states:

I must conclude that [Hannah] Arendt's conception of the *banality of evil* comes closer to the truth than one might dare imagine. The ordinary person who shocked the victim did so out of a sense of obligation—a conception of his duties as a subject—and not from any peculiarly aggressive tendencies.

This is, perhaps, the most fundamental lesson of our study: ordinary people, simply doing their jobs, and without any particular hostility on their part, can become agents in a terrible destructive process. Moreover, even when the destructive effects of their work become patently clear, and they are asked to carry out actions incompatible with fundamental standards of morality, relatively few people have the resources needed to resist authority. A variety of inhibitions against disobeying authority come into play and successfully keep the person in his place.

A lack of compassion was not a particularly salient personality factor in the acts of obedience performed by the followers of Hitler, Jim Jones, and the subjects in the Milgram experiments. Nazi soldiers were capable of decent human behavior toward their friends and family. Some, too, see an irony in that Hitler himself was a vegetarian. The People's Temple members seemed more compassionate and humanitarian than many, and yet they forced their own children to partake of a drink laced with cyanide. Those shocking the victims in the Milgram experiments exhibited signs of compassion both toward the experimenter and to the persons that they thought were receiving the shocks. In fact, Milgram finds that:

7. ADOLESCENCE AND YOUNG ADULTHOOD

It is a curious thing that a measure of compassion on the part of the subject, an unwillingness to "hurt" the experimenter's feelings, are part of those binding forces inhibiting disobedience . . . only obedience can preserve the experimenter's status and dignity.

Milgram's subjects showed signs of severe physiological tension and internal conflict when instructed to shock. Presumably, these signs of psychic pain and tortured indecision were a manifestation of an underlying attitude of compassion for the victim, but it was not sufficient to impel them to openly break with, and therefore embarrass, the experimenter, even though this experimenter had no real authority over them. One of Milgram's subjects expressed this dilemma succinctly:

I'll go through with anything they tell me to do . . . They know more than I do . . . I know when I was in the service (if I was told) "You go over the hill and we're going to attack," we attacked. So I think it's all based on the way a man was brought up . . . in his background. Well, I faithfully believed the man [whom he thought he had shocked] was dead until we opened the door. When I saw him, I said: "Great, this is great!" But it didn't bother me even to find that he was dead. I did a job.

The experiments continued with thousands of people —students and nonstudents, here and abroad—often demonstrating obedience behavior in 60 to 65 percent of the subjects. When the experiments were done in Munich, obedience often reached 85 percent. Incidentally, Milgram found no sex differences in obedience behavior. Though his sample of women shockers was small, their level of obedience was identical to that of the men. But they did exhibit more symptons of internal conflict. Milgram concluded that "there is probably nothing the victim can say that will uniformly generate disobedience," since it is not the victim who is controlling the shocker's behavior. Even when one of the experimental variations included a victim who cried out that he had a heart condition, this did not lead to significantly greater disobedience. In such situations, the experimenter-authority figure dominates the subject's social field, while the pleading cries of the victim are for the most part ignored.

Milgram found that the authority's power had to be somehow undermined before there was widespread disobedience, as when the experimenter was not physically present, when his orders came over the telephone, or when his orders were challenged by another authority. Most importantly, subjects became disobedient in large numbers only when others rebelled, dissented, or argued with the experimenter. When a subject witnessed another subject defying or arguing with the experimenter, thirty-six out of forty also rebelled, demonstrating that peer rebellion was the most effective experimental variation in undercutting authority.

This social orientation in which the authority dominates one's psyche is attributed by Milgram to a state of mind which he terms "the agentic state." A person makes a critical shift from a relatively autonomous state into this agentic state when she or he enters a situation in which "he defines himself in a manner that renders him open to regulation by a person of higher status."

An extreme agentic state is a likely explanation for the scenario at Jonestown, where even the cries of their own children were not sufficient to dissuade parents from serving cyanide. Despite some ambiguity as to how many Jonestown residents were murdered and how many committed suicide, there remains the fact that these victims had participated in previous suicide rehearsals. Jim Jones, assured of their loyalty and of their critical shift into an agentic state, then had the power to orchestrate the real thing. The supreme irony, the likes of which could only be imagined as appearing in the *Tralfamadore Tribune* with a byline by Kurt Vonnegut, was the picture of the Guyana death scene. Bodies were strewn about beneath the throne of Jones and a banner which proclaimed that those who failed to learn from the lessons of history were doomed to repeat them.

How many of us have made the critical shift into an agentic state regarding international relations, assuming that our leaders know best, even though they have repeatedly demonstrated that they do not? Stanley Milgram predicts that "for the man who sits in front of the button that will release Armageddon, depressing it will have about the same emotional force as calling for an elevator . . . evolution has not had a chance to build inhibitors against such remote forms of aggression."

We should recognize that our human nature renders us somewhat vulnerable. For one thing, our own mortality and that of our loved ones is an unavoidable fact underlying our lives. In the face of it, we are powerless; and in our insecurity, many reach out for sure answers. Few choose to believe, along with Clarence Darrow, that not only are we not the captains of our fate, but that we are not even "deckhands on a rudderless dinghy." Or, as someone else has stated: "There are no answers. Be brave and face up to it." Most of us won't face up to it. We want our answers, solutions to our plight, and we want them now. Too often truth and rational thought are the first casualties of this desperate reach for security. We embrace answers from charlatans, false prophets, charismatic leaders, and assorted demagogues. Given these realities of our nature, how can we avoid these authority traps to which we are so prone? By what criteria do we teach our children to distinguish between the charlatan and the prophet?

It seems that the best armor is the rational mind. We must insist that all authorities account for themselves, and we need to be as wary of false prophets as we are of false advertising. Leaders, political and spiritual, must be subjected to intense scrutiny, and we must insist that their thought processes and proclamations measure up to reasonable standards of rational thought. Above all, we must become skilled in activating our inner resources toward rebellion and disobedience, when this seems reasonable.

The power of socialization can conceivably be harnessed so as to develop individuals who are rational and skeptical, capable of independent thought, and who can disobey or disagree at the critical moment. Our society, however, continues systematically to instill exactly the opposite. The educational system pays considerable lip service to the development of self-reliance, and places huge emphasis on lofty concepts of individual differences. Little notice is taken of the legions of overly obedient children in the schools; yet, for every overly disobedient child, there are probably twenty who are obeying too much. There is little motivation to encourage the unsqueaky wheels to develop as noisy, creative, independent thinkers who may become bold enough to disagree. Conceivably, we could administer modified Milgram obedience tests in the schools which detect hyper-obedience, just as we test for intelligence, visual function, vocational attributes and tuberculosis. When a child is found to be too obedient, the schools should mobilize against this psychological crippler with the zeal by which they would react to an epidemic of smallpox. In alcoholism and other mental disturbances, the first major step toward a reversal of the pathology is recognition of the severity of the problem. Obedience should be added to the list of emotional disturbances requiring therapy. Disobedience schools should be at least as common as military schools and reform schools.

The chains on us are not legal or political, but the invisible chains of the agentic state. We have all gotten the message that it is dangerous and requires exceptional courage to be different.

If we are to gain control of our lives and minds, we must first acknowledge the degree to which we are not now in control. We must become reasonable and skeptical. Reason is no panacea, but, at the moment, it is all that we have. Yet many in our society seem to have the same attitude about rationality and reason that they do about the poverty program—that is, we've tried it and it doesn't work.

Along with worrying about the S.A.T. scores and whether or not Johnny can read, we must begin to seriously question whether Johnny is capable of disobedience. The churches and cults, while retaining their constitutional right to free expression, must be more regularly criticized. The legitimate religions have been treated as sacred cows. Too often, criticism of them is met with accusations of religious bigotry, or the implications that one is taking candy from a baby or a crutch from a cripple. The concept of religious tolerance has been stretched to its outer limits, implying freedom from criticism and the nonpayment of taxes. Neither patriotism nor religion should be justification for the suspension of reason.

And, on a personal level, we must stop equating sanity with conformity, eccentricity with craziness, and normalcy with numbers. We must get in touch with our own liberating ludicrousness and practice being harmlessly deviant. We must, in fact, cease to use props or other people to affirm our normalcy. With sufficient practice, perhaps, when the need arises, we may have the strength to force a moment to its crisis.

THE SIBLING BOND
A Lifelong Love/Hate Dialectic

VIRGINIA ADAMS

The link between brothers and sisters is in some ways the most unusual of family relationships. It is the longest lasting, often continuing for 70 or 80 years or more, and the most egalitarian. It is also the least studied. Researchers have been more interested in the relationship between parents and offspring than in sibling interaction. Even when psychologists have focused on the children in a family, they have paid attention chiefly to the effects of birth order on personality and intelligence, or to the role young siblings play in early development when they act as caretakers for still younger brothers and sisters. From the nature of most studies, one might almost deduce that sibling relationships end with childhood.

Over the past few years, however, a growing number of researchers have begun to ask questions about brothers and sisters in adulthood. Do siblings drift apart when they leave home, or do they stay in touch? If they maintain a real relationship, what is it like, and how long is it likely to last?

Some answers have already appeared in print or been reported at professional meetings. Others will be published over the next few months in three book-length studies now in preparation. Most of the new work falls under one of three headings: *fervent sibling loyalty* (never before systematically studied) that arises only under certain family conditions and is so extreme as to produce unexpectedly negative effects in some cases; *sibling rivalry* that can persist into adulthood and even into old age; and *sibling solidarity*, a sense of closeness that leads some siblings to turn to each other for understanding.

Overall, the latest research makes it clear that siblings can exert an important mutual influence, for good or ill, throughout the life span; this is sometimes so even when they are geographically separated. Some psychologists expect sibling relationships to become even more significant as the divorce rate increases, the number of one-parent families grows, and family size declines.

Research confirms what common sense suggests: that the degree and nature of sibling interaction varies greatly, not only from one set of siblings to another but within the same pair at different times in their lives. Some brothers and sisters become and remain best friends; others heartily detest each other.

The most interesting fact, although it is not often noted by researchers or acknowledged by siblings, is that love and hate may exist side by side, in very uneasy equilibrium. One writer speaks of "the delicate balance of competition and camaraderie among all sisters." Her words would seem to apply equally well to brothers. Whether the balance can be maintained, or whether it tips dramatically to one side or the other, depends partly on parental behavior and attitudes and partly on certain critical experiences in the lives of siblings themselves.

Fervent Sibling Loyalty

Among the most innovative of the psychologists now studying adult brothers and sisters are Stephen Bank, adjunct associate professor at Wesleyan University, and Michael D. Kahn, associate professor at the University of Hartford. In a seven-year period they have audiotaped more than 100 interviews with siblings in pairs or in larger groups, and in the course of their work as psychotherapists they have studied many other siblings. Bank and Kahn will describe some of their findings in *Sibling Relationships: Their Nature and Significance Across the Lifespan*, a book edited by Michael E. Lamb and Brian Sutton-Smith that will be published next summer. They will give a more detailed account of their work in their own book, *The Sibling Bond*, due out next February.

Bank and Kahn are particularly interested in "Hansels and Gretels," siblings as intensely loyal as the fairytale children whose loving concern for each other saved them from the wicked witch when their father and stepmother turned them out into the forest to starve. The two psychologists have observed the Hansel and Gretel phenomenon in numerous sibling groups, and they are at pains to distinguish it from sibling solidarity, the kind of cohesiveness that leads adult siblings to offer each other a modest degree of support in time of trouble. In an ordinary relationship, one sister might invite another in the throes of divorce to come and visit for a weekend. An extremely loyal sister, Bank said, might take a sibling into her home indefinitely, acting as "a kind of parent to someone who has been wounded."

Extreme loyalty, Bank says, "involves an irrational and somewhat blind process of putting one's sibling first and foremost," with a willingness to make enormous sacrifices. Pointing out that loyalty comes from the French word *loi* ("law"), he notes that intense loyalty is an unwritten contract providing that "we will stick together."

One of two brothers in their 20s told Bank, "If I ever got in any trouble, I wouldn't go to my wife—I wouldn't go to a friend—I wouldn't go to my boss; I'd go to my brother." One of four brothers aged 36 to 45 said of the other three, "If I knew they needed it, I'd give any one of these guys my last buck, despite my obligations to my wife and children."

The adult Hansels and Gretels that Bank and Kahn studied tried to be together as often as they could and were unhappy when necessity kept them apart. The two young brothers, students at the same university, shared a fantasy that they would not let even marriage part them. They talked about buying a joint homestead, "where their wives and children would blend with them into a big, happy household."

Some of the extremely loyal siblings were bound together by a private code—a special way of exchanging glances, for example, or a word or phrase that meant nothing to outsiders. "The four brothers," Bank said, "repeatedly broke into raucous laughter after one had made what seemed to the interviewer a perfectly neutral comment. They sarcastically 'apologized' for the 'silly' behavior of their brother as if to say, 'This is our sense of humor; *you'll* never understand it, since you didn't grow up with us.' "

How do such intense attachments develop? The critical factor is a kind of family collapse as the children are growing up, with the parents becoming actually or psychologically unavailable. In each of the families studied, the parents were hostile, weak, or absent, and the external circumstances of the children's lives were threatening. "Confronted with a hostile environment," Bank said, "they clung together as the only steady and constant people in one another's lives." The parents of the two college-age brothers died within two years of each other, when the boys were nine and 11 and then 11 and 13, and both boys were seriously abused by a psychotic foster mother. The mother of the four middle-aged brothers died when they were teenagers, and their father was emotionally exhausted, sometimes abusive, and unable to form a close relationship with his children. The family was in a slum in which physical danger was a constant.

Parents who want their children to become and remain fervently loyal to each other do not understand the paradox that Bank underscores: "The *deep* bond between siblings will not develop if parents are real good parents; with good parents, you'll get caring, and solidarity, but not intense loyalty, because there's not much need for it."

Bank's assertion that intense sibling loyalty presupposes a kind of parental abandonment draws indirect support from a number of studies. One of these was done in 1965 by Albert I. Rabin, a Michigan State University psychologist, who compared kibbutz children in Israel with children raised in traditional families there. He found much less sibling jealousy in the kibbutz children, whose parents were with them only two hours a day, than in youngsters whose parents were available to them around the clock.

Even more persuasive evidence comes from the classic study by Anna Freud of six unrelated children orphaned by the Nazis before they were a year old. Kept together in the Terezin concentration camp and, after the war, in a succession of hostels, they had no chance to form consistent relationships with caring adults. When they were about three, they were flown to England and looked after in a special nursery for months to help them make the transition to a more normal life. For a long time, they appeared almost incapable of jealousy.

"It was evident," Anna Freud wrote, "that they cared greatly for each other and not at all for anybody or anything else. They had no other wish than to be together and became upset when they were separated. . . . There was no occasion to urge the children to 'take turns'; they did it spontaneously. . . . They were extremely considerate of each other's feelings. . . . When one of them received a present from a shopkeeper, they demanded the same for each other child, even in their absence. On walks they were concerned for each other's safety in traffic, looked after children who lagged behind, helped each other over ditches, turned aside branches for each other to clear the passage in the woods, and carried each other's coats. In the nursery they picked up each other's toys. . . . At mealtimes handing food to the neighbor was of greater importance than eating oneself."

Among many examples of the children's mutual concern, this one is typical: "John, daydreaming while walking, nearly bumps into a passing child [who does not belong to the group]. Paul shouts at the passerby: '*Blöder Ochs; meine John; blöder Ochs Du*' ('Stupid fool; that's my John; you stupid fool)."

There can, of course, be too much of even such a good thing as loyalty. If parents thought more about it, they might worry a bit less about signs of rivalry and a bit more about excessively close ties. They might look up the story of Dorothy Wordsworth, sister of the poet, who gave her life entirely to her brother William, renouncing any independent existence of her own. They might also remember the reclusive Collyer brothers, who gave their lives to each other—and cannot have had much joy from the sacrifice. In the less extreme cases that Bank and Kahn studied, intense loyalty often imposed "the burden of an obligatory responsibility to one another." Sometimes it stifled friendships with outsiders whom siblings did not happen to like. Often it conflicted with loyalty to spouse and children.

Some very loyal brothers had difficulty admitting wives to their group at all and would submit their choice of spouse to their brothers for approval as if to a review board. At times the system worked well. The first wife admitted to the four-brother group mentioned earlier was a physical-education major who proved eminently suited to life among these macho brothers. "She had grown up with brothers," Bank said. "She knew how to handle brothers. She adapted by becoming 'one of the boys,' accepting their male humor and participating in sports with them. They, in return, adopted her almost as a sister. It was the perfect fit." In other cases, though, wives were made to feel "as if they were on the outside of a very exclusive club."

There were times when extreme loyalty made difficulties for the siblings themselves. The younger of the two college-age brothers felt he was a nobody who existed only as an extension of his brother. Despite the warmth the two felt for each other, Bank said, he decided to get away by himself for a few years so he could discover his own identity. When loyalty was one-way rather than reciprocal, with one sibling doing most of the giving, recipients often felt resentful—despite their gratitude—at being subjected to a kind of domination, while perpetual givers often found themselves shunting aside their own legitimate interests. "If you keep helping and helping, that is in a sense neurotic," said Bank.

But he emphasized that he "would not want to be quoted as calling these people only neurotic." Many siblings reported that their loyalty brought them advantages: never feeling entirely alone, learning skills from each other, having a chance to practice parenthood.

Loyal siblings can be deeply sensitive to each other's needs. According to Bank, "They show some of the qualities of a good psychotherapist; they know when to shut up and when to push." And, he went on, "I don't want to put down the absolute altruism we've seen in these very loyal siblings, because there is such a thing." Citing Jane Goodall's observation that when parent chimpanzees die, a "sibling" takes over as caretaker, Bank suggested that somewhat analogous behavior, having important survival value, may occur in human beings.

Rivalry in Adulthood

Bank and Kahn believe that the degree of rivalry between siblings has been vastly exaggerated and that what looks like rivalry often masks other feelings, for instance, dependency, or simply a need for some kind of intense relationship with a brother or sister. Most psychologists, however, argue that some degree of sibling rivalry is almost inevitable in childhood, and there is plenty of research evidence to support that view. Many people assume that this early rivalry dissipates in adulthood, but on that point the evidence is far from conclusive. Last September two psychologists, both associate professors of education at the University of Cincinnati, made a strong case for the persistence of rivalry into adulthood in a paper presented to the American Psychological Association. Helgola G. Ross and Joel I. Milgram (the brother of psychologist Stanley Milgram) recruited 65 subjects aged 25 to 93 from a Midwestern university community, two senior citizen centers in town, and a Methodist retirement home in the suburbs.

Most of the data came from interviews that were tape-recorded, transcribed, and content-analyzed so that recurrent topics could be noted. In group interviews, all the subjects met in age groups (20s, 30s, 40s, 50s, or 60s) of four to six people. Individual follow-up interviews were conducted with 10 of the subjects so that rivalry could be explored in depth.

Seventy-one percent of the subjects reported that they had sometimes felt rivalrous toward brothers or sisters. Of those, 36 percent claimed that they had been able to overcome the feelings in adolescence or adulthood, but 45 percent admitted that the feelings were still alive.

Of subjects who admitted to experiencing rivalry at some time, 40 percent said it had begun in childhood, 33 percent believed it had not surfaced until adolescence, and 22 percent said they could not remember feeling it at all until adulthood.

About half of the subjects said that adults had initiated their rivalrous feelings when they were children. Parents were most often cited, although grandparents came in for their share of blame, and so, at times, did teachers.

The key stimulus for rivalry was favoritism. Some parents openly compared siblings, asking a child to match standards of behavior, skill, or personal characteristics achieved by a favored brother or sister. In other cases the comparison was covert; the child observed the parents' preferential treatment of one or more favored brothers or sisters.

Roughly 50 percent of all rivalry was said by the subjects to have been initiated by siblings, in half of these cases by brothers, in a third by sisters, in a tenth by the subjects themselves.

When competition began in childhood, Ross and Milgram said in their paper, it appeared to be "a vying for the parents' attention, recognition, and love, but also a more general juggling for power and position among the siblings." Sometimes the precipitating factor was the experience of being cared for by an older sibling. (According to Brenda K. Bryant, a psychologist at the University of California at Davis who has studied siblings as caretakers, sibling babysitters are perceived by their charges as rougher disciplinarians than parents. Young babysitters, Bryant suggests, seem to pattern themselves after the comic strip character Lucy, "whose approach to caretaking with peers such as Charlie Brown is to resort to the sweet reason of the mailed fist.")

Ross and Milgram found that a pattern of rivalry sometimes took shape when a young sibling accepted an older one as mentor and then discovered that the mentor was less disposed to praise the younger person's accomplishments than to depreciate them.

The two experimenters emphasized that inequalities in levels of accomplishment were not by themselves sufficient to generate rivalry. Siblings did not necessarily believe that more is better unless parents thought so.

Some subjects indicated that rivalry could actually be fun. Others said it served to motivate them. "If siblings have the ability to live up to high standards, comparative expectations are not necessarily debilitating," said Ross and Milgram. Many subjects told them that when they managed to find their own areas of expertise, competitiveness gave way to pride in the former rival's successes.

More often than not, however, rivalry was destructive. Some subjects felt deeply hurt. "Many siblings felt excluded from valued sibling or family interactions and the sense of wholeness they can provide," Ross and Milgram said. "Some dissociated themselves psychologically and geographically from particular siblings or the family on a semi-permanent basis; two broke relations completely."

If competition was hurtful, what kept it going? Most often, parental favoritism continuing into adulthood. Also mentioned frequently was provocatively competitive behavior by the siblings themselves.

At times, the difficulty was family gossip that made too much of differences between brothers and sisters, assigning constricting roles or labels. Yet another problem was that siblings rarely talked about their rivalry. Successful siblings often did not even know that their achievements were causing envy, and siblings who felt inferior did not want to say so. "Admitting sibling rivalry may be experienced as equivalent to admitting maladjustment," Ross and Milgram believe. Besides, "To reveal feelings of rivalry to a brother or sister who is perceived as having the upper hand increases one's vulnerability in an already unsafe situation."

Psychologists often observe that even intense feelings of rivalry can coexist with feelings of affection and solidarity. One of the most telling examples is the case of William James, the philosopher-psychologist, and his brother Henry, the novelist. Henry's perceptive biographer, Leon Edel, gives a fascinating account of what he calls their "long-buried struggle for power," which began from the moment of Henry's birth and did not end until William's death.

Whatever praise William voiced over the years for Henry's writing was overbalanced by barbed criticism, especially of Henry's complex literary style. The older brother called Henry "a curiosity of literature." In a letter to the younger man he once wrote, "for gleams and innuendos and felicitous verbal insinuations you are unapproachable" and exhorted him to "say it *out*, for God's sake, and have done with it." William was not quite at ease with his own sharp words, though, and he urged Henry not to answer "these absurd remarks."

On that occasion, Henry complied, but to another critical letter he wrote, "I'm always sorry when I hear of your reading anything of mine, and always hope you won't—you seem to me so constitutionally unable to 'enjoy' it." Yet he did not appear resentful. Inscribing a gift copy of his novel *The Golden Bowl* to his brother, he called himself William's "incoherent, admiring, affectionate Brother."

Henry never knew of William's unkindest remarks about him. They were written in 1905, when William's election to the Academy of Arts and Letters followed Henry's by three months. Petulantly, William declined the offer. He was led to that course, he explained in a letter to the academy secretary, "by the fact that my younger and shallower and vainer brother is already in the Academy and that if I were there too, the other families represented might think the James influence too rank and strong."

As Edel points out, William was being both untruthful with himself

and inconsistent: "He had not considered that there was a redundancy of Jameses when he and Henry had been elected to the Institute [of Arts and Letters, the Academy's parent body] in 1898. He was having this afterthought only now, when Henry had been elected before him to the new body." Indeed, William's conscience was again not quite clear about his own behavior; his letter acknowledged that he was being "sour."

The relationship of the brothers was more complex than these quotations suggest. When William was dying, Henry wrote to a close friend, "At the prospect of losing my wonderful beloved brother out of the world in which, from as far back as in dimmest childhood, I have so yearningly always counted on him, I feel nothing but. . .weakness. . .and even terror."

To someone else, Henry confided that "William's extinction changes the face of life for me." Yet the change was perhaps not wholly unwelcome. Henry "had always found himself strong in William's absence," Edel writes. "Now he had full familial authority; his nephews deferred to him; his brother's wife now became [in effect] wife to him, ministering to his wants, caring for him as she had cared for the ailing husband and brother. Henry had ascended to what had seemed, for 60 years, an inaccessible throne."

Sibling Solidarity

Even rivalrous brother-sister relationships may show a good deal of the quality known as sibling solidarity, or cohesiveness. One of the most important investigators of this quality is Victor G. Cicirelli, a professor of developmental and aging psychology at Purdue University. The burden of Cicirelli's 12 years of sibling research is that brothers and sisters remain important to each other into old age, often becoming more important in time.

In a study published last year in the *Journal of Marriage and the Family*, Cicirelli compared the feelings of 100 college women toward their parents and toward their brothers and sisters and found that his subjects felt significantly closer to siblings than to fathers. For the most part, they also felt closer to siblings than to mothers;

that difference was not significant.

For a study of siblings in midlife, Cicirelli visited 140 Midwesterners, most of them aged 30 to 60, and interviewed them about their relationships with their 336 siblings. His results, as yet unpublished, show little conflict between brothers and sisters.

More than two-thirds of the subjects (68 percent, to be exact) described their relationships with siblings as close or very close; only 5 percent said they did not feel at all close. Sisters felt closer than brothers or cross-sex pairs. In general, cohesiveness increased slightly with age.

Asked how well they got along with brothers and sisters, 78 percent said well or very well; 4 percent said not very well or poorly. As to the degree of satisfaction the relationship brought, 68 percent called it considerable or very great, while 12 percent reported little or no satisfaction.

Other questions brought fewer positive responses. Asked how much interest they thought their siblings had in their—the subjects'—activities, 59 percent said they perceived the interest as very great or moderate, but 21 percent saw little or none. Well under half—41 percent—said they felt free to discuss personal or intimate matters with a sibling; 36 percent said they rarely or never did so. Only 8 percent usually or frequently talked over important decisions with a sibling; 73 percent rarely or never did so.

Cicirelli found little overt conflict: 93 percent maintained that they never or only rarely felt competitive; 89 percent asserted that their siblings were never or only rarely bossy; and 88 percent said they never argued with siblings, or did so only once in a while. Speculating on the far higher percentage of people admitting to rivalry in the Ross-Milgram study, Cicirelli suggested that the small-group interview method those psychologists used "might be more likely to stimulate self-disclosure about rivalry than traditional interview methods."

Cicirelli has administered questionnaires to 300 men and women over 60 in order to learn about sibling relationships in later life. Some 17 percent of his elderly subjects saw the sibling with whom they were in closest touch at least once a week, while 33 percent saw that sibling a minimum of once a month. There was some drop in the frequency of visits with increasing age, but the diminished contact did not seem to lessen closeness.

Overall, 53 percent used the words "extremely close" to describe their relationship, and another 30 percent characterized it as "close." The 83 percent for the two responses combined is striking, compared with the 68 percent reported above for the middle-aged. Cicirelli believes that as older people witness the aging and death of parents and see the effects of the years on themselves and their siblings, their sense of belonging is threatened. Strengthening ties with those who remain, he theorizes, may be partly an attempt "to preserve the attachment to the family system of childhood."

Even so, when Cicirelli asked his elderly subjects which family member they turned to for emotional support and practical assistance, he found that most relied mainly on their children. Only about 7 percent said a sibling was a major source of psychological support.

In a very different kind of study, Cicirelli administered a projective measure, the Gerontological Apperception Test (GAT), to 64 men and women aged 65 to 88 to elicit deep attitudes and feelings of a kind that are not likely to emerge in interviews. The GAT is made up of 14 pictures depicting ambiguous situations such as an older woman on a park bench observing a young couple. Subjects are asked to make up a story about each picture, telling "what the people are thinking and feeling and how it will come out."

From his analysis of responses, Cicirelli learned that sisters had a greater impact on the feelings and concerns of siblings than did brothers. He also discovered that sisters had different effects on sisters than they did on brothers. More often than not, the stories told by female subjects who had one or more sisters showed concern about keeping up their social skills and relationships with people outside the family, helping others in the community, and being able to handle criticism from younger people. Cicirelli interpreted that to mean that "sisters appear to stimulate and challenge the elderly woman to maintain her social activities, skills, and roles."

The stories told by male subjects with sisters tended to be happier in tone than the stories of men with brothers, and they revealed less worry about money, jobs, family relationships, and criticism from the young. "The more sisters the elderly man has, the happier he is," Cicirelli concluded. "Sisters seem to provide the elderly male with a basic feeling of emotional security."

Solidarity may be of major significance to some siblings, but to others it counts for nothing; it may not even develop. Michael Kahn says that if siblings have "low emotional access" to each other—if there are no important interactions early in life because the siblings are many years apart in age, or for other reasons—they cannot become truly important to each other.

Nor are they likely to be close as adults when sibling incest has occurred in the early years. "Feelings of guilt, shame, and anxiety linger on," Kahn says, and sisters, more deeply affected by incest than brothers, experience profound feelings of betrayal, sometimes even if they were willing participants in the experience. "They don't trust their brothers, ever. Sometimes they don't want to be in the same room with them."

Bank also observes that siblings may avoid each other as adults if there are unresolved conflicts left over from childhood because the parents were "conflictophobes," people who forbade sibling quarrels. "If you've been taught that it's dangerous to fight with your sibling, you may freeze around him when you're adults and sit in the same room hating his guts."

Both Bank and Kahn stress the lasting harm of "frozen misunderstandings" developed in childhood. These are distorted perceptions of a sibling that brothers and sisters may carry uncorrected into their adult relationship. A sister whose brother often hit her when they were children may infer, and believe for the rest of her life, that he did it because he was hateful. But it may be that hitting was the only way he could make the physical contact for which he was emotionally starved. And even if a child's image of a sibling was correct in youth, it may need later revision. "We change, but we may transact our relationship as if there had been no change," Bank says.

Adult Turning Points: Satisfactions and Stresses

Circumstances create a great diversity of adult sibling relationships, from hatred to detachment to love. "I have always experienced her as a vicious individual," a 28-year-old woman said of her sister in the course of a study by Elizabeth Fishel, author of *Sisters*. Helene S. Arnstein, author of *Brothers and Sisters, Sisters and Brothers*, quotes a sister who said of her brother: "If we weren't siblings I'd never see David again, because I don't care for the kind of person he has become."

History is filled with examples of more satisfying sibling relationships than that. Freud and his brother, Alexander, enjoyed each other's company all their lives. In his 1957 book on his father, *Glory Reflected*, Freud's son Martin wrote, "My father and Alexander could not have been more different in their outlook on life, but they were always good friends." Alexander often went along on holiday trips with Freud and his wife and children, and was Freud's frequent companion on climbing expeditions. The brothers had a particularly good time swimming together during an Adriatic holiday in 1895. "Uncle Alexander and father were seldom out of the water," wrote Martin. "When, as sometimes happened, they refused to come ashore even for lunch, a waiter would wade or swim out to them balancing a tray with refreshments and even cigars and matches."

In her autobiography *Blackberry Winter*, Margaret Mead tells of her lifelong warm relationship with her sister Elizabeth. "Her perceptions . . . have nourished me through the years. Her understanding of what has gone on in schools has provided depth and life to my own observations on American education. And her paintings have made every place I have lived my home."

Thinking of the women in her mother's family, Mead was struck by the way in which, generation after generation, pairs of sisters became good friends. "In this," she said, "they exemplify one of the basic characteristics of American kinship relations. Sisters, while they are growing up, tend to be very rivalrous, and as young mothers they are given to continual rivalrous comparisons of their several children. But once the children grow older, sisters draw closer together and often, in old age, they become each other's chosen and most happy companions."

If sibling relationships sometimes grow closer as siblings age, they may also founder at certain of life's turning points. Bank and Kahn say that siblings are best studied when under stress, because that is when the frequently submerged dynamics of their relationship are most likely to come to the surface, exposing the true nature, the real strength or fragility, of the bond. It is also at such moments that the "delicate balance" of a sibling relationship may tip decisively.

One turning point that can disrupt a previously close tie comes with a sibling's discovery that a brother or sister is getting a divorce, drinking too much, or going into psychotherapy. Under such circumstances, presumably untroubled siblings are often afraid that they will develop the same difficulty. Their reaction may be to make themselves feel better by offering the troubled sibling hostile advice on "how he could improve himself," Bank says. The mechanism, he explains, is projective identification: "The reason I don't like you is that I see in you what I know all too well I have inside me."

Of course, when a sibling goes into therapy, it does not inevitably damage a relationship. The outcome may be quite the opposite. "Under carefully arranged conditions, siblings can learn to cooperate with each other to resolve important conflicts in family relationships," Kahn and Bank write in a case report to be published in the journal *Family Process*. Called "In Pursuit of Sisterhood," the paper tells of Maureen, a 29-year-old nurse who entered treatment out of a kind of general unhappiness. Among other things, she hated the fact that her family had never taken her seriously.

Eventually the therapist called a series of "sibling rallies," meetings of the four sisters. The three older women told Maureen that she had never let them know how she felt. They all agreed that none of them was as close to the others as they wanted to think they were. Over a period of weeks, the four managed to thaw some of the

"frozen misunderstandings" from their common past and to become a cohesive group; for the first time, Maureen considered herself the equal of her sisters. After a while, Kahn and Bank said, the group confronted their mother and father and "successfully parried their parents' attempts to avoid discussing feelings." The ultimate result: "Faced with a unified sibling group, the parents were forced to redress old grievances by being helpful and accepting."

Physical illness is another situation that sheds light on sibling relationships. One illness that has been studied by many psychiatrists and medical sociologists is end-stage renal disease, in which doctors may be considering a kidney transplant.

Genetically, the best donors are brothers and sisters, because they provide the closest tissue match, reducing the danger that the patient's body will reject the transplanted organ. Yet siblings volunteer to donate a kidney less often than do parents. In a study of more than 300 relatives of transplant patients, Roberta G. Simmons, a sociologist at the University of Minnesota, found that 52 percent of the patients' sisters and 54 percent of their brothers did not volunteer. In contrast, only 14 percent of the patients' parents and only 33 percent of their children failed to volunteer.

Norman B. Levy and Jorge Steinberg, New York psychiatrists who have interviewed many pairs of kidney donors and recipients, have described a patient who got a kidney from her mother and told a researcher, "My sister was better matched, but she got pregnant." The two psychiatrists believe that the pregnancy very likely came about "accidentally on purpose" because the sister did not really want to donate.

Another case Levy and Steinberg described illustrates the apparent power of the emotions siblings feel for each other. The family wanted the needed kidney to be given by the patient's sister, who was not on good terms with the family, because they imagined that the gift would bring family harmony. The patient let herself be persuaded to accept her sister as donor, but she told the psychiatrists that she was afraid her sister would somehow find a way to make her pay for the gift. The prospective donor was no more

enthusiastic. "After I give her the damned kidney, I never want to see her again," she said. In short, Levy said, "They hated each other," and he and Steinberg predicted—correctly—that the patient's body would soon reject the hated sister's kidney.

The dependency, illness, and death of parents provide other possible turning points in the relationship of brothers and sisters. Cicirelli finds that siblings' negative feelings toward each other often emerge as an elderly parent becomes increasingly dependent. "Why should I look after my father?" a sibling thinks. "Let my brother do it."

On the other hand, there are siblings who compete to do the most for a dependent parent. "A common observation for those of us who deal with families of aged people," writes Martin A. Berezin, a psychoanalyst, "is the irrational, hostile attitude that siblings express to each other as they quarrel about what should be the proper care for an aged parent. Each accuses the other of negligence, lack of sympathy, or avoidance of responsibility. They challenge each other with questions of who telephoned or visited how many times." In such cases, according to Barbara Silverstone, a social worker, each sibling claims to have the parent's welfare uppermost, "but the underlying motive is winning out in a family contest."

Many researchers say that sibling bonds often weaken or disintegrate entirely when the last parent dies. "The wounds that are given and received during a parent's illness and dependency," Silverstone says, are often "only the final blows ending a relationship which has been distant or seething for years." A brother or sister who sacrificed too much for a parent may resent those who did too little. Or, Silverstone suggests, "Siblings may resent the martyred caretaker who stood between them and their dying mother, making it impossible for them to share her final days."

In some instances, Bank and Kahn say, a parent's death "can set off fratricidal feelings and struggles as siblings jockey for positions of leadership within the adult sibling group." Yet there are times when a parent's illness brings siblings together. One sister said to Silverstone of her brother, "I never realized what a great person he'd turned into until I spent all that time at home when Mother was sick. We're real friends now." Generally speaking, Bank believes, previously bad sibling relationships get worse after the parents' death; good ones improve.

An Inescapable Bond

Bank and Kahn call the sibling relationship "a lifelong process, highly influential throughout the life cycle." Even when brothers and sisters drift apart as they enter careers and begin to raise families, they are likely to renew their relationship eventually. "It is extremely rare for siblings to lose touch with one another," Cicirelli says. As the British sociologist Graham Allan puts it, it is "permissible to forget about a neighbor or other associate, but less appropriate to 'drop' a sister or brother."

Throughout life, siblings are likely to serve each other as models, spurs to achievement, and yardsticks by which to measure accomplishments. In the best of circumstances, they become sources of practical help, nonjudgmental advice, and true intimacy. Indeed, Cicirelli remarks, "The nature of the sibling relationship is such that intimacy between siblings is immediately restored even after long absences."

"One of the main theses of our book," Kahn says, "is that siblings are becoming more and more dependent upon one another in contemporary families because of the attrition in family size. If you have only one sibling, that one becomes enormously important." The decreased availability of parents in one-parent homes and in households where both mother and father work also makes siblings count for more with each other. The rising divorce rate, too, intensifies adult sibling relationships by making them more significant than a husband or wife in an ephemeral marriage.

Bank and Kahn also predict that the rising divorce rate will lead to increased sibling conflict, "with adolescent and adult siblings taking sides with each parent in a kind of proxy war." But with or without divorce, conflict occurs between many—some would say most—brothers and sisters.

"One can conceive of rivalry as a feeling which is always latent, appearing strongly in certain circumstances while closeness is elicited in others," Cicirelli said recently. "Possibly there is a balance between the functional and dysfunctional aspects of the sibling relation which is manifested in solidarity versus rivalry. There may be a love-hate dialectical process throughout the life span that leads to new levels of maturity or immaturity in sibling relationships."

Some siblings are well aware of their ambivalence. "My mixed feelings toward my sister, my irradicable love-hate mix, have been almost the longest-standing puzzle of my life," a woman of 30 told Elizabeth Fishel.

Even siblings who prefer not to confront their ambivalence probably do not wholly escape its effects. Whatever its nature in particular cases—warm or cool—the sibling relationship remains alive in some sense. Mental images of each other, vivid or half forgotten, can influence brothers and sisters even when they are oceans apart and not consciously preoccupied with each other. "It's a relationship for life," Bank says. "It's forever."

SINGLE PARENT FATHERS:
A New Study

Harry Finkelstein Keshet and Kristine M. Rosenthal

Harry Finkelstein Keshet, Ph.D., and Kristine M. Rosenthal, Ed.D., are co-directors of the Parenting Study at Brandeis University. Dr. Keshet, a divorced father with joint custody of his son, is an instructor in the Parenting Program at Wheelock College, Boston, and clinical director of the Divorce Resource and Mediation Center, Inc., Cambridge, Mass. Dr. Rosenthal, a single parent of three children, is an assistant professor at Brandeis University.

This article is about single-parent fathers who are rearing their young children after marital separation. It discusses what fathers do for and with their children and how being a single parent affects their lifestyles and work responsibilities.

The article is based on interviews, conducted by trained male interviewers, with 49 separated or divorced fathers who live in the Boston area and have formal or informal custody of their children.[1] At the time of the interviews, each father's youngest child was between the ages of three and seven and there were no more than three children in any of the families. Each father had been separated from his wife for at least one year.[2]

Over half of the fathers in the sample (53 percent) were legally divorced; the remainder were separated informally. Most of the men had been married and living with their spouses for a minimum of five years.

The majority of fathers were highly educated; barely a fourth had had less than a full college education and nearly half had either completed or were in the process of completing graduate or professional training. More than half of the fathers were in professional or semi-professional occupations, 20 percent were in business or administration and 20 percent had blue-collar jobs or worked as craftsmen. Another six percent of the sample were students;

two percent were unemployed.

Seventy-six percent of the fathers worked full-time or longer at their occupations; the others worked half time or less. A majority of the fathers earned relatively high incomes: 15 percent earned over $25,000, 25 percent earned between $15,000 and $25,000 and 30 percent earned between $10,000 and $15,000.

We felt that residential stability and housing were important factors in childrearing. Therefore, we were interested in the types of housing arrangements the men had made. Over half of the fathers (54 percent) occupied houses which they owned, while the others lived in apartments or rented homes. Most of the men showed a high rate of residential stability—50 percent had lived in the same dwelling for two years or longer and 12 percent had lived in the same place for more than a year.

Half the men lived alone with their children, 35 percent shared housing with other adults and the remaining 15 percent lived with their lovers. Only a few men in the sample lived with extended families.

Almost all of the fathers reported that their children were attending day care centers or schools or were cared for by hired babysitters when the fathers were working.

Until recently, little research on the role of the father in marriage or after marital separation has been reported. Men have been studied as workers and professionals but not as fathers and husbands. For a married man in modern society, the qualities of being a "good husband" have been similar to those of being a "good father." The good father-husband is an economic provider. He forms a relationship and bond with his wife but not necessarily with his child for the relational tie between child and father has not been deemed essential.

There is a very strong cultural bias that women, biologically, psychologically and temperamentally, are best suited for child care and that mothering rather than parenting is the primary ingredient in child development.[3] This reflects the traditional view that male and female roles are based on traits that are essentially different and complementary. Women bear and rear children, and men support and provide for their wives and children. Men are not expected to be active in childrearing in or outside of marriage.

When a marriage ends in divorce, a father not only separates from his wife but is likely to be separated from his children as well. As E. E. LeMasters said: "The father's parental role in the U.S. is particularly tied to the success or failure of the pair bond between himself and his wife."[4]

The failure of his marriage is also likely to mean the loss of child custody for fathers. Most men do not seek custody of their children and those who do may experience sex role bias on the part of the judiciary. In 1971, for example, it was reported that mothers received custody of their children in over 90 percent of the custody decisions in United States Courts.[5]

Information available on the number of single-parent fathers indicates that more men are now being awarded custody of their children and that the men in our study are part of a small but growing number of fathers who are rearing their children after marital separation. Census data show that from 1960 to 1974 the number of male-headed families, with children and no spouse present, increased from 296,000 in 1960[6] to 836,000 in 1974.[7] The number of families headed by males with children under six and no spouse present also increased, from 87,000 in 1960[8] to 188,200 in 1970.[9]

In our study, we were particularly interested in the following four aspects of parenting: entertainment of children outside of the home, homemaking activities, child guidance and nurturance and child- and parent-oriented services. We also asked the fathers if they felt they had been prepared to carry out these activities and whether they received any help from other adults in conducting them.

Entertainment

Our earlier research had shown that fathers of young children in nuclear families and in separated families often entertain their children. In this study, all the fathers reported that they frequently relied on structured recreational activities for their children. A majority of fathers said they took their children swimming and to playgrounds, museums, restaurants and, less frequently, to child-oriented shows and movies. Finding activities for children seemed to be a major part of their roles as recently-separated fathers. One father reported:

"At first, I had to entertain them . . . bowling, movies, swimming, trips to relatives. I often would do reading and roughhousing with them also . . . I thought it was my responsibility to provide some structured activities."

"I went to every park, museum, playground, movie, zoo and what-have-you imaginable . . . I was constantly looking for things to do," another said.

Recreation and entertainment are "doing" activities and as such they are consistent with the parenting activities and male socialization role found within the nuclear family experience. Separated and divorced fathers often played with their children and felt prepared to perform this aspect of the single-parent role. Although most fathers reported that they received little help with their recreational parenting activities, they also said that when they became involved with women their women friends often accompanied them and their children on recreational ventures. It seems that the recreational aspects of parenting served as a comfortable way for them—and other adults—to interact with their children.

Homemaking

Homemaking seemed to us to be an essential part of the single-parent role. We asked the men about meal preparation and general household management.

Over 90 percent of the fathers reported that they frequently performed the homemaking functions of housecleaning, preparing meals and food shopping and, with the exception of housecleaning, most men felt that they had been prepared to perform these homemaking activities when they separated from their spouses. Most of the fathers also said that they did not receive significant help in homemaking, except in food preparation. Here, nearly half the fathers reported getting some assistance from other adults who shared their dwellings.

Child Guidance and Nurturance

The guidance and nurturance aspect of parenting was defined as direct father-child interaction, with the men giving care or direction to their young. We asked questions concerning discipline and the setting of limits; mealtime, bedtime and bathing routines; dealing with children's feelings; and shopping with children for clothing.

The most frequently reported guidance and nurturing activities were discipline, serving meals, bathing children and dealing with children's feelings and emotional upsets. More than 95 percent of the men reported that they frequently performed these activities on a regular basis. In contrast, less than half of the men said they bought clothing for their children.

The need to provide guidance and nurturance seemed to be more of a problem for the fathers than their entertainment and homemaking roles. Most fathers felt prepared to discipline or bathe their children but were less prepared for dealing with their emotional upsets. As one father explained:

"I felt unprepared for dealing with the children's emotional needs and being open to what they were feeling. I was very inadequate in that I couldn't deal with my own feelings very well. I was afraid, I didn't know how to listen or respond to them."

"Dealing with feelings openly or

expressing them, especially negative feelings, like grief or anger . . . those were the areas where I felt totally handicapped," a second said.

The socialization of males in our society does not prepare them to be nurturant and sex role definitions of childrearing in the nuclear family are likely to emphasize emotional responsiveness as a feminine trait and therefore a part of the mother's role. Boys learn instrumental skills that prepare them to be workers and family providers and they are neither expected nor encouraged to develop emotional skills which may interfere with their achievement.[10]

The more successfully a man has been socialized in instrumental behavior, the more likely he is to lack effective interpersonal skills. For example, a positive and significant relationship between a father's high achievement motivations and his feelings of inadequacy in the father role, and a negative orientation toward preschool children, has been reported by Veroff and Feld.[11] The authors also suggest that consideration of the emotional needs of young children may be incongruent with what fathers have previously learned about good masculine performance.

Our sample of fathers expressed difficulty in relating to their children's emotional needs, yet they attempted to respond to the requirements of their new roles, sometimes with the help of others. They received the most assistance in those areas where they felt least prepared. For example, 44 percent reported receiving help from others in dealing with children's feelings.

Services for Children and Parents

Being a single parent required acting as the child's sole agent in relationship to agencies and professionals. Nearly 75 percent of the men reported calling for babysitters themselves and over 90 percent were actively involved in all the other common child-oriented activities, such as taking a child to a doctor or dentist and talking with their children's teacher and the parents of their friends. Most of the men (80 percent) felt prepared to perform these functions after separation. The calling of babysitters was performed with a lower level of confidence—nearly half

(45 percent) felt unprepared to do this. A majority of the men reported performing these activities with no help from others.

All in all, however, it appears that the fathers in our sample were very active in all the aspects of parenting that we explored. They were most prepared for performing the recreational, homemaking and child-oriented service aspects of parenting, and least prepared for the child nurturance aspects. They received help from other adults in large measure for nurturant activities and certain homemaking activities, while help was not frequently reported for recreational and child-oriented service activities.

Time as Role Salience

More than half of the fathers (53 percent) reported spending half or more of their time at home interacting with their children. Others spent some time doing so and only four percent said that they spent very little time with their children.

We had expected that at least in the families in which the fathers lived with other people, the children would spend a significant amount of time interacting with other adults. However, we found that only nine percent of the fathers said that their children spent half or more of their time with other adults and 43 percent reported that their children spent very little or no time with other adults.

This finding that the father was the adult in the home with whom the children most frequently spent their time is also supported by data on the kinds of childrearing help fathers received from their lovers or dates. Women lovers rarely took or were permitted to take a major child care role with the father's children. Most frequently they accompanied fathers and children in out-of-home recreational activities.

These findings suggest that the single-parent role was a highly salient one and that fathers even protected their children from the influence of other adults.

Compared to the hours spent with other adults, children spent a much greater part of their time with peers: 45 percent of the fathers reported that their children spent half or more of their time playing with other children; 37 percent reported that they

spent some time doing so and only 18 percent reported that they spent very little or no time playing with peers.

In summary, fathers and children's peers were those most involved with the children in terms of time. Other non-related adults, even when they were available, spent less time with the father's children. The new time relationship with their children was described by one father this way:

"Spending time alone with the children came as a change after separation . . . They were alone with me all the time . . . It was the first time I'd been with them alone. I was putting them to bed, getting them dressed, getting their meals . . . I began to see I could do it."

Another explained: "One of the big problems was how being a full-time parent would influence my time. How I could put this whole thing together in terms of spending a lot of time with them; at the same time, having time to do other things. I didn't know how it would work out."

Role Strains

Three-quarters of the men we talked to said that they felt closer to their children as a result of their change in marital status and new parenting role. Their relationship to the children was more direct, no longer being filtered through the conflicts of an ailing marriage. Yet being a single parent had its own set of difficulties. Many fathers expressed serious concerns with the time strains they experienced after the marital breakdown.

Like most men, the fathers in our sample seemed to fit their time structures into the demands of the workplace, school or other organization in which they participated. As single parents, the needs of their children had priority over the requirements of external organizations, especially when the children were ill or had school vacations. At such times, fathers had to take time off from work or arrange for others to care for their children. Role strain often resulted from conflicts between their child care responsibilities and social needs and work responsibilities.

Work and Child Care

As noted previously, the majority of men in our sample were in demanding

professional and semi-professional occupations. In order to ascertain how they perceived the management of child care and work responsibilities, we listed a set of work-related behaviors and asked the men how they felt their child care obligations limited their work activities.

Job mobility was the area cited as being most limited. Fathers reported their work life was hampered in terms of working hours (63 percent), work priorities (62 percent), earnings (55 percent) and job transfer (52 percent). Within the work setting itself, limitations were noted in the areas of type of work (66 percent), promotions (38 percent), relations with co-workers (28 percent) and supervisors (20 percent).

Single parenting clearly limited earnings, hours and work relations. Work identity, a cornerstone of male self-definition, was challenged by an emerging parenting identity. One father said:

"Taking care of her (his daughter) was a real conflict with work. I was trying to build my business, but had to stop to pick her up at 3:00. It was rough starting something new, not making much money and limiting my working hours . . . Often, what I gave her was given grudgingly because I was feeling terribly limited by the schedule and by the tremendous pressures I was feeling."

In an open-ended question, we asked fathers to indicate the kind of services they felt would help make parenting and work more compatible. Over half (55 percent) said that time flexibility at work would help alleviate the constraints of these often conflicting responsibilities. One father for whom this was already a viable arrangement said:

"The advantages of where I work is that my hours are flexible. There is no problem in going home in the late afternoon to be with the kids . . . I make up the time by coming to work at 7:30 a.m. and every other weekend. Sometimes I bring my kids back to work if there is some great necessity, but I don't want this to happen often."

Fathers were also aware of the need for both long-range and immediate services for themselves and their children. Many felt that their parenting roles had not been accorded social legitimacy. Their employers, co-workers and peers often lacked appreciation or understanding of their needs as single parents and offered them little support. The single-parent father's dilemma has been stated clearly: "Economic efficiency is given so much priority in our society that it is difficult to imagine an American father neglecting his job or refusing a promotion out of deference to the needs of his children . . ."[12]

Social Life and Parenting

Our findings indicated that a major limitation for single-parent fathers was the decrease in the amount of time spent alone with other adults. Over half of the fathers studied said that they were involved with a woman as a sexual partner. The majority of these women (61 percent) were either divorced or separated and almost half of them had children from a previous marriage. The majority of the fathers reported that their lovers were helpful with the children.

The most frequently reported role for women friends (62 percent) was companionship with fathers and children in recreational activities away from home. A more active and direct child care role at the father's home was not frequently reported. For example, only 31 percent said that their lovers watched their children at home when the father was present and only 15 percent stated that their lovers cared for children alone for substantial periods of time. This suggests that the single-parent fathers in our sample were somewhat reluctant and guarded in allowing their women friends to share substantially in caring for their children.

When questioned about any conflict with lovers concerning a father's relationship to children, over half (56 percent) indicated that this was an issue because of the limitations placed on their social life by child care obligations. Eighty-five percent of the men said that their social life would be different were the children not present. The areas in which fathers felt their lives to be limited by parenting responsibilities were: social life, mentioned by 63 percent of the fathers; sports, noted by 17 percent; and home activities, referred to by 15 percent.

Being a Single Father

Our analysis showed that being a single parent required a major shift in lifestyle and priorities for most men. The bond between parent and child became a new focal point for self-definition and set the criteria for organizing the more traditional spheres of male functioning at work and in social life. The men in our study limited work and social activities to meet the needs of their children and, in doing so, they felt they had developed a closer relationship with them.

The experience of marital separation had brought the men into the sphere of what is commonly considered the woman's world—of being responsible for children's growth and satisfying children's needs. The men responded by restructuring their daily lives in order to care directly for their dependent children. As a result, fathers felt more positive about themselves as parents and as individuals. A majority of the men reported that being a single parent had helped them to grow emotionally. They felt they had become more responsive to their children and more conscious of their needs, a responsiveness they reported as reaching out to other adults as well.

[1]The interviews were part of a larger study, partially supported by a grant from the Rockefeller Foundation.

[2]The criteria used in the selection of the sample of families (the age of the youngest child, the limit on the number of children in each family and the minimum period of separation of parents) circumscribed the range of time of separation experienced among those interviewed. The mean number of years of separation was 2.6. Half (51 percent) of the fathers in the sample had been separated for no more than two years; over a third (36 percent) had been separated for three to four years, and the remainder, for over four years.

[3]John Bowlby, Maternal Care and Mental Health, Geneva, World Health Organization, 1951.

[4]E.E. LeMasters, Parents in Modern America, Homewood, Illinois, Dorsey Press, 1971.

[5]Robert R. Bell, Marriage and Family Interaction, Homewood, Illinois, Dorsey Press, 1971.

[6]U.S. Department of Commerce, Bureau of the Census, Family Characteristics, 1960 Census.

[7]U.S. Department of Commerce, Bureau of the Census, Current Population Reports, Ser. P-20, no. 271, 1974.

[8]U.S. Department of Commerce, Bureau of the Census, Family Characteristics, 1960 Census.

[9]U.S. Department of Commerce, Bureau of the Census, Family Characteristics, 1970 Census.

[10]Lenard Benson, Fathering, A Sociological Perspective, New York, Random House, 1968.

[11]Joseph Veroff and Sheila Feld, Marriage and Work in America, New York, Van Nostrand Reinhold, 1970.

[12]E.E. LeMasters, op. cit.

DOES PERSONALITY REALLY CHANGE AFTER 20?

ZICK RUBIN

Zick Rubin, who says he hasn't changed much in recent years, is Louis and Frances Salvage Professor of Social Psychology at Brandeis University and a contributing editor of *Psychology Today*. He is the coauthor (with the late Elton B. McNeil) of *The Psychology of Being Human*, Third Edition, an introductory psychology textbook that has been published by Harper & Row.

I n most of us," William James wrote in 1887, "by the age of 30, the character has set like plaster, and will never soften again." Though our bodies may be bent by the years and our opinions changed by the times, there is a basic core of self—a personality—that remains basically unchanged.

This doctrine of personality stability has been accepted psychological dogma for most of the past century. The dogma holds that the plaster of character sets by one's early 20s, if not even sooner than that.

Within the past decade, however, this traditional view has come to have an almost archaic flavor. The rallying cry of the 1970s has been people's virtually limitless capacity for change—not only in childhood but through the span of life. Examples of apparent transformation are highly publicized: Jerry Rubin enters the 1970s as a screaming, war-painted Yippie and emerges as a sedate Wall Street analyst wearing a suit and tie. Richard Alpert, an ambitious assistant professor of psychology at Harvard, tunes into drugs, heads for India, and returns as Baba Ram Dass, a long-bearded mystic in a flowing white robe who teaches people to "be here now." And Richard Raskind, a successful ophthalmol-ogist, goes into the hospital and comes out as Renée Richards, a tall, well-muscled athlete on the women's tennis circuit.

Even for those of us who hold on to our original appearance (more or less) and gender, "change" and "growth" are now the bywords. The theme was seized upon by scores of organizations formed to help people change, from Weight Watchers to est. It was captured—and advanced—by Gail Sheehy's phenomenally successful book *Passages*, which emphasized people's continuing openness to change throughout the course of adulthood. At the same time, serious work in psychology was coming along—building on earlier theories of Carl Jung and Erik Erikson—to buttress the belief that adults keep on developing. Yale's Daniel Levinson, who provided much of Sheehy's intellectual inspiration, described, in *The Seasons of a Man's Life*, an adult life structure that is marked by periods of self-examination and transition. Psychiatrist Roger Gould, in *Transformations*, wrote of reshapings of the self during the early and middle adult years, "away from stagnation and claustrophobic suffocation toward vitality and an expanded sense of inner freedom."

The view that personality keeps changing throughout life has picked up so many adherents recently that it has practically become the new dogma. Quantitative studies have been offered to document the possibility of personality change in adulthood, whether as a consequence of getting married, changing jobs, or seeing one's children leave home. In a new volume entitled *Constancy and Change in Human Development*, two of the day's most influential behavioral scientists, sociologist Orville G. Brim, Jr., and psychologist Jerome Kagan, challenge the defenders of personality stability to back up their doctrine with hard evidence. "The burden of proof," Brim and Kagan write, "is being shifted to the larger group, who adhere to the traditional doctrine of constancy, from the minority who suggest that it is a premise requiring evaluation."

And now we get to the newest act in the battle of the dogmas. Those who uphold the doctrine of personality stability have accepted the challenge. In the past few years they have assembled the strongest evidence yet available for the truth of their position—evidence suggesting that on several central dimensions of personality, including the ones that make up our basic social and emotional style, we are in fact astoundingly stable throughout the course of adult life.

The 'Litter-ature' on Personality

Until recently there was little firm evidence for the stability of personality, despite the idea's intuitive appeal. Instead, most studies showed little predictability from earlier to later times of life—or even, for that matter, from one situation to another within the same time period—thus suggesting an essential lack of consistency in people's personalities. Indeed, many researchers began to question whether it made sense to speak of "personality" at all.

But whereas the lack of predictability was welcomed by advocates of the doctrine of change through the life span, the defenders of stability have another explanation for it: most of the studies are lousy. Referring derisively to the "litter-ature" on personality, Berkeley psychologist Jack Block estimates that "perhaps 90 percent of the studies are methodologically inadequate, without conceptual implication, and even foolish."

Block is right. Studies of personality have been marked by an abundance of untested measures (anyone can make up a new "scale" in an afternoon), small samples, and scatter-gun strategies ("Let's throw it into the computer and get some correlations"). Careful longitudinal studies, in which the same people are followed over the years, have been scarce. The conclusion that people are not predictable, then, may be a reflection not of human nature but of the haphazard methods used to study it.

Block's own research, in contrast, has amply demonstrated that people *are* predictable. Over the past 20 years Block has been analyzing extensive personality reports on several hundred Berkeley and Oakland residents that were first obtained in the 1930s, when the subjects were in junior high school. Researchers at Berkeley's Institute of Human Development followed up on the students when the subjects were in their late teens, again when they were in their mid-30s, and again in the late 1960s, when the subjects were all in their mid-40s.

The data archive is immense, including everything from attitude checklists filled out by the subjects to transcripts of interviews with the subjects, their parents, teachers, and spouses, with different sets of material gathered at each of the four time periods.

To reduce all the data to manageable proportions, Block began by assembling separate files of the information collected for each subject at each time period. Clinical psychologists were assigned to immerse themselves in individual dossiers and then to make a summary rating of the subject's personality by sorting a set of statements (for instance, "Has social poise and presence," and "Is self-defeating") into piles that indicated how representative the statement was of the subject. The assessments by the different raters (usually three for each dossier) were found to agree with one another to a significant degree, and they were averaged to form an overall description of the subject at that age. To avoid potential bias, the materials for each subject were carefully segregated by age level; all comments that

referred to the person at an earlier age were removed from the file. No psychologist rated the materials for the same subject at more than one time period.

Using this painstaking methodology, Block found a striking pattern of stability. In his most recent report, published earlier this year, he reported that on virtually every one of the 90 rating scales employed, there was a statistically significant correlation between subjects' ratings when they were in junior high school and their ratings 30 to 35 years later, when they were in their 40s. The most self-defeating adolescents were the most self-defeating adults; cheerful teenagers were cheerful 40-year-olds; those whose moods fluctuated when they were in junior high school were still experiencing mood swings in midlife.

'Still Stable After All These Years'

Even more striking evidence for the stability of personality, extending the time frame beyond middle age to late adulthood, comes from the work of Paul T. Costa, Jr., and Robert R. McCrae, both psychologists at the Gerontology Research Center of the National Institute on Aging in Baltimore. Costa and McCrae have tracked people's scores over time on standardized self-report personality scales, including the Sixteen Personality Factor Questionnaire and the Guilford-Zimmerman Temperament Survey, on which people are asked to decide whether or not each of several hundred statements describes them accurately. (Three sample items: "I would prefer to have an office of my own, not sharing it with another person." "Often I get angry with people too quickly." "Some people seem to ignore or avoid me, although I don't know why.")

Costa and McCrae combined subjects' responses on individual items to produce scale scores for each subject on such overall dimensions as extraversion and neuroticism, as well as on more specific traits, such as gregariousness, assertiveness, anxiety, and depression. By correlating over time the scores of subjects tested on two or three occasions—separated by six, 10, or 12 years—they obtained estimates of personality stability. The Baltimore

researchers have analyzed data from two large longitudinal studies, the Normative Aging Study conducted by the Veterans Administration in Boston and the Baltimore Longitudinal Study of Aging. In the Boston study, more than 400 men, ranging in age from 25 to 82, filled out a test battery in the mid-1960s and then completed a similar battery 10 years later, in the mid-1970s. In the Baltimore study, more than 200 men between the ages of 20 and 76 completed test batteries three times, separated by six-year intervals. Less extensive analyses, still unpublished, of the test scores of women in the Baltimore study point to a similar pattern of stability.

In both studies, Costa and McCrae found extremely high correlations, which indicated that the ordering of subjects on a particular dimension on one occasion was being maintained to a large degree a decade or more later. Contrary to what might have been predicted, young and middle-aged subjects turned out to be just as unchanging as old subjects were.

"The assertive 19-year-old is the assertive 40-year-old is the assertive 80-year-old," declares Costa, extrapolating from his and McCrae's results, which covered shorter time spans. For the title of a persuasive new paper reporting their results, Costa and McCrae rewrote a Paul Simon song title, proclaiming that their subjects were "Still Stable After All These Years."

Other recent studies have added to the accumulating evidence for personality stability throughout the life span. Gloria Leon and her coworkers at the University of Minnesota analyzed the scores on the Minnesota Multiphasic Personality Inventory (MMPI) of 71 men who were tested in 1947, when they were about 50 years old, and again in 1977, when they were close to 80. They found significant correlations on all 13 of the MMPI scales, with the highest correlation over the 30-year period on the scale of "Social Introversion." Costa and McCrae, too, found the highest degrees of stability, ranging from .70 to .84, on measures of introversion-extraversion, which assess gregariousness, warmth, and assertiveness. And Paul Mussen and his colleagues at Berkeley, analyzing interviewers' ratings of 53 women who were seen at ages 30 and 70, found significant correlations on such aspects of introversion-extraversion as talkativeness, excitability, and cheerfulness.

Although character may be most fixed in the domain of introversion-extraversion, Costa and McCrae found almost as much constancy in the domain of "neuroticism," which includes such specific traits as depression, anxiety, hostility, and impulsiveness. Neurotics are likely to be complainers throughout life. They may complain about different things as they get older—for example, worries about love in early adulthood, a "midlife crisis" at about age 40, health problems in late adulthood—but they are still complaining. The less neurotic person reacts to the same events with greater equanimity. Although there is less extensive evidence for its stability, Costa and McCrae also believe that there is an enduring trait of "openness to experience," including such facets as openness to feelings, ideas, and values.

Another recent longitudinal study of personality, conducted by University of Minnesota sociologist Jeylan Mortimer and her coworkers, looked at the self-ratings of 368 University of Michigan men who were tested in 1962–63, when they were freshmen, in 1966–67, when they were seniors, and in 1976, when they were about 30. At each point the subjects rated themselves on various characteristics, such as relaxed, strong, warm, and different. The ratings were later collapsed into overall scores for well-being, competence, sociability, and unconventionality. On each of these dimensions, Mortimer found a pattern of persistence rather than one of change. Mortimer's analysis of the data also suggested that life experiences such as the nature of one's work had an impact on personality. But the clearest message of her research is, in her own words, "very high stability."

Is *Everybody* Changing?

The high correlations between assessments made over time indicate that people in a given group keep the same rank order on the traits being measured, even as they traverse long stretches of life. But maybe *everyone* changes as he or she gets older. If, for example, everyone turns inward to

about the same extent in the latter part of life, the correlations—representing people's *relative* standing—on measures of introversion could still be very high, thus painting a misleading picture of stability. And, indeed, psychologist Bernice Neugarten concluded as recently as five years ago that there was a general tendency for people to become more introverted in the second half of life. Even that conclusion has been called into question, however. The recent longitudinal studies have found only slight increases in introversion as people get older, changes so small that Costa and McCrae consider them to be of little practical significance.

Specifically, longitudinal studies have shown slight drops over the course of adulthood in people's levels of excitement seeking, activity, hostility, and impulsiveness. The Baltimore researchers find no such changes in average levels of gregariousness, warmth, assertiveness, depression, or anxiety. Costa summarizes the pattern of changes as "a mellowing—but the person isn't so mellowed that you can't recognize him." Even as this mellowing occurs, moreover, people's relative ordering remains much the same—on the average, everyone drops the same few standard points. Thus, an "impulsive" 25-year-old may be a bit less impulsive by the time he or she is 70 but is still likely to be more impulsive than his or her agemates.

The new evidence of personality stability has been far too strong for the advocates of change to discount. Even in the heart of changeland, in Brim and Kagan's *Constancy and Change in Human Development*, psychologists Howard Moss and Elizabeth Susman review the research and conclude that there is strong evidence for the continuity of personality.

People Who Get Stuck

The new evidence has not put the controversy over personality stability and change to rest, however. If anything, it has sharpened it. Although he praises the new research, Orville Brim is not convinced by it that adults are fundamentally unchanging. He points out that the high correlations signify strong associations between measures, but not total constancy. For example, a .70 correlation between

scores obtained at two different times means that half of the variation (.70 squared, or .49) between people's later scores can be predicted from their earlier scores. The apostles of stability focus on this predictability, which is all the more striking because of the imperfect reliability of the measures employed. But the prophets of change, like Brim, prefer to dwell on the half of the variability that cannot be predicted, which they take as evidence of change.

Thus, Costa and McCrae look at the evidence they have assembled, marvel at the stability that it indicates, and call upon researchers to explain it: to what extent may the persistence of traits bespeak inherited biological predispositions, enduring influences from early childhood, or patterns of social roles and expectations that people get locked into? And at what age does the plaster of character in fact begin to set? Brim looks at the same evidence, acknowledges the degree of stability that it indicates, and then calls upon researchers to explain why some people in the sample are changing. "When you focus on stability," he says, "you're looking at the dregs—the people who have gotten stuck. You want to look at how a person grows and changes, not at how a person stays the same."

Brim, who is a president of the Foundation for Child Development in New York, also emphasizes that only certain aspects of personality—most clearly, aspects of social and emotional style, such as introversion-extraversion, depression, and anxiety—have been shown to be relatively stable. Brim himself is more interested in other parts of personality, such as people's self-esteem, sense of control over their lives, and ultimate values. These are the elements of character that Brim believes undergo the most important changes over the course of life. "Properties like gregariousness don't interest me," he admits; he does not view such traits as central to the fulfillment of human possibilities.

If Brim is not interested in some of the personality testers' results, Daniel Levinson is even less interested. In his view, paper-and-pencil measures like those used by Costa and McCrae are trivial, reflecting, at best, peripheral aspects of life. (Indeed, critics suggest that such research indicates only that people are stable in the way they fill out personality scales.) Levinson sees the whole enterprise of "rigorous" studies of personality stability as another instance of psychologists' rushing in to measure whatever they have measures for before they have clarified the important issues. "I think most psychologists and sociologists don't have the faintest idea what adulthood is about," he says.

Levinson's own work at the Yale School of Medicine (see "Growing Up with the Dream," *Psychology Today,* January 1978) has centered on the adult's evolving life structure—the way in which a person's social circumstances, including work and family ties, and inner feelings and aspirations fit together in an overall picture. Through intensive interviews of a small sample of men in the middle years of life—he is now conducting a parallel study of women—Levinson has come to view adult development as marked by an alternating sequence of relatively stable "structure-building" periods and periods of transition. He has paid special attention to the transition that occurs at about the age of 40. Although this midlife transition may be either smooth or abrupt, the person who emerges from it is always different from the one who entered it.

The midlife transition provides an important opportunity for personal growth. For example, not until we are past 40, Levinson believes, can we take a "universal" view of ourselves and the world rising above the limited perspective of our own background to appreciate the fullest meaning of life. "I don't think anyone can write tragedy—real tragedy—before the age of 40," Levinson declares.

Disagreement Over Methods

As a student of biography, Levinson does not hesitate to take a biographical view of the controversy at hand. "To Paul Costa," he suggests in an understanding tone, "the most important underlying issue is probably the specific issue of personality stability or change. I think the question of *development* is really not important to him personally. But he's barely getting to 40, so he has time." Levinson himself began his research on adult development when he was 46, as part of a way of understanding the changes

he had undergone in the previous decade. He is now 60.

Costa, for his part, thinks that Levinson's clinical approach to research, based on probing interviews with small numbers of people, lacks the rigor needed to establish anything conclusively. "It's only 40 people, for crying out loud!" he exclaims. And Costa doesn't view his own age (he is 38) or that of his colleague McCrae (who is 32) as relevant to the questions under discussion.

Jack Block, who is also a hardheaded quantitative researcher— and, for the record, is fully 57 years old— shares Costa's view of Levinson's method. "The interviews pass through the mind of Dan Levinson and a few other people," Block grumbles, "and he writes it down." Block regards Levinson as a good psychologist who should be putting forth his work as speculation, and not as research.

As this byplay suggests, some of the disagreement between the upholders of stability and the champions of change is methodological. Those who argue for the persistence of traits tend to offer rigorous personality-test evidence, while those who emphasize the potential for change often offer more qualitative, clinical descriptions. The psychometricians scoff at the clinical reports as unreliable, while the clinicians dismiss the psychometric data as trivial. This summary oversimplifies the situation, though, because some of the strongest believers in change, like Brim, put a premium on statistical, rather than clinical, evidence.

When pressed, people on both sides of the debate agree that personality is characterized by *both* stability and change. But they argue about the probabilities assigned to different outcomes. Thus, Costa maintains that "the assertive 19-year-old is the assertive 40-year-old is the assertive 80-year-old...*unless something happens to change it.*" The events that would be likely to change deeply ingrained patterns would have to be pretty dramatic ones. As an example, Costa says that he would not be surprised to see big personality changes in the Americans who were held hostage in Iran.

From Brim's standpoint, in contrast, people's personalities—and especially their feelings of mastery, con-

trol, and self-esteem—will keep changing through the course of life... *unless they get stuck.* As an example, he notes that a coal miner who spends 10 hours a day for 50 years down the shaft may have little opportunity for psychological growth. Brim believes that psychologists should try to help people get out of such ruts of stability. And he urges researchers to look more closely at the ways in which life events—not only the predictable ones, such as getting married or retiring, but also the unpredictable ones, such as being fired or experiencing a religious conversion—may alter adult personality.

At bottom, it seems, the debate is not so much methodological as ideological, reflecting fundamental differences of opinion about what is most important in the human experience. Costa and McCrae emphasize the value of personality constancy over time as a central ingredient of a stable sense of identity. "If personality were not stable," they write, "our ability to make wise choices about our future lives would be severely limited." We must know what we are like—and what we will continue to be like—if we are to make intelligent choices, whether of careers, spouses, or friends. Costa and McCrae view the maintenance of a stable personality in the face of the vicissitudes of life as a vital human accomplishment.

Brim, however, views the potential for growth as the hallmark of humanity. "The person is a dynamic organism," he says, "constantly striving to master its environment and to become something more than it is." He adds, with a sense of purpose in his voice, "I see psychology in the service of liberation, not constraint."

Indeed, Brim suspects that we are now in the midst of "a revolution in human development," from a traditional pattern of continuity toward greater discontinuity throughout the life span. Medical technology (plastic surgery and sex-change surgery, for example), techniques of behavior modification, and the social supports for change provided by thousands of groups "from TA to TM, from AA to Zen" are all part of this revolution. Most important, people are trying, perhaps for the first time in history, to change *themselves.*

Some social critics, prominent

among them Christopher Lasch in *The Culture of Narcissism*, have decried the emphasis on self-improvement as a manifestation of the "Me" generation's excessive preoccupation with self. In Brim's view, these critics miss the point. "Most of the concern with oneself going on in this country," he declares, "is not people being selfish, but rather trying to be better, trying to be something more than they are now." If Brim is right in his reading of contemporary culture, future studies of personality that track people through the 1970s and into the 1980s may well show less stability and more change than the existing studies have shown.

The Tension in Each of Us

In the last analysis, the tension between stability and change is found not only in academic debates but also in each of us. As Brim and Kagan write, "There is, on the one hand, a powerful drive to maintain the sense of one's identity, a sense of continuity that allays fears of changing too fast or of being changed against one's will by outside forces. . . . On the other hand, each person is, by nature, a purposeful, striving organism with a desire to be more than he or she is now. From making simple new year's resolutions to undergoing transsexual operations, everyone is trying to become something that he or she is not, but hopes to be."

A full picture of adult personality development would inevitably reflect this tension between sameness and transformation. Some aspects of personality, such as a tendency to be reclusive or outgoing, calm or anxious, may typically be more stable than other aspects, such as a sense of mastery over the environment. Nevertheless, it must be recognized that each of us reflects, over time, both stability and change. As a result, observers can look at a person as he or she goes through a particular stretch of life and see either stability or change or—if the observer looks closely enough—both.

For example, most people would look at Richard Alpert, the hard-driving psychology professor of the early 1960s, and Ram Dass, the bearded, free-flowing guru of the 1970s, and see that totally different persons are here now. But Harvard psychologist David McClelland, who knew Alpert well, spent time with the Indian holy man and said to himself, "It's the same old Dick!"—still as charming, as concerned with inner experience, and as power-oriented as ever. And Jerry Rubin can view his own transformation from Yippie to Wall Streeter in a way that recognizes the underlying continuity: "Finding out who I really was was done in typical Jerry Rubin way. I tried everything, jumped around like crazy with boundless energy and curiosity." If we look closely enough, even Richard Raskind and Renée Richards will be found to have a great deal in common.

Whether a person stays much the same or makes sharp breaks with the past may depend in large measure on his or her own ideas about what is possible and about what is valuable. Psychological research on adult development can itself have a major impact on these ideas by calling attention to what is "normal" and by suggesting what is desirable. Now that researchers have established beyond reasonable doubt that there is often considerable stability in adult personality, they may be able to move on to a clearer understanding of how we can grow and change, even as we remain the same people we always were. It may be, for example, that if we are to make significant changes in ourselves, without losing our sense of identity, it is necessary for some aspects of our personality to remain stable. "I'm different now," we can say, "but it's still me."

As Jack Block puts it, in his characteristically judicious style: "Amidst change and transformation, there is an essential coherence to personality development."

For further information read:

Block, Jack, "Some Enduring and Consequential Structures of Personality" in A.I. Rabin et al. eds., *Further Explorations in Personality*, Wiley-Interscience, 1981, $24.50.

Brim, Orville G., Jr., and Jerome Kagan, eds., *Constancy and Change in Human Development*, Harvard University Press, 1980, $27.50.

Costa, Paul T., Jr., and Robert R. McCrae, "Still Stable After All These Years: Personality as a Key to Some Issues in Adulthood and Old Age," in Paul B. Baltes and Orville G. Brim, Jr., eds., *Life-span Development and Behavior* Vol. 3, Academic Press, 1980, $35.

Levinson, Daniel J., et al., *The Seasons of a Man's Life*, Alfred A. Knopf, 1978, $10.95; paper; Ballantine Books, 1979, $5.95.

244

Development During Maturity and Old Age

There are two extreme points of view concerning the latter part of the life span. The first point of view, "disengagement," argues that the physical and intellectual deficits associated with aging are inevitable and should be accepted at face value by the aged. The second point of view, "activity," acknowledges the decline in abilities associated with aging, but also notes that the aged can maintain satisfying and productive lives if they desire.

Extreme views in any guise suffer from the problem of homogeneity. The problem of homogeneity involves stereotyping all individuals within a category or class as having the same needs and capabilities. Whether one's reference group is racial, ethnic, cultural, or age-related, stereotyping usually leads to counter-productive, discriminatory social policy which alienates the reference group from "mainstream" society.

Evidence compiled during the past decade clearly illustrates the fallacy of extremist views of maturity and aging. Development during the later years of life is not a unitary phenomenon. Although the nature of the problems that must be negotiated at various age periods may differ, individual differences in how problems are resolved are as much a part of later development as they are of early development.

Erik Erikson was among the first to draw serious attention to the conflicts associated with each of the age periods in the life cycle, including adulthood and aging. Subsequent investigators have drawn attention to the major problems of middle age—that period of development which marks the transition to maturity and old age. To date, however, far less is known about development during the adult years than is known about the years of infancy, childhood, and adolescence.

Behavioral gerontology remains a specialization within human development that is absent from most graduate programs in developmental psychology. On the other hand, efforts to encourage training and research in this area are increasing. One can only hope that during the 1980s significant advances will be made in our knowledge of the later years of development.

Looking Ahead: Challenge Questions

How does contemporary American society view the aged? What events may be changing our attitudes toward this period of development?

What are the crises and conflicts which are characteristic of middle-age and old-age? How are they different from crises of earlier age periods of development? How are they the same?

What are the physiological mechanisms underlying aging? How do they affect physical functioning? Intellectual functioning? Emotional functioning?

What are your views on the American way of death? Have you experienced the death of an older person in your family? Would you like your own death to be handled similarly?

Late motherhood: Race against Mother Nature?

ELISE VIDER

Elise Vider, a free-lance writer from West Hartford, previously has written for Better Health, *and regularly for* Connecticut *magazine and the* Hartford Advocate.

Having a baby is no longer an automatic part of a young woman's life.

Medical and sociological changes of recent years have so increased the options for today's young women that the decision of when--and in rare cases whether--to have a baby can be tormenting.

The women born in the immediate post-war baby boom entered their child-bearing years in the late 1960s, a time of social and political upheaval that challenged many of the assumptions of the previous generation. Many women of this new generation chose to pursue careers, often as an alternative to the tradition of marriage and having a family. Many have had more than one marriage; many have not married at all or married much later than their mothers. Add to that the effects of newly developed birth control measures and the changed moral climate that made them acceptable.

The result? A dramatically reduced birth rate in the 1970s. Births for every 1,000 people in Connecticut hovered between 11 and 12 during the late 1970s, compared to its peak of nearly 24 in 1957. Hospitals, including the Hospital of St. Raphael, reduced their numbers of maternity beds. Some hospitals even closed entire wings because of the drop in births. Having babies was simply not a priority for many young women.

Racing the biological clock

But as that generation of baby boom women enters its 30s, a sense of growing urgency about having children has entered the minds of many. No longer do the reproductive years span a seemingly endless horizon. Many women have become acutely aware of the ticking of a biological time clock that limits the number of years they can have children. Many doctors, counselors and other experts have observed anxiety among women in their 30s and early 40s who have not had children. These women, the experts say, feel pressured to have a baby before it's too late to take their turn grabbing what one physician has called "the brass ring."

Not surprisingly, the number of births is once again on the rise, although not nearly at 1950s and '60s levels. Still, the sudden spurt of births has prompted many to refer to "the baby boomlet." Many of these new mothers are markedly older than their own mothers. The U.S. Department of Health and Human Services recently reported that the birth rate for women ages 25 to 29 now equals that for those ages 20 to 24, and that first births are up sharply among women in their early 30s. The National Center of Health Statistics reports that from 1975 to 1978, there was a 37 percent increase in women ages 30 to 34 who had their first child and that for women ages 35 to 39, the increase was 22 percent.

Mothers ages 35...and older

In Connecticut, about five percent of all live births were to mothers ages 35 and older in 1977 and 1978, the last years for which figures are available. That percentage is now starting to increase, a state statistician said. The figure hit its all time high in the mid-1950s when women tended to have more children and therefore continued to bear them later in life. In 1977, over 1,500 Connecticut babies were born to women ages 35 to 39 and 134--or about 10 percent of those whose birth order is known--were first babies. Women ages 40 to 44 had 262 babies, about 9 percent as first babies. Ten babies were born to mothers 45 to 49, one of which was a first baby.

The spate of middle-agers as first time mothers is still too new for experts to determine all the implications. Still, there are medical, sociological and psychological factors that are receiving closer scrutiny. Articles abound and several books have been published on the subject. A recent, three-year study by the Wellesley College Center for Research on Women reported that the decision to have children, the pregnancies, perceptions of childbirth and performance as parents were significantly different for older mothers. Psychologists who did the study reported to the NEW YORK TIMES that when they started their research in 1976, their definition of a "late timing" parent was a 30-year-old. Surprisingly, by the time they finished interviewing in 1979, the boundaries had shifted so dramatically that they ended up focusing on an age range of 37 to 44.

Late motherhood: how risky?

Medical considerations--the health and well-being of mother and child--are the most pressing concern about older mothers. Obstetricians classify first-time mothers aged 35 and older as "elderly primaparas" and as high risk patients. By the late 30s and early 40s, it is agreed, the potential for a problem pregnancy and delivery, difficulty conceiving, maternal mortality rates and the danger of chromosomal abnormalities all increase.

Doctors stress that medical advances make it easier and safer than ever before for women to bear children later in life. Dr. Brian Rigney, chairman of obstetrics and gynecology at St. Raphael's, said the oldest first-time mother in his practice was a 43-year-old who had "beautiful twins" with no problems.

The authors of UP AGAINST THE CLOCK, a 1979 book on deferred childbirth, contacted numerous physicians and health professionals to assess the risks for older mothers and their babies. Their finding: "All in all, the news is good. There is growing consensus among health professionals that the medical risks to both the older mother and her baby have been exaggerated and that healthy women of any child-bearing age who have access to fine medical care can look forward to a normal pregnancy and a healthy baby. These optimistic conclusions, however, should not be taken to mean that there are no drawbacks, from a medical standpoint, of postponing children."

Perhaps the biggest risk for older mothers is the increased probability of bearing a child with a birth defect caused by chromosomal abnormality, particularly Down's Syndrome or mongolism. Dr. Rigney reported that up to age 30, the chances of a mother

bearing a child with Down's Syndrome are less than one in a thousand. At age 35, the odds increase to one in 100 to 120. And by age 45 or older, the odds are one in 45. The probabilities are the same, he added, regardless of whether the woman has had previous children.

The reason for the dramatically increased risks of chromosomal abnormalities, it is believed, is the age of the mother's egg cells. Unlike men, who produce new sperm, women are born with lifetime supplies of oocytes, or immature egg cells. As women age, these cells can be affected by exposure to medications, pollutants, infections and radiation so that, by mid-life, the chances are higher of an oocyte producing an egg with an abnormal number of chromosomes. When the number of chromosomes in the fetus varies from the normal 46, the result can be a serious birth defect, including Down's Syndrome.

Some recent research also indicates some risk among older fathers. Men over age 55 seem more likely than their younger counterparts to contribute a sperm with an incorrect chromosome count. Twenty-three chromosomes from father and 23 from mother provide the normal 46 count in the fertilized egg or fetus.

It is now possible to screen for chromosomal abnormalities through the process of amniocentesis, in which the doctor inserts a hollow needle into the abdomen of a pregnant woman and withdraws a small amount of the amniotic fluid that surrounds the fetus. By examining the fluid, doctors can determine if the fetus has any serious chromosomal birth defect.

Usual practice at St. Raphael's on this sensitive subject is to make expectant mothers in their late 30s aware that such screening is possible and to perform amniocentesis if the patient wishes, according to Dr. Rigney. But, he added, "If a couple does not want an abortion, there is very little sense in the test." Only a small percentage of hospital patients choose amniocentesis, he said.

If a patient opts for amniocentesis, it is crucial that it be handled by a doctor and laboratory well acquainted with the delicate procedure because of its inherent risks. Amniocentesis carries a one in 200 chance of damage to the fetus or loss of pregnancy, said Dr.

Rigney. In some cases, the extracted cells do not grow in the laboratory, necessitating a repeat of the entire procedure, he added.

Age and fertility

Older women have more difficulty conceiving in the first place, doctors say, because fertility drops in the late 30s as satisfactory and regular ovulation begins to fail. The existence of fibroid tumors, which are benign growths in or around the uterus, and endometriosis, in which cells of the uterine lining adhere to the tubes and ovaries, are more likely in older women and both are linked to infertility. For women who choose to defer childbearing until middle age, difficulties in conceiving can be particularly upsetting because of the deadline pressure. And, unfortunately, doctors still have no reliable way to gauge fertility of a patient before she attempts pregnancy.

"If you were to ask at age 26 if you were going to be fertile at age 36, you couldn't possibly be told," said Dr. Rigney. Patients with a history of irregular menstrual periods or abdominal inflammation can be infertile at any age. Fertility is simply less likely with progressive age, he said.

Medical problems such as increased frequency of diabetes and high blood pressure among older women can make pregnancy and delivery difficult. Hypertension, in particular, is linked to toxemia in newborns, a condition that can cause an infant to be stillborn. Vaginal deliveries are sometimes more difficult and, said Dr. Rigney, "It is fair to say that there is an increase in Cesarean sections with the age of the mother." Chances of miscarriage increase after age 40, Dr. Rigney also said. Maternal mortality rates rise with the age of the mother, increasing tenfold between ages 20 and 45, he added.

Despite all the potential problems, there is agreement that medical advances, especially the sophisticated monitoring during pregnancy with devices such as ultrasound, which provides a picture of the developing fetus, amniocentesis, fetal monitoring and other modern tests, make it more likely than ever that a healthy woman in her 30s or early 40s will have a normal pregnancy and healthy baby.

And, who knows for sure what will

happen in the future? For an increasing number of life-threatening problems, medical and sometimes even surgical treatment is becoming possible for babies not yet born. The latest and most exciting treatment for specialists is actual direct treatment of the fetus while still in the mother's womb. An unborn baby girl in California was recently saved from probable death by giving her mother large doses of a vitamin called biotin in the final trimester of pregnancy.

Is there a "best" time?

Nor is biological age the only factor in determining the optimal age for having babies. When social and emotional considerations are taken into account, the benefits of later parenthood increase. A girl of 16--and sometimes even younger--is biologically capable of having a child, Dr. Rigney pointed out, but to do so usually has emotional repercussions from the interrupted adolescence. "Ideally, the age for motherhood is a combination of both biological and emotional maturity. And that's probably going to be in the mid 20s," said Dr. Rigney. The actual average age of first-time mothers at St. Raphael's, by Dr. Rigney's observation, is in the late-teens to early-20s.

When it comes to the quality of parenting, the emotional maturity and stability of many older women are no small consideration. Many experts feel that older adults make better parents because they are often more mature and capable and because, having waited so long, they are exceptionally enthusiastic. The marital stability of older couples "whose relationship has withstood the test of time" is another bonus, according to Dr. Ronald Angoff, a St. Raphael's pediatrician who specializes in child development.

Most of the older parents he sees in his practice are couples who have been married awhile and pursued separate careers, intentionally deferring childbirth. For these couples, the decision to have children does not come lightly.

"What it means to have children and give up the independence you've had before and the ability to do what you want to do, when you want to do it, to have a newborn infant who is totally dependent on you, I think, takes a degree of maturity," said Dr. Angoff.

Some older mothers tend to better educate themselves about pregnancy and childbirth and are "good consumers. And consumerism is healthy in medicine as in anything else," said Dr. Angoff. As a result, the "potential for these infants to do better and

develop better is there." But the flip side of that is that older parents can be excessively anxious and preoccupied with the fear that something is wrong with the child. Overall, he said, it is important to note that "the trend is relatively new. We don't really know how these kids will turn out."

The Wellesley study of older mothers reported a number of advantages. Generally, the older mothers were more financially secure and many, having already had careers, were more settled and secure about having experienced professional success. The women in the study reported less resentment than their younger counterparts about having to sacrifice for their children. One participant was quoted as saying, "I've had 20 years now of going out to the opera and how many more do you have to see?"

For many women in their 30s, however, who have not reached a decision about whether to have children, the dilemma can be fraught with agony and anguish. The fear of a change of heart after it's too late is a strong factor. Still, warned Dr. Angoff, for the woman who reasons that she is in her late 30s and had better have a child although she doesn't feel ready, "That's the wrong reason to have a baby."

The Dynamics of Personal Growth

Linda Wolfe

Linda Wolfe is a writer specializing in psychology and behavior. She is married to a New York psychologist and has one child.

Does psychological growth continue during adulthood or are our life courses fixed, determined by a handful of crucial childhood years? I first began thinking hard about this question in 1972 when, in the course of my work as a behavioral reporter, I covered a meeting of a distinguished psychological research organization, the Life History Society, and witnessed an astonishing event.

Psychologists and psychiatrists are cool people, cautious about over praising their peers, careful and measured in their responses. But at that meeting the assembled researchers seemed, down to the last man and woman, to be overwhelmed with admiration for a report given by one of their members, Dr. Daniel Levinson, professor of psychiatry at Yale, who spoke for a mere 45 minutes or so on his research into the then barely named phenomenon, the midlife crisis.

Levinson had studied a group of men in their late thirties and early forties and had found a remarkable pattern: No matter whether his subjects were having difficulties in their careers or were riding high, were receiving professional slights or high-paying promotions, they felt psychologically struck down when they entered the late-thirties to early-forties period. They became wavering where they had been firm, became frightened where they had been bold. The men he had studied (forty in all, interviewed in detail over a span of several years) came from a variety of economic backgrounds—some were blue-collar workers, some were white collar, some were scientists, and some were writers. But regardless of their class or profession, and even of their personalities, Levinson discovered that into each of these men's lives, some time in the middle years, there was an identifiable and perhaps even predictable time of trouble—the midlife crisis, a painful growth period from which no man or woman is immune.

The response given by his colleagues was immediate and intense. Dr. Joseph Zubin, then a major research director at Columbia University's Psychiatric Institute, stood up and said, "I feel you have just given my biography," and seventy-nine-year-old Dr. Henry Murray, deviser of the widely used Thematic Apperception Test, said, "I had a midlife crisis when I heard you speak. Is it retrogression? To be so pleased to hear your golden words?"

The reason for the enthusiasm was simple. For years psychology had concentrated on childhood and adolescence. Most of us who follow the field had the impression that whatever happened to a person in his or her early formative years (some even went so far to say the first three years of life, or even the first two), laid a groundwork, a barrier so impassable, that the rest of life was but its shadow, a constant repetition and reiteration.

Common sense and life experience seemed to indicate the opposite. All of us know individuals whose happy, easy childhoods should have served, were the psychological theories correct, to inure those people forever from the kinds of life tragedies they later experienced: the divorces, career casualties, psychological depressions, and failures of nerve. And, conversely, all of us know people who, starting life with dreadful memories of childhood, perhaps as orphans or refugees, nevertheless manage to ride out life's storms and land on islands of contentment. Yet most psychologists gave little thought to adulthood. As Levinson says, "We act as if we believe that if you know the really important things that happened in a man's childhood, the rest of his life will be more or less predictable, which of course it isn't. Psychologists speak as if development goes on to age six, or perhaps eighteen. Then there's a long plateau in which random things occur, and then, at around age sixty or sixty-five, a period of decline sets in to be studied by gerontologists. The assumption I am making is that there is something called adult development, an unfolding, just as there is earlier."

Since 1972, the study of adulthood has finally come into its own, with an outpouring of new research and new insights. Across the country, researchers have been studying growth patterns and attempting to delineate normal adult development. While it is not yet possible to say that we know as much about adult development as we do about child development, we are beginning to get some answers to haunting questions: Yes, adults continue to grow psychologically; yes there are relatively predictable cycles of psychological stability and crisis; and yes, a person's adjustment during childhood may be less revealing of his or her adjustment in adulthood than psychology of the fifties and sixties had us believe. Psychology in the seventies is the psychology of wait and see. A life must be fully lived before we can judge the man or woman who lived it as psychologically sound or failed. In that sense, while the seventies' psychology is new, it is also exquisitely ancient and sage, for it is said that the Greeks had carved onto one of their temples the motto, "Call no man lucky till his life is lived."

It was Erik Erikson, author of *Childhood and Society* and father of the 1960s obsession with adolescents and their "identity crises," who was the most influential figure in the current burst of enthusiasm for studying adulthood. Erikson believed that all individuals *travel* through their lives, mastering psychosocial age-related tasks in order to arrive at a psychic place of contentment and fearlessness in old age. In adolescence, the task is *identity*. In the twenties through the forties, the task is *intimacy*. Later still, people must learn to master *generativity*, the concern with "establishing and guiding the next generation." Those age-related tasks accomplished, men and women in late adulthood, which Erikson placed at after sixty-five, can enter the stage of life he called *integrity*. It is, he says, "the acceptance of one's one-and-only life cycle as something that had to be and that, by necessity, permitted of no substitutions. . . . In such final consolidation, death loses its sting."

Erikson was the philosopher of adult development. Those who have followed in his footsteps have been less concerned with the *purpose* of growth than with delineating more closely the various adult stages. Although different researchers give the different stages different names and slightly different age breakdowns, all seem to agree on an age twenty-to-sixties progression that might be listed this way:

20 to 30   Early adulthood: The mastery of intimacy

Around 30   Early adulthood: Transition or crisis

30 to 40   Intermediate adulthood: The mastery of careers

Around 40   Intermediate or midlife transition or crisis

40 to 50   Middle adulthood: Reorganization and renaissance

50 to 65   Later adulthood: Increased awareness of mortality

Ages are approximate: Transitions or crises occur within a year or two of the dates given. The explication of these stages which follows is drawn chiefly from the work of three leading life-span researchers—Dr. Levinson of Yale, Dr. George Valliant of Harvard (who has reported on a major study of close to 100 men whose lives were examined at regular intervals from the time they were undergraduates at an Ivy League college until they reached age fifty), and Dr. Roger Gould of the University of California School of Medicine at Los Angeles (who has reported on a study of several

hundred men and women between the ages of eighteen and sixty).

Early adulthood: The mastery of intimacy (20 to 30)

All the researchers seem to agree, perhaps taking their cue from Erikson, that men and women in their twenties are chiefly concerned with mastering intimacy. This is the time of life when spouses are wooed and wed, and when adolescent friendships are cast off if they no longer seem desirable, or consolidated if they seem worthy of future investment. Dr. Valliant says of the men he studied that no sooner did they "win autonomy from parents than they learned how to entrust themselves to others."

Levinson calls a part of this period GIAW—Getting Into the Adult World—and certainly young adults do not feel quite part of the grown-up life of responsibility and duty. They must get into it, must learn its rules, and experiment with adult roles. While some of them choose and start on careers, it is usually with only a modicum of commitment: others delay, devoting themselves first to sexual and social goals. Many have children now and are intensely absorbed in parenting. Roger Gould finds this period of life a relatively calm one, at least until young adults reach their late twenties. "Extreme emotions are . . . guarded against. . . . The emphasis is on modulating the emotional tone in an experimental effort to learn the proper tone for adult life."

Early adulthood: Transition or crisis (Around 30)

Once marriage and career lines have been established, young adults frequently enter a turbulent period marked by a questioning of values and commitments. Around age thirty, young married men and women may first consider and even have extramarital affairs. The men may switch careers and the women become more serious about having one. The focus of the earlier years is gone, and the direction of the next decade unclear. Gould says the trouble lies in the fact that at this point young adults feel that "some inner aspect is striving to be accounted for." Levinson finds that men around thirty question their career lines because they feel their occupational choice is too constraining or is "a violation or betrayal of a dream they now had to pursue." Similarly, men whose lives in their twenties were unstable and unconstrained may now feel a need for order and responsibility. "This brief transitional period," says Levinson, "may occasion considerable inner turmoil—depression, confusion, struggle with the environment and within oneself—or it may involve a more quiet reassessment and intensification of effort."

Intermediate adulthood: The mastery of careers (30 to 40)

The transitional period behind them, men and women now begin to make a strong commitment to accomplishments outside the personal. Levinson calls this period "Settling Down." People in their thirties seem to want stability, a niche in society, and they also want to "make it," to move toward major occupational goals. The Ivy League men whom Valliant studied were, at this time in their lives, very outerdirected, and not introspective. "They tended to sacrifice playing and instead worked hard to become specialists. Rather than question whether they had married the right woman, rather than dream of other careers, they changed their baby's diapers and looked over their shoulders at the competition."

In their later thirties, men in particular are involved in what Levinson calls BOOM—Becoming One's Own Man. A curious phenomenon of this urge toward selfhood is that people may now begin to pull away from and even turn against formerly cherished professional mentors. The phenomenon has been chiefly documented in men, but at Yale it has begun to be studied in the lives of professional women as well. The mentor, usually ten to fifteen years older than the late-thirties individual, once served as benevolent guide to the depths and shallows of a profession. But in the course of becoming one's own man or woman, a late-thirties striver must abandon or at least separate from the mentor. Levinson's subjects chose physical distance: Scientists set up their own labs in far-off universities; writers changed editors; blue-collar workers and executives switched to new companies.

By thirty-nine, the decks are clear and men—most life-span studies have followed them—begin to focus on key events in their careers, something through which the world will signal whether or not they are valued. It may be the promotion to a new job, the reception of a new book, the acceptance of a piece of research. This signal takes time to receive, so that late-thirties men often seem to be waiting, suspended.

Intermediate or midlife transition or crisis (Around 40)

Slowly, inexorably, whatever the outcome of the awaited event, the midlife crisis now begins. The man around forty may have attained his life's dream—the critics may have lauded his novel, the National Institutes of Health may have granted him bountiful research funds—but still he is thrown into a state of bemused or even rageful introspection. Men at this age feel their achievements are hollow, their bodies are declining, and they hear death whispering to them for the first time in their lives. It is at this point that many men translate *angst* into action: they divorce, take a trip to Europe, change the direction of their research or their style of writing. Even when they don't make such dramatic changes, when the externals of their lives show no such extreme variation, the men themselves are somewhat different. Having held onto their life structures, their personalities change.

Is the crisis harmful as well as painful? Levinson believes that it often enriches and strengthens men for their later years. He himself, at forty-six, changed the direction of his research entirely and left Harvard for Yale, and he points to other more famous examples of men whose lives took a turn for the better after they had weathered a troubled midlife crisis: men like Freud, Gandhi, Frank Lloyd Wright. In fact, says Levinson, "Many men who don't have a crisis at forty become terribly weighted down and lose the vitality they need to continue developing in the rest of the adulthood stages."

Middle adulthood: Reorganization and renaissance (40 to 50)

For most men and perhaps—but we do not know yet—most women, the midlife crisis reaches its peak sometime in the early forties, so that the period which follows is relatively calm. Levinson calls it a time of "restabilization": Gould, one in which there is "a relief from the internal tearing apart of the immediately previous years." Men and women in their forties display a more stable loyalty to their mates. "They very actively look for sympathy and affection from their spouses, who in many ways they seem to be dependent on in a mode similar to that of their former dependency on parents." But Valliant points out that this period is not altogether an untroubled one. For men, at least, it is often marked by depression and renewed questioning of the meaning of life and work. Still, he suggests that despite an increased awareness of unhappiness, the man in middle adulthood is nevertheless calm. Even though men in their forties are conscious of more depression, he writes, "they also in midlife appeared far more able to accept their own tragedy and that of others."

Later adulthood: Increased awareness of mortality (50 to 65)

Gould calls this a period of "mellowing and warming up." Men and women are truly "adults" now, for they no longer see their parents as the cause of their life problems, nor view their spouses as controlling them, and thus resembling parents. According to Gould, "the children's lives are now seen as potential sources of warm comfort and satisfaction. . . . The spouse is now seen as a valuable source of companionship in life." There is, of course, an increased awareness of mortality and this may make men and women in later adulthood seem petty, become focused on their health or their anxieties, or yet again express doubts about the value of their contributions to the world. But, curiously, according to Gould, this is not a time of diminishing interest in human contact. Rather it is a period of life in which "there is a hunger for personal relationships." John A. B. McLeish, an eminent Canadian educator, has documented this in a recent book, *The Ulyssean Adult: Creativity in the Middle and Late Years,* and indicated the remarkable personal and professional accomplishment of many people in late adulthood.

8. MATURITY AND OLD AGE

These, then, are the major life cycles that researchers have recently identified. If they seem to describe men's growth and crisis more often than they do women's, it is because most life-span research was in the past directed toward the lives of men. This is changing. The Yale study of the role of mentors in women's lives will soon be completed, and so far it appears that women, too, must at some intermediate adulthood stage separate from their mentors if they are to continue to grow. At New York's City University, a study of young women and their mothers being conducted by Jill Allen indicated a similar phenomenon: At an early adulthood stage, women who do not separate themselves in their attitudes and aspirations from their mothers seem less capable than others of psychological growth and professional mastery. In the years to come, we will surely know as much about women as we do about men.

While the life cycles already described are interesting, revealing patterns all of us can readily recognize in ourselves, our friends, and parents, they do not seem to me as fascinating as another more subtle aspect of the study of adult growth. Life-span research has indicated that personality itself may alter as adults grow, has revealed that we are not bound to the course we are set on in our early childhood years. Levinson talked of how men's personalities changed depending upon how they weathered their midlife crises. Valliant came to the surprising conclusion concerning the Ivy League men that "a stormy adolescence, per se, was no problem to the normal progression of the adult life cycle. In fact it often boded well," and he noted that : "The men who were judged by the staff to have negotiated adolescence best were not always the men with the best midlife adjustments."

Similarly, midlife adjustments and midlife styles do not always form a reliable guide to what a man or woman can expect in old age. In a recent monumental study of 142 men and women whose lives were examined over a forty-year period—from age thirty to age seventy—California researchers Henry Maas and Joseph Kuypers came up with some surprising information. They noted that while for most people, early adulthood lifestyles set the tone for old-age styles, for certain individuals this was not the case. These people were more often women than men and, astonishingly, they were women who, in their thirties, had been considered by the researchers to have had the lowest energy levels, to have been the most depressed, dissatisfied, and unambitious. Yet these women sometimes changed radically for the better when their dissatisfactions reached a crisis level as a result of a divorce, the death of a spouse or a child. They began to work outside their homes, developed hobbies and friendships, and became in their later sixties "highly involved as guest or visitor or worker."

Some of us, then, continue to grow throughout our adult lives, even during those final decades when, presumably, our choices have been seriously narrowed. As to the rest of adult life, it is, as we now know, not a single unity but a time marked off by distinctive age-related growth cycles. Rooted in our families, we unfold and grow: we have fallow periods and seasons of sudden spurts of change; we are not the same today as we were yesterday, and we will have gone, before we age, through at least three or four distinctively different growth periods.

Stress

*Sometimes harmful, sometimes helpful, it plays
a crucial role in life. Mastery of its demands, says this
expert, is a major key to improving the quality of life.*

Hans Selye, M.D.

*Director of the Institute of Experimental Medicine and Surgery at
Canada's University of Montreal, Dr. Selye holds doctoral degrees in
philosophy and science as well as in medicine. He is the author of
"The Stress of Life" (1956) and "Stress Without Distress" (1974).*

A WOMAN giving birth and her baby being born—
both are under considerable stress. A child on that
first terrifying day at school and a student cram-
ming for a very important exam—both experience a great
deal of stress. A trapeze artist concentrating on a death-
defying jump and a painter producing his best work in a
moment of brilliant inspiration are under stress.

A businessman worrying continually about office
problems while he is at home, or about home problems
while he is at the office, is also under stress. A doctor
treating patients all day and answering sick calls at night
undergoes stress. His patients fighting to regain their
health are also under severe stress.

Nobody can escape stress because to eliminate it
completely would mean to destroy life itself! We are
confronted by stressful stimuli (stressors) innumerable
times a day. The technical definition of stress implies the
body's response to *any* demand. Stress is essential in our
day-to-day living. It is the spice of life. It is a driving force
in every man, woman, and child, urging them on to
creativity in the arts, sports, science, business, and
virtually every other field of human endeavor. An artistic
accomplishment, a triumphant athletic victory, a successful business transaction, winning a lottery: these are all
extremely pleasant and stimulating stresses, also known
as *eustress.* But stress can also be extremely unpleasant and
harmful. A large business loss, physical overexertion,
death of a family member, or even fear can be most
distressing, because they make the body work in an
unusual way. *Distress,* of course, is the opposite of
eustress.

Stress is a state in which a chain of glandular and
hormonal reactions takes place to help our body adapt
itself to changing circumstances and conditions in its
physical and emotional environment. These reactions are
not necessarily destructive. They prepare our body to
withstand extreme weather changes, key us up to accomplish difficult and sometimes apparently impossible
tasks, help us resist physical and emotional shock, and
heal our wounds.

The human species has used these deeply rooted
reactions since prehistoric times. Primitive man had to
survive in a natural and brutal world, so he depended
upon instant, automatic physical responses generated by
the complex biochemical mechanisms of his brain.

Imagine a caveman who suddenly sensed a lurking
saber-toothed tiger. That instant realization of danger
sent a screaming message to his brain, and it in turn
mobilized his body with a tremendous surge of reactions
traveling along the various biochemical circuits to several
glands. Adrenaline poured into his blood, giving him
instant, much-needed energy through the use of sugar
and other stored energy supplies. His blood pressure rose
sharply with his increased respiration and quickened
pulse rate. Certain functions prepared him for "fight or
flight." (For example, his blood coagulation mechanism
went on alert, accelerating his blood clotting time, in case
he should suffer injury.) At the same time, other functions
stopped instantaneously to allow his system to make an
undivided demand on his vital "adaptation energy" so
that he could meet this crucial moment of survival. When
he realized that his chances were slim if he stayed to fight,
he sped away to the safety of his den.

He slowly came down from his hormonal crisis, his
body unwinding from the tension. He felt tired. He had
drawn deeply on his vital adaptation energy, and rest was
the best way to heal the tiny scars inflicted by the "wear
and tear" of those crucial moments.

But the 20th century is a world quite different from the
one inhabited by our prehistoric ancestors. The crises of
our lives are different too. Our adaptive mechanisms keep
up their defensive maneuvers with that same sheer
intensity, but all this is elicited by more subtle, invisible
stressors that have quietly invaded contemporary life.

Modern life—particularly urban life—exposes us to
the everyday stressors of complex interpersonal relationships, job responsibility, richness of diet, and many
associated mental or emotional stimuli. We try to cope as
best we can, and usually we do so quite well. But not

always. We often overreact. Many financial worries can weigh us down by constantly demanding our adaptation energy to defend us against this kind of stress. Bad news from the tax collector, the stock exchange, or commodities market has driven thousands of people to physical and mental rack and ruin.

A three-piece suit is no protection against the stress of life in the business world. Some executives unknowingly become stressed by a curt memo from a superior, by a change in company policy, by a failure to be promoted, or by a smaller than expected raise. The list of stressors in business and professional life is long.

In a 1973 article in The Business Quarterly, John H. Howard, a leading specialist in management stress, wrote: ". . . The stressful aspects of *responsibility* and *decision making* are perhaps best portrayed in the 'executive monkey' experiments. These experiments used four pairs of rhesus monkeys in a 'behavioral-stress' situation. In each pair of monkeys, one animal was designated the experimental animal and the other the control animal. In the experiment both monkeys in each pair were exposed to an electric shock, but only the experimental animal could learn to avoid the shock by pressing a lever at specific intervals. The lever was not available to the control animal. If the lever was pressed in time no shock was administered to either the experimental animal or the control animal. Thus, both animals were shocked an equal number of times, but only the experimental animal could determine whether or not the shock would occur.

"After a few weeks the experimental animal in all four pairs died because of gastrointestinal ulcerations. The control animals that received the same amount of electric shock but were not faced with the decision of controlling it all lived, were examined, and no ulcers were found.

"This experiment tends to convey the possibility for extreme stress when there is responsibility for making decisions on important issues. In real life the experimental situation is often compounded by assigning the responsibility while at the same time limiting the power needed to fulfill the obligation."

These experiments and the conclusions drawn from them are not unanimously accepted by all scientists and, besides, they cannot give us entry into the individual world of each and every business executive. We cannot fully share and feel everything that a businessman experiences, his job priorities, his relationships with his employees, family demands on his invariably limited time, and the many other exacting situations common to his daily schedule.

Picture an executive who receives a memo from his president or director, expressing harsh criticism over his decision in an important business transaction. The businessman's brain registers the censure, eliciting guilt, then perhaps outrage and repressed anger. His pulse rate accelerates, his blood pressure rises, hormones pour into his blood. His "system" mobilizes for instantaneous action but none is forthcoming. Unlike the caveman, our executive cannot run to his cave. Nor can he fight.

Anger turns inward, his own emotions become the target, and minute wounds are inflicted as a result of yet another hormonal crisis. If this type of stressful situation becomes chronic, irreparable damage is caused to his adaptive mechanisms through overwork. The executive is setting himself up as a prime candidate for a heart attack, mental breakdown, hypertension, a peptic ulcer, migraine headaches, or even cancer. These attacks on the physical chemistry of his body, in turn, join the vicious circle of stress, if the stressful situation continues unabated.

Fortunately, most stress-producing situations in everyday life are transient. But each episode of stress always causes chemical scars, which accumulate. Recurring exposure to intense stress, such as that experienced by air traffic controllers, depletes the body's finite resource of adaptation.

In 1936, I was working in the biochemistry department of McGill University in Montreal, trying to find a new ovarian hormone in extracts of cattle ovaries. All the extracts, no matter how prepared, produced the same syndrome, characterized by enlargement of the adrenal cortex, gastrointestinal ulcers, and involution of the thymus. It became evident from these and subsequent animal experiments that the same set of organ changes caused by my toxic, damaging, impure glandular extracts were also produced by cold, heat, infection, trauma, hemorrhage, nervous irritation, and many other stressors.

This reaction became known as the first phase in what I described as the "general adaptation syndrome" (G.A.S.) or the biological stress syndrome. The three phases of the G.A.S. are: the alarm reaction, the phase of resistance, and the phase of exhaustion.

Because of its great practical importance, it should be pointed out that the triphasic nature of the G.A.S. gave us the first indication that the body's adaptability to stress is finite.

Animal experiments have demonstrated without doubt that exposure to any type of stress can be withstood just so long. After the initial alarm reaction to physical as well as mental stress, the body becomes adapted and begins to resist, the length of the resistance period depending upon the person's innate adaptability and the intensity of the stressor. Yet, eventually, exhaustion sets in.

In everyday language we could say that stress is the rate of wear and tear in the human machinery that accompanies any vital activity and, in a sense, parallels the intensity of life. It is increased during nervous tension, physical injury, infections, muscular work, or any other strenuous activity, and it is connected with a nonspecific defense mechanism which increases resistance to stressful or "stressor" agents.

An important part of this defense mechanism is the increased secretion by the pituitary gland of ACTH (adrenocorticotrophic hormone) which in turn stimulates the adrenal cortex to produce other hormones. Various derangements in the secretion of these hormones can lead to maladies which I call "diseases of adaptation" because they are not directly due to any particular pathogen (disease producer), but to a faulty adaptive response to the stress induced by some pathogen.

An example from daily life will illustrate in principle how diseases can be produced indirectly by our own inappropriate or excessive adaptive reactions. If you meet a drunk on your way home from work and he showers you with insults, nothing will happen if you go past and ignore him. However, if you fight or even only prepare to fight him, the result may be tragic. You will discharge adrenaline-type hormones, which increase blood pressure and pulse rate while your whole nervous system will become alarmed and tense in preparation for a "fight."

If you happen to be a coronary candidate (because of age, arteriosclerosis, a high blood cholesterol level, obesity), the result may be a fatal brain hemorrhage or heart attack. Who killed you? You committed biologic suicide by choosing the wrong reaction.

Quite a few misunderstandings about the nature of stress arise from the fact that everybody does not react in the same way while under stress. This should not be surprising, however, since there are no two identical individuals. Each of us is conditioned by *endogenous* and *exogenous* factors.

By endogenous factors, we refer to internal, inherited traits, such as familial diseases, proneness to certain maladies, or weaknesses of specific organs. Exogenous factors encompass various external conditions in the environment, including social, intellectual, and psychological elements as well as climatic and physical surroundings prevailing at the time of stressful experiences. These endogenous and exogenous factors combine to make each individual different from the next.

Think, for instance, of a chain placed under physical tension—that is, stress. No matter what pulls on the chain and no matter in which direction, the result is the same— the chain is faced with a demand for resistance. Just as in the chain the weakest link (or in a machine, the least resistant part) is most likely to break down, so in the human body there is always one organ or system which, owing to heredity or external influences, is the weakest and most likely to break down under the condition of general biologic stress. In some people the heart, in others the nervous system or the gastrointestinal tract, may represent this weakest link. That is why people develop different types of disease under the influence of the same kind of stressors.

Let's extend my earlier definition of stress to "the *nonspecific* response of the body to *any* demand." It can result from tensions within a family, at work, or from the restraining influence of social taboos or traditions. In fact, any situation in life that makes demands upon our adaptive mechanism creates stress. From a psychological point of view, the most stressful experiences are frustration, failure, and humiliation—in other words, distressing events. On the other hand, we derive a great deal of energy and stimulation, considerable force and pleasure, from victories and success. These give us ambition for work, a feeling of youth and of great vitality. As the saying goes, nothing generates success like success, or failure more than failure. We are encouraged and invigorated by our victories, whereas constant defeat eventually deprives us even of the motivation to try.

Although these obvious differences exist between the effects of pleasant and unpleasant experiences, in biologic terms both have a common effect—they cause stress. Even such happy sensations as great joy or ecstasy cause stress, for we must adapt to *any* demands made upon us, whether favorable or unfavorable. *Distress* is much more likely to cause disease than *eustress*, although there is evidence that, in excess, both can be harmful under certain circumstances. There is still no compelling scientific proof of this, but perhaps eustress (for instance, ecstasy) is less likely to be harmful because it rarely equals the intensity and duration of suffering. Besides, it is merely a matter of conditioning that determines whether we perceive a particular experience as pleasant or unpleasant. The event itself is always a stressor causing the same stress. Conditioning can only modify our perception of the event as being either pleasing or displeasing. A thermostat can produce either cold or heat, using the same amount of energy.

There can be no doubt that both eustress and distress have certain measurable biochemical and nervous elements in common. The following example will explain this apparently paradoxical fact: a mother receives a telegram announcing that her only son has been killed at war, and one year later he steps into her living room in perfect health because the news was false. The first experience is extremely painful, the second fills her with joy. Her tremendous grief is significantly different from her great pleasure. If you ask this mother if both experiences were the same, she will obviously answer, "No! Quite the contrary!"

Nevertheless, from the viewpoint of nonspecificity, the demands made upon her body have been essentially the same: a need to adjust to a great change in her life. From a medical and biochemical point of view, the concept of stress implies only an adaptive reaction to change, whether change is for the better or worse.

There are various mechanical instruments—known as "stress meters" or "stress polygraphs"—which measure such physical characteristics as cannot be directly appraised by the layman; however they can register only a few indicators which are usually, but not always, characteristic of stress. Of course, the more signs and symptoms are measured, the more reliable the general picture becomes; yet even the best of these procedures miss the crucial difference between eustress and distress. Moreover, beyond this distinction there lies a still more significant fact: it is our ability to cope with the demands made by the events in our lives, not the quality or intensity of the events, that counts.

What matters is not so much what happens to us but the way we take it.

That is why I recommend that, in everyday normal life, you should judge your own stress level, by easily recognizable signs which tell how you are taking the stress of your life at any particular moment [for some of those signs, see below].

Each of us must learn to recognize *overstress* (hyperstress), when we have exceeded the limits of our adaptability; or *understress* (hypostress), when we suffer from lack of self-realization (physical immobility, boredom, sensory deprivation). Being overwrought is just as bad as being frustrated by the inability to express ourselves and find free outlets for our innate muscular or mental energy.

It is sad to see that many of us today have either lost or abandoned our role as guardians of our own health. We are our own best physician, but we have instead entirely entrusted this responsibility to health-care professionals, hoping that their wonder drugs and sophisticated surgical interventions will save our lives in time of crisis. We must realize that many of our troubles arise only because of neglect and self-abuse. Some diseases, even those caused by stress, are beyond our control, but generally we victimize ourselves through our own bad habits, unhealthy lifestyles, and high-risk working conditions.

Business-oriented careers generally tend to have their own particular type of stressors. Challenges and changes are inherently embedded in every business-oriented career. But they are also great sources of stress. Every businessman has to constantly adapt to changing finan-

Distress signals

A NUMBER of manifestations of stress, particularly of the more dangerous *distress*, are not immediately evident, not constant symptoms that we should monitor throughout life. Depending upon our conditioning and genetic makeup, we all respond differently to general demands. But on the whole, each of us tends to respond with one set of symptoms caused by the malfunction of whatever happens to be the most vulnerable part in our machinery. Learn to heed these signs:

● General irritability, hyperexcitation, or depression associated with unusual aggressiveness or passive indolence, depending upon our constitution.

● Pounding of the heart, an indicator of excess production of adrenaline, often due to stress.

● Dryness of the throat and mouth.

● Impulsive behavior, emotional instability.

● Inability to concentrate, flight of thoughts, and general disorientation.

● Accident proneness. Under great stress (eustress or distress) we are more likely to have accidents. This is one reason why pilots and air traffic controllers must be carefully checked for their stress level.

● Predisposition to fatigue and loss of *joie de vivre*.

● Decrease in sex urge or even impotence.

● "Floating anxiety"—we are afraid, although we do not know exactly what we are afraid of.

● Stuttering and other speech difficulties which are frequently stress-induced.

● Insomnia, which is usually a consequence of being "keyed up."

● Excessive sweating.

● Frequent need to urinate.

● Migraine headaches.

● Loss of or excessive appetite. This shows itself soon in alterations of body weight, namely excessive leanness or obesity. Some people lose their appetite during stress because of gastrointestinal malfunction, whereas others eat excessively, as a kind of diversion, to deviate their attention from the stressor situation.

● Premenstrual tension or missed menstrual cycles. Both are frequently indicators of severe stress in women.

● Pain in the neck or lower back. Pain in the neck or back is usually due to increases in muscular tension that can be objectively measured by physicians with the electromyogram (EMG).

● Trembling, nervous ticks.

● Increased smoking.

● Increased use of prescribed drugs, such as tranquilizers or amphetamines; or increased use of alcohol.

—HANS SELYE, M. D.

cial or economic situations. The bigger the change, the greater is the stress factor.

The many technologic innovations and the social changes in family structure, in the respective rights and duties of men and women, and in the type of work now in demand because of urbanization, present society with unprecedented requirements for constant adaptation. The task of balancing our biologic and psychologic forces is becoming a heavy burden.

That is one reason why my colleagues and I at the University of Montreal are trying to develop accurate stress tests and stress-reducing programs drawn up on the basis of our 40-odd years of research. We hope that our recently created International Institute of Stress will be a center for documentation, research, and dissemination of information about all problems related to stress, which touches virtually every aspect of modern society.

I think that with the information now available through our own work, and that of innumerable scientists throughout the world who have become interested in this topic, the stress concept has been sufficiently clarified to reach the "critical mass" necessary for its continued growth.

My decades of research into the causes and effects of stress have afforded me an opportunity to observe the wide spectrum of this phenomenon. Through my studies I have also developed my own philosophy about stress and the nature of life itself. By way of an admittedly unscientific conclusion, let me share some of these thoughts with you, the Rotarians of the world.

The idea of the *vis medicatrix naturae* (the "healing power of nature") is extremely old. Man has long known that there is a built-in mechanism in all living beings designed to help them right themselves, to restore their integrity when they become wounded or damaged in any way. The principle of the stability of the *milieu intérieur* has been known for centuries, and along with it stress has long been considered as some kind of tension, fatigue, or suffering. But it took many carefully conducted and often tedious experiments to obtain objective, measurable data,

which led to observations—all of which had to be critically evaluated—before we could arrive at the stress concept in its present form, including its implications upon human behavior.

Some of the results of the stress encountered in daily life can be treated with standard medicines, others cannot. In order to meet the social stresses of our time, suggestions have been made for better laws or more severe law enforcement, arms limitations, and so on. To counteract the effects of mental distress, people try out various techniques which they claim eradicate the ill effects or at least help us to cope with them. This is why so many people turn to psychotherapy, meditation techniques, tranquilizers, *ginseng*—the number of stress pills and psychological antistress techniques grows every day. When nothing seems to work, some find an outlet for their energy only in violence, drugs, or alcohol, thus creating still more problems for society. To my mind, the root of the problem lies in the lack of a proper code of motivation that gives our lives a purpose which we can respect.

Viewed from the pinnacle of the eternal general laws governing nature, we are all surprisingly alike. Nature is the fountainhead of all our problems and solutions; the closer we keep to it the better we realize that, despite the apparently enormous divergencies in interpretation and explanation, natural laws have always prevailed and can never become obsolete. The realization of this truth is most likely to convince us that, in a sense, all living beings are closely related. To avoid the stress of conflict, frustration, and hate, to achieve peace and happiness, we should devote more attention to a better understanding of the natural basis of motivation and behavior.

Wouldn't it be better if we could find a way to prevent the stress of life, rather than simply means to help us fight a hostile world? I believe that this should be our highest priority in formulating the policies of governments to meet the stresses of modern civilization, and in devising our personal philosophies of life.

Coping with the Seasons of Life

A British study of the varieties of human experience

John Nicholson

John Nicholson is the author of "Seven Ages" (Fontana Paperbacks). This article is excerpted from the social science weekly "New Society" of London.

There are two very different views about the importance age plays in people's lives. There is the Shakespearean tradition, encapsulated in the "Seven Ages of Man" speech in *As You Like It*, which suggests that our lives fall into distinct phases and that people change as they get older, generally for the worse. Other writers take the view that we are "as old as we feel." André Gide remarked at the age of seventy-three, "If I did not keep telling myself my age over and over again, I am sure I should hardly be aware of it." Which view is right? In an attempt to answer this question, some 600 men, women, and children between the ages of five and eighty were interviewed last winter in the Colchester Study of Aging. The aim of this survey was to build up by objective means a subjective picture of the human lifecycle.

A few years ago you got the "key to the door" at twenty-one. But only one in eight of the adults we interviewed gave twenty-one as the age at which they considered they became adult. Most men judged that they became adult sometime between eighteen and twenty-one, while women gave more varied answers, with a significant proportion of them designating the mid-to-late-twenties.

What are young adults like? Though some functions—notably the performance of our hearts and circulatory systems—reach a peak slightly earlier, the years between twenty and twenty-five represent the pinnacle of our biological development. Physically and intellectually, we have never before been so good and never will be again.

On the negative side we found young adults to be self-centered and still naïve in their views on what life is all about. In psychological terms many of them were still in the throes of the identity crisis that had begun to disturb them in adolescence. Particularly among the unmarried, there seemed to be an internal conflict between wanting to establish a position which commands respect and not wanting to get into a rut. They were afraid of slipping into the habits of their boss, for example, by taking a briefcase home in the evening. And yet many were anxious to carve out a niche for themselves and enjoy the status and economic advantages of their jobs.

On the positive side they enjoyed the feeling of "no longer being a kid." Some—particularly the married—expressed pleasure in mapping out the future. Others complained that they were required to make once-and-for-all decisions, that their options were closing, and that they were anxious about their ability to cope with the future.

For some young adults the feeling of emotional insecurity far outweighs the self-confidence which comes from standing on their own two feet. Psychiatrists consider that many early marriages are "take-care-of-me contracts," entered into in a spirit of panic rather than out of conviction that one has found the ideal partner for life.

This is only one of the dangers presented by the freedom of young adulthood. For some people freedom becomes an obsession, to be guarded at all costs. As one example of this, many young adults we spoke to in Colchester were strongly opposed to the idea of having children, mainly because of the restrictions they felt it would impose on their sense of freedom.

People tend to look back on their early twenties with pleasure. When we asked our sample what age they would most like to be if they had the choice, the twenties proved to be a popular decade. Those who favored it did so because that was when life had been most enjoyable.

Perhaps the most striking feature of young adulthood is its exclusiveness with regard to age. At no other period of life do we spend less time in the company of people older or younger than ourselves; and it may well be that never again are we so sensitive to the difference of even a year.

People who reach young adulthood without having had any sexual experience are now in a minority (though barely). Since most adolescents say that they are opposed to casual sex, and seem to feel quite strongly that sex ought to occur only in the context of an established loving relationship, it may well be that many of today's teenagers have at least one such relationship before young adulthood.

In some respects, the intimate relationships young adults try to form are similar to earlier relationships with their parents and friends. They may well have loved their parents and felt deeply committed to best friends. The new dimension is being in love, and the problem which has taxed young adults since time began is how to distinguish between being in love and liking, depending on, or being infatuated.

"One of the pleasures of middle age is to find out that one was right, and that one was much righter than one knew at, say, seventeen or twenty-three." Or so said the poet Ezra Pound.

People in their forties are sandwiched between two demanding and often—in their eyes — unreasonable generations, both of which rely on them for psychological and practical support. At no other stage in their lives do so many people depend on them, and that realization causes people in their forties to brood darkly on the cyclical nature of life and the passing of time. They also ask themselves more mundane questions like: Should a woman/man of my age really be wearing slit-skirts/jeans? So although the forties may not precisely be middle age, they are an age-conscious decade.

Our appearance does begin to change in the forties. The full effects of these changes are not yet apparent, but we are beginning to look different. The balance between physical improvement and physical degeneration wavers during the thirties, then tips toward degeneration.

During our survey in Colchester we asked people in their forties what was the worst thing about life now compared with ten years ago. Their main complaints were the feeling of being so much older and worries about their appearance and health. But they were not aware of any significant change in their sex life. Most of those we interviewed disagreed strongly with the suggestion that sexual relationships are more important to young people.

The people we talked to seemed to be more concerned about physical than mental decline, and they were right, because in this period of our lives there is little cause for alarm. Any fears about declining mental faculties are more likely to be imaginary than real.

Our survey asked if any age since adolescence had seemed particularly difficult, and we were astonished to find that none of the people in the forty-to-forty-five age group described their present age in these terms. Our results clearly support those who deny existence of a midlife crisis, and we must conclude that if there is such a thing as the male menopause, it affects only a small group.

Unless we decide to tear up our roots and start again, we have to alter the emphasis in our personal relationships from sexuality to sociability, particularly in marriage. It is in the forties that husbands and wives start sizing each other up and wondering how they will adjust to living as a couple again without the shared responsibility of parenthood to bind them. Most couples of this age realize that an enormous emotional vacuum is about to open up, and it is interesting to see how they prepare to deal with it. In Colchester, we found that more women had jobs in the thirty-five-to-forty-nine age bracket than in any other.

The importance of women's jobs rose steadily from one age group to the next. It overtook money as a priority at forty, and by the end of the fifties ranked almost equal with friends. The fact that this was the only change in life-priorities shown by either sex between twenty-nine and fifty-nine clearly establishes this as one of the most significant changes in the forties.

How do men prepare themselves for the future? Although we didn't find that the majority of men in their forties in Colchester were becoming more interested in hobbies, some clearly were prepared to run the gamut of their wives' and children's sneers and were developing new interests to take their minds off worries and prepare for retirement.

Psychologists describe the qualities we need to develop in the forties as mental flexibility and the ability to broaden our emotional investment to include new people, activities, and roles. One person in ten of those in their early forties in Colchester said they found it more difficult than ten years ago to adapt to change or to accept new ideas, whereas three times this number said they found it less difficult.

When the Colchester survey asked its participants to list the things they worried about, we found that concern about children among people of thirty-six to forty-five was the largest single worry of either sex at any age. Nine out of ten of the women we interviewed, whose eldest child was between sixteen and nineteen, said they were worried about their child's future. The men in our survey were much less likely than their wives to worry about their children. Their most frequent worry was money.

People in their forties wonder whether permissiveness hasn't gone too far, and increasingly find themselves identifying with their own parents and defending their values rather than rebelling against them. So as middle age approaches, there seems to be a clear shift in our loyalties and attitudes, which is part of a growing awareness that we are about to join the older generation.

Perhaps the most uncertain feature of life is its length. The uncertainty makes it inevitable that thoughts about death should color the final stage of our lives. But the prospect of dying does not seem to destroy our ability to make the best of whatever age we happen to be. An old woman in Colchester said, "Some people thought I was doing too much and I ought to slow down a bit. I said, 'I've retired from work, I didn't retire from life'."

Researchers have constructed a table of events old people find most stressful. The death of a spouse comes at the top of this list, followed by being put into an institution, the death of a close relative, major personal injury or disease, losing a job, and divorce. Being widowed seems to affect men more severely than women. A recent British study found that the death rate among men and women during the year after they are widowed is ten times higher than among people of the same age who are still married.

As we approach the end of our lives we become less interested in the outside world and more concerned with ourselves. The psychological task that becomes increasingly important is to come to terms with ourselves, to find some justification for our lives, and to reconcile ourselves to the fact that it is going to end. Paradoxically, the person who believes that his or her life has been most worthwhile seems to have the fewest qualms about the prospect of its coming to an end.

What is the recipe for a successful old age? Some people say that unless old people keep themselves active and engaged, they will become a misery to themselves and a burden on others. Others take the view that an old person who continues to devote his energy to worldly matters cannot possibly have time to solve the psychological problems of old age. We now know that there is no single pattern which can be recommended as the recipe for a successful old age.

Perhaps the most useful characteristics we can possess at this stage of our lives are flexibility and acceptance. Changes need to be made, and the unpalatable fact of our own mortality has to be accepted. There is some evidence that we can predict in advance how easily an individual will come to terms with the final realities on the basis of how well he or she has made adjustments earlier in life. For example, people who had problems resolving the intimacy-versus-isolation crisis in their twenties, and who found it difficult to maintain an intimate relationship at age thirty, seem to have difficulty coping with the psychosocial crisis that comes with old age.

Younger people often complain about the way the old ramble on about the past. But research shows that the tendency some old people have to review the events of their lives can actually be therapeutic. It increases their chances of facing death with equanimity.

Some of the strongest fears of death expressed in our interviews came from people in their teens and twenties. Among the old, advancing age seems to produce a greater interest in death, but certainly no greater fear. One old woman said, "I feel that from the day you're born your life is mapped out for you. As for death, well, whatever way that comes, I can't stop it, so it doesn't worry me." An eighty-year-old said, "Dying doesn't worry me. The only thing I do worry about is if I go unexpectedly and leave a muddle for everyone to clear up."

Two major conclusions can be drawn from our Colchester Study of Aging. First: We ought to bury the notion of universal, age-related life crises. Many people never experience any discernible psychological crisis. And when crises do occur they tend to be caused by events which just happen to fall at roughly the same point in most people's lives, rather than because a person has reached a certain age.

Second: Age has remarkably little effect at any stage in life on how we think of ourselves or on how we view the world. As as eighty-year-old Colchester woman said, "I don't put things into blocks. If you just think, 'Well, I'm only one day older than I was,' you don't really feel very much different."

Must Everything Be a Midlife Crisis?

The various stages in a person's life may well be normal turning points.

Bernice L. Neugarten

Bernice L. Neugarten, Ph.D., is a professor of human development at the University of Chicago. This is adapted from an article published in the American Journal of Psychiatry.

The media have discovered adulthood. Gail Sheehy's *Passages,* Roger Gould's *Transformations,* and a dozen other popular books have all accorded adulthood the treatment of high drama, drawing from Erik Erikson's writings, George Vallant's *Adaptation to Life,* Daniel Levinson's *The Seasons of a Man's Life* and other studies. Journalists and psychologists make news by describing a "midlife crisis" as if it were the critical turning point between joy and despair, enthusiasm and resignation, mental health and illness. People worry about their midlife crises, apologize if they don't seem to be handling them properly, and fret if they aren't having one.

From these books, one would predict the imminent arrival of Dr. Gesell or Dr. Spock guides for grown-ups: *The Adult from 40 to 45, The Parent from 55 to 60,* and the like. But the public should not be too quick to assume that adult life changes—whether mild "transitions," or dramatic "transformations," or full-fledged "crises"—define what is normal, or to think that the person who has *not* gone through such changes is somehow

off the track. Many of my colleagues and I feel that media coverage of this subject has come too soon, is based on too little evidence, and generalizes too broadly from a highly select few to all of us. Levinson's study was based on only 40 men, for example, and Vaillant's on only 95 (Harvard graduates, at that).

Stage theories of adult development imply a fixed, one-way progression of issues and conflicts that occur at chronological intervals (such as Sheehy's "The Trying Twenties" and "Passage to the Thirties"). These notions appeal to us because they seem logical, and because they suggest a common pattern for the swirling, complex changes in our lives. But a Gesell for adults is unlikely to emerge—not because adults are changeless, but because adults change far more, and far less predictably, than the oversimplified stage theories suggest.

For many years, my students and I have studied what happens to people over the life cycle. The primary consistency we have found is a lack of consistency. We have had great trouble clustering people into age brackets that are characterized by particular conflicts; the conflicts won't stay put, and neither will the people. Choices and dilemmas do not sprout forth at ten-year intervals, and decisions are not made and then left behind as if they were merely beads on a chain.

It was reasonable to describe life as a set of discrete stages when most people followed the same rules, when major events occurred at predictable ages. People have long been able to tell the "right age" for marriage, the first child, the last child, career achievement, re-

tirement, death. In the last two decades, however, chronological age has moved out of sync with these marking events. Our biological time clocks have changed: the onset of puberty for both sexes is earlier than it used to be just a generation ago; menopause comes later for women; most significantly, more people than ever now live into old age.

And our social time clocks have changed, too. New trends in work, family size, health, and education have produced phenomena that are unprecedented in our history: for example, a long "empty nest" period when the children have left home and the parents are together again. A growing prevalence of great-grandparenthood. An enormous number of people who are starting new families, new jobs, and new avocations when they are 40, 50, or 60 years old.

Old? Even that word has no chronological boundary anymore. Researchers now distinguish the young-old from the old-old, and it is not a given number of years that marks the dividing line. The young-old are a rapidly growing group of retired people who are physically and mentally vigorous and whose major opportunity is the use of leisure time. They are healthy, financially comfortable, increasingly well-educated, politically active, and are eager consumers. They use their time productively, for self-fulfillment and community benefit. This sizable group of people is probably the first real leisure class America has ever had.

We are, in short, becoming an age-irrelevant society. We are already familiar with the 28-year-old mayor, the 30-year-old college president, the 35-year-old grandmother, the 50-year-old retiree, the 65-year-old father of a preschooler, the 55-year-old widow who starts a business, and the 70-year-old student. "Act your age" is an admonition that has little meaning for adults these days.

What gives people the feeling that they are going through "life crises" is not the events of adulthood themselves, but the fact that the events can no longer be keyed to our biological and social clocks. The stage theorists argue that crises occur at regular intervals, set off by natural changes in work, family, and personality.

I would argue precisely the opposite. Leaving the parents' home, marriage, parenthood, work, one's own children leaving home, menopause, grandparenthood, retirement—these are *normal* turning points, punctuation marks along the lifeline. They produce changes in self-concept and identity, of course, and people accept these events with various degrees of aplomb or difficulty; but whether or not they create *crises* usually depends on when they happen. For the majority of women in middle age, the departure of teenage children is not a crisis, but a pleasure. It is when the children do *not* leave home that a crisis occurs (for both parent and child). For the majority of older men, retirement is a normal event. It is when retirement

occurs at 50 instead of 65 that it can become a crisis. Grandparenthood does not set off a "crisis" of worry about old age and death; it is more likely to be a problem when it comes too early, or simultaneously with a new, young family, or when it doesn't come at all.

The popular stage theories are inadvertently replacing one set of expectations about the life cycle with another, and that concerns me. Whether people think their lives consist of three, six, or ten stages, they develop a package of expectations about what will or should happen to them, and when. Men and women compare themselves constantly to their friends, siblings, colleagues, and parents to decide whether they are doing all right. They do not worry about being 40, 50, or 60, but about how they are doing for their age. They do not want to be young again, but they want to grow older in ways that are socially acceptable and, at the same time, personally satisfying.

Certainly there are some psychological preoccupations that are more powerful at one age than another. A study of reminiscence, for example, found that middle-aged people consciously select elements from their past to help them solve current problems; the very old put their store of memories in order, dramatizing some and rearranging others for consistency—creating a coherent life history. But most intra-psychic changes

evolve very slowly, and they may or may not be linked with specific events. Divorce may provoke painful soul-searching for one woman, while another sails through the experience with a minimum of stress.

Most of the themes of adulthood appear and reappear in new forms over long periods of time. Issues of intimacy and freedom, for example, which are supposed to concern young adults just starting out in marriage and careers, are never settled once and for all. They haunt many couples continuously; compromises are found for a while, then renegotiated. Similarly, feeling the pressure of time, reformulating goals, coming to grips with success (and failure)—these are not the exclusive property of the 48-to-52-year-olds, by any means.

The new stage theories do not take into account the power of memory. Adults have a built-in dimension of thought that interprets the present relative to the past; the blending of past and present is our psychological reality. Our experiences with friends, lovers, children, parents, jobs become the fabric of our lives, and we are constantly reweaving these threads into new patterns. They are the stuff of which memories are made, the past reinvented, and the future planned. Whether we choose to celebrate our diversity and range, or confine ourselves to 10-year stages, is, of course, up to us.

LIVING LONGER

RICHARD CONNIFF

The subject, a healthy white male, is 29 years old, an age he would like to maintain more or less forever. At the moment, he is sitting in front of a computer terminal in a private office in Corona del Mar, California, about an hour south of Los Angeles. He is about to learn electronically, on the basis of key physical indicators, whether he is still in his prime or if his body is already slipping into the long decline to senescence. Forget what the mind feels, or what the calendar and a searching look into a mirror might suggest; this machine aims to tell how much the body has actually aged—and therefore how long the subject is likely to live.

The test takes 45 minutes. It requires no syringes, blood pressure belts or strapped-on electrodes. In fact, the whole experience is less like a medical examination than a stint at a computerized arcade game. Not quite Space Invaders, but then, the subject doesn't want to be Luke Skywalker. He's just a guy who'd like to live to 120, and never feel older than 29.

In all, there are more than a dozen tests. They measure such indicators as the highest pitch the subject can hear (known to decline significantly with age), the ease with which the eye adjusts its focus between near and far objects (lens and supporting structure become less elastic over the years) and the subject's ability to identify incomplete pictures of common objects (perceptual organization begins to fall off at about age 30).

Then comes a brief whirring of electronic thought and, ladies and gentlemen, the computer's analysis: The subject has the hearing of a 21-year-old, skin as sensitive as a child's and, overall, a body that just barely passes for 32. So much for the sweet bird of youth.

But not so fast. It turns out that there may be loopholes in the standard existential contract of birth, aging and death. Even people now approaching middle age can expect a bonus of perhaps several years of extra living, thanks to continuing medical progress against cancer, heart disease, stroke and other killers.

At the same time, gerontologists—specialists in the science of aging—are piecing together the details of diet, exercise, personality and behavior that make it practical to shoot for 80, or even 114—the longest human life span reliably recorded.

Finding the Fountain of Youth

There is hope even for people who fear diet and exercise almost as much as an early death. At the most minute level, scientists are now deciphering the basic biologic mechanisms of aging and of rejuvenation. Gerontologists are so confident about fulfilling the promise of their discoveries that a healthy young man aiming to live 120 years begins to appear reasonable. Writing in The American Journal of Clinical Pathology, one researcher recently predicted: "The discipline of gerontology is now advancing at such a rapid rate, with so much overflow from other fields . . . that I rather confidently expect a significant advance in maximum

life-span potential to be achieved for the human species during what is left of the present century. . . ." One probable means of life extension is already available, in a tentative form, and it suggests that a 16th-century Italian named Luigi Cornaro was far closer than Ponce de Leon to the fountain of youth. More about him later.

The so-called life-extension revolution couldn't come at a stranger time. American society is already on the brink of startling change. It is growing up. Whether longevity improves by only a few more years, as some expect, or by whole decades, mature people will for the first time predominate. Between 1970 and 2025, the median age in the United States will have risen almost 10 years, from 27.9 to 37.6—as substantial an age difference for a nation as for an individual. The number of people 65 and over will double, from 25 million now to 51 million in 2025, and there will be 85 million people over the age of 55. All without major new increases in longevity. In earlier societies, so few people managed to reach old age that they were deemed special, endowed with magical powers to ward off the demons of aging. In the United States over the coming decades, the elderly will be commonplace, and possibly more: a powerful, organized political and economic force.

Elderly Astronauts

What will it mean to live longer in such a society? One public-relations man, a product of the post-World War II baby boom, finds himself caught up in his generation's frantic competition for good homes and the best jobs. He suffers nightmares of an old age in which winning admission to a nursing home will prove harder than getting into Stanford or Yale. Cemeteries will be standing room only. Others worry that an increase in longevity will merely mean an increase in the time they'll spend bedridden, senile, catheterized.

Not so, say the gerontologists. They argue that even if there is no significant medical breakthrough, today's young and middle-aged can still look forward to a more youthful old age than their parents or grandparents. Instead of applying for early admission to the local nursing home, these future elderly may acknowledge their dotage merely by switching from downhill to cross-country skiing, or from running to jogging. Rather than worrying about death or about overcrowding at Heavenly Rest Cemetery, they may instead be considering the personal implications of a study, recently begun by the National Aeronautics and Space Administration, to determine how well 55- to 65-year-

---

LINUS PAULING
Scientist
Born: 1901

"I think people should try to keep healthy. I'm especially interested in vitamin C. I continue to recommend that everyone get a good supply. I take 10 grams a day."

---

old women withstand the stress of spaceflight. K. Warner Schaie, research director of the Andrus Gerontology Center at the University of Southern California, cites three well-known but rarely noted reasons for optimism about future old age:

● **The control of childhood disease.** Aging is cumulative. Instead of simply healing and going away, the minor assaults suffered by the body from disease, abuse and neglect can have "sleeper" effects. Chicken pox in a child, for example, can lead much later in life to the hideous itching affliction known as shingles. But vaccines and other wonders of modern medicine have largely eliminated such time bombs. Says Schaie, who is 53 and has lived through whooping cough, measles and mumps: "Most people who will become old 30 or 40 years from now will not have had childhood diseases. Most people who are now old have had them all. That's an important difference."

● **Better education.** Where a grade-school background was typical for the older generation, more than half of all

Americans now 30 or 40 years old have completed at least high school, and studies show that people with more education live longer. They get better jobs, suffer less economic stress and tend to be more engaged with life and more receptive to new ideas, which may help explain the third factor.

● **The fitness revolution.** "We really have changed our habits with respect to diet and exercise and self-care," says Schaie. Per capita consumption of tobacco has dropped 26 percent over the past 15 years, and the drop is accelerating, promising a decrease in chronic obstructive pulmonary disease and lung cancer. Life-style changes and improved treatment of hypertension have already produced a dramatic national decrease in cerebrovascular disease, one of the major chronic problems of old age.

Gerontologists say that these same future elderly will also benefit from the increasingly accepted idea that aging is partly a matter of choice. Speaking before a recent meeting of the American Academy of Family Physicians, Dr. Alex Comfort, the author and eminent gerontologist, argued that 75 percent of so-called aging results from a kind of self-fulfilling prophecy. "If we insist that there is a group of people who, on a fixed calendar basis, cease to be people and become unintelligent, asexual, unemployable and crazy," said Comfort, "the people so designated will be under pressure to be unintelligent, asexual, unemployable and crazy." Changing the image of the elderly may change the way the elderly behave.

Just as important as the image itself is the way individuals react to it. Performance declines on average with age, but individuals can practice not to be average. A 75-year-old man whose joints should be stiffening into immobility can run the marathon. An 80-year-old woman whose capacity for work is undiminished can model herself after Sir Robert Mayer, who declared at age 100: "Retire? Never! I intend to die in harness." (Now 101, Mayer is still arranging concerts, and recently remarried).

STROM THURMOND
United States Senator
Born: 1902

"I think the advantage of my age is the wisdom and knowledge that comes with experience. For those looking ahead to a long life, I offer this advice: Read the Constitution and demand that public officials, legislators and judges abide by it."

GEORGE GALLUP
Pollster
Born: 1901

"There's no substitute for experience. As I look back on my life, I wonder how I could have been so stupid.

"Intellectual curiosity is important, too. A lot of people die just from boredom. I have a whole program that will keep me going until age 100, at least. Incidentally, some years ago we did a study of people over the age of 95. We interviewed over 450 individuals; 150 of them were older than 100. What we found was that those who live a long time *want* to live a long time. They are full of curiosity, alert and take life as it comes."

This reliance on choice and something more—spunk, exuberance, a positive mental attitude—may sound romantic, and is, admittedly, an exotic notion in some of its permutations. At the State University College in Geneseo, New York, for example, Lawrence Casler, a psychologist, is convinced that aging is entirely psychosomatic. In 1970, to break through "brainwashing" about life span, he gave "an extremely powerful hypnotic suggestion" to 150 young volunteers that they will live at least to 120. "We're planning a big gala champagne party for the year 2070," says Casler. "I'm looking forward to it." He also gave a hypnotic suggestion for long life to residents of a nursing home who were already 80 or older. Casler says that the suggestion appears to have reduced serious illness and added two years to life span in the experimental group.

But even medical specialists in aging take the role of choice seriously. Dr. James F. Fries of the Stanford University Medical Center writes in The New England Journal of Medicine that "personal choice is important—*one can choose not to age rapidly in certain faculties.* . . ." The italics are added, but Fries himself writes that the biologic limits are "surprisingly broad." With training, experimental subjects have repeatedly reversed the pattern of decline in testing for intelligence, social interaction, health after exercise, and memory—even after age 70. Fries, Schaie and every other gerontologist interviewed for this article reached the same conclusion: To stay younger for longer, you must stay physically and mentally active. As Fries put it, "The body, to an increasing degree, is now felt to rust out rather than to wear out." What you don't use, you lose.

8. MATURITY AND OLD AGE

The Gray Revolution

A society dominated by old people will inevitably look different. Without a medical breakthrough, even vigorous old people cannot avoid slowing down. Schaie's interpretation is positive: "Young people . . . make many more errors of commission than omission, but the reverse is true for the elderly." Caution. The wisdom of the aged. To accommodate it, traffic lights, elevators, the bus at the corner will also have to slow down and become more patient.

Housing will be redesigned. But that doesn't necessarily mean handrails in the bathroom, wheelchair lifts on the stairway or any of the other depressing impedimenta of old age. "You're thinking of facilities for the ill elderly," says Schaie. Instead, redesign may mean more one-story garden apartments (stairs waste human energy). Homes will be smaller and require less care. House cleaning, home maintenance, dial-a-meal and dial-a-bus services will proliferate. Condominiums for the elderly will be built not near hospitals—an outdated idea, according to Schaie—but near libraries, colleges and shopping and athletic facilities.

The work place will also change, because more old people will stay on the job. As early as 1990, the baby bust of the past two decades will yield a shortage of young workers. Older workers will gain as a result, and there are few things better for an older person's spunk, exuberance or positive mental attitude than a sense of continuing worth in the marketplace and the paycheck that goes with it. The change is already beginning. The personnel department of one high-technology company, unable to find enough job candidates under 30, recently hired Schaie to convince its own top management that gray-haired engineers are just as able to keep the company on the cutting edge of innovation. Schaie says that it will pay to update and retrain older workers, not just for their expertise, but because they are less prone to accidents and absenteeism than their younger counterparts. In the Information Society, automation and robotics will make youth and strength less significant; the premium will be on knowledge and experience.

Indeed, James E. Birren, who runs the Andrus Gerontology Center in Los Angeles, predicts an era of "the experimental aged." A 75-year-old female lawyer came out of the audience once when he was explaining this idea, took his chalk away and told him in detail, with notes, why he was being too restrained. At 62, it seems, Birren still suffered the inhibitions of the conservative young. By contrast, he says, the elderly are past child-rearing and mortgage-paying, and they are also often beyond worrying about what the boss or the family thinks. But they have a much better sense of what they themselves

think, and it is often surprising. In a 1971 Gallup poll, for example, substantially more older people thought that the Vietnam War was a mistake than did those in the 21- to 29-year-old group. If they did not do much about it, perhaps it was because they accepted the youth culture's image of old people as doddering and ineffective. But given respect, independence, a steady paycheck, the prospect of continued vigor and the knowledge of their own numbers, the old may replace the young, says Birren, as the experimenters, innovators and all-around hell-raisers of the world.

Which brings us to the question of sex and the elderly. Future elderly will enjoy, and perhaps enforce, a more tolerant public attitude toward their romantic activities. Dr. Leslie Libow, medical director at the Jewish Institute for Geriatric Care in New Hyde Park, New York, blames decreasing sexual interest among older men and women at least partly on the traditional popular expectation that interest ought to decrease. But that expectation shows signs of changing. One company not long ago introduced a line of cosmetics specially designed for older women and met with 400 times the response it had predicted. And a New York-area motel offering X-rated movies recently began advertising a senior-citizens discount. Future elderly will also benefit from subtler changes. Bernice L. Neugarten, a psychologist at the University

---

LATE, GREAT ACHIEVERS

Herein, proof that life begins—or at least continues—after 70.

Konrad Adenauer (1876-1967) 73 when he became the first Chancellor of the Federal Republic of Germany. Resigned 14 years later.
Walter Hoving (b. 1897) Chairman of Tiffany & Company for 25 years; recently left to start his own design-consulting firm at 84.
Pope John XXIII (1881-1963) Chosen Pope at 77; brought the Catholic Church into the 20th century.
Jomo Kenyatta (c. 1894-1978) Elected Kenya's first President at 70. Led the country for 14 years.
Henri Matisse (1869-1954) In his 70's did a series of sprightly paper cutouts that were exhibited at New York's Museum of Modern Art.
Golda Meir (1898-1978) Named Prime Minister of Israel at 71; held the job for five years.
Cathleen Nesbitt (b. 1889) At 92, revived the role she created on Broadway 25 years ago: Professor Higgins's mother in "My Fair Lady."
Pablo Picasso (1881-1973) Complet-ed his portraits of "Sylvette" at 73, married for the second time at 77, then executed three series of drawings between 85 and 90.
Anna Mary Robertson Moses (1860-1961) Was 76 when she took up painting as a hobby; as Grandma Moses won international fame and staged 15 one-woman shows throughout Europe.
Dr. John Rock (b. 1890) At 70 he introduced the Pill; spent the next 20 years as its champion.
Artur Rubinstein (b. 1887) Was 89 when he gave one of his greatest performances at New York's Carnegie Hall.
Sophocles (c. 496-406 B.C.) Wrote "Electra" and "Oedipus at Colonus" after 70, held office in Athens at 83.
Giuseppe Verdi (1813-1901) Was 74 when "Otello" added to his fame; "Falstaff" followed four years later.
Frank Lloyd Wright (1869-1959) Completed New York's Guggenheim Museum at 89; continued teaching until his death.
Adolph Zukor (1873-1976) At 91, chairman of Paramount Pictures.

of Chicago, suggests that as they put anxiety-producing family and career decisions behind them, men and women are likely to relax more with one another. Sex stereotypes and the arguments they provoke will decline in importance as the years go by. As for male impotence, Libow notes that 20 percent to 40 percent of men continue to have active sex lives well into their 70's even now.

Finally, what about the specter of a long and vigorous life ending wretchedly in a nursing home? The average admission age today is 80, and only 5 percent of the elderly now endure such institutions. Even so, at current rates, the number of nursing-home residents will increase by 57 percent, from 1.3 million now to 2.1 million in 2003. But there are alternatives to terminal "convalescence." Most people now die without what Neugarten calls "a final deterioration that erases individuality." Birren believes that even more people will die "in harness" in the future. Instead of dwindling away, they will remain vigorous longer, then drop away quickly. Temporary "respite care"—an innovation now being imported from Scandinavia—will be available, say, for an octogenarian down with a bad cold. Schaie suggests that old people who are burdened with large homes may also invite friends to come live with them for mutual support. Such communes will give them independence they could never hope for in a nursing home.

Supergenes

So far, all of this assumes that there will be no breakthrough in human longevity, no anti-aging pill or fountain of youth. People will become wrinkled and gray and die at roughly the biblical threescore and ten or fourscore years. But that is no longer a safe assumption.

Dr. Roy Walford is one of the leading gurus of the life-extension move-

ment, and he looks the part. His head is hairless, except for a gray moustache that thickens down past the corners of his mouth. He can seem ferocious on film, but in person he is benign, almost shy. When he tours his complex of laboratories at the School of Medicine of the University of California at Los Angeles, he keeps his elbows close by his sides, and his hands in front of him, tucked like a monk's into the sleeves of his white lab jacket. His colleagues around the country tell you first that he is odd, and second that they respect him more than any other researcher in the field.

Walford believes that he has identified a single supergene that controls much of the aging process. Since 1970, he has been studying a small segment of the sixth chromosome in humans called the major histocompatibility complex. This is the master genetic control center for the body's immune system, and it is a logical suspect in the aging process. The ability to fight off disease peaks in most people during adolescence, and then falls off to as little as 10 percent of its former strength in old age. At the same time, a kind of perversion of the immune system occurs, in which the workhorses of self-defense lose some of their ability to distinguish between friendly and foreign cells. They attack the body's own organs, leading to such characteristic diseases of old age as diabetes and atherosclerosis. The phenomenon is called autoimmunity, and it means that the body is making war on itself.

Walford established the supergene's additional role in aging by comparing 14 strains of mice. The study (conducted by his associate, Kathy Hall) demonstrated significant differences in life span among mice that were genetically identical *except* at key locations in the major histocompatibility complex. Additional study by Walford, Hall and others tied that single genetic variable to two of the most important factors in the aging process as it is now understood. The long-lived mice had improved DNA-repair rates and increased protection against cellular damage from bodily substances known as free radicals. But the most profound idea, first suggested by Richard Cutler of the National Institute for Aging and substantiated by Walford, was simply that genes control longevity. Mice with "good" genes lived longer; the strains with "bad" genes had lives that were nasty, brutish and, above all, short.

Walford points out that there may be other genes controlling longevity. But none has been located so far, and separate studies on the nature of hu-

man evolution suggest that there may be, at most, only a few such supergenes. All of which makes Walford's discovery about the major histocompatibility complex more important. What to do about it? Altering the genetic information in the nucleus of every cell of an adult animal is, at least for now, impossible. Instead, scientists are trying to find other supergenes and identify the mechanisms by which they work. If these supergenes have some regulatory mechanism in common, it may then be possible to manipulate the mechanism rather than the gene.

One promising school of thought theorizes that aging results mainly from accumulating errors in the complex, tightly coiled strands of DNA that are each cell's blueprint for accurately reproducing itself. The damage comes from many sources: ultraviolet radiation, viruses, free radicals, toxic chemicals, even the body's own heat. Repair enzymes correct some damage, but some persists. Over the years, tiny breaks and infinitesimal wart-like bulges accumulate, the DNA coils loosen and the cell begins to malfunction.

The theory suggests that those animals most efficient at DNA repair will live longest and age least. And so it happens in nature. The white lip monkey has a low rate of DNA repair, and lives to be only 12 or so years old. Humans have a very high rate of DNA repair, and live longer than any other primate. Differences in the DNA repair rate appear even among members of the same species, along with corresponding differences in longevity. Walford and Hall's long-lived mice displayed a higher rate of DNA repair than their short-lived counterparts. Walford himself has an unusually high rate of DNA repair. This may be why at 56, he still has pink unwrinkled skin, and looks 45.

What if science could boost other people's DNA-repair rates to the same level as Walford's—or perhaps triple them? The theory is that enhanced repair, probably in tandem with other therapies, would delay the accumulation of errors, slow the aging process and extend human life span. It might even be possible to correct old errors, patching breaks that had become a part of the genetic blueprint, excising bulges, retightening the coils of DNA. In a word, rejuvenation.

MOTHER TERESA
Missionary
Born: 1910

"At the hour of death, when we come face to face with God, we are going to be judged on love—not how much we have done, but how much love we have put into our actions."

Keeping Paramecia Young

Researchers may already have achieved just that in lower animals. Joan Smith-Sonneborn, a professor of zoology and physiology at the University of Wyoming, chose to work with paramecia because the microorganism's single cell in many ways resembles the cells of more complex animals. In an experiment first reported in 1979, she damaged the DNA of paramecia with ultraviolet radiation—the same light waves that cause human skin to tan or burn and, eventually, to age. The damaged animals died sooner than the untreated ones, presumably because they could not repair all the breaks or bumps in their DNA. Damage. Malfunctioning. Old age. Death.

With another group of paramecia, Smith-Sonneborn first induced DNA damage, and immediately stimulated, or "photoreactivated," a repair enzyme known to respond to a particular wavelength of visible light. To her surprise, the photoreactivated paramecia lived not merely as long as untreated counterparts, but substantially longer. They were already at midlife, but somehow achieved a 296 percent increase in their remaining life span, and a 27 percent increase in overall life span. In gerontology, the so-called Hayflick limit represents the maximum life span of each species. It has to do with how many times the animal's cells can divide before they die. There is a Hayflick limit for paramecia and another for humans,

both supposedly inescapable, unmovable. The ultimate deadline. "What we did," says Smith-Sonneborn, "was break through the Hayflick limit for paramecia."

To explain how this happens, Smith-Sonneborn uses the analogy of a sinking ship. When the damage occurs, an S O S goes out. But by the time outside help arrives on the scene, the ship's own crew has plugged the leak and pumped out the bilges. Having no emergency repair to do, the outside help goes to work anyway, overhauling engines and tightening rivets, and may even stay on the scene long afterwards for maintenance. By analogy, the photoreactivated enzymes leave the paramecium more youthful—and youthful for longer—than it was before the damage occurred.

It is, of course, a huge leap from paramecia to human beings. But other scientists have already demonstrated photoreactivation of repair enzymes in human skin. Smith-Sonneborn quickly adds a caveat: No one knows how much ultraviolet damage must be induced in human skin, or with what consequences, before photoreactivation will work as a rejuvenating treatment. So it does not make sense to sit for hours under a light bulb in the hope of unwrinkling or rejuvenating the skin. Then what does any of this matter? The photoreactivation work is important because it demonstrates the possibility of enhancing other DNA-repair mechanisms and other longevity-determining processes elsewhere in the human body. The enhancement may slow aging. And if it is possible to break through the Hayflick limit for paramecia, why not also for people?

MARGARET HICKEY
Public affairs editor
Ladies' Home Journal
Born: 1902

"With age comes serenity, the feeling of being satisfied with what you have done while still looking forward to what you can do. You begin to like yourself."

The Anti-Aging Diet

Another method of enhancement— the only one that is practical now— takes the life-extension story back to the 16th century and to Luigi Cornaro. Cornaro, a Paduan, led such a profligate youth that by the time he was 40, he found his constitution "utterly ruined

by a disorderly way of life and by frequent overindulgence in sensual pleasures." Told to reform, he entered upon the *vita sobria*, a life of moderation in all things. He restricted himself in particular to just 12 ounces of food daily and lived in robust health until he was 98. Since then, scientists have repeatedly demonstrated that undernutrition—as distinguished from malnutrition—extends the lives of experimental animals. But they mistakenly thought that undernutrition had to start at weaning, when it's riskiest. As Walford puts it, "You can't starve babies and you can't have 10 percent mortality rates in order to have a few people live to be 180." When the underfeeding was delayed until midlife, the experimental animals often died prematurely, Luigi Cornaro to the contrary.

I.F. STONE
Author
Born: 1907

"There are great joys in one's later years—as many as there are in one's youth. One of them is learning and studying. The things you study have much more significance; you understand them more fully. I'm studying ancient Greek language and civilization. It's difficult work, but very rewarding.

"My advice is to persist. The mind is like a muscle—you must exercise it."

But by refining the technique of previous researchers, Walford and Richard Weindruch, a co-worker, recently achieved a 30 percent increase in maximum life span for test animals whose diet was restricted after they became adults. The crucial refinement seems to have been moderation. Where other experimenters began adult underfeeding abruptly, Walford and Weindruch gradually reduced the intake of their animals, down to about 60 percent of their normal diet. In humans, Walford says, it would work out to a loss of a quarter to a third of body weight over a six-year period—a big loss, but a slow one.

Walford has not yet developed the optimal human diet for longevity through underfeeding, but says he intends to over the next year or so. For himself, he now fasts two days a week and has shed five or six pounds from his already spare frame. But because undernutrition can easily turn to malnutrition, especially outside a doctor's care, he doesn't recommend that people follow his example. In any case, would such a diet be worth the sacrifice? Walford evidently thinks so. One reason is that experimental underfeeding did not merely delay death, it delayed aging. The lack of extra calories forestalled the development, and hence the decay, of the immune system. Autoimmunity, the body's war on itself, actually decreased. And in other underfeeding experiments, the cancer rate dropped markedly.

As for the sacrifice, it may eventually prove unnecessary. At Temple University Medical School in Philadelphia, Arthur Schwartz is working with an adrenal-gland product called dehydroepiandrosterone, or DHEA. His experiments "suggest that DHEA treatment may duplicate the anti-aging and anticancer effects of caloric restriction."

140 and Still Kicking

It is just this piling on of discoveries and developments in all areas of life-extension research that causes Walford, Cutler, Smith-Sonneborn and others to predict a dramatic increase in human longevity. Walford, in fact, believes that it will become possible within this decade to extend maximum human life span to somewhere between 130 or 140 years.

What such a life will be like is anybody's guess. George Bernard Shaw imagined that extraordinarily long-lived people would quickly abandon mating for more mature pursuits, such as higher mathematics. Aldous Huxley wrote of a brave new world in which old men spend their time "safe on the solid ground of daily labor and distraction, scampering from feely to feely, from girl to pneumatic girl."

The possibilities raised by extreme longevity are as numerous as the additional days people will supposedly live. Walford argues that the very vastness of the change is one reason conservative experts on aging prefer to deny the likelihood of significant life extension. "It blows the data base on which they make their projections," he says. A doubling of human life span would, of course, blow anybody's data base. To cite a single example, a young couple starting out with two children, and passing on their belief in zero population growth to their offspring, could wind up at the healthy middle age of 90 with 44 direct descendants and descendant-spouses, and at age 150, with six or seven living generations in the family. Alternately, if women were to age much more slowly and reach menopause much later, it might be possible for some future generation to put off parenthood and the whole career-family dilemma until age 50.

Whole new sets of questions arise, and at first they seem to have an absurd Woody Allen quality: If you're going to live to be 150, who's going to pay the rent? But if life extension is as near as

---

MICKEY ROONEY
Actor
Born: 1920

"A long time ago I looked around and saw the apathy that the elderly had fallen into. They had nowhere to go, no one to turn to. They were eliminating themselves from the atmosphere of potential. This became a deep concern of mine. So I started a project called Fun Filled Family for people over 45. We have trips, discounts, life insurance policies and gathering places. We have no political or religious ties. We don't use the word age at all, but talk only in terms of experience."

---

gerontologists suggest, perhaps it is time to begin thinking seriously about seemingly absurd possibilities. Does it make sense, for example, to buy life insurance at 30 if you will be living to 150 and if, barring accidents, medicine makes the time of death ever more predictable? Should you plan now for a second career? (Walford intends to be a researcher in artificial intelligence.) Will longer life encourage a flowering of abilities? Is marriage "till death do us part" practical when there is a good chance that you'll see your 100th anniversary? What if the husband pursues the life-extension therapy and the wife rejects it? What will it mean if marriages and generations become asynchronous? What will happen—in fact, is already happening—to the drilled-in timetables of schooling, mating, child-rearing, income-producing, retirement and death?

To many people, the promise and the possibilities suggested by life extension are truly wonderful. "I see humanity as a fragile organism that has been evolving all along into a more complex, autonomous, intelligent species," says F. M. Esfandiary, an author who lectures at the University of California at Los Angeles, "and I believe that in our time this ascent will rapidly accelerate, propelling us into entirely new dimensions. I also believe that our mortality, the very fragility of life, has for eons drastically impaired the quality of life. It is not just the imminence of one's own death that is so cruel, it is the ever-present fear of losing all the people one loves. In the future, people will not be as programmed to finitude and mortality. Instead, they will see a whole avalanche of new options. . . ."

On the other hand, it may be worth remembering that Moses lived to be 120, and still never entered the promised land.

COPING WITH DEATH IN THE FAMILY

The key to minimizing the impact of death in the family, say experts on the subject, is more honesty on the part of everyone involved. The patient should be told—as much as possible—about the seriousness of the illness. And the rest of the family should openly express their own feelings.

The dramatic news today is that there are experts on death and dying and a large and growing body of knowledge about the subject. Thanks to the work of such people as Chicago psychiatrist-author Dr. Elizabeth Kübler-Ross, the burden of coping with death—from the shock of learning that a spouse or a child is terminally ill, through the death, funeral, and subsequent adjustment period—is being lightened for many families. "There's a healthier, more open attitude today toward death and how to handle it," says Dr. Austin H. Kutscher, who teaches a new course called thanatology (the study of death) at the Columbia University medical school. "People are getting help they couldn't have gotten two or three years ago."

Kübler-Ross, who has written extensively on the subject, is one of many experts who travel widely today lecturing on death at schools, hospitals, and religious institutions around the country. Dr. Mary Cerney, a psychologist at the Menninger Foundation in Topeka, says she has just finished a lecture series on death at a Kansas high school. More and more hospitals and university psychology departments offer seminars on coping with death. The Candlelighters, with 39 chapters around the country, provides counsel to the families of children stricken with cancer, and the National Foundation for Sudden Infant Death tries to help grieving parents.

Orville Kelly was a newspaper editor in Davenport, Iowa, when he was stricken with lymphatic cancer in 1973. He launched Make Today Count for other cancer patients a year later and now counts 54 chapters. "As a cancer patient, I can now share my frustrations and anger with other patients," says 45-year-old Kelly. "We talk, we have speakers, we have hope."

Perhaps the biggest change has come among those who deal with terminal patients in hospitals (where 80% of all Americans die). Dr. Stephen E. Goldston, a psychologist with the National Institute of Mental Health in Rockville, Md., finds hospital personnel increasingly aware that "the dying need to be dealt with honestly, without pretense, and safeguarded against the greatest pain of death, which is terrible loneliness."

Goldston warns the family of someone approaching death that false good cheer only serves to deepen the patient's fear and anxiety. The survivors, in turn, require a period of unrestrained grief in order to adjust to a loved one's death, and that goes for children as well as adults.

"Children," says Goldston, "need to be told the truth and not lied to about a death in the family." Sugar-coating death, say Goldston and others, can cause greater pain—and sometimes serious guilt feelings—among children. "Families have a chance to educate their growing children about death," says Edward K. Leaton, president of a New York City consulting firm, whose two sons died of muscular dystrophy. "This need not be frightening. The death of a pet or a bird gives you a chance to explain and remove fears."

The patient

Step one in coping with death is dealing with the patient, and here the old approach of telling the patient nothing is giving way to a new realism. Patients are being told more today—often to the point of telling a person that he or she is going to die. "Ideally, the patient should be the first to know," advises Dr. Herman Feifel, professor of psychiatry at the University of Southern California medical school. "The only question is how to tell the person. Some must be told gradually."

There are practical reasons for revealing the truth to a terminal patient. The patient, if "protected" by family and friends, suffers through hours of pointless, trivial talk. Another reason for telling the truth is that the patient will then feel less guilty in demanding services or attention or openly crying and expressing grief. Grieving is vital, since it enables the patient to accept the reality of death more easily.

Admitting the likelihood of death to a terminal patient is not a step to be taken lightly. Hope alone can sometimes prolong life; denial of hope can sometimes prod a person to suicide. Obviously, the matter should be discussed thoroughly with the patient's doctor and probably with a clergyman, as well. Several important questions must be answered: Is the doctor absolutely certain that the person is going to die? Does the patient really want to know the truth? "Some people decide to tell the dying person out of a compunction to be honest or a need to absolve their own guilt," says Dr. Kenneth D. Cohen, psychiatrist and clinical director of the Philadelphia Psychiatric Center.

Some people should not be told at all. A strong, relatively independent person can handle the news far better than someone who tends to be weak and dependent. The patient who still hopes for recovery probably should not have those hopes dashed. Rather, the matter should usually be postponed until the patient raises the question of whether he or she is going to die.

"You do not rush the person into a recognition of their condition," says Columbia's Kutscher. "You listen carefully for hints of a desire to discuss the possibility of death." Cohen agrees that "the telling must be approached delicately." A basic point to get across, he feels, is that the patient is not alone and will not be alone.

"Above everything, don't make pretenses," advises a New York City advertising executive whose wife died of cancer. "When you feel that the suffering person wants the truth—tell it. Pretense keeps two people from ever really talking. I regret that before my wife died, we never got to the stage of real frankness. It made her death harder for both of us."

HONESTY IS BEST, WITHIN REASON

Kutscher and others suggest that the wishes of the dying be heeded when it comes to deciding on desperate life-prolonging measures. "Most terminal patients at this point know they will die," says Kutscher. "I suggest that you listen to them. They will probably make their wishes felt about this last-minute fighting to stay alive."

Sometimes, the word of impending death simply can't be concealed from the patient—when there is a will or other legal or business papers to be signed, for instance. Some people are anxious to get their affairs in order, and doing it can help them face death. "But do this gently," Kutscher suggests. "Have the family lawyer present the will. He can be more casual about it than you would be."

Most experts agree on the need to tell children as much as they can handle about a parent's illness. "If a parent has a terminal illness, even a small child should be told what this means—in terms the child can understand," says Dr. Edward Rydman, a Dallas psychologist. "Families tend to shy away from dealing with this," he adds. Says Dr. David A. Switzer, a Southern Methodist University psychologist: "The children need to know that their father or mother is going away, and not just for 15 minutes or overnight."

Most psychologists agree and point out that in explaining the terminal illness of a parent, the child should be told in such a way that he or she can fully comprehend that death is permanent, that every effort was made to save the parent's life, that the loss will be painful, that it is good to cry or get angry to express grief, and that the parent's duties will be taken care of.

"In addition," says Goldston of NIMH, "I would make it clear that the rest of the family will remain together—assuming this is true." He notes that if the terminal patient is a brother or sister, it is important to let the other children know that "any evil wishful thinking they have had about the sick child doesn't really count for anything." This, he explains, can ease often dire guilt feelings among children, who frequently do "wish" bad things on siblings—the wish that a brother or sister would disappear forever, for example.

Don't resort to fairy tales. Telling a child that "your mother went to sleep and will forever rest peacefully" can be frightening, not reassuring. "They will be afraid to go to bed," says Sandi D. Boshak, of Dix Hills, N.Y., whose son died in infancy. Boshak,

who is active in group meetings of the National Foundation for Sudden Infant Death, feels that honesty is vital in explaining death to small children, as well as to teen-agers.

Another common mistake, say both psychologists and parents, is to use a religious explanation of a death in a family that has had no religious background. "Such an explanation will mean little to the child, may create fear, and can cause the child to later react negatively to religion," Goldston explains.

A child's role

Honest explanations to children in cases of a dying parent can also have a positive effect on the parent. Some children secretly fear that they are to blame for their parent's illness. Freed of this anxiety, the child will relate more closely to the ill parent and make the parent feel better. A child should rarely be kept apart from a parent who is dying. "Even though our daughter was only three years old, we kept her in contact with her mother," says a New York businessman whose wife died of cancer several years ago. "She fed her mother cracked ice," he adds. "This made her understand that her mother was very sick, and helped both of them when death came."

When a parent or sibling dies, even children as young as seven or so can and should take part in the mourning ritual observed by the family, say psychologists. "Ask yourself if a young child is capable of participating along with the adults and older children," says Cohen, of the Philadelphia Psychiatric Center. "Most children are capable of this, but some families want to protect and keep the child away from the funeral." This is a mistake, he feels, because the child is denied the outpouring of grief that helps lead to acceptance of the death.

"Equally important," advises Cohen, "is that the remaining parent, or both parents if a sibling has died, be with the child afterward to help the youngster with his or her feelings and talk about the death if the child wants to talk. And there is no need to be a 'stout fellow' about it. It's better to cry."

A number of specialists say today that even a child should be told—very cautiously—about his or her own terminal illness. Says Kübler-Ross: "Children can deal with death sometimes much better than grownups. The worst thing you can do is lie to them."

But the news—if it is told at all—must be greatly softened. David M. Kaplan, director of clinical social work at Stanford University, feels that "you might use such statements as 'you are very sick and we hope the doctor can help.' " But children as young as five or six usually sense that they are seriously ill, he explains, and pretense and lying only loses you their trust and isolates them, making their loneliness more difficult.

"With a child," advises Kutscher, "the task is to be honest—but not frightening. A parent will want to discuss these points with the doctor, clergyman, and possibly other family members, before deciding what to tell a child." But all the warnings about how—and when—to tell a dying patient, apply doubly with a child.

The household routine should be kept as normal as possible when a child is at home with a terminal illness. "That's important to the sick child and the other children," says Spencer Smith, an Atlanta businessman whose four-year-old daughter died after an 18-month fight against cancer.

"We did our best not to change our lifestyle, and treated Margaret as normally as possible," Smith adds. "Cobalt treatment caused her hair to fall out, and we got a wig for her. For 6 of her last 10 months, things were pretty normal and Margaret was a happy child." Smith speaks of his other small children: "We fully explained her illness to the others, ages 5 and 8, telling them about the cancer and that there was a good chance she would not survive. When Margaret died, they understood and were not shocked. They accepted it."

How to handle the five stages of dying*

*Based on the five stages of death outlined in *On Death and Dying* by Dr. Elizabeth Kübler-Ross, psychiatrist and author

PATIENT BEHAVIOR	STAGE	FAMILY RESPONSE
	DENIAL	
In effect, the patient says, "It cannot be true." Patients often search frantically for a favorable diagnosis.		Understand why the patient is grasping at straws. Patience and willingness to talk are important.
	ANGER	
The patient says, "Yes—but why me?" Deep anger follows, and the patient may bitterly envy those who are well and complain incessantly about almost everything.		Consider that the patient is angry over the coming loss of everything: family, friends, home, work, play. Treat patient with understanding and respect, not by returning the anger.
	BARGAINING	
The patient says, "Maybe I can bargain with God and get a time extension." Promises of good behavior are made in return for time and some freedom from physical pain.		If the patient's 'bargain' is revealed, it should be listened to, not brushed off. This stage passes in a short time.
	DEPRESSION	
The patient grieves, and mourns approaching death.		Attempts to cheer up or reassure the patient mean very little. The patient needs to express sorrow fully and without hindrance.
	ACCEPTANCE	
The patient is neither angry nor depressed, only quietly expectant.		News of the outside world means little, and few visitors are required. There will be little talk, and it is time merely for the presence of the close family.

Facing yourself

Your own personal adjustment to a death or terminal illness in the family is made easier to the extent that you can open up to others. "People who share the struggle, who are open rather than secretive, seem to do better," says James Gibbons, director of chaplaincy at University of Chicago hospitals. But "talking it out" may not be easy.

"The well person may have a deep fear of death," notes Menninger Foundation psychologist Cerney. "If a husband has a terminal illness, the wife may freeze and deny the coming death, pretending that it won't happen." The wife might then remain aloof from the husband and cause more pain for both.

Cerney suggests that the spouse of a terminal patient who is concerned about his or her own inability to cope should first try to come to grips with the fear of death.

"The books of Kübler-Ross [such as *On Death and Dying*] offer real help," she advises. "Or phone your local university and ask someone in the psychology department about one-day seminars on thanatology. There are more and more of these—and they aren't morbid."

Hearing a good talk and being a part of a question-and-answer session can help lift some of the unreal fears of death and send a seminar participant home with a clearer idea of how to relate to the terminal patient.

"I dreaded the death seminar idea," confesses a Columbus (Ohio) businessman whose wife had terminal cancer. "But it seemed I was the one pulling us down, while she was the one having the chemotherapy. I went to the sessions at the hospital and came away feeling tolerably better about those difficult talks with my wife."

In Tiburon, Calif., near San Francisco, Nora E. Grove, a lawyer's wife, tells how attending Candlelighters Society meetings helped her family adjust to having a child at home with terminal cancer. "You meet once a month in a church hall for about two hours with 30 to 40 other parents of cancer children," she explains. "You have coffee and then listen to a speaker and have group discussions. Personally, I got some sanity. I felt I would cave under, but meeting with the other parents helped me to know about physically caring for the boy and got me through the sadness."

Most people are psychologically unprepared for their own reaction to the death of a spouse or child, or in some cases, a parent. "The death of a parent usually is manageable," says Goldston, "but losing your wife or child is a different story." And a sudden, unexpected death is the most difficult to face.

There are many psychosomatic side effects from grief, from an inability to concentrate at work to trouble in making even routine decisions. There may be headaches or insomnia, and psychologists note that some people even take on the symptoms of the deceased—stomach pains, where the death was from stomach cancer, or chest pains following a heart attack death.

HOSTILITY IS ALSO A STAGE OF GRIEF

Anger and hostility, a stage of dying, are also a stage of grief and mourning,

PUTTING YOUR HOUSE IN ORDER

There are a number of practical matters that must be dealt with despite the emotional pain and distraction of a loved one dying. A terminally ill person will often feel better knowing that his or her affairs are being put in order. The details can be handled by a spouse, if able to, or by a family lawyer.

DURING ILLNESS

Will: Make necessary changes. Examples: specific financial provision for children's education and provision for personal guardianship of children. This is the time to make any changes in the choice of executor (often spouse or relative) or co-executor (lawyer, bank). A letter apart from the will can cover non-mandatory instructions to the executor—for example, naming children's schools.

Estate tax: Consider buying "flower bonds" to pay any estate taxes. These Treasury issues can be bought at a discount and used to pay estate tax at the full face amount. Discuss with the lawyer a tax-saving trust for the children.

Bank account: Provide funds to carry the family immediately after the death by setting up (or increasing) a savings account in the surviving spouse's name. Bank funds in the patient's name will be frozen upon death. Funds in a joint account are not frozen, but since any jointly owned property becomes part of a decedent's taxable estate, any money withdrawn still would be subject to the estate tax.

Investments: Alter instructions to patient's broker where necessary—"stop orders," for example, that ought to be dropped. It may be wise to sell shares to buy flower bonds or to take the family out of speculative situations.

Gifts: Consider making gifts of the patient's property to family members. If the gift-tax exemption has been used up, there will be a tax. But the taxpayer will be entitled to a credit against estate tax; there may be a net tax saving.

Insurance: Make sure that life insurance beneficiary clauses are up to date. A gift of ownership of the policy itself (right to convert, borrow, and

such) will not take benefits out of the patient's taxable estate because of the three-year contemplation-of-death rule.

Power of attorney: Consider a "durable, general" power of attorney, with the power assigned by the patient so the surviving spouse can perform all transactions. This is particularly necessary where a patient becomes incapacitated.

Benefits: Contact the employee benefits department of the patient's company to ensure that desired beneficiaries are named. If an individual is named, noncontributory benefits (mainly pension and profit-sharing) are kept out of the taxable estate.

Funeral: Consider making advance arrangements for the funeral. A funeral director will handle purchase of a cemetery plot and bill you later along with other charges (assuming death is imminent). If you deal directly with a cemetery company, it will want full payment for one grave and will finance the balance if you wish.

AFTER DEATH

Lawyer-executor: Arrange to see the family lawyer and executor or co-executor together, preferably in the lawyer's office, within two or three days after the funeral. Speed is important because the lawyer starts probate of the will.

Papers: Send military discharge papers of deceased to the funeral director so he can file a claim with the Veterans Administration for routine $400 cash death benefit ($800 if a service-connected death). Provide Social Security number for a similar $255 benefit if the deceased was covered.

Miscellaneous: Phone life insurance companies to request claim forms. Notify insurer carrying deceased's homeowner's and auto insurance policies, employee benefits department, banks used by decedent (thus freezing accounts), and stockbrokers (automatically nullifying any brokerage instructions and freezing accounts). Notify credit card companies, and destroy or surrender the cards.

8. MATURITY AND OLD AGE

and awareness of this is especially important when a child dies. "Some men suffer more than their wives when a child dies," says Sandi Boshak whose son died in infancy. "From formal meetings with other parents of deceased children, I've seen that the husband has a harder time with his grief. He tends to bottle it up, while his wife cries openly." She explains that feelings of guilt can bog a parent down, particularly following the death of a first child. It is then that meetings of such groups as the National Foundation for Sudden Infant Death, with chapters in 40 cities, can help most.

An excessive grief reaction may justify psychotherapy if daily life has become impaired, at home or the office, and you realize your own conduct is hurting others. But avoid rushing into therapy. "See a therapist only if the level of grief is getting beyond you," says Dr. Harold Visotsky, chief of psychiatry at Northwestern University medical school.

If deep grief lasts more than two months or so, it may be wise to seek help, preferably using a psychologist, psychiatrist, or social worker who has had experience with similar problems. "Just any 'qualified' person won't do," says a Washington (D.C.) psychologist. "You must have somebody who knows thanatology." Most major hospitals will provide such a lead.

Be careful in using drug sedation to cope with excessive grief. "The risk of sedation is that you will lose the impact of the personal loss suffered, and your grief will only be delayed or prolonged," advises Cohen.

Adjusting after a death

Managing a household as a lone surviving parent, and "dating" to start new relationships, is the last phase of adjustment to the death of a spouse.

A formula for handling the children, including teen-agers to age 15, is suggested by an executive who is a widower: "First, if you are a working widow or widower, get a good housekeeper, if possible. She must be kindly and concerned about the kids, but still firm. And she must have the authority she needs to run things. If this is unrealistic, then make contacts with close married friends who have children the same age. Ask for their advice and help. At the least, you'll need some after-school-to-suppertime supervision for the kids."

And there is this cardinal rule: Maintain a spirit of discipline among the children. "The death of a parent should be no excuse for drifting into excessive permissiveness," says the executive.

A structured daily home life should be continued for the children's sake as well as your own. The older children should be given greater responsibilities to help make the household operate smoothly. But it is unwise to treat a teen-ager as a peer. This instant-adult treatment is a common mistake, say the specialists.

Dating again, after a spouse has died, can cause problems. "First of all, don't date to get over your grief," warns USC's Feifel. "Once you start dating," he explains, "it should be a signal that you have worked the mourning through." Catching a new person "on the rebound"—to override your grief—is a common mistake usually leading to new relationships that are shaky, at best.

Conversely, if you experienced a long period of grief before your spouse died and mourning has been fulfilled, then a new successful relationship can be approached fairly soon. "Some people are ready even to remarry after two months," says Kübler-Ross.

The children's reaction to a dating parent, however, poses a difficult problem. Cohen notes that "it is very difficult for the kids to accept another woman or man." Keep family communications open and try to talk matters through with the children.

One point is certain: A parent who waits for the children's approval of a proposed new marriage partner runs a good risk of staying single.

272

WHO'S WRITING AND WRITTEN ABOUT

INDEX

Credits/Acknowledgments

Cover design by Charles Vitelli

1.   Perspectives
Facing overview—UNITED NATIONS/Y. Nagata. 13—Professor Thomas Bouchard, Psychology Department, University of Minnesota. 27—M.E. Challinor.
2.   Prenatal Period
Facing overview—WHO/photo. 45—CRM Books, Random House. 46—Joe Baker, The Image Bank. 47-48—Dr. Robert Rugh and Landrum B. Shettlers from *Conception to Birth,* Harper & Row, 1971. 49—From Fitzhugh, Mabel Lum and Newton, Michael; Posture in Pregnancy, Am. J. Obster. Gynecol. *85*:1091-1095, 1963. 50—Everett Davidson. 76—*American Scientist,* March/April 1979.
3.   Infancy
Facing overview—EPA Documerica. 89—Jason Laure/Woodfin Camp, courtesy of Professor Lewis P. Lipsitt and Tryg Engen of Brown University. 91—Aidan Macfarlane. 92—Professor Lewis P. Lipsitt of Brown University. 98—From *Human Nature,* October 1978. Copyright ©1978 by Human Nature, Inc. Reprinted by permission of the publisher.

4.   Childhood
Facing overview—EPA Documerica. 127—Clemens Kalischer, Image Photos. 131-132—Farrell Greham. 148—Reprinted from *Psychology Today* Magazine. 152—Reprinted from *Psychology Today* Magazine.
5.   Education and Child Development
Facing overview—UNITED NATIONS/Photo by L. Solmssen. 162—From *Phi Delta Kappan,* September 1979.
6.   Child Rearing and Child Development
Facing overview—UNITED NATIONS/Y. Nagata.
7.   Adolescence and Young Adulthood
Facing overview—WHO/photo.
8.   Maturity and Old Age
Facing overview—From *The Depression Years,* Arthur Rothstein.

WE WANT YOUR ADVICE

ANNUAL EDITIONS: HUMAN DEVELOPMENT 82/83

Article Rating Form

Here is an opportunity for you to have direct input into the next revision of this reader. We would like you to rate each of the 56 articles listed below, using the following scale:

1. **Excellent: should definitely be retained**
2. **Above average: should probably be retained**
3. **Below average: should probably be deleted**
4. **Poor: should definitely be deleted**

Your ratings will play a vital part in the next revision. So please mail this prepaid form to us just as soon as you complete it.
Thanks for your help!

Rating	Article	Rating	Article
	1. The Unique You		29. They Learn the Same Way All Around the World
	2. The Instinct to Learn		30. Navigating the Slippery Stream of Speech
	3. Twins: Reunited		31. Infant Day Care: Toward a More Human Environment
	4. Ethnic Differences in Babies		32. Diet and School Children
	5. How Children Influence Children: The Role of Peers in the Socialization Process		33. The Mismatch Between School and Children's Minds
	6. The Japanese Brain		34. Learning Right from Wrong
	7. Just How the Sexes Differ		35. Toward a Nonelitist Conception of Giftedness
	8. New Genetic Findings: Why Women Live Longer		36. A New Look at Life with Father
	9. The More Sorrowful Sex		37. When Mommy Goes to Work . . .
	10. Pregnancy: The Closest Human Relationship		38. Suffer the Children
	11. Embryo Technology		39. The Curse of Hyperactivity
	12. Test Tube Babies: Solution or Problem?		40. The Children of Divorce
	13. Prenatal Psychology: Pregnant with Questions		41. Go Get Some Milk and Cookies and Watch the Murders on Television
	14. A Perfect Baby		42. How I Stopped Nagging and Started Teaching My Children to Behave
	15. Premature Birth: Consequences for the Parent-Infant Relationship		43. Adolescents and Sex
	16. Biology Is One Key to the Bonding of Mothers and Babies		44. Pregnant Children: A Socio-Educational Challenge
	17. The Importance of Mother's Milk		45. The Many Me's of the Self-Monitor
	18. What a Baby Knows		46. Why Johnny Can't Disobey
	19. Attachment and the Roots of Competence		47. The Sibling Bond: A Lifelong Love/Hate Dialectic
	20. Are Young Children Really Egocentric?		48. Single Parent Fathers: A New Study
	21. If Your Child Doesn't Get Along with Other Kids		49. Does Personality Really Change After 20?
	22. The Myth of the Vulnerable Child		50. Late Motherhood: Race Against Mother Nature?
	23. American Research on the Family and Socialization		51. The Dynamics of Personal Growth
	24. Mood and Memory		52. Stress
	25. Erik Erikson's Eight Ages of Man		53. Coping with the Seasons of Life
	26. Piaget		54. Must Everything Be a Midlife Crisis?
	27. Learning About Learning		55. Living Longer
	28. 1,528 Little Geniuses and How They Grew		56. Coping with Death in the Family

(continued on back)

About you

Name _____ Date _____

Address _____

City _____ State _____

Zip _____ Telephone _____

---

1. What do you think of the Annual Editions concept?

2. Have you read any articles lately that you think should be included in the next edition?

3. Which articles do you feel should be r_____ the ne_____ _n? Why?

4. In what other areas would you like to see an Annual Edition? Why?

HUMAN DEVELOPMENT 82/83

BUSINESS REPLY MAIL
First Class   Permit No. 84   Guilford, Ct.

Postage Will Be Paid by Addressee

Attention: Annual Editions Service
The Dushkin Publishing Group, Inc.
Sluice Dock
Guilford, Connecticut 06437-0389

NO POSTAGE
NECESSARY
IF MAILED
IN THE
UNITED STATES